Commercial Airplane Design Principles

Commercial Airplane Design Principles

Pasquale M. Sforza

University of Florida

AMSTERDAM • BOSTON • HEIDELBERG • LONDON
NEW YORK • OXFORD • PARIS • SAN DIEGO
SAN FRANCISCO • SINGAPORE • SYDNEY • TOKYO

Butterworth-Heinemann is an imprint of Elsevier

Butterworth-Heinemann is an imprint of Elsevier
The Boulevard, Langford Lane, Kidlington, Oxford OX5 1GB, UK
225 Wyman Street, Waltham, MA 02451, USA

First edition 2014

Notice
No responsibility is assumed by the publisher for any injury and/or damage to persons
or property as a matter of products liability, negligence or otherwise, or from any use
or operation of any methods, products, instructions or ideas contained in the material
herein. Because of rapid advances in the medical sciences, in particular, independent
verification of diagnoses and drug dosages should be made.

British Library Cataloguing in Publication Data
A catalogue record for this book is available from the British Library

Library of Congress Cataloging-in-Publication Data
A catalog record for this book is available from the Library of Congress

ISBN: 978-0-12-419953-8

For information on all Butterworth-Heinemann
publications visit our web site at store.elsevier.com

14 15 16 17 18   10 9 8 7 6 5 4 3 2 1

Working together
to grow libraries in
developing countries

www.elsevier.com • www.bookaid.org

Contents

v

Preface

This book grew out of a handbook originally prepared to support a one-semester senior undergraduate course devoted to airplane design. It is intended primarily to involve students in the preliminary design of a modern commercial turbofan or turboprop transport. The technical content is arranged to illustrate how the material covered in aerospace engineering analysis courses is used to aid in the design of a complex system. In addition, the format of the book is aimed at being amenable to self-education in the topic and to distance-learning online applications.

Commercial aircraft design is a relatively mature field and sufficient reference material is available to provide a secure mooring for student research and study. An industrial approach is taken in order to help instill the spirit of the design process, which is that of making informed choices from an array of competing options and developing the confidence to do so. In the classroom setting this design effort culminates in the preparation of a professional quality design report and an oral presentation describing the design process and the resulting aircraft. This report and presentation provides evidence of both the design and communication skills of the author(s) and can be of significant value in job interviews as well as in the developing report-writing skills for graduate theses.

The book is arranged in a manner that facilitates team effort, the usual course of action, but also provides sufficient guidance to permit individual students to carry out a creditable design as part of independent study. Emphasis is placed on the use of standard, empirical, and classical methods in support of the design process in order to enhance understanding of basic concepts and to gain some familiarity with employing such approaches which are often encountered in practice. Though some popular computational methods are described, none are specifically used in this book. Students may choose to use available codes for particular applications, and have done so, with varying degrees of success. Spreadsheet skills are generally sufficient to support the preliminary design process and such skills are quite valuable to those setting out on industrial careers. Because CAD courses are generally required in engineering programs, their use in preparing drawings is encouraged.

My experience in teaching design over the years has led me to embrace the use of basic analyses and reliable empiricisms so that students have the opportunity to learn some of those applied aeronautical engineering skills that have been edged out of modern curricula by reductions in the total credit hours required as well as the perceived need for broadening skills in other areas. Indeed, it is often the case that many students studying aerospace engineering have only a passing acquaintance with the airplane as a system that integrates many of the individual technologies they have been studying in analysis courses. Class meetings in a university setting rarely provide more than 40 contact hours for explaining the design process and for conferring with the instructor. Thus there must be a substantial amount of time spent outside class in preparing the aircraft design.

This book represents cumulative efforts of the author over a number of years of offering this course at the University of Florida, and before that, at the Polytechnic Institute of New York University in Brooklyn. Because of the wide diversity of material and techniques employed, errors are bound to appear. The responsibility for such errors is mine and I would appreciate learning of any so they may be corrected.

I would like to acknowledge the contributions to these notes over the years made by the late Professors A.R. Krenkel, B. Erickson, and G. Strom and to the students who have participated in my course over the years and often provided suggestions and corrections that proved to be of great value. Finally, I am delighted to once again thank my wife, Anne, for her continuing encouragement and support for my efforts in writing yet another book.

Pasquale M. Sforza
Highland Beach, FL
September 2, 2013

Introduction and Outline of an Airplane Design Report

Preliminary design

The preliminary design of a commercial aircraft is a feasibility study aimed at determining whether or not a conceptual aircraft is worthy of detailed design study. Only six statements are needed to define the mission specification and initiate a preliminary design study for a civil aircraft: type (private, business, or commercial), powerplant (reciprocating, turboprop, or turbofan), passenger capacity, cruise speed, cruise altitude, and maximum range in cruise. The information sought in a preliminary design study is described in the following list:

1. market survey: an assessment of the competing aircraft that have characteristics similar to those of the concept aircraft,
2. initial sizing and weight estimate: development of preliminary weight information based on range requirements and empirical data,
3. fuselage design: development of cabin design layout based on mission specification,
4. engine selection: terminal operations and cruise requirements set the range of engine possibilities and the wing size,
5. wing design: wing planform and airfoil selection, high lift devices, maximum lift capability, cruise drag including compressibility effects,
6. tail design: tail surface planform and airfoil selection, longitudinal and lateral static stability considerations
7. landing gear design: loads on landing gear, clearance issues, shock absorbers, tires and wheels, braking systems,
8. refined weight estimate: center of gravity location, control surface sizing, final layout of aircraft with all components properly placed,
9. drag estimation: careful analysis and accounting of the drag contribution of all components throughout the contemplated speed and altitude envelope
10. performance analyses: take-off, climb, cruise, descent, and landing performance of the final configuration,
11. economic analysis: capital cost of the airframe and engines, direct operating cost, and revenue generation,
12. final report and presentation: synthesis of the preliminary design in a report suitable for management, customers, and technical staff, version for oral presentation.

Final design

The tasks for final design are much more detailed and involve many more engineers than those for preliminary design. The more important areas of interest for final design are described in the following list:

1. detailed aerodynamic studies: computational fluid dynamics (CFD) analyses of the aircraft configuration, wind tunnel tests for verification of CFD studies and to assess particular aerodynamic issues, performance analyses throughout the flight range, preparation of aerodynamic models for stability and control analyses, handling qualities studies and simulations,
2. detailed structural design: detailed layout of structural components of the airframe, computational assessment of structural integrity, detailed weight estimation,
3. loads and dynamics: determination of aerodynamic, inertial, and other loads imposed upon the aircraft in all flight and ground operations, determination of the natural frequencies of the airframe and all important modes under excitation from engines, control surfaces, and other aerodynamics forcing functions, determination of unsteady loads and flutter effects,
4. design of subsystems: electrical, hydraulic, auxiliary power, avionics (guidance, navigation, and control), cockpit layout and human factors engineering, selection of equipment and vendors,
5. manufacturing considerations: detailed CAD drawings of all manufactured parts, assembly drawings, installation drawings, component interference evaluation, design of manufacturing tools, dies, and jigs, and assembly line design and preparation,
6. economics and sales studies: marketability analyses, customer feedback, cash flow analysis, capitalization,
7. prototype model: construction of demonstrator aircraft for ground, flight, and structural tests.

Preparation of technical reports

One of the most important tasks facing every engineer is the preparation of a technical report. This may be a document like a proposal, which seeks to engage the interest of a sponsor to financially support the technical task proposed, or one describing the work that has been carried out in completing a technical task. Typically, engineers enjoy performing the technical work required to solve the particular problem at hand, but often dread the planning, writing, and preparation of the technical document that describes the work.

A report is intended to present information clearly, and in a manner that is both self-contained and interesting. Conceptually, report preparation is rather simple, being in essence an edited log of the work that has been, or is proposed to be, performed.

Thus, it is convenient to keep a good journal of the work done along with the relevant background and illustrative material used. Though the technical work done may be well understood and appreciated by the engineer who carried it out day by day, this is not necessarily the case for other people who also need to know about that work. If the reader finds the report difficult to understand because the presentation is poor, then the engineer has wasted all the technical work done because the information cannot get beyond the person who actually did the work. So it is important to be sure that some basic requirements are met by the design report, such as the following: the reader should not have to search for important facts, the technical content should not be obscured by poor writing, and ambiguity should be avoided.

There is always some concern about the perspective of the report, that is, who is the reader? For design reports there are generally three classes of readers: business and sales executives, technical managers, and technical staff engineers. To satisfy this broad group with one report, it is common to include an executive summary, a main text, and detailed appendices. Executives generally read the brief executive summary to clearly understand the general approach and results of the study. Technical management personnel read the executive summary and the main text so as to be able to guide the executive group as necessary. The technical staff needs all three sections since they may be called upon to review detailed questions from the other two groups who are involved in making major business decisions.

Outline of the design report

Aircraft companies have standard formats for reports used both within and outside the company. Though the details may vary from company to company, there is a general outline that tends to be followed. The particular format to be followed here is shown in outline form as follows:

Chapter Number	Chapter Heading
	Title page
	Executive Summary
	Table of contents
1	Introduction and Market Survey
2	Preliminary Weight Estimate
3	Fuselage Design
4	Engine Selection
5	Wing Design
6	Tail Design
7	Landing Gear Design
8	Refined Weight Estimate
9	Drag Estimation
10	Performance Analysis

Chapter Number	Chapter Heading
11	Airplane Pricing and Economics
12	Conclusions and Recommendations
	References
	Appendices

The report should be prepared using a computer for word processing, for production of graphs, and for production of drawings using CAD systems, like Pro/Engineer. It is likely that a well-prepared and -presented report will be of great value in preparing similar reports and in improving one's design and communication capabilities.

Title page, executive summary, and table of contents

The title or cover page should be informative and attractive to the prospective reader. Therefore it should be laid out in a logical and interesting fashion and should give the title of the report, the author, the organization issuing the report, the name of the organization for which the report was prepared, and the date of issue of the report.

The executive summary of the report is a brief, but complete, narrative description of the report. Its length will vary, depending upon the size and complexity of the report, but in most cases should be kept to one or two pages. This summary, as its name implies, is intended for a prospective reader in the executive ranks of a corporation or government agency, one who needs to know the nature and outcome of the report but cannot devote the time to read through the whole report. In this sense it is perhaps the most important section of the report since it will be read by executives who can say "yes" to a proposal of this sort. Though many echelons of industrial personnel may say "no," the approval of a major decision-maker will make these technical critics pay close attention to the details within the body of the report before bringing a "no" to the upper levels. Therefore it is important to make the best possible case for your proposed design and to include some of the most interesting and compelling results the report contains. Remember that the objective of writing the report is to communicate your ideas to others and to persuade them of the value of your work to them. In the particular instance of design reports, the author is actually carrying out part of a sales activity, where the author is trying to get the reader, or reader's organization, to financially support the proposed design.

The table of contents provides still more information to the reader by laying out the chapter headings, subsection headings, etc. and their page numbers in the report. One may learn more about the details of the format requirements described here by studying the textbook reference materials, and other publications. There are generally accepted norms for the layout of technical reports and one should keep an observant eye on how this is done by others in the field.

Main text

The main text of the design report parallels the outline of this book. Chapter 1 is the introduction to the design case studied and describes the mission profile in detail. The market survey portrays details of the competing aircraft that most closely meet the specified mission requirements. The preliminary weight estimate is covered in Chapter 2 and illustrates how the aircraft is sized as to total weight, empty weight, fuel weight, etc. Analytical techniques and historical correlations are used to generate the weight make-up of the proposed aircraft that will be used to carry the design along through more detail. Fuselage design comes next in Chapter 3 and planning and sizing the overall configuration culminates in the generation of three-view drawings providing the outline of the proposed aircraft. Using the gross size and weight characteristics from the previous chapters permits the selection of appropriate engines for the proposed aircraft and this comprises the content of Chapter 4. The take-off, landing, and cruise constraints on the design are applied to narrow the choice of engines among those that are currently available. Chapter 5 addresses the design of the wing by utilizing methods that form an industry-wide standard basis for comparative aerodynamics and stability and control evaluations. After the design is taken this far, the stability and control surfaces of the proposed aircraft, i.e., the horizontal and vertical tails, are sized and positioned on the aircraft in Chapter 6. The requirements for the size and location of the landing gear and the associated ground handling characteristics of the aircraft are developed in Chapter 7. Then, having determined the size and placement of all the major components of the aircraft in the previous chapters, a refined weight estimate and center of gravity location are carried out in Chapter 8. In Chapter 9, a careful drag estimate is carried out, again using common industrial approaches. Chapter 10 presents an analysis of the performance potential of the aircraft in all aspects of its mission profile, as carried out using standard techniques. An assessment of the cost of producing and operating the proposed aircraft is carried out in Chapter 11 using a variety of techniques developed by government agencies and industry associations. Finally, there should be a Chapter 12 which presents the conclusions and recommendations arrived at in carrying out the design of the proposed aircraft for the specified mission. Following that would be a list of the references quoted and the appendices that may have been cited within the main text.

Suggestions for report preparation

Report preparation should proceed in a timely manner; in a university setting this should be carried out regularly during the course presentation and not be delayed to the end of the semester. The reports should be prepared on a word processor and attention should be paid to regularly backing up files. Drawings are to be made on any CAD system available. Again it is important to remember to back up all files. The report should be submitted as a spiral bound or stapled document printed on standard white 8.5 × 11 inch paper. The pages must be numbered consecutively using a standard system throughout the report and should be consistent with the table of contents.

It may be useful to review some of the common errors in report preparation which are listed as follows:

1. Title page omitted or incomplete.
2. Table of contents omitted or paginated incorrectly.
3. An important section of the report, like the executive summary or a main test chapter, is omitted.
4. References enumerated properly at the end of chapters, but not cited in the text of the chapter, or vice versa.
5. Figures captioned properly as they appear, but not cited in the text, or vice versa.
6. Pages are not numbered sequentially, or are omitted entirely.
7. Three-view drawings of the aircraft, and similar supporting drawings are omitted or are of poor quality.
8. Printing is of poor quality or non-standard font employed.
9. Spelling, grammar, and punctuation are poor.
10. Material presented indicates that the author(s) poorly understands it.
11. Extensive quoting, or even plagiarizing, of previously published material.
12. Improper use of appendices; repetitive material, tables, etc. are to be incorporated into appendices so as to keep the flow of ideas smooth in the main text.
13. Improper inclusion of calculations in the main text; there should be sufficient information, such as equations, in the report to permit another engineer trained in the art to reproduce the results shown, but no actual incorporation of numbers therein.
14. Uneven emphasis among chapters, usually indicating varying degrees of effort among contributors.
15. Too much repetition or discussion of irrelevant material.
16. The conclusions reached by carrying out the work are not clearly stated.

Suggestions for preparing graphs and other figures are as follows:

1. Figures should be introduced as part of the text, with appropriate numbering and captioning.
2. The figures should be placed near to where they are described, so that they can be easily found.
3. Figures should be self-contained so that, with the descriptive caption, their content can be easily grasped.
4. The ordinate and abscissa of a graph must be clearly labeled, including units and scale divisions; a somewhat heavier line width may be used for the axes.
5. Scales must be chosen appropriately so that the behavior being described in the text is apparent; auto-scale features of graphing packages must be scrutinized in this regard.
6. Analytic results should be indicated by lines that are distinguishable without color since reports are often Xeroxed or faxed without benefit of color; different line styles, such as dotted, dot-dash, etc. may be employed.

7. Experimental data should be represented by discrete symbols; again avoid using symbols of different color.
8. Maintain some degree of uniformity among the symbols and lines used on graphs dealing with essentially the same subject for ease of interpretation.
9. Extrapolations and interpolations should be indicated by a change in the line style, such as changing from solid to dotted lines.
10. Major grid lines in both coordinate directions should be shown.

Market Survey and Mission Specification

1.1 A growing market for commercial aircraft

The fleet of commercial aircraft is aging and new market pressures portend a large and growing market for new aircraft. Approximate average ages of the fleets of representative US and international airlines as of the end of 2011, according to Airfleets (2012), are shown in Figure 1.1.

The escalation of fuel prices, increased concern over environmental effects, and changes in the nature of airline services to passengers are pushing airline operators into renewing their fleets. Manufacturers are therefore putting intense efforts into designing and producing aircraft that will provide superior economic and environmental performance. Sales of US commercial aircraft rose from a low point in 2003

Commercial Airplane Design Principles. http://dx.doi.org/10.1016/B978-0-12-419953-8.00001-2

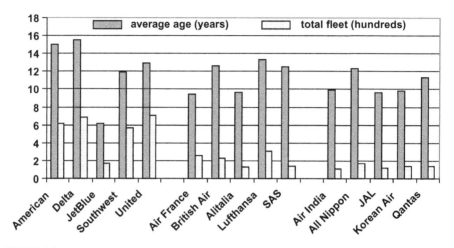

FIGURE 1.1

Size and average age of the fleets of US international airlines as reported by Airfleets (2012).

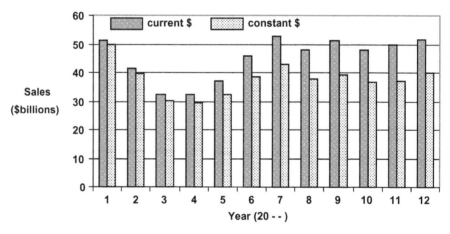

FIGURE 1.2

Annual sales of US civil aircraft over the period 2001–2012 as reported by the Aerospace Industries Association (2012). The numbers for 2012 are estimates. The deflator used by the AIA assumes 2000 as the base year.

up to a fairly constant value starting in 2007, as can be seen in Figure 1.2, as reported by the Aerospace Industries Association (2012).

Airbus has had some setbacks, in both the new A380 and the developing A350, and now has somewhat more than 40% of market share, a figure that is often taken

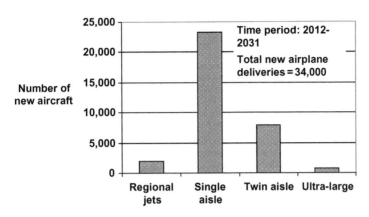

FIGURE 1.3

Number of new aircraft deliveries forecast by Boeing Current Market Outlook (2012) for the period 2011–2030 grouped according to aircraft type.

as an indicator of serious competition. The Airbus figures owe a great deal to the continued success of the narrow-body A320. However, in the wide-body field it is losing out to Boeing, whose 777, 747-8, and the new 787 are all doing very well in terms of orders, in part because of A350 development delays and A380 operational problems, like wing cracks, described by Flottau (2012). Though it is anticipated that sales will grow somewhat, as shown in Figure 1.2, there is concern that the world economic situation in general, and the rapidly escalating fuel prices in particular, may reduce these estimates. The former suggests possible drops in demand while the latter forces fleet upgrades to newer, more fuel-efficient aircraft. The net result is currently uncertain, though precipitous drops in sales are unlikely. In any event, the backlog of orders is currently so large that if there is a drop-off in sales the effects won't be felt for several years, giving manufacturers an opportunity to revise their business strategies.

The Boeing Current Market Outlook (2012) calls for about 34,000 new commercial aircraft to be delivered over the next 20 years, as shown in Figure 1.3, constituting a doubling of the fleet in service. The major segment of this new growth in airplane deliveries is predicted to be in single-aisle, or narrow-body, aircraft, a segment expected to more than double in size. This reflects Boeing's confidence in the move toward more direct flights, rather than hub and spoke flights, and this outlook is felt to be in response to passenger demand, as well as the prospect for reduced fuel burn by using more direct flights.

This potential growth is also encouraging new aircraft manufacturers. China, Russia, and Japan are all developing 70–90-seat regional jets providing direct competition to current regional jet leaders Embraer and Bombardier. However, the regional jet market is seen to be relatively weak and all these manufacturers are contemplating moves toward larger size (100 plus seats) jets where they will be starting to compete

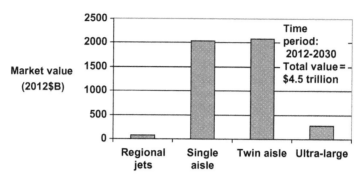

FIGURE 1.4

Market value of new airplane deliveries forecast by Boeing Current Market Outlook (2012) for the time period 2012–2030 grouped according to aircraft type.

with Boeing and Airbus. The short-range, or regional, market is slowly returning to turboprop-powered aircraft, largely because of the better fuel economy they offer.

However, another passenger demand is increased comfort and this is best filled by twin-aisle, or wide-body, aircraft. Not only are pressures from business travelers making international carriers expand their business class sections, but now seats that fold flat into beds are becoming a major factor in airline competition. The long-range capabilities of newer aircraft make these upgrades increasingly important. It is not unusual to have 16- and even 18-hour flights, like Singapore Airlines' direct flight from Singapore to Los Angeles and to Newark. Although the projected number of twin-aisle aircraft to be delivered in the 20-year period is only about one-third that of single-aisle aircraft, the market value of the twin-aisle aircraft is about the same as that of the single-aisle aircraft, as illustrated in Figure 1.4. The Boeing Current Market Outlook (2012) forecast for the change in the size and make-up of the fleet is summarized in Figure 1.5 which shows the fleet complexion in 2011 compared to that projected for the year 2030.

The Boeing Current Market Outlook (2012) also highlights the rapid growth and industrialization of the Asia-Pacific region as illustrated by the market value share breakdown according to region shown in Figure 1.6. China is expected to lead in capacity growth over the period considered, at a rate of about 8% and by 2027 will have half the traffic of the North American market. Currently it has less than 20% of that market. Expansion of the European Union and the Commonwealth of Independent States of the former Soviet Union is expected to result in their domestic traffic overtaking the North American domestic traffic in the near term, then falling back to a lesser value by 2026.

The International Civil Aviation Organization (ICAO) released preliminary figures for the year ending 2011 indicating that passenger traffic on the world's airlines increased by about 6.5% over 2010, in terms of passenger-kilometers performed, according to ICAO (2011). Passengers carried on scheduled services grew by around

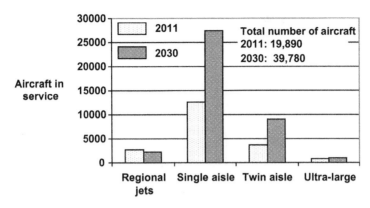

FIGURE 1.5

Forecasts for the change in the size and make-up of the fleet by the year 2030 grouped according to aircraft type.

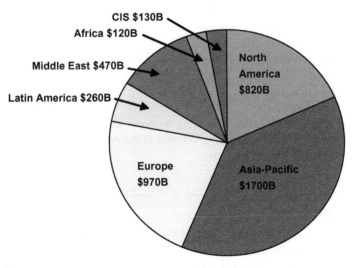

FIGURE 1.6

Forecast of the market value of the purchases of new aircraft during the time period of 2012–2031. Total market value = $4.5 trillion.

5.6% to 2.7 billion; the trend over the past 10 years is illustrated in Figure 1.7. Statistics supplied by ICAO's 190 Contracting States also show an overall increase of about 4.3% over 2010 based on metric ton-kilometers performed; this is a measure which includes passenger, freight, and mail traffic. The global passenger load factor was 78%, about the same as in 2010. Freight carried worldwide on scheduled

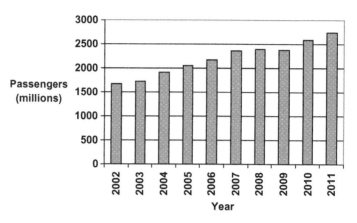

FIGURE 1.7

Airline passengers carried over the 10-year period 2002–2011 as reported by the ICAO (2011).

services in 2011 went up to approximately 51.4 million metric tons compared with about 50.7 million metric tons in 2010.

The ICAO points out that the continued high growth rates in passenger traffic across all regions in a difficult world economic environment was a positive for 2011. They forecast 3.5% growth in the world economy over the next 2 years resulting in a 6% growth in revenue passenger-kilometers over the same period. However, the uncertainty regarding the weakly growing economy in the United States and Europe, which could expand to other regions, coupled to the erratic behavior of growing fuel prices makes predictions questionable.

1.2 Technology drivers

Airliner design is a rather mature field and it is unlikely that improvements will involve breakthroughs. Instead, the persistent application of small cumulative advances will provide competitive advantages for the manufacturer that can successfully integrate them into their product. The areas of greatest promise center around reductions in weight, drag, and engine emissions, as well as noise abatement techniques. Wall (2006) reported on the initiation of a "Clean Sky" program by the European Union as a major part in the Aeronautics and Air Transport portion of its 7-year, $69 billion, Seventh Framework Program. The objective was to research an array of emissions-reducing technologies that could be ready enough to be incorporated into designs for aircraft entering service around 2020. This aeronautical research program, which included demonstrator engines, was funded at a level of about $2 billion and now a second, larger phase, to run through the 2014–2016 period is being drawn up, according to Warwick (2012).

In the summer of 2006 jet fuel prices rose to over $2.20 per gallon in New York and remained fairly stable through the summer of 2007. However, by the summer of 2008 the price almost doubled, increasing to almost $3.85 per gallon. During the global economic slump of 2008–2009 the price dropped back down to about $1.25. By 2011 the price had slowly climbed back up to around $3.00 and has remained around that value through the middle of 2012. Crude oil futures costs are volatile and are expected to average $115 per barrel, leaving most manufacturers feeling the pinch to improve fuel economy. After a brief period of profitability following the losing years between 2002 and 2006, the airlines are again facing the prospects of substantial losses and are seeking new ways to reduce costs. Boeing's emphasis on designing the new 787 as a particularly fuel-efficient airliner has proven to be a remarkably good decision. Airbus has looked to the A350 to provide similar attractiveness to the airlines. In the high-volume single-aisle narrow-body aircraft category the fuel efficiency mantra has been at the center of the heated battle between the B737Max and the A320Neo, even though both are still in the development stage, as noted by Norris (2012). The high oil prices have also served to cool off the market for regional jets as airlines look to larger-capacity aircraft which have lower fuel costs per seat.

Environmental concerns are pushing airlines to demand "greener" and more fuel-efficient aircraft. This is of particularly acute concern since airline fleets are aging and next-generation replacements for the immensely popular single-aisle aircraft like the Boeing B737 and Airbus A320 are now being designed. Operators would like to phase out older aircraft for new ones that burn less fuel, can meet more stringent environmental regulations that are looming in the future, and can spend more flight time between overhauls. Currently the fleet average fuel economy is about 20 km/l (48.2 mpg) per passenger and this is expected to rise to about 34.5 km/l (83.4 mpg) as new products with performance like the B787 and A350 are fully in service with the airlines.

Reducing the perceived and actual environmental impact of aircraft operations is being sought both by manufacturers and operators. Although a major effort in such areas is bound to involve political solutions ranging from taxes for exceeding tightening standards to sophisticated emissions brokering for trading credits on pollutants, the returns from technological improvements in all aspects of aircraft design, manufacture, and operation will likely be more rewarding. The effects of gaseous pollutants from engines, noise from engines and airframes and terminal maneuvers, and toxic materials used in production are the major issues under study.

1.2.1 Fuel efficiency

Aboulafia (2012) presents an interesting forecast of the airline business based on the changes brought by the operational and structural improvements forced by intense competition over the last decade. He suggests that the deciding factor for airline success is in the exploitation of technological advances that reduce fuel burn. The figure of merit for fuel efficiency is the specific fuel consumption which is defined

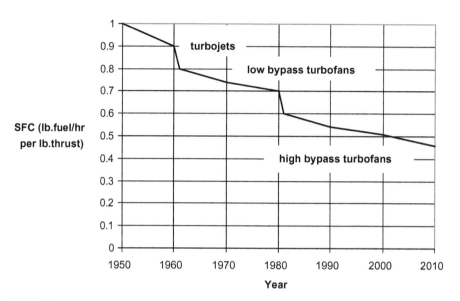

FIGURE 1.8

Trend in specific fuel consumption (SFC) since the introduction of the jet engine into commercial service.

as the weight of fuel consumed per hour per unit of thrust. Improvements in specific fuel consumption over the time period since the introduction of jet engines into commercial service are indicated by the trend line shown in Figure 1.8. The major contribution to the improvements arises from the introduction and development of the turbofan engine as discussed by Sforza (2012).

Airline operators have seen fuel costs overtake labor costs as the largest item in the breakdown of operating costs as we will show in Chapter 11. US airlines spent over $50 billion on fuel alone in 2011, according to Anselmo (2012), helping keep airline profit margins down to 2%. Though airlines have grown sophisticated in hedging the cost of crude oil purchases, there is an additional cost premium for refined jet fuel. The volatility of these premiums has caused one airline, Delta, to develop a subsidiary, Monroe Energy, with its own fuel refinery in an attempt to have additional control over fuel prices. The promising capabilities of new aircraft, like the competing single-aisle offerings of Boeing's B737Max and Airbus's A320Neo, will push production up as airlines move to upgrade their fleets. The larger two-aisle wide-body aircraft are less technology-dependent, Aboulafia contends, and carriers can lean on equipment upgrades, particularly in the area of premium seating accommodations and cabin refinements.

Another turnabout caused by the rise in fuel costs is the revival of interest in turboprop aircraft. The common vision in the airline industry was that turbofan aircraft were more attractive to travelers than turboprops in the short-range regional market.

However, the unrelenting increase in fuel costs prompted a second look at turbo-props, especially because they can save as much as 30% in fuel consumption per seat compared to jets. Although projections for the growth of the near-term market for turboprop aircraft currently appear modest, the outlook for a decade into the future seems to be brightening. That this is the case is fortified by the news that five aircraft manufacturers are considering developing new, larger (90-seat) turboprop aircraft, according to Perrett (2013). The companies are widely spread internationally, with Asian entries from China, India, and South Korea, as well as the current Western manufacturers ATR in Europe and Bombardier in Canada.

1.2.2 Weight reduction

Weight reduction and control is a central theme in airplane design since it is the major determinant in performance. Part of the performance is connected with fuel burn since a more fuel-efficient aircraft needs less volume to store the fuel, reducing the empty weight of the airframe. Furthermore, requiring less fuel per trip means a reduced takeoff weight, and reduced total weight requires less thrust in takeoff and climb. Naturally, burning less fuel per trip also reduces the emissions of CO_2, NO_x, and other pollutants, making the airplane more environmentally efficient. With fuel costs becoming increasingly important in revenue operations, the ability to reduce fuel usage then reduces both costs and environmental impact.

The new Boeing B747-8, a successor to the B747-400, has a market niche in between the 365-seat B777-300ER and the 525-seat Airbus A380. Norris and Flottau (2012) report that Lufthansa's first revenue transatlantic flight of the B747-8 showed about 3% lower fuel burn than would have been achieved with their B747-400 aircraft. Planned performance improvement packages for the larger General Electric GEnx-2B engine will help reduce fuel burn, but in addition, weight reductions of up to 5000 lb are planned, as are aerodynamic enhancements. That weight reductions are seen as very important in this regard is made clear when it is realized that this planned weight reduction is quite small, amounting to only about 1% of the empty weight of around 470,000 lb, which, in turn is about 48% of the planned maximum takeoff weight of 987,000 lb. Other relatively small, but cumulatively important weight savings accrue from details like reduced galley equipment and cargo container weight, reducing the potable water carried, maintaining control of the amount of contingency fuel carried, and using lighter seats.

The move toward composite materials in airframe, engine, and cabin equipment is part of the drive to reduce aircraft empty weight. The last decade has seen the weight fraction of composite material used in commercial aircraft rise from about 10% to over 50% in the search for weight reductions. This deep penetration of composites into major structural members is characteristic of new programs. For example, the next-generation Boeing 777X is planned to have all-composite wings which are expected to yield substantial weight savings. The B787 uses composite structures for the fuselage where their increased resistance to hoop stresses allows substantial cabin pressure increases. Previously, aircraft cabins were pressurized to the equivalent of

8000–9000 ft altitude. Now the B787 can safely provide a cabin pressure equivalent to a 6000 ft pressure rating which results in increased comfort for passengers.

However, uncertainties concerning the effects of long-term exposure to extreme environmental conditions on the structural and functional properties of polymer matrix composites have thus far been a deterrent to their widespread use. There are challenges in engineering polymer matrix composites and the cost of testing and quality assurance costs can limit the benefits of their use. Because the movement is continually in the direction of increasing use of composite materials, the aircraft manufacturers, airline operators, and regulatory bodies are cooperating on the standardization of materials, techniques, and training certification.

1.2.3 Drag reduction

Drag reduction and overall improvements in flow control are expected to yield continuing improvements in performance. Aerodynamic improvements of airliner wings are focused on wingtip treatments, both in planform shape and in continually evolving winglet designs aimed at reducing lift-induced drag in cruise. Both the B737Max and the A320Neo are using the latest version of winglet which incorporates two fins, one tilted up and one down, on each wingtip. Vortex generators are used on wing surfaces to energize boundary layers thereby avoiding separation and extending the range of attached flow. Greater attention to detail, such as improving seals and minimizing gaps on aerodynamic surfaces, has become a necessity in the drive to reduce drag. Similarly, simple operational improvements like keeping aircraft surfaces washed and free from the roughness caused by dust, insects, etc., can pay a dividend in reduced drag.

A new area of practical drag reduction involves exploiting many years of research in boundary layer control, particularly techniques to maintain laminar flow over large surface areas. Boeing is introducing hybrid laminar flow control on the horizontal and vertical stabilizers on their new stretched B787-9 in an attempt to reduce profile drag by 1%. If this technique, which has been flight-tested, can be effectively employed on the wings, fuel burn may be reduced by as much as 20% on typical flights. Research is also being carried out in relatively new areas such as synthetic microjets for localized flow management and plasma generators embedded in aircraft skin which can locally ionize the flow and achieve electromagnetic control of the flow.

1.2.4 Engine design

The availability of new engines that promise greatly improved fuel efficiency is the basis for the advanced single-aisle B737Max and A320Neo aircraft. In mid-2008 Pratt & Whitney introduced the PW1000G, the geared turbofan, which incorporates the results of research initiated a decade earlier. This engine improves fan performance by matching its speed to the driving turbine more effectively as well as by incorporating a variable area fan nozzle. The PW1000G, which has completed over one hundred hours of flight testing, is the first in a line labeled "Pure Power" engines by Pratt & Whitney.

CFM International, a joint venture by GE Aviation and SNECMA that brought out the successful CFM56 series, is bringing out the new LEAP-X (Leading Edge Aviation Propulsion) engine. This engine utilizes GE's advanced combustor technology along with composite, rather than titanium, fan blades to provide its performance improvements.

One of the other approaches for improved fuel efficiency is the revisiting of the unducted fan, now often called the open rotor, which sacrifices some turbofan speed for improved fuel economy. This engine, which lies between a conventional turboprop engine and a turbofan engine, was extensively studied in the 1980s when inflation and increased oil costs made fuel economy a major issue. The higher propulsive efficiency of a propeller is achieved at the cost of maximum speed, as shown by Sforza (2012) and the unducted fan may be an optimal solution for some airplanes. In this regard it is interesting to note that in 2004 the turboprop fleet constituted 23% of the world passenger fleet and is expected to continually drop over the next 20 years to the point where they will comprise only 1% of the world fleet. There is also seen a continuing trend of reduced travel on 300-mile routes, apparently indicating a growing preference to drive rather than fly over those distances. But skyrocketing fuel prices are forcing manufacturers to reconsider turboprops in new airplane designs. Turboprops are much more fuel-efficient than turbofans and are well-understood propulsion systems, but after weaning the public away from propellers to jets the industry faces the challenge of winning them back to propellers. The widespread public awareness of environmental issues, as well as the economic factor, will be helpful in this regard.

Engine design is aimed at producing the least polluting strategies for combustor design over the entire operating regime of the engine. This is part of a much larger research program covering combustion in all its guises, including automobiles, trucks, off-road equipment, and stationary power plants. At the heart of the matter is the extent to which the combustion of hydrocarbon fuels produces two important gases: CO_2 and NO_x, both of which are associated with atmospheric degradation. The former is a so-called greenhouse gas whose molecules are effective in trapping the infrared radiation emitted by the surface of the Earth and reducing the possible cooling effect. The latter is a chemically active gas involved in complex reactions with other gases in the atmosphere, and can contribute to depletion of ozone in the upper atmosphere that otherwise would help reduce the ultraviolet radiation from the sun from reaching the Earth's surface. Unfortunately, the production of CO_2 and NO_x is closely related and when the concentration of one is reduced, that of the other is increased. Therefore, very strict combustion control is necessary to find an optimal middle ground for the acceptable production of each. Some current research programs in these areas are reported by Chang et al. (2013).

1.2.5 Carbon footprint

The entire aviation industry is considered to produce around 2% of the CO_2 emissions produced by fuel-burning activities in the world (for comparison, the electricity and heating sector produces about 32%) and is becoming keenly aware of its role in terms of its "carbon footprint." The European Union has developed rules concerning the inclusion of aviation in its emissions trading scheme and it is likely that the US, and ultimately the rest of the world, aviation community will come under some similar kind of regulatory system. As emissions trading protocols involve economic penalties for producing more than some baseline amount of emissions, there is considerable scrutiny of the means to achieve the required regulatory goals. The carbon footprint for an individual passenger flying between two cities may be estimated using the online calculator available from ICAO (2012). For example, one passenger flying a round trip from JFK airport in New York to FCO airport in Rome generates an average of about 900 kg of carbon dioxide. The methodology used to form the estimate is also described. Information based on modeling analyses of this type is used in determining carbon offsets for use in different environmental trading programs.

1.2.6 Biofuels

Fuel cost and availability concerns caused increased attention to be paid to synthetic hydrocarbon fuels as an alternative to standard Jet-A fuel. Synthetic paraffinic kerosene (SPK) is generally made by converting coal or natural gas into a liquid fuel (often designated as CTL or GTL products, respectively) using the Fischer-Tropsch process, a technique developed in Germany in the 1920s and used to produce fuel during World War II. However, such fuels, and others derived from ethanol, are often less energetic than Jet-A, and suffer from other operational disadvantages, such as their behavior at the low temperatures typically met at operational altitudes. Tests at Tinker AFB in Oklahoma on a TF33 engine burning a half-and-half blend of synthetic kerosene and JP-8 performed the same as 100% JP-8. The combustion of the synfuel blend resulted in a 50% drop in sulfur emissions, since the Fischer-Tropsch process produces no sulfur in the synthetic kerosene. This elimination of sulfur is one of the major advantages of CTL and GTL fuels since coal and natural gas are generally quite rich in sulfur, while a disadvantage is that the process itself is energy-intensive. The test was carried out using fuel produced by Syntroleum, a Tulsa, Oklahoma refinery. Flight tests using different fuels have been carried out at an increasing rate and the degree of success in these tests may influence the commercial sector in turning to a blend of Jet A and synfuels, and ultimately perhaps to pure synfuels. South African Airways has been flying blends of CTL and Jet A for many years. The CTL is produced by Sasol in Secunda, South Africa.

Because of the growing number of possible synthetic or biofuel production mechanisms the technical standards organization ASTM International has been active in streamlining processes for approving new fuels to replace the standard jet fuel, known as Jet A, including CTL and GTL synthetic fuels, and examining the possibility of including biofuels. The ASTM has developed the D7566 specification for synthetic jet fuels including Fischer-Tropsch fuels and hydroprocessed esters and fatty acids (HEFA) fuel. In any event the inclusion of those fuels will generally only be permitted blends of up to 50% with ordinary petroleum-based fuel. Tests of biofuels produced from vegetable oils (Generation 1 biofuel) have been carried out, but concern about diversion of food stocks to fuel has turned emphasis to pursuing non-food sources, such as cellulose waste, non-food plant oils, and algal oils (Generation 2 biofuel). Current interest centers on non-food plants like Jatropha in Africa, Pongamia in India, and the castor oil plant in Brazil, while algal oils are considered to be later entries into the alternative fuel competition. Warwick (2012a) presents a brief but comprehensive review of the many directions biofuel producers are considering.

1.2.7 Alternative fuels and power sources

Hydrogen, which is often touted as an alternative to hydrocarbon fuels because of its clean-burning characteristics and high energy content per unit mass, suffers from an inadequate infrastructure for producing, distributing, and storing the product for use on a large scale. In addition, because of its low density, hydrogen suffers from relatively poor energy content per unit volume, a characteristic that necessitates larger tanks, and therefore larger and heavier aircraft. There is also interest in fuel cells, which still require fuel, generally hydrogen, with all the problems mentioned earlier. However, for the airline industry the application of fuel cells will likely be for onboard electrical power generation rather than propulsion. In this manner the propulsive fuel burn will be reduced since the main engines will not need to provide for all the auxiliary power demands. For example, electricity provided by fuel cells can power electric motors mounted in the landing gear and would be used to taxi the aircraft.

Surface treatments of airplanes to control corrosion often include toxins, like cadmium and chromium, which manufacturers would like to replace with environmentally benign substances. Even at the end of an aircraft's life there is concern about its environmental impact and manufacturers are seeking ways to improve the recycling capability of its aircraft. Boeing has teamed with a group of companies to form the Aircraft Fleet Recycling Association whose goal is to establish standards for disposing of aircraft. Wall (2008) reported on efforts carried out by Airbus which showed the possibility of converting 85% of the aircraft weight into salable products, typically as secondary raw material. The large percentage of aluminum in aircraft currently reaching the end of their life cycle is particularly valuable since its production is energy-intensive and recycling it uses 90% less energy. Composite materials are less of a concern at the moment because they are just entering the fleet and have considerable lifetime remaining before their disposal is imminent.

1.2.8 Noise and vibration

Noise and vibration control is another area of strong environmental concern and involves engine and airframe design, as well as operational factors such as terminal maneuvers and trajectories. Quieting the cabin during long cruise periods is of great concern to passengers, while sideline noise and flyover noise footprint is important to airport staff and residents in the airport operational area. To deal with these issues, research by engine manufacturers is ongoing in improving the design of inlets, fan and compressor blades, and nozzles, while airframe builders are studying landing gear and flap design, and airline operators are examining flight operations and trajectories that would alleviate some of the noise problems.

The largest European Commission research project dedicated to aircraft noise reduction, SILENCE(R) was started in April 2001 with the French engine manufacturer Snecma acting as project coordinator. By completion of the project in late 2007, improvements in noise reduction technology had been achieved in several areas. Because engine and airframe each generate about equal contributions to the total noise produced by an aircraft both contributions were studied. In the area of airframe noise, an important practical improvement was made on acoustic liners for engine nacelles by the development and assessment of zero-splice inlet liners. Since conventional liners are manufactured in several pieces (typically three), splices are necessary to join the pieces together. It is well known that these splices have a negative effect on the performance of the liners. This is not only because of the reduced area covered by the lining material, but also because the splices cause the prevailing circumferential acoustic modes to scatter to other modes, which are less effectively attenuated. Large-scale research on zero-splice inlet liners has been conducted on the Rolls-Royce Trent engine. Airbus has already patented the zero-splice inlet and installed it on all A380 engines, contributing to the remarkably low-noise levels registered during the aircraft's acoustic certification process. In future, it will be fitted as standard on all new Airbus aircraft and may well be adopted by other manufacturers as well. The extensive airframe noise tests also focused on technologies to reduce landing gear noise and noise generated by high lift devices. Flight tests were carried out on an Airbus A340 with landing gear fitted with low-noise fairings. For future applications, more comprehensive changes to the landing gear configuration may be needed to reduce noise.

Envia (2012) describes NASA's current and future engine noise reduction research, focusing on three areas: alterations to the engine cycle, application of noise reduction materials, and reducing engine noise propagation by judicious shielding of the engine by airframe components. The trend of noise reduction with year of aircraft certification is shown in Figure 1.9. The noise level is given in terms of effective perceived noise level measured in decibels (EPNdB) which according to FAR Section A36.4 is a single number evaluator of the subjective effects of airplane noise on human beings. It consists of the instantaneous perceived noise level corrected for spectral irregularities and for duration. The stage levels shown in Figure 1.8 correspond to the maximum noise levels permitted as described by the FAA Part 36 regulations. The research goals for the 2020–2025 time frame are also shown in Figure 1.9.

The most effective engine cycle change for reducing noise involves increasing the bypass ratio, that is, the ratio of mass flow of air passing through the engine fan to the mass flow of air passing through the engine core where the fuel is added and burned. The turbojet engine, which has a bypass ratio equal to zero, became the engine of choice for airliners in the mid-20th century because of its ability to maintain high thrust at high speeds, unlike propellers. However, incorporating a fan into

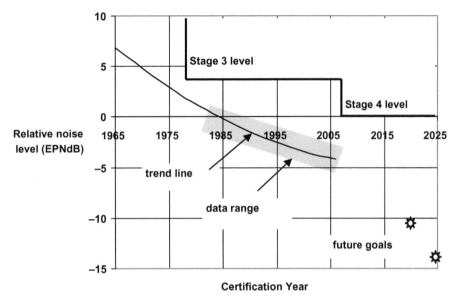

FIGURE 1.9

Noise levels for jet transport aircraft as a function of the year of aircraft certification relative to Stage 4 airplane requirements. Also shown are the goals of current research.

the turbojet engine can provide substantial improvements in fuel efficiency without significant reductions in thrust at high speed, as shown by Sforza (2012). The turbofan engine became the engine of choice in the latter part of the century accompanied by continuing increases in the bypass ratio. Currently the largest turbofan engines, the GE90 and Rolls-Royce Trent 1000, have bypass ratios between 9 and 10. The improvements in fuel efficiency as the bypass ratio grew also provided a reduction in the engine noise level. Carrying the high bypass ratio to even higher levels brings one to the realm of ultra-high bypass turbofans and ultimately to the open rotor, or unducted fan, as indicated schematically in Figure 1.10.

Reduction and containment of engine noise may be achieved by constructing the fan casing of a sound-absorbing material. Foamed metal material like cobalt or an iron-chromium-aluminum alloy is used within the casing surrounding the fan. The pores in the metal are instrumental in dissipating the acoustic energy of the noise. Another engine structural modification aims at mitigating noise associated with the interaction of the aerodynamic wake of the moving fan rotor with stationary guide vanes. The so-called soft vane concept is based on providing hollow chambers within the stationary vane, each of which is sized to act in damping specific frequencies of the spectrum of time-varying pressure field seen by the vane.

Noise blocking by appropriate integration of the airplane structure with the engine placement is another approach. To be most effective, the engines must be placed above the wing or fuselage so that those structures can block some of the

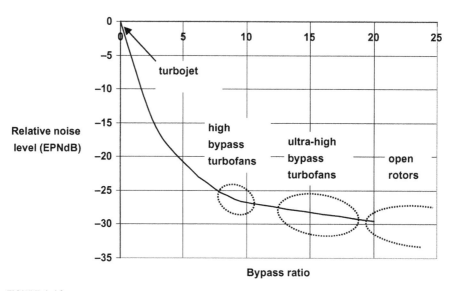

FIGURE 1.10

Noise reduction compared to turbojet for increasing bypass ratio.

radiated engine noise from reaching the ground. In a similar manner tail surfaces may serve to block sideline engine noise. Of course, such alterations in the aerodynamic shape must be carefully analyzed to ensure that achieving noise reduction does not cause degradation of the overall airplane performance. For example, a negatively scarfed, or raked-back inlet face, can change the directional pattern of the radiated engine noise, so that more noise will be directed upward, and less noise downward. However, application of the concept to existing aircraft requires extensive redesign and is unlikely to be economically viable.

1.3 Cargo aircraft

Globalization and the attendant increase in international trade make air freight an increasingly important component for the movement of goods. In addition, growing concerns about environmental impact make fleet modernization a necessity. The market for cargo aircraft is estimated to be in the range of $100 billion and this is driving aircraft manufacturers to be competitive. The Boeing B747-8F is a new freighter aircraft that was first delivered in late 2011. It is the counterpart of the international passenger version, the B747-8I. Operators of the B747-8F are reporting fuel burn about 1% better than expected in the first group of aircraft in service. But Boeing is targeting additional improvements in performance as production continues. The Airbus A380F was in development but has been put on hold in order to concentrate on issues with the A380 passenger version that were considered more pressing. The

A330-200F, a freighter based on the A330 family, entered service late in 2010. A substantial fraction, around 20%, of earlier wide-body passenger aircraft that are being retired in favor of newer, more fuel-efficient aircraft, will be converted to freighters. These include earlier aircraft like the following: McDonnell Douglas MD10 and MD11, Boeing B747-200, and Airbus A300B, and later aircraft like Boeing's B747-400 and B767 and Airbus A300-600. This will provide additional work for maintenance, repair, and overhaul companies (MROs) around the world.

More than half of international air cargo shipments are carried on pure freighters with the remainder carried aboard passenger aircraft and this fraction is expected to grow in coming years. The impetus for shipping by air is delivery speed, but this is tempered by the increased cost, and if costs can be trimmed there will be even greater demand for air cargo transport. Although freight operations constitute an important segment of the aircraft industry, the study of some of the specialized aspects of freighter aircraft design is outside the scope of this book.

1.4 Design summary
1.4.1 Mission specification

The objective of the academic design project is to develop a preliminary design for a commercial aircraft which will best satisfy a particular mission requirement. The mission is basically defined in terms of the range over which a specific payload is to be transported at a desired speed. In the case of commercial airliners the payload is generally stipulated in terms of the number of passengers to be carried. Generally, each student or group of students is assigned a different mission to be satisfied. Data for 46 operational commercial jet transport aircraft are shown in Figure 1.11 illustrating the relationship between the two basic mission specifications, range R and the number of passenger seats N_p. It is obvious from Figure 1.11 that there is a broad $R-N_p$ design space available to airline operators and a wide array of aircraft from which to choose. For ranges greater than around 5000 mi aircraft cabins are typically configured for three classes: first, business, and economy, while for ranges less than 5000 mi the cabins are mostly two class configurations. Although there is a broad $R-N_p$ space, the same is not true when we consider the relationship of the maximum gross weight $W_{g,\max}$ to the number of passengers N_p. The correlation between $W_{g,\max}$ and N_p is quite close, as shown in Figure 1.12, and may be approximated by the following equation:

$$W_{g,\max} \approx 221.5 N_p^{1.361}$$

1.4.2 The market survey

Prior to designing any new aircraft, it is essential to see what the competition is offering. If your aircraft is not decidedly better than what exists in the market, it would be foolish to "bet your company" on developing a new design. By the same token, if

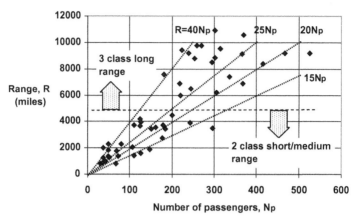

FIGURE 1.11

Range versus number of passenger seats for 46 operational commercial jet transports.

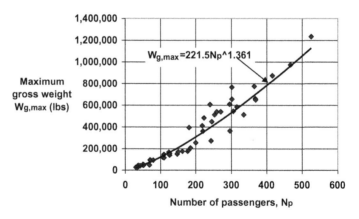

FIGURE 1.12

Take-off weights versus number of passenger seats for 46 operational commercial jet transports.

it would serve a market that doesn't show promise of growth, or at least robustness, a new design would be unlikely to succeed. For example, Singapore Airlines (SIA) was the first operator of the super-jumbo Airbus A380 which entered service with the airline in late 2007. They acquired these airplanes because they felt that new planes have very strong passenger appeal, and their Boeing B747-400 aircraft were aging. The A380 delivers 12–15% lower seat-mile costs while its large size permits revenue growth in dense, heavy routes with slot-constrained airports. In 2012 there were 75 A380 aircraft in commercial service. Passenger handling issues in very large aircraft pose some challenges and airlines often configure their A380 aircraft to seat fewer than the nominal 555 passengers, perhaps around 480 seats. This has also been the case with B747 flights, where seating is often kept in the range of 380–385 rather than the nominal 440. But the airlines are keenly aware that they need to sell more seats rather than fewer to prosper. The difficulties faced by Airbus in bringing the very large A380 into production and service have been exploited by Boeing deciding to build the B747-8, the third-generation 747. For the purposes of a preliminary design, the best business case must be made by the design teams to identify a suitable niche exists for the designs assigned, and the purpose of the exercise is to make the best aircraft to meet the specifications.

There is a second reason for reviewing existing aircraft designs. It is usually not necessary to reinvent the wheel. Designers do not start with a blank sheet of paper. It is generally useful to follow what others have done in a preliminary feasibility study. Similar aircraft will suggest the type of engines, flaps, airfoils, wing geometries, fuselage sizes, etc. to be used as first cut designs.

The market survey should rigorously examine three or four existing aircraft for which information is available and which most closely satisfy the assigned mission. Major sources of detailed technical information may be found in Janes' All the World's Aircraft (2012), *Aviation Week and Space Technology*, especially the annual inventory issue, like Aerospace (2012), and other sources, such as books and trade journals, as well as on-line sources; for example, see the websites of aircraft manufacturers like Boeing, Airbus, Embraer, Bombardier, etc. It is not expected that all required information will be easily available for all aircraft. It is valuable to gather three-view scale drawings, so that data may be estimated from the diagrams. Scale drawings and photographs can provide much useful data not generally presented in tabular form. In addition, models of all commercial aircraft, typically in the scale of 200:1, are readily available and inexpensive. Purchasing a model of one or more of the market survey aircraft to aid in visualizing subtle design features not noticeable in drawings can also be quite helpful. These other aircraft are the "competition" and the proposed design should turn out to be "better." Collecting copies of photographs and three-view drawings of these aircraft for use in the final design report, always adhering to the applicable copyright rules, is important. There are a fairly large number of photographs of jet transports which are in the public domain and can be used freely. It is a good idea to keep an eye out for aircraft details when flying on commercial airlines and taking photographs for use in the design report when permission is granted by the flight crew.

Thus Chapter 1 of a design report should introduce the reader to the mission specification, the competitor aircraft and their characteristics, and the special attributes which the proposed design aircraft will possess. The detailed quantitative data for the competitor aircraft should be presented in tabular form, along with three views, in an Appendix for easy reference. However, photos of competitor aircraft should appear in the body of Chapter 1 along with the descriptions of these aircraft. Note that all tables should be numbered and titled, as should all figures.

Keep in mind that the United States government, through the Federal Aviation Agency (FAA), establishes airworthiness requirements to ensure public safety in aviation. The FAA, which is part of the US Department of Transportation, issues Federal Aviation Regulations (FAR) and FAA Advisory Notes which lay down rules for aircraft and their operation. The FAR are published as Title 14 of the Code of Federal Regulations and are available online in Federal Air Regulations (2012).

References

Aboulafia, R., 2012. Jetliner demand: the changing landscape. Aerospace America, 50 (5), 22–25.

Aerospace, 2012. Aviation Week and Space Technology, 23–30.

Aerospace Industries Association, 2012. <www.aia-aerospace.org/assets/Table 1.pdf>.

Airfleets, 2012. <www.airfleets.net/home/>.

Anselmo, J.C., 2012. Airlines to wall street: where is the love?. Aviation Week & Space Technology, 12.

Boeing Current Market Outlook, 2012. <www.boeing.com/commercial/cmo/index.html>.

Chang, C.T. et al., 2013. NASA Glenn combustion research for aerospace propulsion. Journal of Aerospace Engineering 26 (2), 251–259. http://dx.doi.org/10.1061/ (ASCE)AS.1943-5525.0000289.

Envia, E., 2012. Emerging Community Noise reduction Approaches, NASA TM-2012-217248.

Federal Aviation Regulations, 2012. <http://rgl.faa.gov/Regulatory_and_Guidance_Library/rgWebcomponents.nsf/>.

Flottau, J., 2012. Wing worries. Aviation Week & Space Technology, 45–46.

ICAO, 2011. <www.icao.int/icao/publications/Documents/9975_en.pdf>.

ICAO, 2012. <www.icao.int/environmental-protection/CarbonOffset/Pages/default.asp>.

Janes' All the World's Aircraft, 2012–2013. IHS Global, Englewood, Colorado (published annually).

Norris, G., 2012. Actual mileage may vary. Aviation Week & Space Technology, 46–47. April 2.

Norris, G., Flottau, J., 2012. Big promises. Aviation Week & Space Technology, 28–31.

Perrett, B., 2013. Field of five. Aviation Week & Space Technology, 38–40.

Sforza, P.M., 2012. Theory of Aerospace Propulsion. Elsevier, New York.

Wall, R., 2006. Cleaner from cradle to grave. Aviation Week & Space Technology, 134–138.

Wall, R., 2008. Waste management. Aviation Week & Space Technology, 44.

Warwick, G., 2012. Coming together. Aviation Week & Space Technology, 31–33.

Warwick, G., 2012a. Spreading the bet. Aviation Week & Space Technology, 42–44.

Preliminary Weight Estimation

2

CHAPTER OUTLINE

2.1 The mission specification

The basic mission specifies that an airplane be designed to carry N_P passengers at a cruising speed of V nautical miles per hour (knots) over a range of R nautical miles using turboprop or turbofan engines. This specification actually contains all the major economic information which will decide whether or not a particular commercial design will be successful. The airplane itself will have an empty weight W_e which may be

Commercial Airplane Design Principles. http://dx.doi.org/10.1016/B978-0-12-419953-8.00002-4

thought of as the "showroom" weight at the time of purchase. This is the weight of the aircraft that the original equipment manufacturer (OEM) presents to the buyer, which, as we shall see in Chapter 11, is proportional to the capital cost of the airplane to the buyer. On the other hand, the operating cost, i.e., the expense incurred by the operator in flying the airplane, is made up of several parts including fuel expense, crew expense, and maintenance expense. The first item is a function of airplane design and engine performance while the last two items are influenced by FAA requirements and tend to be dependent on the size of the airplane. As mentioned in the previous chapter, fuel has become the major operating expense item for airline operators. Therefore, at this first step in the design process, major attention will be paid to determining the amount of fuel necessary to operate the airplane in accomplishing the mission specification safely. This is readily expressed as W_f, the weight of fuel which must be carried by the airplane. Finally, there is the positive factor of income generation by the airplane which is accomplished by charging a fee for each passenger. The transport of cargo by the aircraft is also a source of revenue for the operator and passenger aircraft usually have cargo space for freight in addition to the baggage requirements of the passengers. Of course, there are freighter aircraft dedicated solely to carrying cargo for revenue. These aircraft are either converted passenger aircraft or derivatives of passenger aircraft. The carriage of passengers and/or cargo constitutes the payload weight W_{pl}. This is the portion of the aircraft weight which contributes revenue and is proportional to the number of passengers that can be accommodated, N_p.

As pointed out in USDOT (2004), accurately calculating an aircraft's weight and center of gravity location before flight is essential to comply with the FAA certification limits established for the aircraft. These limits specifically include both weight and center of gravity location. By complying with these limits and operating under the procedures established by the airplane manufacturer, an operator is able to meet the weight and balance requirements specified in the aircraft flight manual. Typically, an operator calculates takeoff weight by adding the operational empty weight of the aircraft, the weight of the passenger and cargo payload, and the weight of fuel. The objective is to calculate the takeoff weight and center of gravity location of an aircraft as accurately as possible so as to ensure safe operation. In this chapter we will form a preliminary estimate of the maximum gross weight of the aircraft and the contributions to that weight from various components.

2.2 The mission profile

It is useful to consider the flight to progress as a series of distinct segments: engine start and warm-up, taxi, take off, climb, cruise to full range, 1 h additional flight at cruise conditions, descent to destination and refused landing, climb, diversion to alternate airport 200 nautical miles distant, descent, landing. These stages are numbered and appear in Figure 2.1 and Table 2.1.

The normal flight plan starts at 0 with the aircraft standing at the gate, fueled and boarded and ready to push back and begin taxiing to the runway. The flight plan

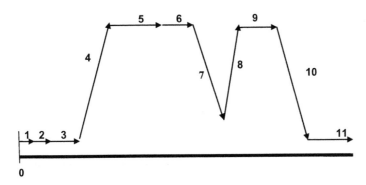

FIGURE 2.1

Mission profile showing the 11 flight stages for a domestic flight ($R<3000\,\text{nm}$). International flights have an additional segment between 10 and 11 calling for a 30 min hold at 15,000 ft altitude and segment 6 is calculated at 10% of the normal cruise time, rather than a blanket 1 h.

Table 2.1 Mission Segments

Segment	Description
1	Engine start and warm-up
2	Taxi
3	Takeoff
4	Climb
5	Cruise to full range
6[a]	Additional flight time at cruise conditions
7[a]	Descent to destination and refused landing
8[a]	Climb
9[a,b]	Diversion to alternate airport 200 nm distant
10	Descent
11	Landing

[a] Denotes flight diversion stages requiring reserve fuel.
[b] International flights include hold time at the alternate airport.

then proceeds through stages 1–5 plus stages 10 and 11, while stages 6–9 represent possible flight diversions due to poor weather or other such situations. These extra stages require the use of the reserve fuel which must always be carried by the aircraft. Operational rules for determining fuel reserve requirements are set by the airline operator to equal or exceed the reserves required by the FAA. The notional mission profile shown in Figure 2.1 is based on general operational rules suggested in ATA (1967). The Air Transportation Association of America (ATA) is now Airlines for America (A4A).

2.3 Weight components

There are many weight definitions that appear in the description of aircraft operations and they are not always entirely consistent between manufacturers, operators, and regulators. We will attempt to keep to the minimum number of weight classifications necessary for carrying out the design process.

2.3.1 Gross, takeoff, and operating empty weights

At the start of the mission the total weight of the aircraft W_0 is often called the maximum gross weight of the aircraft, $W_{g,\max}$, and may be written as the sum of the major weight components as follows:

$$W_0 = W_{g,\max} = W_{oe} + W_f + W_{pl} \qquad (2.1)$$

After taxiing and waiting in a queue for permission to take off some fuel will have been used and the aircraft weight has diminished. When calculating takeoff performance the aircraft is considered to be at the takeoff weight W_{to}, which is defined as the weight of the aircraft at the point of the start of the takeoff ground roll. The difference between the maximum gross and takeoff weights is typically around 2%. Though imprecise, it is common for $W_{g,\max}$ and W_{to} to be used interchangeably in the literature. During the remainder of the flight more fuel will be used until finally the aircraft will touch down at landing weight W_l. These weights are mentioned here because they generally will have maximum values that are not to be exceeded and these values can be determined for a given aircraft. Of course, other weight definitions arise and they will be treated in subsequent sections as they arise.

The fuel weight W_f and the payload weight W_{pl} appearing in Equation (2.1) have been broadly defined previously and it only remains to define the operating empty weight W_{oe} of the airplane as follows:

$$W_{oe} = W_e + W_c + W_{tfo}$$

The operating empty weight of the airplane is the weight of the airplane in a condition ready to fly, but with no fuel or payload yet taken on board. It therefore includes the empty weight of the airplane W_e, the weight of the trapped fuel and oil (that is, the fuel and oil left in lines and at the bottom of tanks, etc., and therefore necessary but unusable) W_{tfo}, and the weight of the crew, W_c. This last term includes the weight of the flight crew, the flight attendants, and all their baggage and this may be determined by the methods described in the following Section 2.3.2.

2.3.2 Passenger and crew weights

The US Department of Transportation in USDOT (2004) gives a chart for estimating the average weight of passengers, and it is reproduced here as Table 2.2.

Table 2.2 Standard Average Passenger Weights from USDOT (2004)

Standard Average Passenger Weight	Weight per Passenger (lb)
Summer weights	
Average adult passenger weight	190
Average adult male	200
Average adult female	179
Average child (2–13 years old)	82
Winter weights	
Average adult passenger	195
Average adult male	205
Average adult female	184
Average child (2–13 years old)	87

Table 2.3 Average Crew Member Weights from USDOT (2004)

Crew Member	Average Weight (lb)	Average Weight with Bags (lb)
Flight crew member	190	240
Flight attendant	170	220
Male flight attendant	180	220
Female flight attendant	160	200
Crew member roller bag	30	NA
Pilot flight bag	20	NA
Flight attendant kit	10	NA

As reported in USDOT (2004), the average passenger weights in Table 2.2 include a 16-pound allowance for personal items and carry-on bags, based on the assumption that:

(a) One-third of passengers carry one personal item and one carry-on bag.
(b) One-third of passengers carry one personal item or carry-on bag.
(c) One-third of passengers carry neither a personal item nor a carry-on bag.
(d) The average weight allowance of a personal item or a carry-on bag is 16 pounds.

Standard average weights for checked bags should be at least 30 pounds. It is reasonable to assume that the one-third of passengers that carry neither a personal item nor a carry-on bag [item (c) above], will have checked a bag.

The number of crew members is usually set by the operator, observing the minimum requirements stipulated by the FAA, while the baggage allowance is set by the operator. The USDOT (2004) gives a chart for estimating the average weight of crew members and their baggage and it is given here as Table 2.3. When the number of crew members is determined, their total weight may be found by using this table. Consideration of the weight of additional payload in the form of cargo freight will be given subsequently. The number of pilots is typically two, with a third pilot required for flights that are likely to last more than 8 h, and these are typically international flights. For ultra-long-duration flights, which are 12 or more hours in duration, four pilots will be necessary. In addition, rest areas for the flight deck officers must be incorporated into long-range aircraft design. FAA regulations stipulate that there must be at least one flight attendant for every 50 seats on an airliner, although the actual number depends upon the operator's preferences. Flight attendants constitute part of the indirect (or overhead) costs of operation so their number on any flight is subject to considerations of economics and competition. It is not unusual to have an average number of 20–30 seats per flight attendant, with the lower range more common on long-range flights and on flights with several levels of service.

2.3.3 Cargo weight

Revenue may also be produced by cargo carried as part of the payload, but this is difficult to specify in the initial design, being dependent upon the priorities placed on cargo by different airline operators. Therefore the payload in the initial design phase will be set by the passenger load alone as described previously. Once the fuselage design is accomplished the volume available in the cargo hold, over and above that necessary to accommodate the checked luggage, can be estimated. Then the additional payload due to freight can be estimated and included in the refined weight estimate which will be carried out in Chapter 8.

2.3.4 Fuel weight

The total usable fuel weight, W_f, may be considered to be the sum of two parts, the nominal weight of fuel, W_{fn}, necessary for the mission range R, and the fuel reserve, W_{fr}, or $W_f = W_{fn} + W_{fr}$. The quantity of reserve fuel carried is generally set by the operator, within the requirements posted by the FAA. The maximum gross weight may then be expressed as follows:

$$W_0 = W_{g,\max} = W_e + W_{plc} + W_{fn} + W_{fr} + W_{tfo} \tag{2.2}$$

We introduce the following definitions for the weight fractions of the various components:

$$M_f = \frac{W_f}{W_{g,\max}}$$

$$M_{tfo} = \frac{W_{tfo}}{W_{g,\max}}$$

$$W_{plc} = W_{pl} + W_c$$

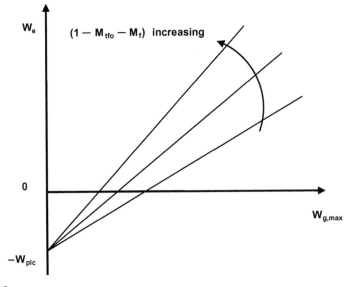

FIGURE 2.2

Empty weights as a function of maximum gross weight.

Equation (2.2) may be solved for W_e and the result written as

$$W_e = \left(1 - M_f - M_{tfo}\right) W_{g,\max} - W_{plc} \tag{2.3}$$

Thus, the equation for the empty weight is that of a straight line, i.e., $W_e = aW_{g,\max} + b$. This result is shown schematically in Figure 2.2 where it is seen that the quantity W_{plc} is the anchor point for the design of the airplane. All the possible results for W_e radiate out from the point $(0, -W_{plc})$ and depend upon the slope, a, of the straight line, and this is primarily dependent upon the fraction of fuel carried by the aircraft.

The remaining problem in the weight estimation process is the determination of the coefficient of the takeoff weight in the equation for empty weight, which is the slope $a = 1 - (M_f + M_{tfo})$. The term M_f clearly depends upon the amount of fuel used in carrying out the mission specification, including the reserve requirements. The total weight of fuel actually used in a nominal flight is the difference between the maximum gross weight and the landing weight, or $W_{fn} = W_{g,\max} - W_{final}$ and therefore

$$M_{fn} = 1 - M_{final} \tag{2.4}$$

This means that the fraction of maximum gross weight that is usable fuel on a nominal flight is given by $M_f = 1 - M_{final}$. During the flight the only weight change comes about from fuel consumption. Thus each stage shows a certain change in aircraft weight between that at the start of the segment and that at the end of the segment. The fuel fraction M_f may be found by applying a chain product of the weight fractions of each of the n stages as follows:

$$M_{final} = \frac{W_{final}}{W_{g,max}} = \frac{W_{11}}{W_0} = \prod_{i=1}^{n} \frac{W_i}{W_{i-1}} \tag{2.5}$$

To calculate the portion of fuel used for the nominal mission and the portion of fuel kept in reserve, we note that the former is given by

$$\frac{W_{fn}}{W_{g,max}} = M_{fn} = 1 - M_{final} - M_{fr} \tag{2.6}$$

Then the landing weight fraction for a nominal mission, i.e., one where no reserve fuel is used, is as follows:

$$\frac{W_l}{W_{g,max}} = M_{final} + M_{fr} \tag{2.7}$$

The second term on the right-hand side of Equation (2.7) is given by

$$M_{fr} = \frac{W_{fr}}{W_{g,max}} = \frac{W_5}{W_0} - \frac{W_9}{W_0} = \left(\prod_{i=1}^{5} \frac{W_i}{W_{i-1}} \right) \left(1 - \prod_{i=6}^{9} \frac{W_i}{W_{i-1}} \right) \tag{2.8}$$

The total fuel, it will be recalled, is given by

$$\frac{W_f}{W_{g,max}} = 1 - \frac{W_{final}}{W_{g,max}} = 1 - M_{final} \tag{2.9}$$

Recall that the normalized maximum fuel load is given by

$$M_f = \frac{W_f}{W_{g,max}} = \frac{W_{fn} + W_{fr}}{W_{g,max}} = M_{fn} + M_{fr} \tag{2.10}$$

For purposes of illustration, this term may be approximated using information on empty weights and numbers of passengers that have been compiled from information in the open literature for 50 operational airliners. First, we make the following approximation:

$$M_f \approx \frac{W_{g,max} - W_e - W_{plc}}{W_{g,max}} = 1 - \frac{W_e + W_{plc}}{W_{g,max}} \tag{2.11}$$

Data using his approximation are shown in Figure 2.3 along with a trend line given by $M_f = 0.0045\sqrt{R}$, where R is measured in statute miles. Note that for very long-range aircraft the total fuel fraction approaches half the take off weight. This information provides a check on the estimates being made for the fuel requirements of the aircraft being designed.

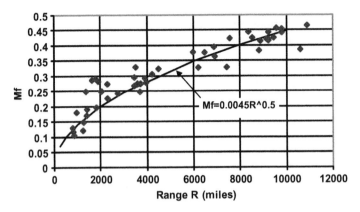

FIGURE 2.3

The total fuel fraction M_f is shown as a function of range as estimated from available information on 50 airliners. The solid line is an approximate curve fit to the data shown.

2.3.5 Fuel consumption by mission segment

The 11 general mission stages are described in Table 2.4, along with Roskam's (1986) suggestions for applicable average weight fractions for turbofan-powered aircraft; turboprop-powered aircraft are treated in a subsequent section. Recall that stages 6 through 9, inclusive, correspond to flight diversions requiring reserve fuel and that the nominal flight includes only stages 1–5, plus 10 and 11. In addition, for long-range international flights, stage 6 is calculated not for 1 h, but for 10% of the nominal flight time. In addition, in stage 9 for international flights there would be included with the diversion to an alternate airport 200 nm away a 30 min hold at 15,000 ft altitude at the alternate airport. The terms used in Table 2.4 are explained more fully later in this chapter.

2.3.6 Fuel consumption in segments other than cruise

The quantity W_i/W_{i-1} is the ratio of the weight of the aircraft at the end of stage i to the weight at the start of stage i (i.e., the end of the previous stage). The weight fractions for segments of the flight which involve acceleration, like stages 1–4 and 10–11, are generally very close to unity because commercial flights spend most of the flight time at essentially constant velocity cruise conditions. Compilations of representative values for these weight fractions for all stages except the cruising stages are those given by Roskam (1986) and are included in Table 2.4 along with cruise-related equations. It may be noted that once the characteristics of the airplane are known in greater detail it will be possible to more accurately determine the weight of fuel used during climb and descent and this is carried out as part of the refined weight estimates discussed in Chapter 8.

Table 2.4 Weight Fractions for the Mission Segments for Turbofan Aircraft (R in nm, V in kts, and C_j in hr^{-1})

Stage	Description	W_i/W_{i-1}
1	Engine start and warm-up	0.990
2	Taxi	0.990
3	Takeoff	0.995
4	Climb	0.980
5	Cruise to full range	$\exp[-RC_j/V(L/D)]$
6[a]	One hour additional flight at cruise conditions	$\exp[-C_j/(L/D)]$
6[b]	Ten percent nominal flight time additional	$\exp[-RC_j/10V(L/D)]$
7[a]	Descent to destination and refused landing	0.990
8[a]	Climb	0.980
9[a]	Diversion to alternate airport 200 nm distant	$\exp[-200C_j/V(L/D)$
9[b]	Diversion to alternate airport 200 nm distant plus 0.5 h hold at 15,000 ft altitude at alternate airport	$\exp[-200C_j/V(L/D)$ $+\exp[0.5C_j/V(L/D)]$
10	Descent	0.990
11	Landing	0.992

[a] *Flight diversion for domestic flights.*
[b] *Flight diversion for international flights.*

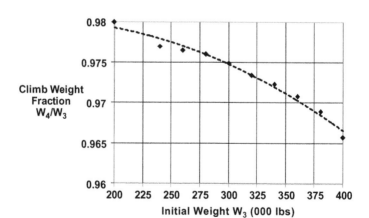

FIGURE 2.4

The weight fraction W_4/W_3 of Table 2.1 for the Douglas DC-10-10 airliner in a climb to 35,000 ft as given by Shevell (1989). Dotted line is added to indicate the trend.

For example, Figure 2.4 shows the ratio of the weight at the end of climb to that at the start of climb as a function of the weight at start of climb; that is, the ratio W_4/W_3 in Table 2.4. The values shown are taken from Shevell (1989), and refer to the fuel to climb for the Douglas DC-10-10 airliner for the particular case of climb to 35,000 ft. In Table 2.4 Roskam's (1986) generic value, $W_4/W_3 = 0.98$, is specified, and this is reasonably close to the more accurate values in Figure 2.4. A discussion of the climb and descent segments of the mission profile is presented in Chapter 10.

2.3.7 Fuel consumption in cruise

A detailed discussion of the Breguet range equation suitable for this stage of the investigation appears in Chapter 10; here it is written as follows:

$$R = \frac{V}{C_j} \left(\frac{L}{D}\right) \ln \frac{W_4}{W_5} \tag{2.12}$$

This form is applicable to aircraft powered by turbofan or turbojet engines; here R is the range, V is the cruise speed, L/D is the lift to drag ratio, and C_j is the thrust specific fuel consumption. The corresponding relation for turboprop aircraft is given in Section 2.7. All the variables in Equation (2.12) are to be evaluated under cruise conditions.

Note that the quantity V/C_j must have the same units as does the range R for Equation (2.12) to be consistent. There are variations in the choice of units between different sectors of the aerospace industry, in spite of the trend toward metrification in many other industries. The common units of horizontal distance and speed in airline operations are still nautical miles, abbreviated nm, and knots, or nautical miles per hour, abbreviated kts, respectively. Vertical distance, that is, altitude, is commonly referred to in feet and the specific fuel consumption C_j is typically given in the units of pounds of fuel burned per hour per pound of thrust. The units for C_j are convenient because they reduce to solely inverse hours or hr^{-1}. Thus, using this set of units the ratio of V/C_j has the unit of nautical miles. Some typical units are shown in Table 2.5. The units used in airline operations, like common US units, make RC_j/V dimensionless with no corrections. The metric version of specific fuel consumption

Table 2.5 Some Units in Use for Range, Speed, and Specific Fuel Consumption

Property	Airline Operations	English Units	Common US Units	Metric Units	Common Metric Units
Horizontal distance	1 nm	6076 ft	1.151 mi	1852 m	1.852 km
Horizontal speed	1 kt	1.688 ft/s	1.151 mph	0.5144 m/s	1.852 kph
Specific fuel consumption	1 lb/lb-hr	$1\,hr^{-1}$	$1\,hr^{-1}$	28.33×10^{-6} kg/N-s	28.33 mg/N-s

is often clumsy to use in the present context because it is a ratio of a mass flow rate to a force.

Solving for the fuel weight fraction expended during cruise from Equation (2.12) yields

$$\frac{W_5}{W_4} = \exp\left[-R\left(\frac{V}{C_j}\frac{L}{D}\right)^{-1}\right] \tag{2.13}$$

In this equation the units must be consistent, for example, R in nautical miles (or miles), V in knots (or miles per hour), and C_j in pounds fuel per hour per pound of thrust, so that the argument of the exponential function is dimensionless. One may now determine a value for W_5/W_4, W_6/W_5, and W_9/W_8 and therefore M_f for the prescribed range if one picks a set of values for the airplane's performance parameters V, C_j, and L/D. A range of representative values for some of these parameters is described in the next section. Of course, wherever possible, it is preferable to use information from the research carried out on the market survey aircraft to improve the estimates for these values.

2.3.8 Selection of cruise performance characteristics

The cruise speed V should be as high as is reasonable, remembering that the drag rises with the square of the speed and the engines selected later on in the design process must be able to provide the thrust necessary to overcome this drag. At the speeds considered for jet transports it is preferable to deal with the Mach number $M = V/a$ rather than the velocity because it is the appropriate similarity variable for the force and moment coefficients. Furthermore, the commercial jet transports considered here generally cruise around the tropopause, the altitude where the stratosphere begins (around 36,000 ft or 11 km) and where the atmospheric temperature, and therefore the speed of sound a, is approximately constant (at 574 kts or 660 mph), so it is also convenient to work with the cruise Mach number. Standard atmospheric profiles may be used in the design process, as shown in Appendix B. Typical cruise Mach numbers are in the range of $0.76 < M < 0.86$ and the actual values for the market survey aircraft may be used as a guide for the present design.

Similarly, the assumed cruise value of the specific fuel consumption should be as low as possible, but should be in keeping with existing or planned engine characteristics. Current turbofan technology has provided cruise-specific fuel consumptions in the range $0.5 < C_j < 0.6$, in the units of pounds of fuel consumed per hour per pound of thrust produced. Again, published data for the engines used by the market survey aircraft can provide a guide for representative values to be used for the current aircraft design. It should be noted that most published engine data show values for takeoff-specific fuel consumption which are in the range of 0.3–0.4 pounds of fuel per hour per pound of thrust, but this applies only to takeoff conditions and should not be used for cruise calculations. To repeat, the value of specific fuel consumption for cruise is considerably larger, namely 0.5–0.6 for current high bypass turbofan engines.

The lift to drag ratio L/D represents the aerodynamic efficiency of the airplane and should be as high as possible. Current commercial jetliners have lift to drag ratios in the range of $14 < L/D < 19$, with the higher values associated with long-range versions. The actual L/D for the design aircraft will ultimately be calculated so it is wise, as usual, to try to make reasonable estimates at every stage of the design. Though the cruise lift for an aircraft is easily determined since it is merely equal to the weight of the aircraft at any stage of the cruise, the drag is not as easily determined. Indeed, the drag is a closely guarded secret of aircraft manufacturers and great efforts are expended to continually reduce the drag of new designs. Some of the most modern techniques used for developing reduced drag designs are quite sophisticated and beyond the scope of this book. Thus the lift to drag ratios to be determined later in the design process will not be as high as achieved in current practice, so it is advisable to be somewhat conservative in the choice of the cruise lift to drag ratio, probably confining the maximum value to 16 or so.

Loftin (1985) discusses the estimated L/D for the Boeing B707 and the Douglas DC-8, which he gives as 19.0–19.5 and 17.9, respectively. He argues that the additional length of the DC-8 fuselage increased the total wetted area of the airplane and therefore the profile drag coefficient C_{D0}, thereby bringing down L/D. The later Boeing B767-200 is said to have $L/D = 18$. Once again, the reduction from the 19 or so of the B707 is attributed to the fact that the ratio of wetted surface area to wing planform area, S_{wet}/S, is larger for the B767-200, although S is comparable for both.

The Boeing B747 is estimated to have $L/D = 18$, like the B767-200. Other wide-body airliners, like the older Lockheed L1011-200 and the McDonnell-Douglas MD 10–30, are estimated to have L/D values between 17 and 17.5. Heffley and Jewell (1972) present data on the characteristics of a number of aircraft in cruise, as well as

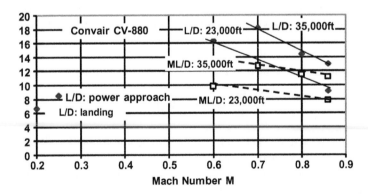

FIGURE 2.5

Data from Heffley and Jewell (1972) show the variation of L/D and ML/D as a function of M for the Convair CV-880 jetliner at two altitudes. The L/D for typical power approach and landing at sea level are also shown.

in power approach and landing configurations. A particular case is that of the Convair CV-880, a medium-size four-jet airliner, similar to and contemporaneous with the B 707 and DC-8 airplanes. The L/D and ML/D behavior with Mach number is shown in Figure 2.5. Note that although the L/D drops quite rapidly with Mach number, the more important quantity for the range equation, $ML/D \sim VL/D$, drops much more slowly.

2.4 Empty weight trends

Our objective at this point is to develop a reasonable estimate for the empty weight of the design aircraft W_e and in this regard it is instructive to first see what historical precedents apply. In keeping with common usage, W_{to} will be considered to be equal to $W_{g,max}$ in the following discussion. Roskam (1986) collected a substantial database on existing aircraft and generated curve fits describing the relationship between W_e and W_{to} and offered a correlation equation of the form

$$\log_{10} W_e = \frac{1}{B} \left(\log_{10} W_{to} - A \right) \tag{2.14}$$

In Equation (2.14) the values of A and B are constants that are different for different classes of aircraft. For jet transport aircraft Roskam (1986) suggests $A = 0.0833$ and $B = 1.0383$, which results in a curve that is very close to a straight line and may be approximated by the equation $W_e = 0.5 W_{to}$. A database for 50 commercial airliners provides the results shown in Figure 2.5 which support this simplified result.

It must be kept in mind that this simple correlation is based on a wide variety of commercial aircraft built over a fairly long period of time and the scatter, though appearing small on Figure 2.6, is in the range of up to tens of thousands of pounds. When the focus is narrowed to the particular class of market survey aircraft considered, the scale of the graph of W_e vs. W_{to} will be larger and deviations from the historical curve more evident. The utility of a correlation of this type is in its ability to provide a guideline for the development of a new design. There is a slight nonlinearity in the relationship of empty weight to takeoff weight that is not apparent in Figure 2.6, but is made clearer in Figure 2.7. There the correlation

$$\frac{W_e}{W_{to}} = 1.59 \left(W_{to} \right)^{-0.0906} \tag{2.15}$$

is shown to fit the actual data better than the simple approximation $\frac{W_e}{W_{to}} = 0.5$.

As mentioned previously the scale of Figure 2.6 is large enough to obscure deviations from the proposed variation of W_e with W_{to}. This should be clear from the results shown in Figure 2.7 where the deviation of the W_e from the average of $0.5 W_{to}$ is substantial for $W_{to} < 100,000$ lb. This is the domain of regional jets and turboprop aircraft which will be discussed subsequently. As an additional indicator of the design range for the new aircraft under consideration one may make use of the characteristics of the three or four market survey aircraft that have a mission similar

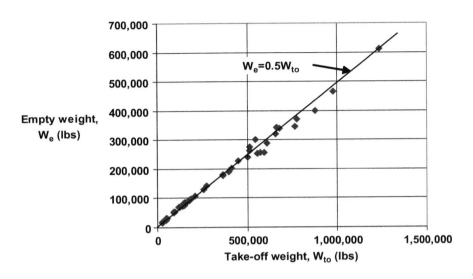

FIGURE 2.6

Empty weights as a function takeoff weight for 50 commercial airliners.

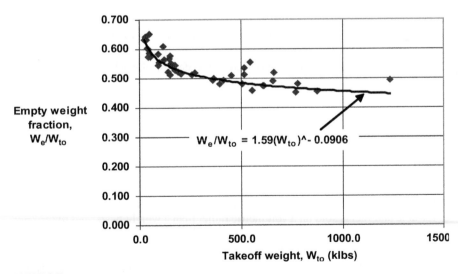

FIGURE 2.7

Variation of the empty weight fraction W_e/W_{to} with W_{to} for 50 commercial airliners illustrating the slight nonlinearity of the relationship.

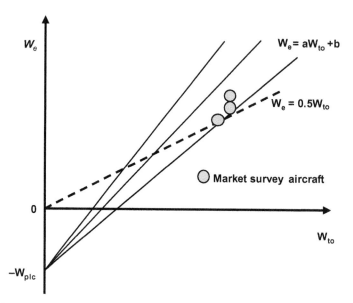

FIGURE 2.8

Estimation of empty weight using empirical weight relations and characteristics of market survey aircraft which are shown as circular symbols.

to that of the proposed aircraft. Since the empty and take off weights are known for these aircraft they may be represented as discrete points on the plot of W_e as a function of W_{to}. Those points, which represent the aircraft most like the design aircraft, serve to further limit the probable region of the design plot where the new aircraft will fall. This is illustrated in Figure 2.8.

2.5 Fuel characteristics

Jet fuel is a hydrocarbon fuel composed primarily of paraffin (approximately 70%) and aromatic (approximately 20%) petroleum compounds. Some characteristics are shown in Table 2.6 as given by Chevron (2000). The most commonly used fuel in the US is Jet A and that fuel will be used in developing the aircraft design. Fuel density is variable and fuel is sold on a volumetric rather than a weight basis, and for our purposes it will be considered sufficient to use the standard density shown. Some of the fuel (and lubricating oil) carried on the aircraft will not be drainable from the tanks and therefore is unusable. The weight of this component has been denoted by W_{tfo} and we need an estimate of this value to calculate the term M_{tfo} used in the relation between W_e and W_{to} used to generate Figure 2.9.

Torenbeek (1982) provides an expression for the weight of trapped fuel and oil in terms of the volume of the fuel tanks on the aircraft. That expression can be converted into an expression involving the weight of the fuel by using the value for the density of Jet A in Table 2.6 which results in the following relation:

Table 2.6 Characteristics of Commonly Used Jet Fuels

General Designation	US Commercial Designation	US Military Designation	Density at 15C in lb/gal	Freezing Point in degrees (C)	Energy Content (Btu/lb)	Energy Content (Btu/gal)
Wide-cut gasoline	Jet B	JP-4	6.36	−50 to −58	18,720	119,000
Kerosene	Jet A, Jet A-1	JP-8	6.76	−40 to −50	18,610	125,800

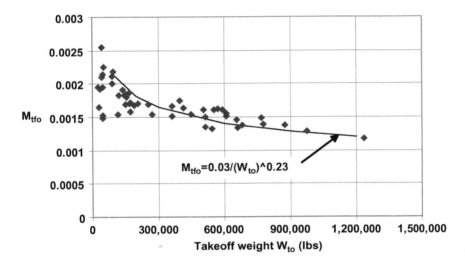

FIGURE 2.9

Data points show results for trapped fuel and oil fraction M_{tfo} for Equation (2.16) when using database for 50 commercial airliners. Also shown is a curve-fit describing the trend of the results.

$$M_{tfo} \approx 0.227 \left(\frac{M_f^2}{W_{to}} \right)^{\frac{1}{3}} \tag{2.16}$$

Applying Equation (2.16) to the database of 50 commercial airliners results in the data shown in Figure 2.9. Also shown on the figure is an approximation given by

$$M_{tfo} = \frac{0.03}{\sqrt[3]{(W_{to})}^{0.23}}$$

This expression for M_{tfo} depends upon W_{to} (in pounds) while in the equation for W_e it was assumed that it is a constant. However, it is seen that M_{tfo} will be small, lying in the range $0.001 < M_{tfo} < 0.005$. Therefore it is suggested that M_{tfo} be

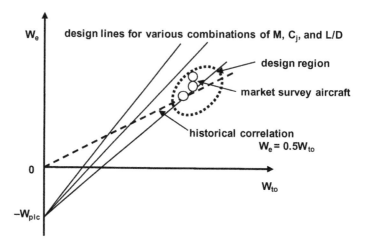

FIGURE 2.10

General design curve illustrates the manner in which the design region is determined for jet aircraft. The final choice for W_e and W_{to} is set by the appropriate combination of values of V, C_j, and L/D for the proposed aircraft.

estimated using the value of M_f calculated and an approximate value for W_{to} based on the values for the market survey aircraft.

2.6 Estimation of the takeoff and empty weights

Equipped with a small set of reasonable performance estimates (V, D/L, and C_j) for the design aircraft, a number of design lines can be entered on the W_e vs W_{to} plot. The confluence of design lines, the historical data correlation, and the market survey data constitute the basic design region and this is illustrated in Figure 2.10. It is within this design region that the design choice exists and in order to continue the design a specific point must be selected. That choice will be one which best defines the initial estimate of the empty weight and the takeoff weight of the design aircraft. These weights will influence the subsequent development of the design and if they are chosen poorly may require substantial alteration and iteration later on. Nevertheless a choice must be made and the major result of this chapter of the design report is the design plot of Figure 2.10 with all its information and the final values for W_e and W_{to} and the associated assumed performance data.

The design lines for the different sets of C_j, V, and L/D should appear and be identified on the design chart as should the market survey aircraft points. The design point chosen for the new aircraft should be clearly shown and a discussion illustrating the criteria by which this choice was made should be included in the final design report. In addition, a weight breakdown table showing the relevant weight components for the market survey aircraft and for the design aircraft should be provided. Of course, all of the work carried out should be used to form a clear narrative of this phase of the design process.

2.6.1 New materials for weight reduction

As mentioned in Chapter 1, the pressure to reduce aircraft weight has manufacturers on a quest for new materials that can outperform conventional airplane construction materials at lower weight. There is extensive work underway to incorporate ever-greater percentages of composite materials into the aircraft structure in particular and such efforts were assiduously pursued in the new Airbus A380 and Boeing 787, as well as aircraft under development, like the Airbus A350, among others. There is still concern about material lifetime, particularly with regard to fatigue, as well as to the ability to affordably accommodate repair and maintenance work. There are other issues surrounding the use of composite regarding porosity, environmental robustness, effect of lightning strikes, and the like. The rapidly growing use of composites in aircraft is demonstrated by the A380 using it for around 25% and the B787 using it for up to 50% of the airframe weight. The material of choice for the Airbus A380 is glass-fiber-reinforced aluminum while that for the Boeing 787 is carbon-fiber-reinforced plastic, known as CFRP. More details are available in the literature; see, for example, NRC (1996). As might be expected, military aircraft are leading the way in this regard with the Bell Boeing V-22 and the Eurofighter already using composites for about 75% of the airframe weight and the F/A-22 and F/A-18E/F using between 50% and 60%. It is imprudent, at this stage of the design, to make optimistic assumptions about the extent to which new composite materials may reduce the empty weight of the design aircraft. One may be guided by the weight trends shown by the later model aircraft in the market survey. The cultivation of an even-handed approach to estimation is important because overly conservative estimates are as burdensome in their own way as overly optimistic ones.

2.7 Weight estimation for turboprop-powered aircraft

The recent rapid rise in fuel prices has forced a revaluation of turboprop-powered aircraft, particularly for regional airline service. As pointed out in Chapter 1, manufacturers of regional jets have been targeting larger aircraft of around 100 or more seats, underscoring the prediction that turbofan-powered airliners will ultimately abandon the regional market to turboprop aircraft. In general, the public has been moved to consider jet aircraft to be the preferred mode of travel, even for regional distances, so that the question remains as to how much emphasis will be placed on returning turboprops to a major role in airline service. At the moment the high and uncertain fuel prices are moving airline operators to seriously consider asking aircraft manufacturers for advanced-design turboprop aircraft.

For short-range applications cruise speed is not as important in keeping travel time brief as it is for longer-range flight. This becomes apparent when one considers that most of the time in a short-range flight is spent in taxiing from the gate, waiting in a queue for takeoff, climbing, descending, and once again taxiing to the gate. The best time advantage for a turbofan compared to a turboprop may be assumed to be in the ratio of the cruise speeds, that is, about 500 mph/350 mph = 1.43, so that a 60-min flight in a turbofan would be about an 80–90-min flight in a turboprop. Of course, a transcontinental flight would be quite different, with a 6-h flight in a turbofan becoming a 9-h flight in a turboprop. Thus for ranges of up to 500 or 600 miles the turboprop can deliver its good fuel economy with relatively little passenger inconvenience.

2.7.1 Fuel weight estimation for turboprop airliners

The determination of the fuel consumption for turboprop airliners follows the procedures discussed previously for turbofan airliners. However, the stage weight fractions now depend on two new variables, C_{tp} and η_p, rather than C_j and V. The Brequet range equation for jet aircraft given in Equation (2.12) is slightly different for propeller aircraft, now being given by

$$R = \frac{326\eta_p}{C_{tp}} \left(\frac{L}{D}\right) \ln \left(\frac{W_4}{W_5}\right) \tag{2.17}$$

All the variables in Equation (2.17) are to be evaluated under cruise conditions. The fuel weight fraction expended during cruise may be obtained from the above equation in the following form:

$$\frac{W_5}{W_4} = \exp \left[-R \left(\frac{375\eta_p}{C_{tp}} \frac{L}{D}\right)^{-1} \right] \tag{2.18}$$

Table 2.4 showing the weight fractions for different mission segments for jet engines is modified for turboprop applications and appears in Table 2.7. The quantity C_{tp} is the specific fuel consumption of the turboprop engine which has the units of pounds of fuel per hour per shaft horsepower (lbs/hr-hp) and the quantity η_p is the propeller efficiency and is dimensionless. The numerical coefficient 326 incorporates unit conversions. Note that the range R is still in nautical miles and the velocity V is the cruise velocity in knots. The cruise speeds of turboprop commercial aircraft are in the range of 250–300 kts while the ratios of lift to drag are in the range of 14–18. Propeller efficiencies are in the range of 82–92% while specific fuel consumption is in the range of 0.5–0.7 lbs/hp-hr. The numerical values for the weight fractions in the other flight stages are taken from Roskam's (1986) suggestions for turboprop aircraft.

2.7.2 Empty weight estimation for turboprop airliners

The weight estimation procedures in Section 2.2 are still applicable with some minor changes specific to regional airliners. At the lower takeoff weights typical of regional airliners the nonlinearity of the relationship between W_e and W_{to} shown in Figure 2.7 must be taken into account. Empty weights for regional turbofan and turboprop airliners are described more closely in Figure 2.11 and it is seen that the

Table 2.7 Weight Fractions for the Mission Segments for Turboprop Aircraft (R in nm, V in kts, and C_{tp} in lb/hp-hr)

Stage	Description	W_i/W_{i-1}
1	Engine start and warm-up	0.990
2	Taxi	0.995
3	Takeoff	0.995
4	Climb	0.985
5	Cruise to full range	$\exp\{-RC_{tp}/[326\eta_p(L/D)]\}$
6	One hour added flight at cruise conditions	$\exp\{-(1\text{hr})(V)C_{tp}/[326\eta_p(L/D)]\}$
7	Descent to destination and refused landing	0.985
8	Climb	0.985
9	Diversion to alternate airport 200 nm distant	$\exp[-200C_p/326\eta_p(L/D)]$
10	Descent	0.990
11	Landing	0.995

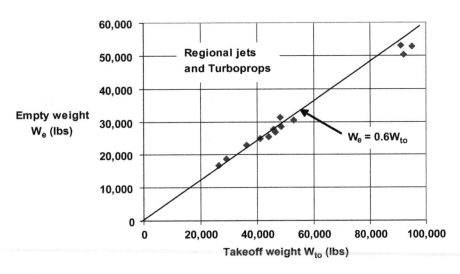

FIGURE 2.11

Empty weights as a function of takeoff weight for 11 regional jet and turboprop airliners. Three heavier commercial airliners are shown to illustrate the initial deviation from the linear correlation.

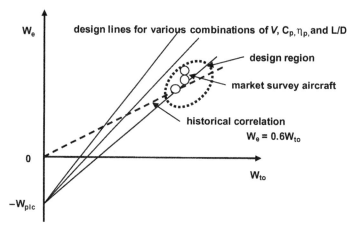

FIGURE 2.12

General design curve illustrates the manner in which the design region is determined for turboprops. The final choice for W_e and W_{to} is set by the appropriate combination of values of V, C_p, η_p, and L/D for the proposed aircraft.

correlation equation is now given more accurately by $W_e = 0.6W_{to}$ in this weight class. Also shown on Figure 2.11 are turbofan aircraft with takeoff weights approaching 100,000 lb, illustrating the start of the deviation of the slope from the 0.6 value. In estimating the empty weight for regional turbofan aircraft the design chart in Figure 2.10 is constructed as before and only the historical empty weight curve changes from $W_e = 0.5W_{to}$ to $W_e = 0.6W_{to}$. However for turboprop regional airliners the factors to deal with are C_p, η_p, L/D, and V rather than C_j, L/D, and M. Thus the design chart in Figure 2.10 changes slightly to that shown in Figure 2.12.

2.8 Design summary

The design choice developed through the steps of this chapter may be best summarized by filling in a table like that shown in Table 2.8. In this fashion, the weights that may be needed in subsequent design developments are readily available.

In addition, the mission profile for the design aircraft should be illustrated. A design graph showing the variation of empty weight W_e as a function of the maximum gross weight $W_{g,max}$, like that in Figure 2.10 or 2.12, should be presented. The market survey aircraft data points and the proposed aircraft design point should be clearly identified on the graph.

Table 2.8 Preliminary Weight Estimates for the Design Aircraft

	Range R (nm)	Number of seats N_p	Cruise speed V (kts)	Cruise C_j or C_{tp} (hr^{-1} or lb/hp-hr)	Cruise L/D
Mission requirement					

Segment Weight	Definition	Weight (lb)
W_0	Maximum gross weight $W_{g,max}$	
W_1	Weight at start of pushback	
W_2	Takeoff weight W_{to}	
W_3	Climb weight	
W_4	Weight at start of cruise	
W_5	Weight at end of cruise	
W_6	Weight at end of 1 h of additional cruise	
$W_6{}^a$	Weight at end of 10% additional cruise time	
W_7	Weight at end of descent to descent to destination	
W_8	Weight at end of climb from refused landing	
W_9	Weight at end of diversion to alternate airport	
$W_9{}^a$	Weight at end of diversion to alternate airport plus 0.5 h hold at alternate airport	
W_{10}	Landing weight W_l	
W_{11}	Final weight W_{final}	
W_e	Empty weight	
W_{oe}	Operating empty weight	
W_c	Weight of crew	
W_{pl}	Weight of payload	
W_{plc}	Weight of payload and crew	
W_{tfo}	Weight of trapped fuel and oil	
W_f	Weight of total fuel	
W_{fn}	Weight of nominal mission fuel	
W_{fr}	Weight of reserve fuel	

^{a} International flights.

2.9 Nomenclature

A	constant
a	sound speed
B	constant
C_j	turbofan thrust specific fuel consumption
C_{tp}	turboprop-power specific fuel consumption
D	drag
L	lift
M	Mach number
M_{final}	final weight ratio
M_f	fuel weight ratio
M_{tfo}	trapped fuel and oil ratio
R	range
V	velocity
W_c	weight of full flight crew
W_e	aircraft empty weight
W_f	total fuel weight
W_{final}	weight of aircraft at end of flight
W_{fn}	fuel weight used in nominal mission
W_{fr}	reserve fuel weight
$W_{g,max}$	maximum gross weight
W_{ln}	aircraft weight at end of nominal mission
W_{oe}	operating empty weight
W_{pl}	payload weight
W_{plc}	weight of payload plus flight crew
W_{tfo}	weight of trapped fuel and oil
W_{to}	takeoff weight
W	weight
η_p	propeller efficiency

2.9.1 Subscript

i	index denoting stage number

References

Chevron, 2000. Aviation Fuels Technical Review. Chevron Products Company FTR-3, Chapter 2, <www.chevron.com/products/prodserv/fuels/aviationfuel/toc.shtm>.

Heffley, R.K., Jewell, W.F., 1972. Aircraft Handling Qualities Data, NASA CR-2144, December.

Loftin, K., 1985. Quest for Performance – The Evolution of Modern Aircraft, NASA SP-468.

NRC, 1996. New Materials for Next-Generation Commercial Transports. National Research Council, National Academy Press. <www.nap.edu/openbook/0309053900/html/R1.html>.

Roskam, J., 1986. Rapid sizing method for airplanes. Journal of Aircraft 23 (7), 554–560.

Shevell, R., 1989. Fundamentals of Flight. Prentice-Hall, Englewood Cliffs, NJ.

Torenbeek, E., 1982. Synthesis of Subsonic Airplane Design. Kluwer Academic Publishers, Dordrecht, The Netherlands.

USDOT, 2004. US Department of Transportation Advisory Circular AC120-27D, August 11.

Fuselage Design

3.1 Introduction

The fuselage is that portion of the aircraft wherein the payload is carried. In jet transports the payload consists of the passengers and their baggage and/or cargo. Primary considerations when designing the airplane's fuselage are as follows:

- Low aerodynamic drag.
- Minimum aerodynamic instability.
- Comfort and attractiveness in terms of seat design, placement, and storage space.

Commercial Airplane Design Principles. http://dx.doi.org/10.1016/B978-0-12-419953-8.00003-6

- Safety during emergencies such as fires, cabin depressurization, ditching, and proper placement of emergency exits, oxygen systems, etc.
- Ease of cargo handling in loading and unloading, safe and robust cargo hatches and doors.
- Structural support for wing and tail forces acting in flight, as well as for landing and ground operation forces.
- Structural optimization to save weight while incorporating protection against corrosion and fatigue.
- Flight deck optimization to reduce pilot workload and protect against crew fatigue and intrusion by passengers.
- Convenience, size, and placement of galleys, lavatories, and coat racks.
- Minimization of noise and control of all sounds so as to provide a comfortable, secure environment.
- Climate control within the fuselage including air conditioning, heating, and ventilation.
- Provision for housing a number of different subsystems required by the aircraft, including auxiliary power units, hydraulic system, air conditioning system, etc.

Though these factors must be taken into account in fuselage design, they are constrained by the need to minimize overall structural weight and aerodynamic drag because they affect performance and initial cost, as well as impacting operating costs. The overall shape and dimensions very much dictate the effectiveness of the design and these are in turn influenced by the mission. A modern approach to aircraft cabin optimization is presented by Nita and Scholz (2010).

3.2 Commercial aircraft cabin volume and pressure

The pressurized cabin of a commercial airliner, as shown schematically in Figure 3.1, includes the flight deck in the nose cone, the passenger cabin and cargo areas, and terminates at the aft pressure bulkhead. The cross-sectional shape of the cabin is generally circular, or close to circular, because of the structural and manufacturing benefits of such a shape. To simplify the analysis attention is focused on the passenger cabin, which comprises the major portion of the fuselage volume. The fuselage width is proportional to the number of seats abreast, their width, and the width of any aisles between the seats. We may approximate the fuselage width w_{fus} by the following equation:

$$w_{fus} \approx c_1 + c_2 w_s N_a \tag{3.1}$$

Here the seat width and the number of seats abreast are denoted by w_s and N_a, respectively. Similarly, the length of the cabin L_c is proportional to the product of the seat pitch P and the number of rows of seats N_r and may be expressed by

$$L_c \approx c_3 \left(PN_r\right)^{1+\varepsilon} \tag{3.2}$$

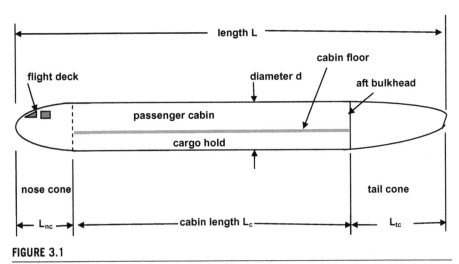

FIGURE 3.1

Schematic diagram of fuselage showing passenger cabin.

The seat pitch P is the distance between a point on the seat and the same point on the seat behind it in the next row, $c_3 \geq 1$ is a constant accounting for the presence of lavatories, galleys, etc. available for the convenience of the passengers, and $\varepsilon \ll 1$ accounts for any slight non-linearity. The number of passengers that can be accommodated is

$$N_p \approx N_a N_r \tag{3.3}$$

The approximation sign is used to note that we tacitly assume the number of seats abreast to be a constant. On some aircraft there may be a few rows that have a different number of seats abreast for various operational or design reasons.

3.2.1 Cabin volume

We may consider two volume measures, the pressurized volume and the free volume. The pressurized volume is defined as the gross volume contained within the pressure shell and for the passenger cabin this is given by

$$V_{press} = \frac{\pi}{4} w_{fus}^2 L_c \approx \frac{\pi}{4} \left(c_1 + c_2 w_s N_a\right)^2 c_3 \left(PN_r\right)^{1+\varepsilon} \tag{3.4}$$

The free volume is defined as that volume readily accessible to the passengers in the cabin neglecting the volume occupied by seats and partitions. Assuming the aircraft is large enough for passengers to readily move around, one may define some average headroom h, and then the free volume is given by

$$V_{free} = w_{fus} L_c h \approx \left(c_1 + c_2 w_s N_a\right) c_3 \left(PN_r\right)^{1+\varepsilon} h \tag{3.5}$$

Table 3.1 Average Fuselage Data for Typical Commercial Aircraft

N_a	c_1	c_2	c_3	$c_{3,pax}$	P (ft)	w_s (ft)	h (ft)	N_r
1	0.71	1.09	1.32	1.08	2.5	1.43	4	1–4
2	2.49	1.24	1.32	1.08	2.5	1.43	5	6–10
3	2.49	1.24	1.32	1.08	2.5	1.43	5.5	10–15
4	2.49	1.24	1.32	1.08	2.5	1.43	6	16–18
5	2.49	1.24	1.08	1.08	2.7	1.43	6	19–31
6	2.49	1.24	1.08	1.08	2.7	1.43	6	31–37
7	0	1.3	1.08	1.08	3	1.5	6	33–37
8	0	1.3	1.08	1.08	3	1.5	6	37–45
9	0	1.3	1.08	1.08	3	1.5	6	45
10	0	1.3	1.08	1.08	3	1.5	6	

If we use the approximation of Equation (3.3) we may form the volume per passenger as follows:

$$\frac{V_{press}}{N_p} \approx \frac{\pi}{4} c_3 P \left(\frac{c_1^2}{N_a} + 2c_2 w_s + c_2^2 w_s^2 N_a \right) \left(\frac{PN_p}{N_a} \right)^{\varepsilon} \tag{3.6}$$

In the same fashion we may determine the free volume per passenger to be

$$\frac{V_{free}}{N_p} \approx c_3 h P \left(\frac{c_1}{N_a} + c_2 w_s \right) \left(\frac{PN_p}{N_a} \right)^{\varepsilon} \tag{3.7}$$

Torenbeek (1982) suggests $\varepsilon = 0.052$ and we find that the last term on the right-hand side of Equations (3.6) and (3.7) is a slowly varying function that lies between about 1.1 and 1.3 for a wide range of typical airplanes, while the remaining terms are all constants particular to a given aircraft. The characteristics of common commercial aircraft like Boeing's 737, 767, and 777 families and the regional turboprop ATR-72 provide some useful information, as shown in Table 3.1.

For aircraft that seat four or more abreast, Equation (3.4) is satisfactory for determining the pressurized volume. For smaller aircraft, that is for $N_a \leq 3$, the length of the pressurized volume should instead be approximated by

$$L_c' \approx c_3' (PN_r)^{1+\epsilon} \tag{3.8}$$

Here we recognize that in the smaller-diameter aircraft the cargo holds are in the cabin forward and/or aft of the passenger compartment rather than below the floor, so the pressurized cabin is longer than the passenger compartment. Thus, for these smaller aircraft the pressurized cabin length Equation (3.2) is used with c_3 replaced by c_3', where $c_3' > c_3$. The range of the number of rows $N_r = N_p/N_a$ given in Table 3.1 is used to estimate the slowly varying function $(PN_p/N_a)^{0.052}$ that appears in Equations (3.6) and (3.7). With all this information we are now able to estimate

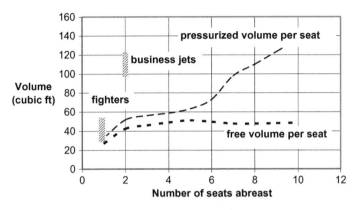

FIGURE 3.2

Variation of the nominal free and pressurized volume per passenger for commercial aircraft. Shown for comparison are typical ranges for business jets and military fighter aircraft.

the likely values of free and pressurized volume per passenger seat for commercial aircraft and these are shown as a function of N_a in Figure 3.2.

The interesting feature is that the free volume per passenger is essentially constant for all commercial aircraft cases, with a range of 35–50 ft³ per passenger. The pressurized volume is about 20% greater than the free volume until the aircraft is large enough that the (circular) cross-sectional area is sufficiently large to permit use of pressurized above- and below-floor space for other uses besides passenger accommodation. Thereafter, as would be expected, the pressurized volume grows linearly with the cabin diameter.

Business jets may have twice the value of the specific volume as commercial airliners since they typically are optimized for comfort. For example, the Gulfstream G200, which carries a maximum of 10 passengers, has a specific free volume of about 87 ft³ per passenger and a specific pressurized volume of about 109 ft³ per passenger. The larger G500, which can carry 18 passengers, provides a specific free volume of 93 ft³ per passenger and a specific pressurized volume of 125 ft³ per passenger. On the other hand, military fighter aircraft have typical cockpit dimensions that suggest a specific free volume of about 40 ft³ per passenger, very much like that of commercial aircraft. The range of volume for these types of aircraft is also indicated in Figure 3.2 for comparison.

3.2.2 Cabin pressure

At the normal stratospheric cruising altitudes of 30,000–38,000 ft, the outside pressure is 0.3–0.2 atm, respectively, while the cabin pressure is maintained at a level equal to that found at altitudes between about 5500 ft and 8000 ft, or between about 0.8 and 0.7 atm. This is normally tolerable for healthy adults without serious difficulties, but on long-duration flights noticeable effects may arise which are similar to what is commonly called mountain sickness. O'Connor (2012) described studies of volunteers subjected to simulated flights with different levels of cabin pressure. As the pressure in the cabin was decreased to levels corresponding to altitudes between 7000 ft and 8000 ft, blood oxygen levels were found to decrease as much as 4.4%. At these lower cabin pressures the volunteers were more likely to complain of headaches, fatigue, muscle cramps, and stomach aches. The general conclusion reached in these studies is that for long-duration flights the cabin pressure should be maintained at a level of about 6000 ft or less to reduce the probability of passenger discomfort. New long-range aircraft, like the Boeing 787, have incorporated increased cabin pressures, at about the 6000 ft altitude level, to address this aspect of passenger comfort.

3.3 General cabin layout

Obviously, the layout of the passenger cabin controls the final length and diameter of the fuselage since the passengers must be completely enclosed within the fuselage. If there are N_P passengers they may be arrayed in a layout with one seat and N_P rows, or N_P seats and one row, or any other combination in between. The first extreme results in a long slender airplane which will be hampered by large bending moments which will require a stronger and heavier structure, as well as by the waste of much room with just one aisle and one seat abreast. The second extreme will encounter high drag due to the large frontal area of the fuselage which must enclose N_P seats abreast. These two extreme cases serve to illustrate the competing roles of the length and diameter of the fuselage. It is to be expected that there will be some optimum ratio of the length to the diameter which will best serve the overall needs of the aircraft's mission. Simple rules of thumb for the typical configuration of passenger cabins of commercial jet transports are that the number of passengers abreast $N_a = \frac{1}{2}\sqrt{N_p}$ and the number of rows $N_r = 2\sqrt{N_p}$, which of course satisfies the requirement that $N_p = N_a N_r$. This is approximately correct for both single- and dual-aisle cabins, but double-deck layouts like the new Airbus A380 may not conform to the rule. However, the actual configuration can only be determined by actually laying out the cabin floor plan to scale. A general elevation view of the fuselage illustrating the important parameters is shown in Figure 3.3. The overall length of the fuselage is L and is made up of a nose cone of length L_{nc}, a passenger cabin of length L_c, and a tail cone of length L_{tc}.

A good rule for fuselage design is that "the fuselage is designed from the inside out." The passenger cabin cross-section is designed first. Next, the plan view of the passenger cabin is laid out. Then the cargo volume is checked for suitability in accepting the requisite number of cargo containers of the standard type chosen. The flight deck may then be designed, although that will not be covered here; it is described in some detail in Torenbeek (1982).

The final step is to design a nose cone and tail cone around the passenger compartment and wrap the fuselage skin around them in an aerodynamically sound fashion. Much of the fuselage design may be carried out prior to determination of the overall configuration so it is advisable to begin this work and conduct it in parallel to the

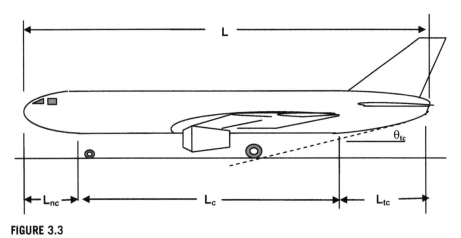

FIGURE 3.3

Schematic diagram of fuselage showing major segments.

market survey and other design activities. To conduct all design tasks serially stretches the work out unacceptably and generally leads to missing completion deadlines.

3.4 Cabin cross-section

The typical passenger cabin is designed to be cylindrical in cross-section as closely as possible. There are a number of reasons for this:

- A circle has the greatest cross-sectional area per unit perimeter. The drag of a typical fuselage, which has a rather large fineness ratio, that is, the ratio of length to diameter, is dominated by the skin friction component, as will be seen later on in the design analysis.
- A circular cross-section is strongest under internal pressure. At the normal stratospheric cruising altitudes of 30,000–38,000 ft, the outside pressure is 0.3–0.2 atm, respectively, while the internal pressure is maintained at a level equal to that found at 8000 ft, or about 0.7 atm. Therefore the pressure difference across the thin skin of the cabin ranges from 0.4 to as much as 0.5 atm, or 6–7 psi (40–50 kPa).
- A circular cylinder can more easily accommodate growth in N_P in terms of manufacturing since cylindrical sections, called plugs, can be readily added to an existing fuselage to create a so-called stretched version of a given aircraft.

There are disadvantages to circular cross-section cabins as well, and these are related to limited space outside the passenger compartment for auxiliary systems and cargo. It is evident that the passenger compartment must be located in the vicinity of a diameter of the circle since this gives the greatest width for seats and aisles. However, this leaves awkward circular sectors above and below the passenger compartment which can serve for luggage and cargo stowage as well as support items

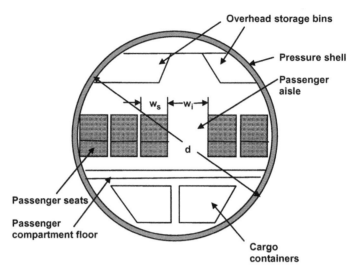

FIGURE 3.4

Schematic diagram of typical passenger cabin cross-section.

like electric cables, ventilating ducts, etc. Many modern designs have expanded the lower portion of the circular cabin into a more rectangular cross-section in the vicinity of the wing root chord to accommodate more internal carriage for landing gear, fuel tanks, and the like. The cabin forward and aft of the wing root is maintained as an essentially circular cross-section, and any stretching to be done to the fuselage will require the plugs to be added in these regions. It is useful to consult the websites of major airframe companies to obtain additional detailed information regarding fuselage layouts. A schematic diagram of a typical cabin cross-section is shown in Figure 3.4.

In order to develop the design of the cabin cross-section it is necessary to choose the style of seating to be employed. As mentioned previously a simple rule of thumb to start the layout of a typical passenger cabin of a commercial jet transport is $N_a = \frac{1}{2}\sqrt{N_p}$ and $N_r = 2\sqrt{N_p}$. Naturally in a detailed layout there may be some rows which have a different number of seats abreast; the former rule of thumb is just a guide. The actual configuration can only be determined by drawing the cabin floor plan. A CAD drawing of a typical cabin floor plan is shown in Figure 3.5. The emergency exits shown in Figure 3.5 are discussed subsequently in Section 3.10.

The diameter of the fuselage is the most important parameter of the fuselage design since it sets the maximum frontal area of the airplane, and it is essential to make it as small as possible while remaining consistent with passenger comfort and safety. Most commercial airlines require aircraft to be easily configured so as to provide at least two cabin seating arrangements, first class and economy class, with the great majority of seats (80% or more) in the latter category and this arrangement will set the diameter of the cabin. Seats, carry-on luggage (and therefore overhead bins),

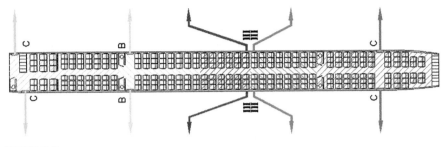

FIGURE 3.5

Typical CAD layout of cabin floor plan showing seating arrangements, lavatories, galleys, and Type B, C, and III exits.

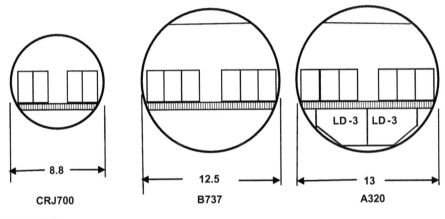

FIGURE 3.6

Cabin cross-sections of typical single-aisle narrow-body aircraft with standard seating. The B737 and the A320 (which can carry a reduced height ULD) are similar in size while the CRJ100 regional jet is substantially smaller. All dimensions are approximate and are given in feet.

cargo containers, and galley carts (and therefore aisles) are standardized so that there is a good deal of constraint on the layout shown in Figure 3.5. Torenbeek (1982) is a good source for detailed information on all these dimensions and these should be used in developing the cabin design. Any increase in cabin interior diameter without a corresponding change in external diameter is a major achievement. The new Boeing 787 is said to have gained an extra inch in interior cabin width because the fuselage's all-composite construction doesn't require a moisture barrier as do fuselages constructed of aluminum, according to Aviation Week and Space Technology (2005).

Cabin cross-sections of single-aisle narrow-body airliners in current service are shown on the same scale in Figure 3.6. Regional jet or turboprop airliners, exemplified here by the Bombardier CRJ700, have only sufficient room for 2-2 seating, while the popular mid-size jets, like the Boeing B737 and the Airbus 320, have space for up

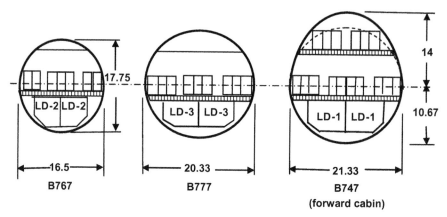

FIGURE 3.7

Cabin cross-sections of the Boeing family of two-aisle wide-body aircraft with typical seating. The forward cabin of the B747 is oval in shape with a second deck. The aft cabin is circular (dashed lines) and has only the main deck. All dimensions are approximate and are given in feet.

to 3-3 seating in the coach section of the cabin. The A320 can accept cargo containers shortened in height but the B737 generally does not carry such containers. These containers are called unit load devices (ULD) and are discussed in more detail later in this chapter. Overhead storage bins have not been included in the diagrams for the sake of clarity. The maximum headroom ranges from about 6 ft in the regional jets to about 7 ft in the mid-size single-aisle, and this value occurs only in the aisles.

Cross-sections of the larger twin-aisle aircraft, leading up to the largest jets, are illustrated by the Boeing B767, B777, and B747 shown which are shown on the same scale in Figure 3.7. Note that the seating can go from 2-3-2 to 3-3-3 to 3-4-3 and that the B747 has a second, upper, deck that runs about one-third the length of the entire main cabin.

The main cabin of the B747 has first class seating in the nose cone of the aircraft, the flight deck being on the upper level. Therefore the B747 has a fuselage cross-sectional shape that changes rather substantially. This can be seen in Figure 3.8, where the pear-shaped nose section appears, as well as the main deck cabin, which is circular in cross-section the rest of the way down from the distinguishing humped shape of the upper deck. For comparison, the largest jet airliner, the Airbus A380, is also shown in Figure 3.8 to the same scale as the B747. The A380 cabin has an oval cross-section throughout its length because the main and upper decks are maintained throughout the fuselage.

3.5 Estimation of fuselage width

As described in the previous section, the number of seats abreast plus the number of aisles desired can be used to estimate the fuselage width. In general, there is one aisle for every three seats abreast; that is, one may have one, two, or three seats abreast

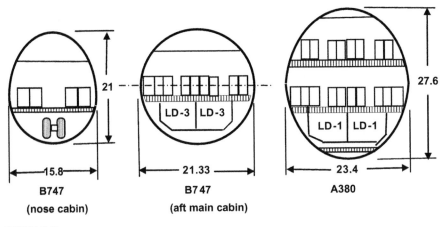

FIGURE 3.8

Cross-sections of the B747 two-aisle wide-body aircraft with typical seating. The first class nose cabin (note the stowed nose wheel) is pear-shaped while the aft main cabin is circular and has only the one main deck. Also shown is the two-deck, two-aisle Airbus A380 illustrating its oval cross-section. All dimensions are approximate and are given in feet.

on either side of one aisle. If four seats are to be joined then there must be an aisle on either side of the four seats. Seating abreast may include the following: 1-1, 1-2, 2-2, 2-3, 3-3, 1-2-1, 2-2-2, 2-3-2, 2-4-2, 3-4-3, etc., where the hyphen denotes an aisle. Then the internal width of the fuselage in the cabin region of the fuselage is proportional to the total width of seats and aisles, or $d_f \approx w_s N_a + w_i N_i$, where w_s is the seat width, N_a is the number of seats abreast, w_i is the aisle width, and N_i is the number of aisles. The equation is written as an approximation because details of seat placement and cabin wall curvature determine the exact width. Typically the cross-section of the fuselage will be circular, but in any case, the actual width or diameter will be obtained from a CAD layout of the cabin cross-sections. Two examples of CAD drawings of fuselage cross-sections with different seating arrangements are shown in Figure 3.9. It is reasonable at this stage in the preliminary design to assume that the distance between the inner and outer cabin surfaces is about 4 in. (10 cm) and is independent of the actual cabin diameter. Similarly, the passenger cabin floor is generally taken to have the same thickness.

3.6 **Estimation of fuselage length**

In the same fashion it is possible to estimate the length of the fuselage cabin section. The fuselage is usually divided up into three sections: the nose cone, the cabin, and the tail cone. Typically the passengers are all housed in the cabin, which tends to be in the shape of a right circular cylinder. This shape is structurally sound, easy to

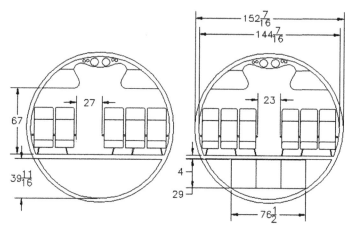

FIGURE 3.9

Typical CAD drawings of fuselage cross-sections with different seating arrangements. All dimensions shown are in inches.

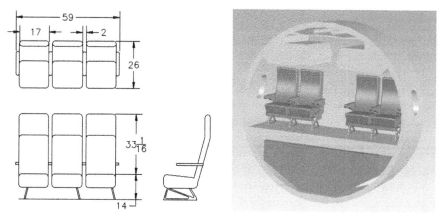

FIGURE 3.10

CAD drawing of a typical three-seat arrangement and a 3-D rendering of a 2-2 seating configuration. All dimensions are in inches.

manufacture, permits increases in length by the addition of "plugs," and has good drag characteristics. In some designs it may be necessary to have several rows of passenger seats extend slightly into the nose or tail cone sections. The length of the cabin then depends upon the number of rows of seats desired and the pitch P of the rows, i.e., the longitudinal distance between adjacent rows. The number of rows is fixed by the number of seats abreast chosen for the design, and is given by N_p/N_a.

Table 3.2 Typical Seating Dimensions

Description	First Class (in.)	Coach (in.)
Seat width w_s	19–21	17–18
Armrest width	2.75	2.5
Aisle width w_i	18–20	16–18
Pitch P	37–42	32–36
Seat height	17	17

FIGURE 3.11

Cabin interior of a typical narrow-body single-aisle aircraft with three-abreast seating

A correlation for the cabin length, based on the discussion in Section 3.2.1, may be put in the following form, where the dimensions are in feet:

$$L_c = 1.08 \left(PN_r\right)^{1.052}$$

This equation is merely for reference and should not be used to set the design. The actual cabin length, like the cabin width, must be obtained from a CAD layout of the planform view of the fuselage. A CAD drawing of a typical seat design is shown in Figure 3.10 along with an isometric rendering of seats placed in the cabin cross-section. Typical dimensions for seats and aisles are shown in Table 3.2. Additional information can be obtained from aircraft manufacturers' and airline operators' websites.

A typical narrow-body single-aisle aircraft with 3-3 seating, the B737-800, is shown in Figure 3.11. The overhead storage bins and the cabin wall curvature are

FIGURE 3.12

Airbus A300 fuselage cross-section illustrating the passenger floor, 2-4-2 seating configuration, overhead bins, and cabin ceiling. Also shown is the cargo hold with unitary load devices (ULD). Attribution: Aaron Siirila.

clearly seen. The photograph illustrates the challenge to the interior designer who is faced with making a long narrow passage seem spacious and comfortable while keeping within the strict restraints of size and weight particular to airplane design. The cross-section of a typical two-aisle wide body, the Airbus A300, is shown in Figure 3.12 illustrating the passenger floor, 2-4-2 seating configuration, overhead bins, and cabin ceiling. Also shown is the cargo hold with unitary load devices (ULD).

3.7 Influence of fuselage fineness ratio
3.7.1 Fuselage effects on drag

Fuselage drag may be estimated by considering the pressure and shear stress on the three major portions of the fuselage as depicted in Figure 3.13. The drag may be considered as the sum of the pressure and friction drag components acting on the three fuselage sections as given by

$$D_{fus} = D_{p,nc} + D_{f,nc} + D_{p,c} + D_{f,c} + D_{p,tc} + D_{f,tc} \tag{3.9}$$

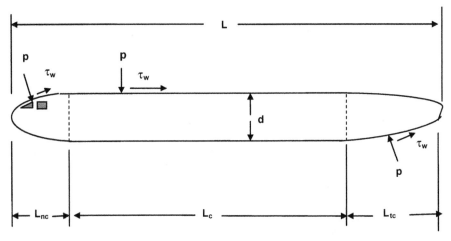

FIGURE 3.13

Schematic diagram of a typical fuselage showing the orientation of the pressure and shear stress on the different fuselage sections.

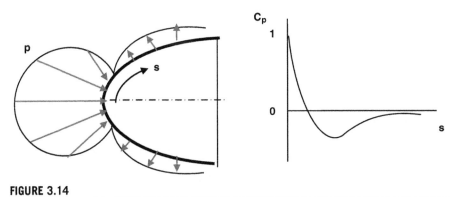

FIGURE 3.14

The pressure distribution about an axisymmetric body at zero angle of attack is shown. The graph shows the pressure coefficient as a function of distance along the body.

The nose cone of the fuselage may be approximated by an axisymmetric body of revolution at zero angle of attack. The pressure distribution about such a body is illustrated in Figure 3.14. The pressure distribution near the stagnation point is in excess of the free stream pressure and produces a drag force while the flow going around the smooth curve of the nose cone develops a pressure less than the free stream pressure and produces a thrust force. This pressure variation is also shown in Figure 3.14 in terms of the distribution of the pressure coefficient as a function of distance along the surface of the nose cone. The pressure coefficient is defined as

$$C_p = \frac{p - p_\infty}{q_\infty} = \frac{p - p_\infty}{\frac{1}{2}\rho_\infty V_\infty^2} = \frac{p - p_\infty}{\frac{1}{2}\gamma p_\infty M_\infty^2} \tag{3.10}$$

The drag and thrust forces due to pressure approximately cancel out, so that a reasonable assumption about the pressure drag of a smoothly curved nose cone is approximately zero, or $D_{p,nc} = 0$. Recalling that the drag is defined as the force in the direction of the free stream velocity permits us to also set the pressure drag on the constant cross-section cabin also equal to zero, since the cabin walls are parallel to the free stream velocity and thus pressure can provide no force in that direction, so $D_{p,c} = 0$. This leaves us with three skin friction terms plus a pressure drag term for the fuselage drag. The three skin friction terms depend on the skin area exposed to friction, or the wetted area, S_{wet}.

The pressure drag of the tail cone is usually called the base drag because it represents the drag due to the inability of the flow to follow the body surface at the base of the fuselage and instead produce an eddying flow there. The drag is usually made dimensionless by forming its ratio with the product of the dynamic pressure of the free stream and some reference area, typically S, the planform area of the wing. For a complete description of the planform parameters of the aircraft see Chapter 5. We may now write the drag of the fuselage and the corresponding drag coefficient for the fuselage as follows:

$$D_{fus} = D_{f,nc} + D_{f,c} + D_{f,tc} + D_{p,tc}$$

$$C_{D,fus} = \frac{D}{q_\infty S_{front}} = C_F(Re, M)\frac{S_{wet}}{S_{front}} + \frac{D_{p,tc}}{q_\infty S_{front}} \tag{3.11}$$

The skin friction coefficient C_F in the drag coefficient on the right-hand side of Equation (3.11) is an integrated value for the whole fuselage and is based on the wetted area of the body surface. The second term on the right-hand side of Equation (3.11) may be written in terms of the tail cone drag coefficient, which is usually defined as

$$C_{D,tc} = \frac{D_{p,tc}}{q_\infty S_{front}} \tag{3.12}$$

It is important to keep in mind that drag coefficients for conventional aircraft are always referred to the wing planform area S, while base drag coefficients are typically referred to the maximum-cross-sectional, or frontal, area of the body, S_{front}. This is an important distinction and lack of attention to the proper non-dimensional factors can lead to significant errors. The pressure drag of the tail cones of smooth axisymmetric bodies like fuselages and nacelles in subsonic flow is found to depend on the speeding up of the flow as it passes around the body and on the location of separation of the boundary layer on the body. Hoerner (1958) suggests the following correlation for the tail cone drag coefficient defined by Equation (3.12):

$$C_{D,tc} = C_F(Re, M)\frac{S_{wet}}{S_{front}}\left(\frac{1.5}{F^{3/2}} + \frac{7}{F^3}\right) \tag{3.13}$$

Here $F = L/d$ is the fineness ratio of the body. Introducing this representation of the tail cone drag coefficient into our fuselage drag coefficient formula of Equation (3.11), and remembering to properly account for the reference areas, yields

$$C_{D,fus} = C_F(Re, M) \frac{S_{wet}}{S_{front}} \left[1 + \frac{1.5}{F^{3/2}} + \frac{7}{F^3} \right] \tag{3.14}$$

If we note that the wetted area of the fuselage surface may be represented by

$$S_{wet} = k\pi L d \tag{3.15}$$

Here the constant $k < 1$ and if the nose cone is approximated by a paraboloid of revolution, the cabin as a circular cylinder, and the tail cone as a right circular cone we may estimate k as follows:

$$k = 1 - \frac{F_{nc}}{3F} - \frac{F_{tc}}{2F} \tag{3.16}$$

As mentioned previously, the parameter F is the fineness ratio of the fuselage $F = L/d$ and for typical modern airliners F is around 8–10. Similarly $F_{nc} = L_{nc}/d$ is the fineness ratio of the nose cone and $F_{tc} = L_{tc}/d$ is the fineness ratio of the tail cone. The fuselage may be considered to be circular in cross-section so that the reference base area is

$$S_{front} = \pi \frac{d^2}{4} \tag{3.17}$$

Now the area ratio in question in Equation 3.13 may be represented by

$$\frac{S_{wet}}{S_{front}} = k\pi d L \frac{4}{\pi d^2} = 4k \left(\frac{L}{d} \right) = 4kF \tag{3.18}$$

Then the drag coefficient, based on frontal area, of the fuselage may be shown to have the following form:

$$C_{D,fus} = 4kC_F(Re, M) F \left(1 + \frac{1.5}{F^{3/2}} + \frac{7}{F^3} \right) \tag{3.19}$$

The nature of this functional form is shown in Figure 3.15, as evaluated for the case of a typical airliner fuselage at an altitude of 35,000 ft and a cruise Mach number $M = 0.85$. The general result is that the minimum drag coefficient occurs at relatively small fineness ratios, about $F = 3$. However, such a small fineness ratio would not result in an effective fuselage for transporting passengers. Typically commercial airliners have fineness ratios in the range $8 < F < 11$ and the contribution of the pressure drag of the tail cone is small. The slope of the drag coefficient curve depends on the value of the integrated skin friction coefficient as can be seen from Equation (3.19). A more detailed description of the drag characteristics will be covered in Chapter 9.

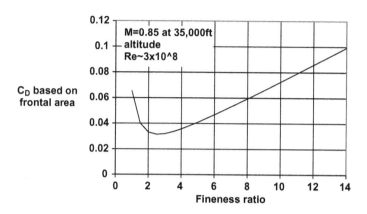

FIGURE 3.15

General trend of fuselage drag coefficient with fineness ratio F.

3.7.2 **Fuselage effect on lift**

The lift distribution on a wing, which is described in some detail in Appendix C, is affected by the presence of the fuselage as a result of the following effects:

- The presence of the fuselage disturbs the longitudinal velocity field in the vicinity of the wing.
- At an angle of attack relative to the free stream, the fuselage also perturbs the flow about the wing in planes normal to the free stream.
- The fuselage has a blocking effect on the flow.

The change in the longitudinal velocity in the vicinity of the wing and, hence, the lift on the wing, is a result of the finite length of the fuselage. The change is not large for a slender fuselage but may be important when the fuselage is relatively bulky, so that a substantial alteration to the local longitudinal velocity may result. The cross-flow caused by the fuselage at angle of attack changes the component of free stream velocity normal to the fuselage axis and affects the downwash flow produced by the wing. The blocking effect of the fuselage is always present even if the fuselage is an infinite cylinder aligned with the free stream, in which case the other two effects are not present. Zlotnick and Diederich (1952) carried out a theoretical analysis which showed that the presence of a slender fuselage does not have an important effect on the lift distribution on an unswept wing of moderate aspect ratio, but a larger change in the lift distribution on a wing in the presence of a fuselage may be anticipated if the wing is swept. Comparisons with experiment supported their results. Pamadi (2004) notes that for configurations with relatively large values of the ratio of wing span to fuselage diameter ($b/d > 2$) the mutual interference effects between the fuselage and the wing are small and can be neglected so that the effects of the body and the wing can be determined individually and summed. In general, experimental results suggest that the maximum lift coefficient for the wing alone and for the wing-body combination typically differs by 5% or less. The lift curve slope is likewise affected little by the fuselage when b/d is large. Roskam (1971) offers the following approximate correction to the lift curve slope of the wing for the wing-body combination:

$$\left(\frac{dC_L}{d\alpha}\right)_{wb} = \left[1 + \frac{1}{4}\left(\frac{d}{b}\right) - \frac{1}{40}\left(\frac{d}{b}\right)^2\right]\left(\frac{dC_L}{d\alpha}\right)_w \tag{3.20}$$

Note that for typical airliners $d/b \sim 0.1$ and therefore the lift curve slope of the wing-body is approximately equal to that of the wing alone. As a result of these considerations it is reasonable to assume that the lift produced by the wing alone is equivalent to that of the wing-body for preliminary design purposes.

3.8 Estimation of nose cone and tail cone length

The shape of the nose and tail cone sections is usually set by a combination of aero-dynamic and operational requirements. The aerodynamics of the situation dictates a smooth contour for the fore and aft sections of the fuselage. The nose cone is approx-imately ellipsoidal in nature while the tail cone tends to be somewhat more conical. Structurally, there is a requirement for good visibility from the flight deck in the nose section and a need for a kick-up of the bottom of the tail cone section to provide ground clearance during takeoff rotation; this is shown as the angle θ_{tc} in Figure 3.3.

Specific data for these fuselage characteristics are shown for 15 commercial air-craft spanning the range of current practice in Figures 3.16 and 3.17. As expected, the

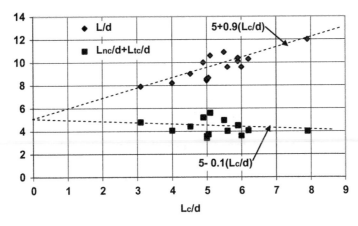

FIGURE 3.16

Correlation of total length with cabin length and correlation of nose plus tail cone length with cabin length (d=fuselage diameter).

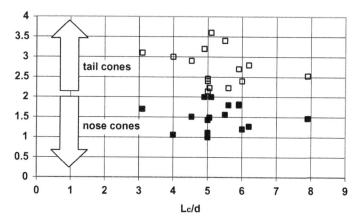

FIGURE 3.17

Correlation of nose cone length and tail cone length with cabin length for various aircraft is shown. Open symbols denote L_{tc}/d and closed symbols denote L_{nc}/d.

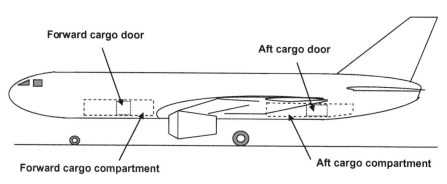

FIGURE 3.18

Typical placements of cargo compartments and doors for airliners.

overall length of the fuselage scales reasonably well with the cabin length, as shown in Figure 3.16. The ratios of nose cone length to fuselage diameter and tail cone length to fuselage diameter are shown as a function of the ratio of cabin length to fuselage diameter in Figure 3.17. Although there is appreciable scatter among the various aircraft it is clear that the ratios of nose cone length to fuselage diameter lie between 1.0 and 2.0, while the ratios of tail cone length to fuselage diameter lie between 2.0 and 3.5.

3.9 Cargo containers

Passenger aircraft must carry passenger luggage and they often carry revenue cargo as well. To facilitate the loading and unloading of such cargo the use of cargo

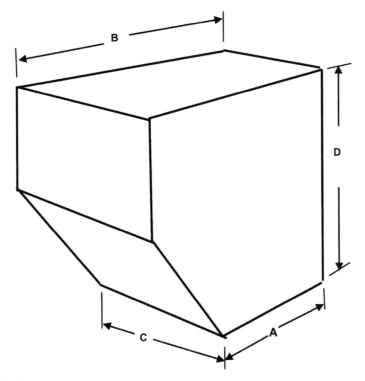

FIGURE 3.19

Typical cargo container, or unit load device (ULD) for airliners.

Table 3.3 Dimensions of Common Cargo Containers (ULD) for Airliners

Container type[a]	A (in.)	B (in.)	C (in.)	D (in.)	Volume (ft³)
LD-1	61.5	92	60.4	64	173
LD-2	47	61.5	60.4	64	120
LD-3[b]	61.5	79	60.4	64	159
LD-6	125	160	60.4	64	316
LD-8	96	125	60.4	64	243

[a] Letters in first row refer to dimensions shown in Figure 3.16.
[b] Most common ULD in use.

containers design to properly fit and be secured within the cargo hold of the aircraft, usually the volume below the floor of the passenger compartment. A typical configuration for the cargo compartments is shown in an elevation view of the aircraft in Figure 3.18. Typical containers are illustrated in Figure 3.19 and dimensions of the most commonly used containers are given in Table 3.3. A photograph of cargo containers being loaded onto a B747 through a front cargo door is shown in Figure 3.20.

FIGURE 3.20

Unit load devices being loaded onto a B747 through a front cargo door.

Detailed information on these containers and pallets, called Unit Load Devices (ULD), including dimensioned drawings may be found in Unit Load Device (2012). As mentioned in Chapter 2, the baggage allowance for passengers can be converted to volume by using an average density of $12.5\,lb/ft^3$. Once the luggage volume requirement is met, additional volume in the cargo hold is available for freight, figured at an average density of $10\,lb/ft^3$. Additional weight due to freight is accounted for in the refined weight estimate in Chapter 8.

The database of 50 commercial airliners provides cargo weights for 44 of the 50. Most aircraft have the capability to carry about 10% of the takeoff weight in the cargo holds as illustrated by Figure 3.21. These figures are for passenger aircraft and do not include freighter aircraft.

3.10 Emergency exits

The material on this subject is cited in detail in the Federal Aviation Regulations (2012). For our purposes only the information most relevant to commercial airliners is discussed and the full regulations should be consulted for completeness. These regulations should be considered carefully in the design process and their implementation should be clearly described in the final report.

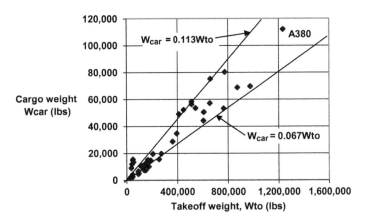

FIGURE 3.21

Cargo weights for 44 commercial airliners.

FAR Section 25.803 requires that each crew and passenger area must have emergency means to allow rapid evacuation in crash landings, with the landing gear extended as well as with the landing gear retracted, considering the possibility of the airplane being on fire. For airplanes having a seating capacity of more than 44 passengers, it must be shown that the maximum seating capacity, including the number of crewmembers required by the operating rules for which certification is requested, can be evacuated from the airplane to the ground under simulated emergency conditions within 90 s and compliance with this requirement must be shown by actual demonstration. To aid in complying with this basic requirement, FAR Section 25.807 defines the types of acceptable emergency exits in detail, and the major types are as follows:

(1) *Type I.* A floor-level exit having a rectangular opening of not less than 24 in. wide by 48 in. high, with corner radii not greater than 8 in.

(2) *Type II.* A rectangular opening of not less than 20 in. wide by 44 in. high, with corner radii not greater than seven in. Type II exits must be floor-level exits unless located over the wing, in which case they must not have a step-up inside the airplane of more than 10 in. nor a step-down outside the airplane of more than 17 in.

(3) *Type III.* A rectangular opening of not less than 20 in. wide by 36 in. high with corner radii not greater than seven in., and with a step-up inside the airplane of not more than 20 in. If the exit is located over the wing, the step-down outside the airplane may not exceed 27 in.

(4) *Type IV.* A rectangular opening of not less than 19 in. wide by 26 in. high, with corner radii not greater than 6.3 in., located over the wing, with a step-up inside the airplane of not more than 29 in. and a step-down outside the airplane of not more than 36 in.

(5) *Type A.* This is a floor-level exit with a rectangular opening of not less than 42 in. wide by 72 in. high, with corner radii not greater than 7 in.

Table 3.4 Maximum Number of Passenger Seats Permitted for Each Exit Type

Exit Type	Number of Seats
A	110
B	75
C	55
I	45
II	40
III	35
IV	9

(6) *Type B.* This is a floor-level exit with a rectangular opening of not less than 32 in. wide by 72 in. high, with corner radii not greater than 6 in.

(7) *Type C.* This is a floor-level exit with a rectangular opening of not less than 30 in. wide by 48 in. high, with corner radii not greater than 10 in.

Each required passenger emergency exit must be accessible to the passengers and located where it will afford the most effective means of passenger evacuation. For an airplane that is required to have more than one passenger emergency exit for each side of the fuselage, no passenger emergency exit shall be more than 60 ft from any adjacent passenger emergency exit on the same side of the same deck of the fuselage, as measured parallel to the airplane's longitudinal axis between the nearest exit edges. The maximum number of passenger seats permitted depends on the type and number of exits installed in each side of the fuselage. The maximum number of passenger seats permitted for each exit of a specific type installed in each side of the fuselage is presented in Table 3.4.

For a passenger seating configuration of more than 110 seats, the emergency exits in each side of the fuselage must include at least two Type I or larger exits. The combined maximum number of passenger seats permitted for all Type III exits is 70, and the combined maximum number of passenger seats permitted for two Type III exits in each side of the fuselage that are separated by fewer than three passenger seat rows is 65. If a Type A, Type B, or Type C exit is installed, there must be at least two Type C or larger exits in each side of the fuselage.

For aircraft with fewer than 110 seats, passenger ventral or tail cone exits, ditching emergency exits for passengers, and emergency exits for flight crew the full text of FAR Section 25.807 should be consulted for corresponding requirements.

3.11 Recent developments in fuselage design

In October 2008, NASA initiated the NASA N+3 program, described in NASA (2008), which seeks advanced concepts for aircraft that would satisfy commercial air transportation needs while meeting specific energy efficiency, environmental and

FIGURE 3.22

Boeing's sub-scale version of the X-48B Blended Wing-Body (BWB) aircraft flies over Rogers Dry Lake at Edwards Air Force Base. *Credit: NASA.*

operational goals in 2030 and beyond. Among NASA's goals for a 2030-era aircraft, compared to aircraft entering service at the start of the program, are:

- Reduction of noise by 71 db below current FAA noise standards to contain objectionable noise within airport boundaries.
- More than a 75% reduction on the International Civil Aviation Organization's (ICAO) standard for nitrogen oxide emissions for improving air quality around airports.
- More than a 70% reduction in fuel burn to reduce greenhouse gas emissions and the cost of air travel.

A number of alternative aircraft designs were developed by various industry and academic teams as can be seen in NASA (2008). One of the approaches of interest here is the blended wing-body (BWB) concept which involves a flattened cabin area that blends smoothly into the wing. A sub-scale flying model of Boeing's X-48 design for a BWB aircraft is shown in Figure 3.22.

Though the BWB concept brings a number of performance advantages there are weight penalties associated with the noncircular fuselage design which will be discussed subsequently. The flattened cross-section of a BWB cabin provides a more familiar setting for passengers who are used to occupying rooms that are essentially

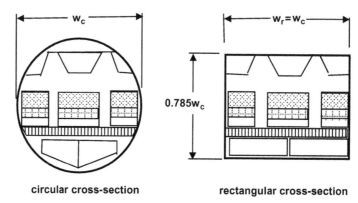

FIGURE 3.23

Circular and rectangular cabins having the same seating and equal frontal area.

rectangular in cross-section. Safety concerns regarding rapid egress from an aircraft in emergency situations make it likely that the allowable seating would be limited to 10 abreast in a two-aisle, 3-4-3 arrangement. For this case the aircraft cabin width w_c is closely approximated by the relation

$$w_c \approx w_s N_a + w_i N_i = w_s N_a + 2w_i$$

Consider two cabins having the same seating arrangement and frontal area, but with different cabin cross-sections, as shown in Figure 3.23. There appears to be somewhat more headroom and overhead bin volume in the rectangular cabin, but the cargo hold will likely have smaller volume and require different types of cargo container. The circular cross-section cabin has some overhead space for electrical and HVAC utilities, while the rectangular cross-section cabin has none.

The equivalence in frontal area provides for both fuselage shapes to have approximately equivalent pressure drag. However the perimeter of the rectangular cross-section cabin is $3.57w_c$ while that of the circular cross-section cabin is $3.14w_c$. If the two cabins have the same total number of seats N_p with the same pitch between rows, then the length of both cabins would be equal. Thus the ratio of fuselage surface wetted area $S_{w,r}/S_{w,c}=1.14$ so that the rectangular fuselage would likely have about 14% greater skin friction drag than the circular fuselage. So there are still advantages to circular fuselage shapes in addition to manufacturability issues. To see the parameter space where circular and rectangular cabin cross-sectional shapes have their advantages, we may examine the ratios of cross-sectional areas A_c/A_r and of perimeters P_c/P_r as follows:

$$\frac{A_c}{A_r} = \frac{\pi}{4} \left(\frac{h}{w_c} \right)^{-1}$$

$$\frac{P_c}{P_r} = \frac{\pi}{2} \left(1 + \frac{h}{w_c} \right)^{-1}$$

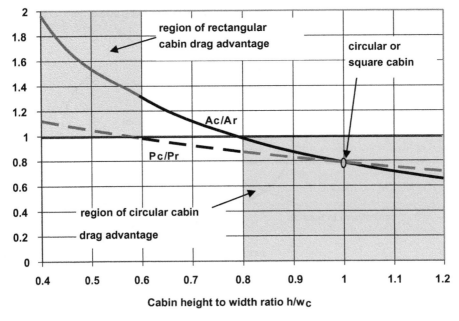

FIGURE 3.24

Cross-sectional area and perimeter ratios for circular and rectangular cross-section fuselage cabins shown as a function of the height to width ratio of the cabin.

The area ratio is a measure of the ratio of the pressure drag while the perimeter ratio is a measure of the skin friction drag. The scale parameter is the ratio of the height of the cabin to the width of the cabin, h/w_c. The width of the cabin w_c is set by the number of seats abreast and must equal the height of the circular cabin, that is, $h/w_c = 1$ for a circular (or square) cabin. However, for a rectangular cabin the height is essentially set by a small multiple of the ceiling height desired for passenger comfort and can be as small as 0.4 for a large number of seats abreast, say 10. The area ratios are depicted in Figure 3.24 and show that for $h/w_c > 0.8$ the circular cross-section should have lower drag while for $h/w_c < 0.8$ the rectangular cross-section should have lower drag. For the range $0.6 < h/w_c < 0.8$ the circular cabin will likely have larger pressure drag but lower skin friction drag.

Mukhopadhyay (2005) explains that high stress levels in the walls of rectangular cross-section cabins lead to weight problem in blended wing-body (BWB) pressurized cabins as compared to the tubular fuselages of traditional aircraft. Consider again the two cabin configurations shown in Figure 3.23 which compares a conventional circular cabin with a rectangular cabin in a BWB aircraft with both operating under the same net internal pressurization level p. In the circular cross-section cabin of diameter w_c and skin thickness t, the pressure is resisted by a uniform hoop stress $\sigma = p(w_c/t)$. However, in the rectangular cabin of the BWB airplane the cabin walls resist the pressure by bending outward. Considering the flat wall as a simply

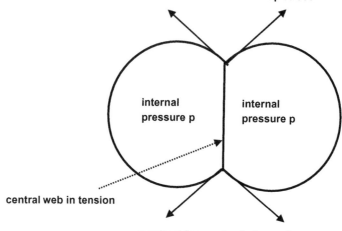

resultant forces due to hoop stress

internal
pressure p

internal
pressure p

central web in tension

resultant forces due to hoop stress

FIGURE 3.25

The double bubble concept for a flattened cabin. Two circular cylinders are attached to each other along a line of intersection and a central web joins the two lines of intersection.

supported plate of length w_c and thickness t, the maximum bending stress is equal to $\sigma_{max} = 0.75p(w_c/t)^2$. Then the ratio of the maximum bending stress to the hoop stress is $0.75(w_c/t)$, which is very high because $w_c/t \gg 1$. In an effectively rectangular cabin of a BWB airplane the bending stiffness of the wall must be increased, and this can lead to significant weight penalties.

An alternative fuselage design which achieves a flattened cabin shape without incurring the large bending stress penalty is the double bubble concept. Drela (2011) describes the D8 aircraft design, one of those which showed promise in satisfying the stringent goals posed by NASA. A distinguishing feature of the twin-aisle cabin is a "double bubble" cross-section. The concept is shown in Figure 3.25 where two bubbles have a center vertical web to give a pure-tension pressure vessel in cross-section. Drela (2011) points out that the web material could be concentrated into slender struts making the cabin appear effectively wide open.

The name double bubble is quite clear because the configuration appears like two soap bubbles joined together by a common membrane. Like the soap bubble, the elements of the cabin structure are all in tension. An appreciation for the appearance of the wide flattened fuselage shape may be gained from Figure 3.26 which shows a model of the D8 mounted in an MIT wind tunnel.

Although these new concepts show great promise they still require a substantial amount of research and development to achieve their potential and further discussion of them is beyond the scope of this book.

The double bubble concept used in the D8, a configuration which is about the size of a B737, expands the horizontal dimension of the cabin. Though this has advantages with respect to passenger comfort, the horizontal double bubble fuselage design contributes to the improvement of a number of other performance issues

FIGURE 3.26

An experimental model of the D8 aircraft with the double bubble fuselage is shown as mounted in an MIT wind Tunnel.

which are discussed by Drela (2011). The double bubble concept may be applied to a smaller airliner and used in a vertical configuration to expand the usable cabin and cargo space. This is the approach taken in the Embraer 175 which seats 78 to 88 passengers. The E175 cabin has the cross-section shown in Figure 3.27 in which two circles of different diameter placed on vertically displaced centers form an inverted almost pear-shaped cross-section. This permits passenger comfort in terms of cabin width, but also adds cargo space in terms of cabin depth.

The oval shape is not readily apparent in a typical photograph of the E175, such as that shown in Figure 3.28. However it is clearly apparent in the case of the Breguet Atlantic Br.1150 of the French Navy shown in Figure 3.29. The Atlantic is a twin-engined, turboprop airplane that also has a vertically configured "double-bubble" fuselage, with the upper lobe comprising a pressurized crew compartment, and the lower lobe housing a 9 m (27 ft 6 in.) long weapons bay.

3.12 **Design summary**

At this stage of the project the proposed fuselage shape should be designed and the floor plan with seating, emergency exits, lavatories, and galleys completed. The cargo hold with the associated ULDs and cargo doors should also be specified. These aspects of the design of the fuselage may be conveniently carried out following the procedure below:

1. Draw the cabin deck floor line as a start in developing the fuselage cross-section.

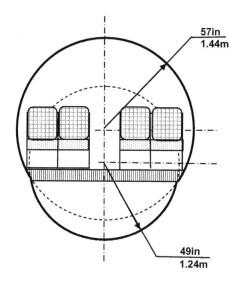

FIGURE 3.27

An illustration of the Embraer 175 double bubble fuselage arrangement. The dimensions shown are approximate and the dotted lines denote the remainders of the two circles comprising the double bubble.

2. Place the seats and aisle locations on the floor line. The typical range of seating dimensions is presented in Table 3.1.
3. Sketch seated and standing passengers to note head clearances and aisle width clearances.

FIGURE 3.28

An Embraer 175 on the ramp. The double bubble shape is not apparent in this typical view.

FIGURE 3.29

A Breguet Atlantic Br.1150 of the French Navy. The double bubble fuselage shape is apparent in this frontal view.

4. Choose a floor thickness to locate ceiling of cargo hold.
5. Place standard cargo containers in cargo hold.
6. Note the extreme positions of all necessary components that limit the internal dimensions of the cabin.
7. Draw the interior contour of the fuselage as tightly around the extreme points as possible, preferably maintaining a circular cross-section for the reasons outlined previously in this chapter.
8. A wall thickness of about 4 in. (10 cm) is found to be generally applicable, independent of cabin diameter, and the exterior boundary of the fuselage is drawn accounting for this thickness.
9. Indicate windows in the cross-sectional view to illustrate passenger's field of view.
10. Lay out the floor plan of the cabin by placing all the seats in rows appropriately pitched for the class of travel and accounting for emergency exits, lavatories, galleys, and wardrobes.
11. Design the shape of the nose cone and tail cone using the cabin diameter and length as guides; an appropriate kick-up angle of the tail cone must be provided to minimize tail strikes on landing and take-off.
12. Lay out the cargo holds and doors and the distribution of the cargo containers.
13. Produce detailed final CAD drawings of the fuselage cross-section and plan and elevation views of the fuselage layout.

3.13 **Nomenclature**

A	frontal area
b	wingspan
C_D	drag coefficient
C_F	skin friction coefficient
C_L	lift coefficient
C_p	pressure coefficient
D	drag
d	cabin outside diameter
d_f	cabin inside width
F	fineness ratio
k	constant
L	fuselage length
L_c	cabin length
L_{nc}	nose cone length
L_{tc}	tail cone length
M	Mach number
N_a	number of seats abreast
N_i	number of aisles
N_p	number of passengers
N_r	number of rows
P	seat pitch or cabin perimeter
p	pressure
q	dynamic pressure
Re	Reynolds number
S	wing planform area
S_{wet}	wetted surface area
S_{front}	fuselage cross-sectional area
V	velocity
w_s	seat width
w_a	aisle width
γ	ratio of specific heats
ρ	density

3.13.1 **Subscripts**

c	cabin or circular cross-section
car	cargo
f	friction
fus	fuselage
nc	nose cone
p	pressure
r	rectangular cross-section
tc	tail cone

References

Aviation Week and Space Technology, 2005. Boeing Commercial Aircraft, November, 21, p. 19. <www.boeing.com/commercial/flash.html>.

Drela, M., 2011. Development of the D8 Transport Configuration. In AIAA Paper 2011-3970, 29th AIAA Applied Aerodynamics Conference, Honolulu, HI.

Federal Aviation Regulations, 2012. <http://rgl.faa.gov/Regulatory_and_Guidance_Library/ rgWebcomponents.nsf/>.

Hoerner, S.F., 1958. Fluid Dynamic Drag. Midland Park, NJ. published by the author.

Mukhopadhyay, V., 2005. Blended-Wing-Body (BWB) Fuselage Structural Design for Weight Reduction. In: AIAA 2005-2349, 46th AIAA/ASME/ASCE/AHS/ASC Structures, Structural Dynamics and Materials Conference, Austin, TX.

NASA, 2008. NASA Advanced Concepts Studies Awardees, <www.aeronautics.nasa.gov/ nra_awardees_10_06_08.htm>.

Nita, M., Scholz, D., 2010. From Preliminary Aircraft Cabin Design to Cabin Optimization. Deutscher Luft- und Raumfahrtkongress 2010, Document 161168.

O'Connor, A., 2012. Flying Can Cause Mountain Sickness, New York Times, Science Times Section, May 29, scitimes@nytimes.com.

Pamadi, B.N., 2004. Performance, Stability, Dynamics, and Control of Airplanes. AIAA, Reston, VA.

Roskam, J., 1971. Methods for Estimating Stability and Control Derivatives of Conventional Subsonic Airplanes. Lawrence, Kansas. published by the author.

Torenbeek, E., 1982. Synthesis of Subsonic Aircraft Design. Kluwer Academic Publishers, Dordrecht, The Netherlands.

Unit Load Devices, 2012. <www.fredoniainc.com/glossary/air.html>.

Zlotnick, M., Diederich, F.W., 1952. Theoretical Calculation of the Effect of The Fuselage on the Spanwise Lift Distribution on a Wing. NACA RM L51J19.

CHAPTER OUTLINE

4.1 Introduction

The engine selection process is dominated by that portion of the mission profile characterized by acceleration: the takeoff phase. This is due primarily for the need to

Commercial Airplane Design Principles. http://dx.doi.org/10.1016/B978-0-12-419953-8.00004-8

accelerate rapidly and climb out from airports where traffic may be quite heavy and noise abatement requirements severe. This is different from the cruise portion of the mission profile, which is mainly influenced by the lift to drag ratio, as was apparent from the discussions in Chapter 2. The ability to takeoff and climb rapidly within a distance compatible with typical airport runway lengths calls for substantial thrust levels while the ability to deliver high fuel economy requires tailoring the engine appropriately for cruise as well as for takeoff.

However, the design actually begins by using the preliminary weight estimates developed in Chapter 2 for analyzing the landing phase. Here, the deceleration levels must be acceptable for a wide variety of passengers, which means that it ordinarily cannot be out of the range of $g/3$–$g/2$. Thus the approach and landing speeds must be low enough so that deceleration on the ground is not harsh, and this makes stringent requirements on the aerodynamic lift capabilities of the aircraft. Once a range of appropriate lift coefficients in landing are selected attention can turn to the takeoff phase. Because lift is proportional to the square of velocity and typical runway lengths are limited, the tradeoff is between high thrust for rapid acceleration to high speeds and high lift coefficients for safe takeoff at low speeds. At the same time the cruise phase also puts practical constraints on the thrust level required. Combining the results of these analyses forms a relatively small design region within which the design aircraft must be placed. A final check on the adequacy of the lift and thrust capabilities of the proposed aircraft is carried out by determining the runway length, the so-called balanced field length, needed to provide safety with one engine inoperative, either by taking off successfully or for bringing the aircraft to a safe stop on the runway.

The engine selection process for turbofan engines is followed by a similar analysis specialized to turboprop-powered airliners. These aircraft are likely to be the predominant type used for regional service as regional jets are phased out because of relatively poor fuel economy with little added time saving over short ranges.

4.2 Landing requirements

In landing it is desirable to have a smooth gradual descent with fairly high drag values; more rapid deceleration can be safely achieved once on the ground by means of thrust reversers and brakes. The landing phase is critical since the aircraft is making a transition from three-dimensional motion to two-dimensional motion at a hard interface—the runway surface. Therefore it is important to land at a reasonable speed, one which is safe but not too slow. It will be seen that too slow a landing puts difficult constraints on either excessive wing surface area S (which is detrimental in cruise because of high skin friction penalties) or $C_{L,max}$ in landing (which would require extensive application of high lift devices). Assuming that the landing may be considered a quasi-static process where the vertical acceleration is very small prior to touchdown, the lift may be equated to the landing weight, or

$$L_l = W_l = \frac{1}{2}C_{L,l}\rho S V_l^2$$

This equation may be solved for the wing loading in landing to yield

$$\left(\frac{W}{S}\right)_l = 3.392 \times 10^{-3} C_{L,l} V_{E,l}^2 \tag{4.1}$$

Here $C_{L,l}$ is the lift coefficient in the landing configuration and $V_{E,l}$ is the equivalent landing speed in knots, usually abbreviated to kts. Aircraft velocity is typically measured in knots (nautical miles per hour) and the conversion factor is $kts = 1.1515\,mph = 1.852\,kph$.

The equivalent airspeed is defined through the definition of dynamic pressure as follows:

$$q = \frac{1}{2}\rho V^2 = \frac{1}{2}\rho_{sl} V_E^2$$

This equation incorporates the effect of local density through the equation $V_E = V\sqrt{\sigma}$, where σ is the ratio of local density to sea level density, that is, $\sigma = \rho/\rho_{sl}$. Thus the equivalent velocity is the velocity at sea level which yields the same dynamic pressure as that at another velocity and altitude. Solving Equation (4.1) for the equivalent landing speed (in kts) yields

$$V_{E,l} = 17.17\sqrt{\frac{(W/S)_l}{C_{L,l}}} = \sqrt{\sigma} V_l \tag{4.2}$$

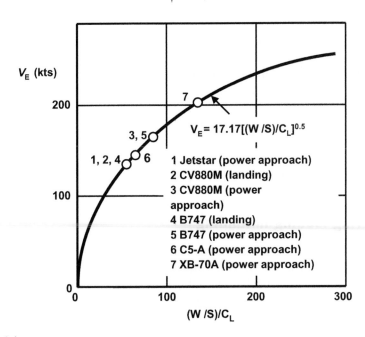

FIGURE 4.1

Equivalent airspeed is shown as a function of the ratio of wing loading to lift coefficient for various aircraft using flight test data from Heffley and Jewell (1972).

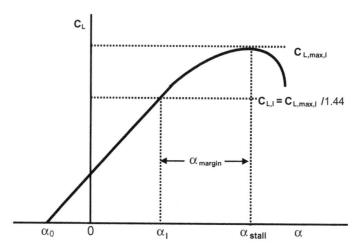

FIGURE 4.2

Sketch of C_L versus α illustrating different operating regimes.

The relation in Equation (4.2) is plotted in Figure 4.1 along with some flight data for several aircraft in both the landing configuration and the power approach configuration using flight test data from Heffley and Jewell (1972). Note that the landing speed is given by $V_{E,l} = kV_{E,S}$ where $V_{E,s}$ is the equivalent airspeed at stall and k is taken as 1.2. This provides a margin of safety against the possibility of encountering tailwind gusts which might reduce the actual airspeed to values below the stall speed in the crucial seconds of touchdown. The stall speed is defined to be the speed pertinent to steady level flight at the maximum lift coefficient for any given configuration, as illustrated in Figure 4.2. In landing, the equivalent stall speed is

$$V_{E,S} = 17.17\sqrt{\frac{(W/S)_l}{C_{L,l\,\max}}} = \sqrt{\sigma}V_S \qquad (4.3)$$

The approach speed is required, by FAR Part 25, to be equal to or greater than 130% of the stall speed, i.e., $V_{E,a} \geq 1.3\ V_{E,S}$.

4.3 Wing loading in landing

An indication of the wing loading for market survey aircraft may be obtained by using their quoted landing speeds and Equation (4.3) to form a graph like that shown in Figure 4.3. It is clear from the data in Figure 4.1 that the landing speeds typical of jet transports result in a ratio of wing loading to lift coefficient which ranges from roughly 50–60. In terms of the ratio of wing loading to maximum lift coefficient in landing, the range figures must be divided by 1.44 to give the spread 35–42, i.e.,

$$35 < \frac{(W/S)_l}{C_{L,l,\max}} < 42$$

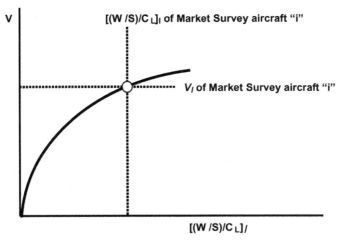

FIGURE 4.3

Sketch of landing velocity V_l versus $[(W/S)/C_L]_l$.

This is so because $C_{L,l,\max} = 1.44 C_{L,l}$ when $V_{E,l}$ is taken as $1.2 V_{E,S}$. Maximum lift coefficient capability for modern jet transports are typically between 2 and 3, depending upon the extent of use of high lift devices, as will be discussed subsequently.

4.4 Landing field length

When an aircraft's configuration is fully defined we may analyze the performance using the approach described in Chapter 10. However, at this point in the design process we have little detailed information with which to work. Therefore, the landing maneuver sketched in Figure 4.4 will be treated using the approximation developed in Section 10.5. The approximate Equations (10.69) and (10.70) give the landing distance as

$$x_l = x_a + x_g \approx \frac{(W/S)_l}{g\rho_{sl}\sigma C_{L,\max,l}} \left[7.11 + \frac{1.44}{\frac{\bar{a}}{g}} \right] \tag{4.4}$$

In Equation (4.4) the landing length x_l is the sum of the air distance in landing (starting from the 50-ft obstacle) and the ground run from touchdown to the stopping point. It is assumed that an average deceleration $\bar{a}$ is applied during the ground run through the use of thrust reversers and brakes; the quantity g is the acceleration of gravity. Equation (4.4) has been plotted in Figure 4.5 as a relation between the non-dimensional deceleration $\frac{\bar{a}}{g}$ and the landing distance x_l; with the quantity $\frac{(W/S)_l}{\sigma C_{L,\max,l}}$ appearing as a parameter. In this figure it is seen that for the normal range of ground run decelerations and the normal range of $\frac{(W/S)_l}{\sigma C_{L,\max,l}}$ the corresponding range of x_l is about 4500–7000 ft. Market survey aircraft decelerations may be estimated by using their quoted values for nominal landing distance and speed, x_l and V_l, respectively.

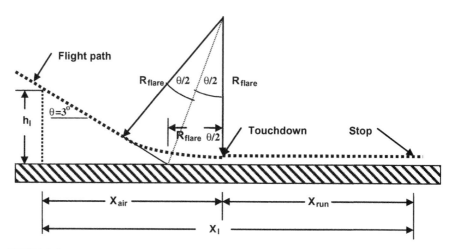

FIGURE 4.4

A sketch of the landing phase showing the important variables.

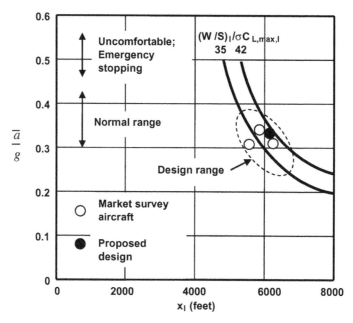

FIGURE 4.5

Variation of non-dimensional average decelerations in the ground run as a function of the landing distance for various values of $\frac{(W/S)_l}{\sigma C_{L,max,l}}$.

First, a value of $\frac{(W/S)_l}{\sigma C_{L,max,l}}$ is deduced from Figure 4.1 by using $V_{E,l} = V_l \sqrt{\sigma}$. Then, with the nominal quoted value of x_l and the estimated value of $\frac{(W/S)_l}{\sigma C_{L,max,l}}$ the market value data may be displayed on a plot like Figure 4.5. It is useful to add the data for

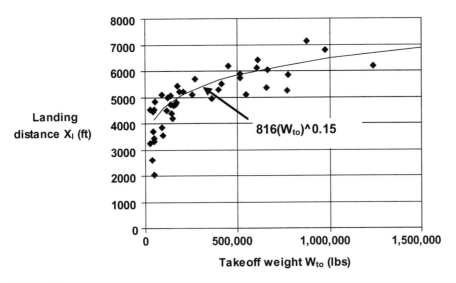

FIGURE 4.6

Nominal landing distances for 50 commercial airliners.

the market survey aircraft to Figures 4.1 and 4.5 because this information provides a clear indication of the competitive range which must be met or bettered. Note that the effects of airport altitude and hot day operation are easily assessed because the density ratio σ appears in both figures.

For a particular choice of average deceleration $\frac{\bar{a}}{g}$ and wing loading to lift coefficient ratio $\frac{(W/S)_l}{\sigma C_{L,max,l}}$ it is possible to estimate the design landing distance and to place a data point for the design aircraft on the plot. It is expected that the design point, shown as a solid circle, will lie in the design range defined by the dashed ellipse shown in Figure 4.5. Thus a method is available for determining the landing characteristics of the design aircraft. For example, a desire to achieve a particular landing distance coupled with a choice of a reasonable deceleration leads to a candidate value for $\frac{(W/S)_l}{\sigma C_{L,max,l}}$.

Nominal landing distances for aircraft in the database of 50 commercial airliners are plotted as a function of takeoff weight in Figure 4.6. An approximation to the behavior of the landing distance is shown in the figure as the correlation $x_l = 816 W_{to}^{0.15}$. Note that for aircraft with $W_{to} > 100,000\,\mathrm{lb}$ the landing distance lies in the range $4000\,\mathrm{ft} < x_l < 7500\,\mathrm{ft}$, which is consistent with the results in Figure 4.5. At lower takeoff weights the aircraft are characterized by lower wing loadings and these move the curves in Figure 4.5 to the left, that is, to lower landing distances.

4.5 Wing loading in takeoff

Having selected the candidate design point for the parameter $\frac{(W/S)_l}{\sigma C_{L,max,l}}$ shown in Figure 4.5 it is possible to determine the corresponding parameter for takeoff conditions by noting that the takeoff wing loading may be written as

$$\left(\frac{W}{S}\right)_{to} = \left(\frac{W}{S}\right)_l \left(\frac{W_{to}}{W_l}\right)$$

The nominal landing weight for the design aircraft is known from the weight estimation results obtained in Chapter 2. However, the nominal design landing weight may not be achieved if there is an in-flight emergency requiring immediate landing prior to completing the full range length. Under these heavier weight conditions the landing speed may be too high to permit landing within a practical field length. Therefore it is necessary to design to some realistic maximum landing weight design condition. Typically, landing weight restrictions are based on (a) airframe structural strength, (b) runway mechanical strength, and (c) landing speed restrictions. Torenbeek (1982) offers a correlation for maximum landing weight in the following form:

$$\phi = \frac{W_{l,\max} - W_{zf}}{W_{to} - W_{zf}} = a + b\exp\left(-\frac{R}{R_r}\right)$$

Here we offer suggestions for the constants a and b that yield results which appear to agree closely with the database for 50 commercial airliners:

$$\phi = 0.36 + 0.72\exp\left(-R/R_r\right) \tag{4.5}$$

In the definition of ϕ, $W_{l,\max}$ is the maximum landing weight, W_{zf} is the zero fuel weight, R is the design range, and R_r is the reference range which is taken to be equal to 1000 nm or 1151.5 mi. This equation may be rearranged, using the fuel weight fraction M_f discussed in Chapter 2, to yield

$$\frac{W_{l,\max}}{W_{to}} = 1 - (1 - \phi)M_f \tag{4.6}$$

Using the database of 50 commercial airliners permits an evaluation of the relationship between landing weight and takeoff weight according to Equations (4.5) and (4.6). This is shown in Figure 4.7 and suggests that a reasonable correlation, for takeoff weight in pounds, is

$$\frac{W_{l,\max}}{W_{to}} = 1 - 0.003W_{to}^{1/3}$$

It is possible to now determine a reasonable and consistent design value for the takeoff wing loading in terms of the maximum landing weight and this is given by

$$(W/S)_{to} = \frac{(W/S)_{l,\max}}{\left[1 - (1 - \phi)M_f\right]} \tag{4.7}$$

Since a design value for $\frac{(W/S)_l}{\sigma C_{L,\max,l}}$ was chosen at the conclusion of the previous section, one may use Equation (4.7) to calculate a corresponding design value for the takeoff phase, i.e., $\frac{(W/S)_{to}}{\sigma C_{L,\max,to}}$ which then defines a takeoff wing loading, $(W/S)_{to}$, for a specified value of both σ and $C_{L,\max,l}$. Using the result obtained previously for the range of $\frac{(W/S)_l}{\sigma C_{L,\max,l}}$ one may-construct the plot of (W/S) vs $\sigma C_{L,\max,l}$ shown in Figure 4.8.

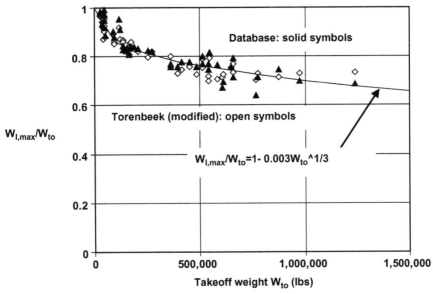

FIGURE 4.7

Variation of the maximum landing weight fraction with takeoff weight as deduced from the database of 50 commercial airliners and from Equation (4.6). Also shown is an approximate equation for $W_{l,max}/W_{to}$.

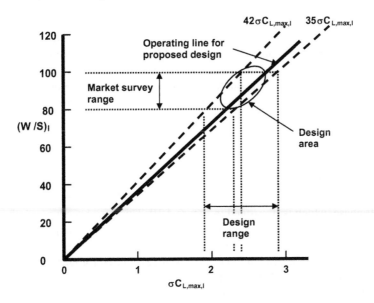

FIGURE 4.8

Variation of landing wing loading with landing maximum lift coefficient for conditions appropriate to jet transports.

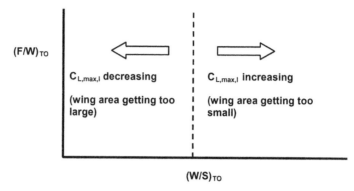

FIGURE 4.9

Plot of takeoff thrust-to-weight ratio vs takeoff wing loading showing design constraints due to $C_{L,max,l}$.

The market survey aircraft may be indicated on the plot and will probably fall within the region designated as the design area. This shows that the quantity $\sigma C_{L,max,l}$ lies between about 1.8 and 2.8. The line marked "operating line" in the figure denotes the design value chosen earlier and aids in narrowing down the choice for the design value of $\sigma C_{L,max,l}$. Then, with this design value or range of values one may determine the corresponding $(W/S)_{to}$ value(s) and draw them on a plot of takeoff thrust to weight ratio, $(F/W)_{to}$, as a function of takeoff wing loading as shown in Figure 4.9. Though the landing phase information doesn't depend on the takeoff thrust to weight ratio and would appear as vertical lines in Figure 4.9, this plot will ultimately have added to it information regarding the takeoff and cruise conditions and it will then be the major design tool for carrying out the engine selection process. Note that as $\sigma C_{L,lmax}$ decreases, the required wing area S grows while as $\sigma C_{L,lmax}$ increases, S shrinks, for a given W_{to}.

4.6 Takeoff distance

Although the thrust does not enter the analysis for landing, it does strongly affect the takeoff phase, which is illustrated in Figure 4.10. This is so because both the takeoff roll and the climb-out are acceleration processes requiring the application of a substantial thrust force. An analysis of the acceleration of the aircraft, assuming it is a point mass, and the determination of the distance required for takeoff is presented in Chapter 10 where a detailed derivation of the equations for takeoff is presented along with some simplified results. At this stage in the preliminary design we may simply use the approximate result derived in Chapter 10 for the total distance from start of the ground roll to climb-out over a takeoff obstacle of height h_{to}, which is given by

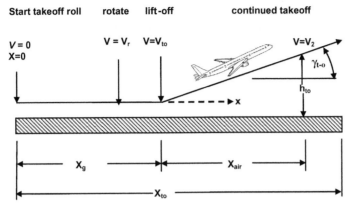

FIGURE 4.10

Takeoff procedure showing ground run, air run, and total takeoff distance.

$$x_{to} \approx \frac{-V_{to}^2}{0.12g} \ln \left[\frac{\left(\frac{F_{avg}}{W}\right)_{to} - 0.075}{\left(\frac{F_{avg}}{W}\right)_{to} - 0.015} \right] + \frac{V_{to}^2 + 20gh_{to}}{20g\left[\left(\frac{F_n}{W}\right)_{to} - 0.075\right]} \qquad (4.8)$$

The first term on the right-hand side of Equation (4.8) is the ground run distance from a standing start to the point of rotation and takeoff. This represents a substantial change in speed from 0 to V_{to}, which is typically around 200–250 ft/s (118–148 kts). Because the net thrust of turbojet and turbofan engines drops off slowly with flight speed, an average net thrust is used to account for the actual variation and this average thrust, derived in Chapter 10, is approximated as follows:

$$\left(\frac{F_{avg}}{W}\right)_{to} = \left(\frac{F}{W}\right)_{to}\left(1 - M_{to}\frac{1+\beta}{3+2\beta}\right)$$

The second term on the right-hand side of Equation (4.8) is the air distance from the takeoff point to climb-out over the takeoff obstacle as illustrated in Figure 4.10. In this phase of the takeoff the speed increases very little above V_{to} so the actual net thrust is used; it is given as

$$\left(\frac{F_n}{W}\right)_{to} = \left(\frac{F}{W}\right)_{to}\left(1 - 2M_{to}\frac{1+\beta}{3+2\beta}\right)$$

Both net thrust terms are defined as functions of the static thrust of the engines F_{to}, the bypass ratio of the engines β, and the takeoff Mach number M_{to}.

The static thrust is defined for the condition of zero forward speed, although it is commonly called the takeoff thrust, hence the notation F_{to}. The static thrust is quoted by engine manufacturers and is readily available from their literature. Similarly, the bypass ratio β, defined as the ratio of the fan mass flow to the core engine mass flow, is available from the engine manufacturer. The fan flow is said to "bypass" the core of the engine wherein fuel is burned. Thus the fan flow is said to be cold and the central core flow is said to be hot. The takeoff Mach number $M_{to} = V_{to}/a$, where V_{to} is the takeoff velocity and a is the sound speed in the ambient atmosphere. Using the atmospheric temperature ratio $\theta = T/T_{sl}$ we may write the local speed of sound as $a = \sqrt{\gamma R T_{sl} \theta}$. Because the vertical acceleration of the aircraft during the airborne phase of the takeoff is negligible with respect to the gravitational acceleration, the aircraft may be considered to be in equilibrium with its weight balanced by its lift, that is,

$$L_{to} = C_{L,to} \left(\frac{1}{2} \rho V_{to}^2 \right) S = W_{to}$$

Then the takeoff velocity may be written as

$$V_{to}^2 = \frac{2 \left(\frac{W}{S} \right)_{to}}{C_{L,to} \rho_{sl} \sigma}$$

Assuming a takeoff velocity equal to 120% of the takeoff stall velocity, so as to provide a stall margin as discussed in Section 4.2, we have

$$V_{to}^2 = (1.2)^2 V_{S,to}^2 = \frac{2.88 \left(\frac{W}{S} \right)_{to}}{C_{L,\max,to} \rho_{sl} \sigma}$$

Now Equation (4.8) provides a relationship between $(W/S)_{to}$, $(F/W)_{to}$, $C_{L,\max,to}$, and x_{to} with bypass ratio (β), airport atmospheric conditions (σ, θ), and takeoff obstacle height (h_{to}) appearing as parameters. The takeoff obstacle height h_{to} is generally assumed to be 35 ft and the airport altitude may be chosen as sea level and one other altitude, say 5000 ft (Denver, Colorado, for example). At this stage of the design process the engines have not yet been selected, therefore one must choose a nominal value for the bypass ratio β based on the market survey results of Chapter 1. With this information it is then possible to generate curves of $(W/S)_{to}$ as a function of $(F/W)_{to}$ for given values of x_{to} and $C_{L,\max,to}$ and plot them on a graph like that in Figure 4.11 whereon the $C_{L,\max,l}$ results had already been plotted. In this fashion one may produce a graph which illustrates the limits on both $(W/S)_{to}$ and $(T/W)_{to}$ and thereby narrow down the design values for these quantities pertinent to the design aircraft.

The design graph as it has developed thus far is shown in Figure 4.11. The vertical dashed lines are lines of constant maximum lift coefficient in the landing configuration and represent the corresponding values of $(W/S)_{to}$. There is no dependence of these results on the thrust to weight ratio because this parameter has little influence in the landing phase. However, the solid lines in Figure 4.11, which represent lines of constant maximum lift coefficient in the takeoff phase, show a definite dependence on

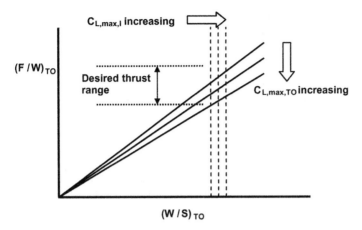

FIGURE 4.11

Design graph showing constraints due to $C_{L,max,to}$ as well as those due to $C_{L,max,l}$.

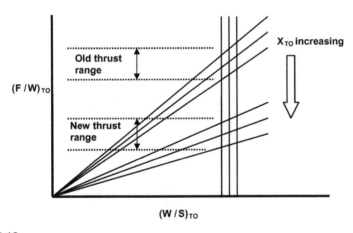

FIGURE 4.12

Design graph showing the effect of changing the takeoff distance on the required thrust level.

the thrust to weight ratio. As a practical matter it should be noted that although these solid lines are shown as straight in Figure 4.11 they may have some slight curvature to them. Furthermore Equation (4.8) is not applicable near the origin and any irregular behavior there may be ignored. The overlap region of the dashed and solid lines defines values of $(W/S)_{to}$, $(F/W)_{to}$, $C_{L,max,l}$, and $C_{L,max,to}$ (for an assumed value of x_{to}) which are mutually consistent with the requirements of both takeoff and landing.

If there are no suitable engines in the desired thrust range as defined in Figure 4.11 one may change the takeoff distance x_{to} while leaving the other parameters

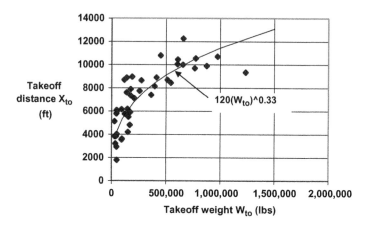

FIGURE 4.13

The nominal values of takeoff distance as a function of takeoff distance taken from the database for 50 commercial airliners.

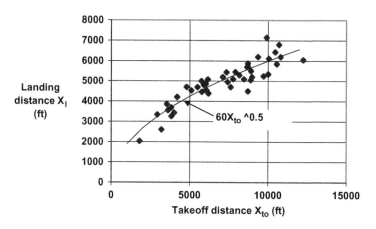

FIGURE 4.14

The landing distance is shown as a function of takeoff distance according to the database for 50 commercial airliners. Also shown is a trend line for the data.

fixed which will alter the design graph as shown in Figure 4.12. The choice of x_{to} can be set by market survey results, the landing distance x_l, or any other reasonable approximation. For example, the database of 50 commercial airliners yields the results for takeoff distance as a function of takeoff weight shown in Figure 4.13. A trend line is shown for reference even though there is a good deal of scatter of the data which is due to the range of values of wing loading, thrust to weight ratio, high lift devices, etc. available to the aircraft design team.

Another possibility is to use the landing distance x_l determined in Section 4.4 to aid in choosing a starting value for the takeoff distance x_{to}. Figure 4.14 shows the relationship between landing distance and takeoff distance according to the database for 50 airliners.

4.7 Cruise requirements

The final basic constraint on the thrust to weight ratio $(F/W)_{to}$ relates to the third major phase of the flight, the cruise stage. The design region delineated thus far is based on landing and takeoff requirements alone. Additional bracketing of the region is provided by noting that in cruise $(L/D)=(F_n/W)^{-1}$, so that at the start of cruise, stage 4 in Chapter 2, $(L/D)_4=(F_n/W)_4^{-1}$. The flight is carried out at close to constant L/D so that the highest thrust required occurs at the heaviest weight, i.e., at the start of cruise. Based on available reported data, e.g., Svoboda (2000), one may show that a simple but reasonable approximation for the net thrust at the typical cruise Mach number (around 0.8) and altitude (around 35,000 ft) is given by $F_n/F_{to}=\delta=p/p_{sl}$. Therefore we concentrate on the start of cruise and note that

$$\left(\frac{L}{D}\right)_4 = \frac{W_4}{F_{n,4}} = \frac{1}{\delta}\left(\frac{W_4}{W_{to}}\right)\frac{1}{\left(\frac{F}{W}\right)_{to}} \tag{4.9}$$

A more accurate representation is described in Section 4.12. Note that by using the weight fractions given in Table 2.4 we find $W_4/W_{to}=0.956$. The L/D ratio may also be expressed as follows:

$$\left(\frac{L}{D}\right)_4 = \frac{\left(\frac{W}{S}\right)_4}{\left(\frac{D}{S}\right)_4} = \frac{\left(\frac{W_4}{W_{to}}\right)\left(\frac{W}{S}\right)_{to}}{C_{D,4}q_4}$$

Equating this expression for L/D with that given by Equation (4.9) yields

$$\left(\frac{W}{S}\right)_{to} = \frac{C_{D,4}q_4}{\delta\left(\frac{F}{W}\right)_{to}}$$

Noting that the dynamic pressure $q_4 = \frac{1}{2}\gamma p_{sl}\delta M_4^2$ permits rewriting the above equation as follows:

$$\left(\frac{W}{S}\right)_{to} = \frac{\frac{1}{2}C_{D,4}\gamma p_{sl}M_4^2}{\left(\frac{F}{W}\right)_{to}} \tag{4.10}$$

The variable here is the cruise drag coefficient which, as discussed in Chapter 9, is given by $C_{D,4}=C_{D,0}+kC_L^2$, where $k=(\pi eA)^{-1}$. The detailed evaluation of the drag is carried out later in the design, but for the moment we may take some typical values as follows:

$$\gamma = 1.4 \qquad A = 8$$
$$M_4 = 0.8 \qquad \delta = 0.24(35,000 \text{ ft altitude})$$
$$e = 0.85 \qquad q = 227.5 \text{ lb/ft}^2$$

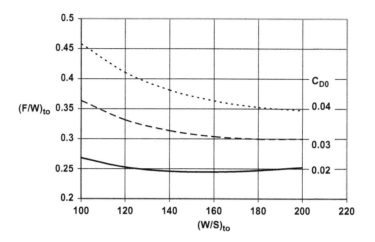

FIGURE 4.15

Takeoff thrust as a function of takeoff wing loading for cruise of a turbofan at $M=0.8$ and 35,000 ft altitude for three profile drag coefficients.

Then the drag coefficient becomes

$$C_{D,4} = C_{D,0} + 0.047\frac{(W/S)_4^2}{q_4^2} = C_{D,0} + 0.047\frac{(0.96)^2 (W/S)_{to}^2}{(227.5)^2}$$

Solving for the thrust to weight ratio yields

$$\left(\frac{F}{W}\right)_{to} = 948C_{D,0}\left(\frac{W}{S}\right)_{to}^{-1} + 7.88 \times 10^{-4}\left(\frac{W}{S}\right)_{to} \qquad (4.11)$$

The lowest values of $C_{D,0}$ are around 0.02 for a clean jet transport below $M=0.6$ while the highest acceptable values are no more than about 0.04. With these two extremes one may bound the allowable $(F/W)_{to}$ values for the design aircraft. The results of Equation (4.11) for the assumed conditions are shown in Figure 4.15. As expected the low-drag coefficient case makes the least demands on engine thrust. Because the fuel consumption is directly proportional to the thrust required, the low-drag coefficient case will deliver better fuel economy. This is the reason for the great effort expended on reducing drag.

4.8 Construction of the engine selection design chart

In the preceding sections, choices were made for the range of lift coefficients in landing and takeoff, along with a choice for the range of cruise zero-lift drag coefficients. Those selections may be combined on one graph of takeoff thrust to weight ratio as

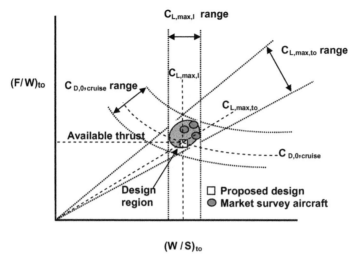

FIGURE 4.16

Design graph showing constraints posed by landing, takeoff, and cruise. The design point fixes the target values for $C_{Lmax,l}$, $C_{Lmax,to}$, $C_{D0,cruise}$, and, implicitly, x_{to}.

a function of takeoff wing loading as shown in Figure 4.16. This design graph shows all the boundaries for the proposed design, the resulting design region, the market survey aircraft operating points, and the proposed design aircraft operating point. Note that the design region in Figure 4.16 encompasses a range of takeoff thrust to weight ratios depending on the takeoff wing loading.

An initial estimate for the takeoff weight of the design aircraft was determined in Chapter 2, so that selecting production engines with specified takeoff thrust performance permits the formulation of available thrust to weight ratios for the design aircraft. The objective is to select a production engine for which the total thrust available, that is, the product of the number of engines and the takeoff thrust of one engine, will permit an available thrust line to pass through the design region, as shown in Figure 4.16. Should the available engines not provide a line passing through the design region, the design region must be expanded, typically by changing, for example, the takeoff distance, as was described in Section 4.6. The design point must lie somewhere on that line but within the design region. Choosing the design point then will fix the takeoff wing loading, $(W/S)_{to}$ and, because the takeoff weight is known, the wing planform area S will also be defined.

In addition, the maximum lift coefficients required for takeoff and landing, $C_{L,max,to}$ and $C_{L,max,l}$, respectively, are determined, as shown in Figure 4.16. Note that these coefficients are connected to the assumption of specific takeoff and landing distances, x_{to} and x_l, respectively. The projected cruise zero-lift drag coefficient $C_{D,0}$ is also fixed by the design point because one curve of $(F/W)_{to}$ as a function of $(W/S)_{to}$ defined by Equation (4.11) must pass through it, as shown in Figure 4.16.

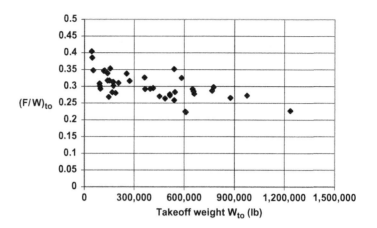

FIGURE 4.17

Takeoff thrust to weight ratio $(F/W)_{to}$ for 43 turbofan-powered airliners is shown as a function of takeoff weight W_{to}.

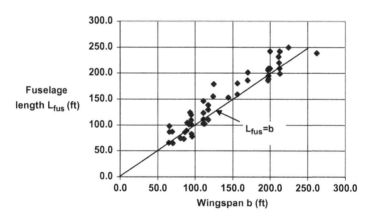

FIGURE 4.18

The relationship between the wingspan and the fuselage length among the 50 airliners in the database is shown here with the line $L_{fus} = b$.

An appreciation for the range of takeoff thrust to weight ratio is given in Figure 4.17 where $(F/W)_{to}$ for 43 turbofan-powered airliners is shown as a function of takeoff weight W_{to}. We will discuss the takeoff situation for turboprop airliners at the end of this chapter.

A first-order estimate of the wingspan b may be obtained using the fuselage length L_{fus} already determined in Chapter 3. The relationship between the wingspan and the fuselage length among the 50 airliners in the database is shown in

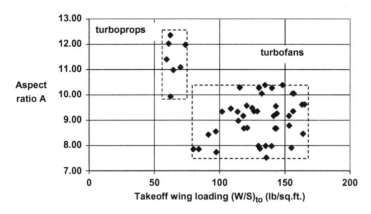

FIGURE 4.19

The relationship between the aspect ratio and the takeoff wing loading among the 50 airliners in the database is shown.

Figure 4.18. The straight line shows that $L_{fus} \sim b$ and indeed $b = 0.94L_{fus}$ within about $\pm 10\%$. Then the aspect ratio may be determined from the definition $A = b^2/S$. As a check on the magnitude of the aspect ratio one may examine Figure 4.19 which collects the published data regarding aspect ratio as a function of takeoff wing loading on 50 commercial airliners. It is clear that the turboprop aircraft are immediately set apart from the jet aircraft because of their relatively low wing loading and high aspect ratio. The low wing loading is a consequence of the relatively low thrust levels achievable with a practical size propeller on a turboprop engine, but this permits good takeoff performance using relatively simple high lift devices. The high aspect ratio is mainly a result of the turboprop aircraft having no appreciable sweepback, and this high aspect ratio pays off in reduced induced drag and therefore high fuel efficiency. As mentioned previously, the cost of this efficient performance is a reduced cruise speed, but this is a defect only if long-range travel is the object.

4.9 Flight test data for landing and power approach

A series of flight tests were performed on a number of aircraft in order to determine their characteristics over various portions of their flight envelopes and the results were reported by Heffley and Jewell (1972). Some of these data may be more readily available in McCormick (1995). For the particular case of terminal operations, like power approach and landing, data have been reproduced in Table 4.1. Some of these data appear in Figure 4.1, which was discussed at the beginning of this chapter. Additional information on performance characteristics of a variety of aircraft is provided by Loftin (1985).

Table 4.1 Landing and Power Approach Data from Flight Tests[a]

Airplane in landing configuration	W_{TO} (pounds)	W_L (pounds)	S (ft²)	V (kts)	C_L	C_D	α (deg)	V/V_{stall}	δ_F/gear (d = gear down)
Jetstar	38,204	23,904	543						
CV880M	155,000	126,000	2000	134	1.05	0.154	5.2		50°/d
B747	636,600	564,000	5500	131	1.76	0.263	8.5	1.2	30°/d
C-5A	654,362	580,723	6200						
XB-70A	384,524	300,000	6298						

Airplane in power approach	W_{TO} (pounds)	W_L (pounds)	S (ft²)	V (kts)	C_L	C_D	α (deg)	V/V_{stall}	δ_F/gear (d: gear down)
Jetstar	38,204	23,904	543	133	0.74	0.095	6.5	1.4	40°/d
CV880M	155,000	126,000	2000	165	0.68	0.080	4.3		30°/up
B747	636,600	564,000	5500	165	1.11	0.102	5.7	1.4	20°/up
C-5A	654,362	580,723	6200	146	1.29	0.145	2.7	1.4	30°/d
XB-70A	384,524	300,000	6298	205	0.33	0.055	7.5		

Airplane in power approach	W_{TO} (pounds)	W_L (pounds)	S (ft²)	A	C_L	C_D	e	$C_{D,0}$	δ_F/gear (d: gear down)
Jetstar	38,204	23,904	543		0.74	0.095			40°/d
CV880M	155,000	126,000	2000	7.2	0.68	0.080	0.85	0.056	30°/up
B747	636,600	564,000	5500	7	1.11	0.102	0.85	0.036	20°/up
C-5A	654,362	580,723	6200	7.7	1.29	0.145	0.85	0.064	30°/d
XB-70A	384,524	300,000	6298		0.33	0.055			

[a] Data taken from Heffley and Jewell (1972); note: (1) Convair CV-880M in landing has speed brakes extended 8°, (2) B747 and C-5A takeoff weights are at 40% less fuel than maximum and XB-70A takeoff weight is at 50% maximum fuel; (3) $C_{L,max}$ for B747 and C-5A is estimated from C_L in landing to be 2.53; (4) Zero-lift drag coefficient estimated from the power approach data assuming a parabolic drag polar given by $C_D = C_{D,0} + kC_L^2$ where $k = 1/\pi eA$ and e is the Oswald span efficiency factor.

4.10 **Turbojet and turbofan engines**

Turbojet and turbofan engines are discussed in detail by Sforza (2012). The complexity of practical aircraft engines is substantial and cutaway drawings of existing engines are often so detailed as to lack clarity in emphasizing the basic design features. To gain an appreciation for the general arrangement of major systems in a complete engine, we will consider cross-sectional diagrams of simple schematic form for typical turbojet and turbofan engines. In addition, we will illustrate the nature of the major flow properties within a specific turbojet engine and a specific turbofan engine, each of which has seen extensive use. The engines illustrated all employ axial flow compressors driven by axial flow turbines; this is typical for aircraft engines, except perhaps at the smallest sizes where centrifugal compressors may sometimes be employed.

4.10.1 **Dual shaft turbojet**

In Figure 4.20 a common turbojet configuration is shown in which a low-pressure compressor and its driving turbine ride on an inner driveshaft while a high-pressure compressor rides on a concentric outer driveshaft connected to a separate high-pressure compressor. In this fashion each compressor-turbine set is free to turn at a different rotational speed appropriate to the desired performance characteristics. In the turbojet the turbines extract sufficient power to drive the compressors with the remaining thermal energy in the gas arising from combustion of the fuel passing out the exhaust nozzle at high-speed producing pure jet thrust.

The Pratt & Whitney (P&W) J57 is a two-shaft engine with a 9-stage low-pressure compressor, a 7-stage high-pressure compressor, a single-stage high-pressure turbine, and a 2-stage low-pressure turbine. The military versions of this engine powered such aircraft as the North American F-100 Super Sabre and Vought F8U Crusader fighter aircraft while the civil versions were flown by the Boeing 707 and Douglas DC-8 commercial airliners. The schematic diagram in Figure 4.20 represents the P&W J57-P-43WB, the military version of the P&W JT3C-6 turbojet. The compressor sections raise the stagnation pressure from the standard atmospheric level of 14.7 psia (101.3 kPa) to about 160 psia (1100 kPa). The fuel is burned at almost constant pressure in the combustor, raising the stagnation temperature to over 2700R (1500 K). The turbines then remove work from the hot gases and the stagnation pressure drops to about 36 psia (250 kPa), or about 2.5 atm and the stagnation temperature drops to about 1440 F (800 K). The converging nozzle accelerates the hot exhaust gases to a speed of about 1700 ft/s (518 m/s) providing a static thrust of 11,200 lbs (49.82 kN).

4.10.2 **Dual shaft high bypass turbofan**

In the turbofan engine, as illustrated in Figure 4.21, the turbines extract power to drive not only the compressors but also the fan. The hot gases of combustion are again accelerated through a nozzle to produce pure jet thrust as in the turbojet.

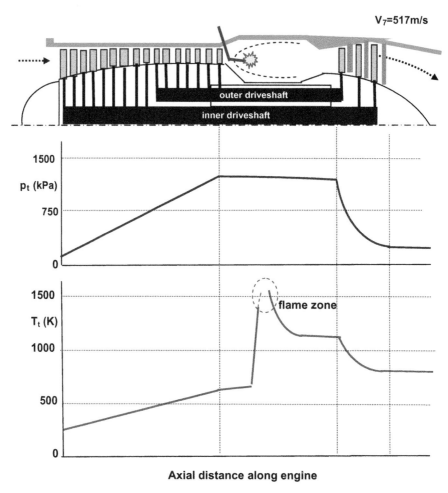

FIGURE 4.20

Schematic diagram of a dual shaft turbojet engine similar to the Pratt & Whitney JT3C showing the axial variation of the major flow properties through the engine; from Sforza (2012).

However, the power supplied to the fan is transmitted to the air and therefore acts much like a propeller. The fan accelerates the air passing through it, called the bypass air, by directly doing mechanical work on the air without any appreciable heating of that bypass air. In this fashion the turbofan combines the high speed capability of the pure turbojet with the fuel efficiency and good acceleration characteristics of the propeller. As a consequence all modern commercial airliner engines are turbofan engines; the major difference between such engines is in the degree of bypass utilized. Turbofans have a bypass ratio describing the ratio of the mass flow passing through the fan (the "cold" flow) to that passing through the core engine (the "hot" flow) and is expressed as follows:

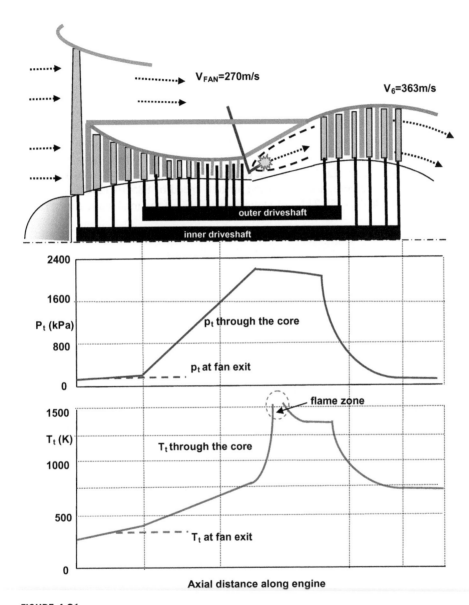

FIGURE 4.21

Schematic diagram of a dual shaft, externally mixed, high bypass turbofan engine similar to the Pratt & Whitney JT9D showing the axial variation of the major flow properties through the engine; from Sforza (2012).

$$\beta = \frac{\dot{m}_{fan}}{\dot{m}_{core}} \tag{4.12}$$

High bypass turbofan engines may be considered to be those with β greater than 4 or 5; current engines have β values as high as 10.

The high bypass turbofan, characterized by a large fan diameter, is illustrated in Figure 4.21. Some high bypass turbofan engines, e.g., the Rolls-Royce Trent 900 that powers the A380, carry this specialization of compressor-turbine pairs further so as to have three concentric drive shafts with the very inner one dedicated to the fan and its driving turbine. The Pratt & Whitney JT9D shown in Figure 4.21 is a two-shaft engine with a 3-stage low-pressure compressor and integral fan, an 11-stage high-pressure compressor, a 2-stage high-pressure turbine, and a 4-stage low-pressure turbine. It is an engine commonly used on wide-body airliners like the Boeing 747 and the Airbus A300. The Pratt & Whitney JT9D is a high bypass ratio ($\beta = 5$) turbofan with separate nozzles for the fan and the jet. The fan, which processes five times as much air as the core, accelerates the air to a speed of 885 ft/s (270 m/s) providing about 34,330 lbs (152.7 kN) of static thrust, 79% of the total static thrust of the engine.

The compressor pressure ratio is 21.5 so that sea level air is compressed to a stagnation pressure of about 316 psia (2170 kPa) and a stagnation temperature of 1340R (744 K). Then the air is fed into the combustor where the stagnation temperature rises to around 2430R (1350 K). The turbines then remove work from the hot gases and the stagnation pressure drops to about 20.9 psia (144 kPa), or about 1.4 atm, and the stagnation temperature drops to about 1310R (727 K). The converging jet nozzle accelerates the hot exhaust gases to a speed of over 1200 ft/s (365 m/s) providing a static thrust of 9130 lb (40.6 kN), or 21% of the total static thrust of 43,500 lb (193.5 kN).

4.10.3 Determination of net thrust

The turbojet and turbofan thrust arise from the change in momentum imparted to the air entering the engine. As shown by Sforza (2012) the gross thrust produced by turbofan or turbojet engine exhausting into an ambient atmosphere of pressure p_0 is given by

$$F_g = \dot{m}_{core} V_{e,jet} + A_{e,jet} \left(p_{e,jet} - p_0 \right) + \dot{m}_{fan} V_{e,fan} + A_{e,fan} \left(p_{e,fan} - p_0 \right)$$

Note that this general result includes the special case of the turbojet because the fan terms are zero in that case. The gross thrust involves only the exit stations of the engine and therefore is the case for zero momentum of the air entering the engine, that is, the case of static thrust where the engine is at rest. If the engine is moving at some flight speed V_0 the momentum of the air entering the engine must be taken into account. This quantity is often called the ram drag and is defined as follows:

$$F_r = \left(\dot{m}_{core} + \dot{m}_{fan} \right) V_0$$

Then the force provided by the engine for propulsion is the net thrust

$$F_n = F_g - F_r = \dot{m}_{core} \left(V_{e,jet} - V_0 \right) + A_{e,jet} \left(p_{e,jet} - p_0 \right) + \dot{m}_{fan} \left(V_{e,fan} - V_0 \right)$$
$$+ A_{e,fan} \left(p_e - p_0 \right)$$

For completeness it is noted that we have assumed that the mass flow of fuel added in the core is assumed to be negligible with respect to the air mass flow entering the core engine. The fuel to airflow ratio, f/a, is typically around 2%. In the case where the pressures at the exit of the jet and the fan are equal to the surrounding pressure, p_0, the net thrust depends solely on the difference between the jet and fan exhaust velocities and the flight velocity. If the fan and jet exhaust velocities are both subsonic, their exhaust pressures are equal to the surrounding ambient pressure. Engines in commercial airliners operate with high subsonic jet and fan exhausts so we may rewrite the net thrust equation as follows:

$$F_n = \dot{m}_{core} \left[V_{e,jet} + \beta V_{e,fan} - (1 + \beta) V_0 \right]$$

The important engine properties to consider in the selection process for preliminary design are takeoff thrust, cruise thrust, and cruise-specific fuel consumption. Takeoff thrust is commonly considered to be the static thrust quoted by the manufacturer. The static thrust is the thrust measured with the engine stationary, as would be the case when the aircraft is initiating the takeoff roll. The net thrust produced drops off with forward speed, with the drop-off less pronounced for turbojets than for turbofans. However, the larger static thrust capability and better specific fuel consumption of the turbofan engine more than offset the larger thrust drop-off with forward speed.

4.10.4 Net thrust in takeoff

A detailed analysis of the variation of net thrust with flight speed is given in Chapter 10. There it is shown that an average thrust level F_{avg} may be defined over a range of flight speeds. The takeoff distance correlation given by Equation (4.8) uses a value for the net thrust which is averaged over the speed range covered during takeoff as given by

$$\left(\frac{F_{avg}}{W} \right)_{to} = \left(\frac{F}{W} \right)_{to} \left(1 - M_{to} \frac{1 + \beta}{3 + 2\beta} \right) \tag{4.13}$$

The takeoff thrust is defined for zero flight velocity so it is equivalent to the gross, or static, takeoff thrust. Then the takeoff thrust to weight ratio is defined as follows:

$$\left(\frac{F}{W} \right)_{to} = \frac{F_{g,to}}{W_{to}} \tag{4.14}$$

This is the takeoff thrust to weight ratio used as the ordinate in the design graph of Figure 4.16. The takeoff weight is known from the estimation procedure carried out in Chapter 2 and the static thrust is determined from the quoted values for operational engines. Note that the static thrust level used is the *total* static thrust produced, that is, the product of the static thrust of the engine and the number of engines used.

Typical commercial aircraft use two, three, or four identical engines. The number of engines selected is often a result of engine availability, although it is generally considered preferable to have only two engines. This is the trend because turbine engines are now so reliable that there is no significant safety advantage to having more than two. In addition, fewer engines mean reduced maintenance and repair costs. For the same reasons, though aircraft may be offered by the airplane manufacturer with a choice of engines, on any one aircraft all the engines are the same. Thus, the number of engines to be used is part of the design selection process. For example, if the only engines available in the thrust class required by the design require the choice of three engines rather than two, it would be wise to see if altering other design choices, like takeoff distance, which would move the design region in Figure 4.16 and make it again possible to use two engines, would be worthwhile.

4.10.5 Net thrust in cruise

The cruise thrust available from the engines helps fix another boundary of the design region, as described in Section 4.7. There it was suggested that ratio of cruise to takeoff thrust could be approximated by the cruise altitude atmospheric pressure ratio $\delta = p/p_{sl}$. This rule of thumb is easy to remember but it tends to give optimistic thrust levels. However, it is necessary to estimate the cruise thrust for the selected engines when carrying out the performance estimates in Chapters 9 and 10. Sometimes the engine manufacturers will quote nominal cruise thrust performance, typically giving the net thrust at a specific speed and altitude, commonly $M = 0.8$ and $z = 35,000$ ft. If such values are given they should be used. If not, a reasonable and simple approximation is that the cruise thrust (in lbs) is given by

$$F_{cr} \approx 0.2 F_{to} \tag{4.15}$$

This result is a correlation of the performance of 26 engines as reported by Svoboda (2000) and data gathered from manufacturers and other sources for 9 other engines. The data are shown in Figure 4.22 along with lines showing $\pm 10\%$ variations illustrating that the correlation is reasonable. Above takeoff thrust levels of 60,000 lbs the correlation becomes overly optimistic and adjusted accordingly.

4.10.6 Specific fuel consumption in cruise

The specific fuel consumption, denoted as C_j, measures the weight flow rate of fuel (lb/hr) used for each unit of thrust (lb) produced and is a major figure of merit for engines. This variable is dependent upon the fuel flow rate for the actual thrust level produced and is often quoted by the manufacturer as evaluated at the static or takeoff thrust level, F_{to}. The value of $C_{j,to}$ is around 0.33 for engines currently in use. However, in order to estimate fuel usage during cruise, when the majority of the fuel is consumed, it is necessary to have the value of the cruise-specific fuel consumption C_j. This quantity, like the cruise thrust itself, is often very difficult to find in the open literature since it is usually considered proprietary information. The database developed in Svoboda (2000) also has some information on cruise-specific fuel consumption and these data are shown in Figure 4.23. A simple equation which illustrates the general trend for those data is given by

$$C_{j,cr} = 0.7 - \left(\frac{F_{cr}}{10^5}\right) \tag{4.16}$$

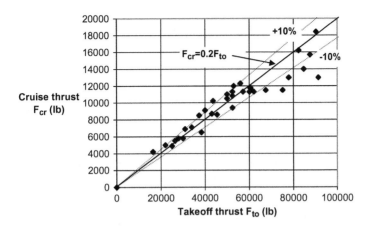

FIGURE 4.22

Correlation of cruise thrust with takeoff thrust for 35 turbofan engines: 26 from Svoboda (2000) and 9 from other sources. The lines representing ±10% variations from the indicated equation are shown for reference.

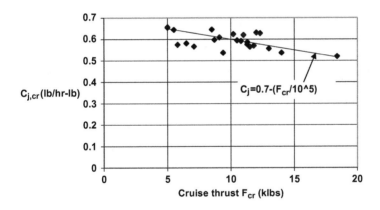

FIGURE 4.23

Correlation of cruise-specific fuel consumption with takeoff thrust for 26 turbofan engines; from Svoboda (2000).

In this equation the units for thrust and for specific fuel consumption are lbs and lb/hr-lb, respectively.

4.11 Turboprops

The shaft power required to drive a propeller can come from different sources: the conventional reciprocating engine which has been the power source since the early days of flight, the electric motor which has been receiving substantial attention in recent years, and the gas turbine engine which we have been concentrating on here. The advantages of the propeller in terms of low speed thrust, high propulsive efficiency, and good fuel efficiency could be coupled to the smooth running, high reliability of the gas turbine while also reaping the benefit of some jet thrust. The only major technical hurdle was mating the two because the rpm of a gas turbine is about ten times greater than that appropriate for a propeller. Thus a gear reduction unit is required between the propeller and the gas turbine. The weight and reliability of such units reduced the attractiveness somewhat, but sufficient advantages still accrued to this propeller-jet combination that their use is widespread today in applications for larger aircraft that need not fly much faster than about 300 kts.

The turboprop started out by adding the propeller and gearbox directly to a gas turbine engine and extracting the work needed for the compressor and the propeller, leaving a smaller percentage of the enthalpy rise for expansion through a nozzle for jet thrust. It became apparent that this was less effective than employing a separate shaft for the propeller assembly that would be driven by a so-called free turbine whose output power and speed could be more efficiently adapted to different shaft power requirements. The compressor-burner-turbine unit would be essentially a hot gas generator for the free turbine. A schematic diagram of a typical free turbine turboprop engine is shown in Figure 4.24.

The engine selection procedure for turboprop-powered airliners follows the same steps as that for turbofan-powered airliners with somewhat different details in places. This was apparent in Section 2.8 where the weight estimation procedure was specialized to turboprop aircraft. In this chapter some special treatment is once again needed for turboprop airliners and that material is discussed in the subsequent subsections.

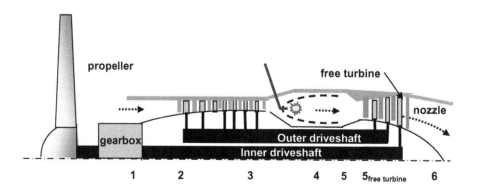

FIGURE 4.24

Turboprop engine showing free turbine driving propeller through gearbox. Compressor-burner-turbine gas generator unit feeds free turbine and nozzle.

4.11.1 Takeoff distance

For preliminary design purposes, where details of the aircraft being designed are not yet accurately known, it is reasonable to employ a simplified analysis. In Chapter 10 an approximate relation for the takeoff distance of a turboprop-powered aircraft was developed and is given by

$$x_{to} = \frac{0.66 V_{to}^2}{g\left[\left(\frac{F}{W}\right)_{to} - 0.015\right]}\left(1 + 2.13\frac{gh}{V_{to}^2}\right) \tag{4.17}$$

Note that Equation (4.17) may be used with any consistent set of units. We may substitute for the takeoff velocity the relation

$$V_{to}^2 = 2\frac{1.44\left(\frac{W}{S}\right)_{to}}{\rho_{sl}\sigma C_{L,max,to}^2}$$

Then the takeoff distance is

$$x_{to} = \frac{1.9\left(\frac{W}{S}\right)_{to}}{\rho_{sl}\sigma g C_{L,max,to}\left[\left(\frac{F}{W}\right)_{to} - 0.015\right]}\left[1 + 2.13\frac{\rho_{sl}\sigma g h C_{L,max,to}}{\left(\frac{W}{S}\right)_{to}}\right] \tag{4.18}$$

The takeoff distance is a function of $(W/S)_{to}$, $C_{L,max,to}$, and the static thrust to weight ratio $(F/W)_{to}$, as well as, of course, the altitude, through the atmospheric density ratio σ. This is the same situation as found in the case of the turbofan-powered airliners in Section 4.6 and therefore the same procedures outlined there apply here. For a selected altitude and takeoff distance, we may use Equation (4.18) to develop a plot of specific static thrust $(F/W)_{to}$ as a function of takeoff wing loading $(W/S)_{to}$ for a specified $C_{L,max,to}$. This curve would be analogous to that described in Figure 4.11. Note that the landing analysis for turboprop airliners is the same as that for turbofan-powered airliners because the thrust characteristics don't enter the picture in landing and therefore don't influence the maximum lift coefficient in landing $C_{L,max,l}$. The takeoff, or static, thrust to weight ratio for seven turboprop airliners is shown as a function of takeoff weight in Figure 4.25. Note that these static thrust to weight ratios are as much as 50% greater than those for the turbofan airliners shown in Figure 4.17.

4.11.2 Turboprop cruise requirements

The final basic constraint on the $(F/W)_{to}$ term relates to the third major phase of the flight, the cruise stage. The design region delineated thus far is based on landing and takeoff requirements. Additional bracketing of the region is provided by noting that in cruise $(L/D) = (F_n/W)^{-1}$, so that at the start of cruise, stage 4 in Chapter 2, $(L/D)_4 = (F_n/W)_4^{-1}$. The flight is carried out at close to constant L/D so that the highest thrust required occurs at the heaviest weight, i.e., at the start of cruise. Therefore we concentrate on the start of cruise and note that

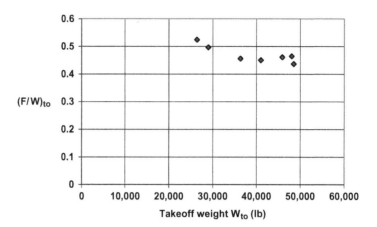

FIGURE 4.25

Takeoff thrust to weight ratio for seven turboprop airliners.

$$\left(\frac{L}{D}\right)_4 = \frac{W_4}{F_{n,4}} = \left(\frac{W_4}{W_{to}}\right)\frac{1}{\left(\frac{F_{cr}}{W_{to}}\right)} \tag{4.19}$$

Note that by using the weight fractions given in Table 2.4 we find $W_4/W_{to} = 0.965$. The L/D ratio may also be expressed as follows:

$$\left(\frac{L}{D}\right)_4 = \frac{\left(\frac{W}{S}\right)_4}{\left(\frac{D}{S}\right)_4} = \frac{\left(\frac{W_4}{W_{to}}\right)\left(\frac{W}{S}\right)_{to}}{C_{D,4}q_4}$$

Equating the two expressions for L/D yields

$$\left(\frac{W}{S}\right)_{to} = \frac{C_{D,4}q_4}{\left(\frac{F_{cr}}{W_{to}}\right)}$$

Noting that $q_4 = \frac{1}{2}\gamma p_{sl}\delta M_4^2$ permits rewriting the above equation as follows:

$$\left(\frac{W}{S}\right)_{to} = \frac{\frac{1}{2}C_{D,4}\gamma p_{sl}\delta M_4^2}{\left(\frac{F_{cr}}{W_{to}}\right)} \tag{4.20}$$

The variable here is the cruise drag coefficient which is given by $C_{D,4} = C_{D,0} + kC_L^2$, where $k = (\pi e A)^{-1}$. The detailed evaluation of the drag is carried out later in the design, but for the moment we may take some typical values for a turboprop airplane as follows:

$$\gamma = 1.4 \qquad A = 12$$
$$M_4 = 0.5 \qquad \delta = 0.356(26,000 \text{ ft altitude})$$
$$e = 0.85 \qquad q = 131.8 \text{ lb/ft}^2$$

Then the drag coefficient becomes

$$C_{D,4} = C_{D,0} + 0.031\frac{(W/S)_4^2}{q_4^2} = C_{D,0} + 0.031\frac{(0.965)^2 (W/S)_{to}^2}{(131.8)^2}$$

Solving for the cruise thrust to weight ratio from Equation (4.20) yields

$$\left(\frac{F_{cr}}{W_{to}}\right) = 132C_{D,0}\left(\frac{W}{S}\right)_{to}^{-1} + 2.19 \times 10^{-4}\left(\frac{W}{S}\right)_{to} \qquad (4.21)$$

Using the results of Section 10.5.4 the cruise thrust may be related to the static thrust as follows:

$$\frac{F_{cr}}{F_{to}} = \frac{326\eta_p P_{cr}}{V_{cr}}\frac{N_{to}d_p}{33000P_{to}\left(\frac{C_T}{C_P}\right)_{to}} = \frac{\eta_p}{J\left(\frac{C_T}{C_P}\right)_{to}}\frac{P_{cr}}{P_{to}}\frac{N_{to}}{N_{cr}} \qquad (4.22)$$

Remember that the aircraft speed V_{cr} is in knots, the propeller speeds N_{cr} and N_{to} are in rpm, and the powers P_{cr} and P_{to} are given in horsepower, but the ratios shown are dimensionless and accept any consistent set of units. The quantity $J = V/(nd)$ is the dimensionless propeller advance ratio which is the ratio between the flight speed and twice the linear speed of rotation of the propeller tip. Because the cruise thrust is generated at the cruise altitude and the static thrust is generated at low altitude, typically close to sea level, we must correct for the altitude difference. For constant engine rpm the propeller rpm is also constant because the propeller-engine gear ratio is fixed. Cycle analysis for a gas turbine engine shows that the power at a given rpm depends upon the total pressure p_t and total temperature T_t entering the engine and has the following characteristic:

$$P \sim \frac{P_t}{T_t}N = \frac{p}{T}\left(\frac{p_t}{p}\right)\left(\frac{T}{T_t}\right)N = \frac{p_{sl}\delta}{T_{sl}\theta}\frac{\left(1 + \frac{\gamma-1}{2}M^2\right)^{\frac{\gamma}{\gamma-1}}}{\left(1 + \frac{\gamma-1}{2}M^2\right)}N$$

Then a reasonable correlation to adjust the shaft power for altitude using the isentropic flow relations for air with $\gamma = 1.4$ is

$$\frac{P_{cr}}{P_{to}}\frac{N_{to}}{N_{cr}} \approx \sigma\left(1 + .2M^2\right)^{2.5}$$

Then the cruise to static thrust ratio may be approximated by

$$\frac{F_{cr}}{F_{to}} \approx \frac{\eta_p\sigma\left(1 + 0.2M^2\right)^{2.5}}{J\left(\frac{C_T}{C_P}\right)_{to}} \qquad (4.23)$$

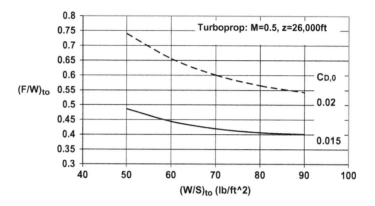

FIGURE 4.26

Takeoff thrust as a function of takeoff wing loading for cruise of a turboprop at $M=0.5$ and 26,000 ft altitude for two profile drag coefficients.

As shown by Sforza (2012) good propeller designs will cruise around $M=0.5$ with typical propeller efficiencies $\eta_p=0.85$ at advance ratios around $J=2$. Under static conditions those propellers will be characterized by values of the ratio of thrust coefficient to power coefficients around $C_T/C_P=2$. Then, assuming good propeller performance, we may use these values to arrive at the following approximation which should be suitable for preliminary design purposes:

$$\frac{F_{cr}}{F_{to}} \approx 0.24\sigma \tag{4.24}$$

The lowest values of $C_{D,0}$ are around 0.015 for a clean turboprop airliner below $M=0.6$ while the highest acceptable values are no more than about 0.025. By substituting Equation (4.24) into Equation (4.21) and using these two extremes for $C_{D,0}$ one may bracket the allowable $(F/W)_{to}$ values for the design aircraft. The results of this process for the assumed conditions are shown in Figure 4.26. As expected, the low-drag coefficient case makes the least demands on engine thrust. Because the fuel consumption is directly proportional to the thrust required, the low-drag coefficient case will deliver better fuel economy. This is the reason that such great effort is expended on reducing drag.

4.11.3 Estimating takeoff thrust and specific fuel consumption

One may use the discussion in Chapter 10 leading to Equation 10.89 to form an equation with which to estimate the takeoff thrust of market survey aircraft, for example

$$F_{to} \approx 8.21 n_e \left(\sqrt{\sigma} P d_p\right)^{\frac{2}{3}} \tag{4.25}$$

In Equation (4.25) the propeller diameter is in feet and the power in horsepower and n_e denotes the number of engines. Unlike the turbofan, the specific fuel consumption for a turboprop, C_{tp}, is essentially constant with flight speed and for current turboprop engines generally varies between 0.45 and 0.55 lbs fuel/hr/hp.

4.12 **Engine-out operation and balanced field length**

It is important to consider how an engine failure during takeoff affects the runway length required to ensure safe operation. In general, it is assumed that a single engine failure during a takeoff is a possible emergency which warrants corrective action, while the failure of more than one engine on a given takeoff is considered so unlikely that it cannot be provided for in design. If an engine were to become inoperative during the takeoff ground run there are two possible actions: either the pilot can brake and bring the airplane to a safe stop on the runway or the pilot can continue the takeoff on the remaining good engine(s). Obviously if the engine failure occurs very early in the takeoff run when the speed is low, the takeoff should be aborted and the aircraft brought to a stop on the runway. If the engine failure occurs near the end of the takeoff run where the velocity is almost equal to the takeoff velocity, then the takeoff should continue on the one good engine. The questions are: what is the speed that separates braking to a stop or continuing the takeoff and how long a runway is needed for safety? A reasonable approach is to seek the aircraft speed for which the distance needed to reject a takeoff and come to a safe stop on the ground is equal to the takeoff distance required for operation with one engine out when the engine failure occurs at a given speed. This speed may be considered a critical speed below which braking is indicated and above which a continued takeoff is indicated. The runway length corresponding to this criterion is called the balanced field length, x_{bf}. A detailed discussion and development of the balanced field length concept is provided in Chapter 10 and once the design has proceeded to that stage the value of x_{bf} should be calculated for the design aircraft.

Basically, the definition of the balanced field length may be put into equation form as follows:

$$x_{crit} + x_{stop} = x_{crit} + x_{cto,1eo} \qquad (4.26)$$

The critical distance x_{crit} is that distance achieved during the ground runup until the critical velocity V_{crit} is reached at which time an engine fails. At that speed the decision is made either to abort and x_{stop} is the distance required to decelerate the aircraft to a safe stop on the ground or to continue the takeoff and $x_{cto,1eo}$ is the distance required to accelerate the aircraft to the takeoff speed, lift off, and clear the 35-ft obstacle.

The variables involved in this equation are illustrated in Figure 4.27. The two sides of Equation (4.26) may be calculated using the methods of Chapter 10 and then plotted on one graph. The typical appearance of such a plot is shown in Figure 4.28

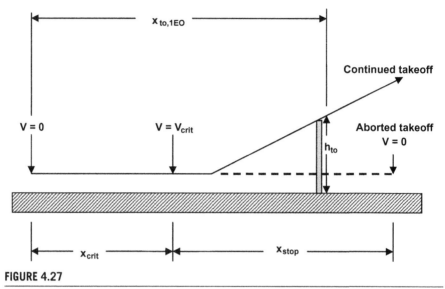

FIGURE 4.27

Schematic diagram of one-engine-out operations.

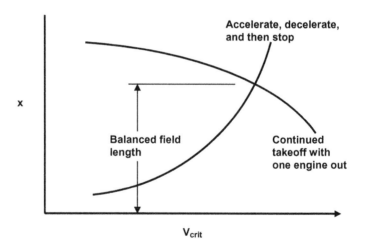

FIGURE 4.28

Schematic illustration of the determination of the balanced field length.

where the cross-over point of the two curves defines the balanced field length. This is the runway length which provides safety for the aircraft in the event of a rejected takeoff or a one-engine-out takeoff.

One advantage of having three or four engines instead of two is that the thrust level is not as severely compromised as it is for twin-engine aircraft. Therefore a steeper climb-out angle may be maintained if one engine becomes inoperative on takeoff. When selecting engines one should keep in mind that it may not be in the best interests of the overall design to choose a lower thrust level engine from those in the competitive range, particularly in the case of a twin-engine configuration. Aircraft with $(F/W)_{to} < 0.25$ are all four-engine aircraft and are not as constrained as twin-engine aircraft. With one engine out, the thrust to weight ratio of the four-engine aircraft is about 0.18, well above that for a twin-engine aircraft, which is likely to be about 0.15.

In the past, the lack of high thrust levels in available engines often called for adding engines. However this tends to be relatively inefficient in terms of purchase and maintenance cost, despite the margin gained in one-engine-out scenarios. In recent years the high thrust available from individual high bypass ratio turbofan engines, coupled with the high reliability of turbine engines in general, has made it possible to develop appropriate levels of safety in twin-engine configurations.

Note that the concept of a critical velocity is introduced to provide an objective means of determining whether to continue a takeoff or to remain on the ground and come to a stop on the runway when an engine goes out during the takeoff run. Obviously, the critical velocity must be less than the takeoff velocity because one is already airborne after the takeoff velocity is reached. Usually the critical velocity is around 90% of the takeoff velocity. In addition, it should be clear that the balanced field length is longer than the normal takeoff distance. As a check, one must be sure that the one-engine-out takeoff distance is longer than the normal takeoff distance, as would be expected.

4.13 **Design summary**

At the end of this stage of the design the landing speed (see Figure 4.1) for the design aircraft should be determined. A plot of the variation of the landing speed V_l as a function of $(W/S)_l/C_{L,l}$ in landing should be prepared, as in Figure 4.3. The variation of the average acceleration in landing $\bar{a}/g$ as a function of landing distance x_l should be shown on a graph like that in Figure 4.5. The landing wing loading $(W/S)_l$ should be illustrated as a function of $\sigma C_{L,\max,l}$ as is illustrated in Figure 4.8. The engine selection graph, $(F/W)_{to}$, as a function of $(W/S)_{to}$ (Figure 4.16) should be displayed and the design point for the proposed aircraft should be shown on this graph. Data points indicating values for the market survey airplanes should be shown and identified on all pertinent graphs. All this information should be summarized in a table like that shown in Table 4.2.

Thus, this chapter provides the reader with additional information concerning the configuration and nominal performance of the design aircraft. There is now a sized aircraft fuselage with specified preliminary weight components, a set of engines, and an estimated wing design with a desired area and likely values of wingspan and aspect ratio. The next phase of the design will be to develop the wing design in more detail so that it can provide the basic lift characteristics targeted by the results of this chapter.

Table 4.2 Design Summary for Engine Selection

Parameter	Value	Units
Landing speed V_l		
Maximum landing lift coefficient $C_{L,max,l}$		
Average deceleration in landing $\bar{a}/g$		
Landing distance x_l		
Landing wing loading $(W/S)_l$		
Maximum takeoff lift coefficient $C_{L,max,to}$		
Takeoff distance x_{to}		
Takeoff wing loading $(W/S)_{to}$		
Wing area S		
Wingspan b		
Aspect ratio A		
Drag coefficient in cruise $C_{D,0}$		
Takeoff thrust to weight ratio $(F/W)_{to}$		
Number, type, and brand of engines		
Takeoff thrust per engine		

4.14 Nomenclature

A	aspect ratio or area
a	acceleration or sound speed
b	wingspan
C_D	drag coefficient
C_F	skin friction coefficient
C_L	lift coefficient
C_j	thrust-specific fuel consumption
C_P	propeller power coefficient
C_p	pressure coefficient
C_T	propeller thrust coefficient
C_{tp}	power-specific fuel consumption
D	drag
d_p	propeller diameter
e	span efficiency factor
F_n	net thrust
F_{cr}	cruise thrust
F_{to}	static, or takeoff, thrust

F_{avg}	average thrust
g	acceleration of gravity
h	obstacle height
J	propeller advance ratio
k	constant
L	lift
L_{fus}	fuselage length
M	Mach number
M_f	fuel weight fraction
N	propeller rpm
n_e	number of engines
P	shaft power
p	pressure
q	dynamic pressure
R	range or gas constant
Re	Reynolds number
S	wing planform area
V	velocity
V_{crit}	critical velocity for takeoff
V_E	equivalent velocity
W	weight
x_{crit}	ground distance to $V=V_{crit}$
x_l	landing distance
x_{stop}	distance to stop in ground run
x_{to}	takeoff distance
α	angle of attack
β	bypass ratio
δ	atmospheric pressure ratio
ϕ	correlation factor
γ	ratio of specific heats
γ_{to}	takeoff climb angle
λ	engine bypass ratio
η_p	propeller efficiency
μ'	effective friction and drag coefficient
ρ	density
σ	atmospheric density ratio
$t(\theta)$	atmospheric temperature ratio

4.14.1 Subscripts

avg	average
cr	cruise
$crit$	critical
cto	continued takeoff

e	exit
l	landing
max	maximum
p	pressure
r	reference
sl	sea level
to	takeoff
zf	zero-fuel
$1eo$	one engine out

References

Heffley, R.K., Jewell, W.F., 1972. Aircraft Handling Qualities Data, NASA CR-2144, December.

Torenbeek, E., 1982. Synthesis of subsonic airplane design. Kluwer Academic Publishers, Dordrecht, The Netherlands.

Svoboda, C., 2000. Turbofan engine database as a preliminary design tool. Aircraft Design 3, 17–31.

McCormick, B., 1995. Aerodynamics, Aeronautics, and Flight Mechanics, second ed. J. Wiley & Sons, New York.

Sforza, P.M., 2012. Theory of Aerospace Propulsion. Elsevier, New York.

Loftin Jr. L.K., 1985. Quest for Performance, The Evolution of Modern Aircraft, NASA SP-468, <http://www.hq.nasa.gov/pao/History/SP-468/cover.htm>.Keywords

Landing weight, Takeoff weight, Landing distance, Takeoff distance, Wing loading, Thrust to weight ratio, Engine selection

Wing Design

CHAPTER OUTLINE

Commercial Airplane Design Principles. http://dx.doi.org/10.1016/B978-0-12-419953-8.00005-X
© 2014 Elsevier Inc. All rights reserved.

5.1 General wing planform characteristics

A general wing planform shape is shown in Figure 5.1. In this case an ogee shape somewhat like that of the Concorde supersonic airliner is depicted. The main parameters of importance are shown in the figure. The overall length of the wing l and the wingspan b are readily apparent and the centroid of the wing area is indicated. The x-axis is an axis of symmetry and the chord length is defined as the distance between the leading and trailing edges of the wing as measured parallel to the centerline of symmetry, that is, the x-axis. Thus the chord length is a function of spanwise distance y and is denoted by $c(y)$ with the root chord being measured along the centerline, $y=0$, and is denoted by $c_r=c(0)$.

The wing area S is defined as the planform area, that is, the projected area on the x-y plane, and it is commonly taken as the reference area for most airplane performance calculations. Part of the wing planform may be obstructed by the fuselage and to complete the full planform the leading and trailing edges are faired in to the centerline in a consistent and reasonable fashion. The area is then given by

$$S = \int_{-b/2}^{b/2} c(y)dy = 2\int_{0}^{b/2} c(y)\, dy \qquad (5.1)$$

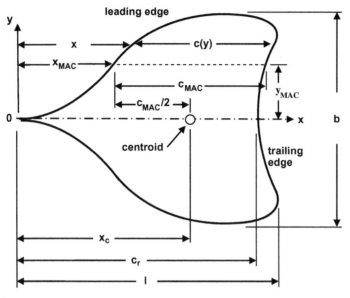

FIGURE 5.1

General wing planform showing various parameters.

The mean aerodynamic chord c_{MAC} is defined as

$$c_{MAC} = \frac{\int_0^{b/2} c^2 dy}{\int_0^{b/2} c\, dy} = \frac{2}{S} \int_0^{b/2} c^2 dy \tag{5.2}$$

The centroid of the wing planform area lies on the centerline of symmetry and its location x_c on that line may be found from the equation for the moment of the area about the origin:

$$x_c S = \int_{-b/2}^{b/2} \left(x_{LE} + \frac{c}{2} \right) c\, dy$$

Here the x_{LE} is the distance from the y-axis to the leading edge of the chord. Solving for the centroid location yields

$$x_c = \frac{2}{S} \int_0^{b/2} \left(c x_{LE} + \frac{1}{2} c^2 \right) dy \tag{5.3}$$

Note that the centroid location may also be expressed as the sum of the distance from the origin to the leading edge point of the mean aerodynamic chord and the mean aerodynamic chord length as follows:

$$x_c = x_{MAC} + \frac{1}{2} c_{MAC}$$

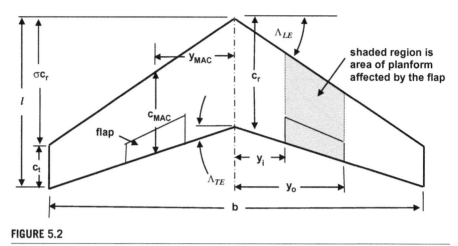

FIGURE 5.2

Straight-tapered wing planform showing various parameters.

Here x_{MAC} is the distance from the origin to the leading edge point of the mean aerodynamic chord. The spanwise location of the mean aerodynamic chord is denoted y_{MAC}. Another important parameter is the aspect ratio of the wing which is defined as

$$A = \frac{b^2}{S} \tag{5.4}$$

The slenderness of the wing is suggested by the aspect ratio, as may be seen by considering the wing to have an average chord length c_{avg} such that $S = bc_{avg}$. Then the aspect ratio $A = b/c_{avg}$ and a large aspect ratio would represent a wingspan much greater than the average chord length. Similarly one may consider the aspect ratio to be the ratio of the area of a square formed by the wingspan, that is, b^2, to the actual planform area S. Another characteristic of the slenderness of a wing is the ratio b/l, where l is the overall length of the wing measured from the apex of the wing to its most aft point.

5.1.1 The straight-tapered wing planform

A conventional wing planform is characterized by straight leading and trailing edges as shown in Figure 5.2. The general wing parameters may be readily calculated for this type of wing.

The defining features of the straight-tapered wing are the sweepback angles of the leading and trailing edge, Λ_{LE} and Λ_{TE}, respectively, and the taper ratio $\lambda = c_t/c_r$. The wing planform area defined by Equation (5.1) may be written as

$$S = \frac{1}{2}bc_r\,(1 + \lambda) \tag{5.5}$$

The mean aerodynamic chord defined by Equation (1.2) becomes

$$c_{MAC} = \frac{2}{3}c_r \frac{1 + \lambda + \lambda^2}{1 + \lambda} \tag{5.6}$$

The centroid location defined by Equation (5.3) is

$$x_c = \frac{c_r}{3}\left(\lambda + \sigma + \frac{1 + \lambda\sigma}{1 + \lambda}\right) \tag{5.7}$$

The quantity σ, depicted in Figure 5.2, is defined as follows:

$$\sigma = \frac{l - c_t}{c_r} \tag{5.8}$$

The aspect ratio, Equation (5.4), becomes

$$A = \frac{2b}{c_r(1 + \lambda)} \tag{5.9}$$

The spanwise location of the mean aerodynamic chord is

$$y_{MAC} = \frac{b}{6}\left(\frac{1 + 2\lambda}{1 + \lambda}\right) \tag{5.10}$$

The sweepback of different chordwise locations is often needed in the design process, particularly the quarter-chord and half-chord positions. The sweepback of any location n, as a fraction of the chord, may be found from the known sweepback at some other position m, also as a fraction of chord, as follows:

$$\tan \Lambda_n = \tan \Lambda_m - \frac{4}{A}(n - m)\frac{1 - \lambda}{1 + \lambda} \tag{5.11}$$

For example, if the leading edge sweepback angle is $\Lambda_{LE} = \Lambda_0$, then the quarter-chord sweepback angle $\Lambda_{c/4} = \Lambda_{0.25}$ and

$$\tan \Lambda_{0.25} = \tan \Lambda_0 - \frac{4}{A}(0.25 - 0)\frac{1 - \lambda}{1 + \lambda} = \tan \Lambda_{LE} - \frac{1 - \lambda}{A(1 + \lambda)}$$

Assuming typical values $\Lambda_{LE} = 30°$, $A = 9$, and $\lambda = 1/3$ we find that $\Lambda_{c/4} = 27.56°$.

5.1.2 Cranked wing planform

It is sometimes advantageous to employ a cranked wing planform, that is, one with discontinuous changes in sweepback angle of the leading and/or trailing edges, like that shown in Figure 5.3. Some commercial airliners have essentially zero sweepback along the trailing edge in the central portion of the wing as shown in Figure 5.4. This design feature, sometimes called a "yehudi," allows for better stowage of landing gear, cargo, and a stronger wing root structure. In addition to these intuitively

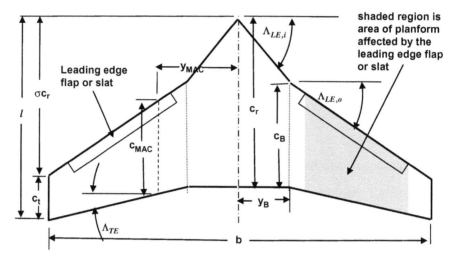

FIGURE 5.3

Wing planform with cranked leading and trailing edges.

FIGURE 5.4

Photograph of a B737 illustrating the cranked wing planform.

obvious attributes, the longer chord length in the vicinity of the fuselage junction provides for improved aerodynamic flow characteristics over the wing.

For the case of a single kink at one spanwise location on the leading and/or trailing edge, as shown in Figure 5.3 or 5.4 we may readily find the important geometrical properties of the wing. The area may be considered to be comprised of the sum of the inner panel area S_i and outer panel area S_o, or

$$S = S_i + S_o = \left[\frac{1}{2}y_B\left(c_r + c_B\right)\right] + \left[\left(\frac{b}{2} - y_B\right)\left(c_B + c_t\right)\right]$$

This may be put in the form

$$S = \frac{1}{2}bc_r\left[(1 - \lambda)\frac{2y_B}{b} + \lambda + \lambda_i\right] \tag{5.12}$$

Here we have introduced y_B, the spanwise distance outboard to the break in trailing edge angle, and $\lambda_i = c_B/c_r$, the taper ratio of the inboard panel. The aspect ratio is

$$A = \frac{b^2}{S} = \frac{2b}{c_r\left[(1 - \lambda)\frac{2y_B}{b} + \lambda + \lambda_i\right]} \tag{5.13}$$

The mean aerodynamic chord may be written in terms of the mean aerodynamic chord of the inner and outer panels as follows:

$$c_{MAC} = c_{MAC,i}\frac{S_i}{S} + c_{MAC,o}\frac{S_o}{S}$$

The mean aerodynamic chord then may be written as

$$c_{MAC} = \frac{2}{3}c_r\frac{\left(\lambda_i^2 + \lambda_i\lambda + \lambda^2\right) + \left[\frac{y_B}{(b/2)-y_B}\right]\left(1 + \lambda_i + \lambda_i^2\right)}{\left(\lambda_i + \lambda\right) + \left[\frac{y_B}{(b/2)-y_B}\right]\left(1 + \lambda_i\right)} \tag{5.14}$$

The spanwise location of the mean aerodynamic chord may be written in terms of the spanwise location of the mean aerodynamic chord in each panel:

$$y_{MAC} = y_{MAC,i}\frac{S_i}{S} + \left(y_B + y_{MAC,o}\right)\frac{S_o}{S}$$

Similarly the spanwise location of the mean aerodynamic chord is given by

$$y_{MAC} = \frac{1}{3}\left(\frac{b}{2} - y_B\right)\frac{\lambda_i\left(1 + 2\lambda_i\lambda\right) + \left[\frac{y_B}{(b/2)-y_B}\right]\left(2 + \lambda_i\lambda\right) + \left[\frac{y_B}{(b/2)-y_B}\right]^2\left(1 + 2\lambda_i\right)}{\left(\lambda_i + \lambda\right) + \left[\frac{y_B}{(b/2)-y_B}\right]\left(1 + \lambda_i\right)}$$

$$\tag{5.15}$$

FIGURE 5.5

Photograph of a B737 illustrating the dihedral angle of the wing.

The wing area centroid is given by

$$x_c = x_{MAC} + \frac{1}{2}c_{MAC} \qquad (5.16)$$

The location of the leading edge of the mean aerodynamic chord is given in terms of the spanwise location of the mean aerodynamic chord in the inner and outer panels as follows:

$$x_{MAC} = y_{MAC,i} \tan \Lambda_{LE} \frac{S_i}{S} + \left(y_B \tan \Lambda_{LE,i} + y_{MAC,o} \tan \Lambda_{LE,o} \right) \frac{S_o}{S} \qquad (5.17)$$

5.1.3 Wing dihedral

The planform views discussed thus far cannot illustrate another common wing characteristic, dihedral. This is the angle that a wing makes with a horizontal line like the ground plane as illustrated in Figure 5.5. It is typically a small angle, on the order of 3–8°. This angle has an influence on the stability of an aircraft in roll and will be the subject of discussion in a later chapter.

5.2 General airfoil characteristics

A cross-section taken through the wing planform in Figure 5.3 at a general spanwise location $y =$ constant reveals a shape called the airfoil section. The historical

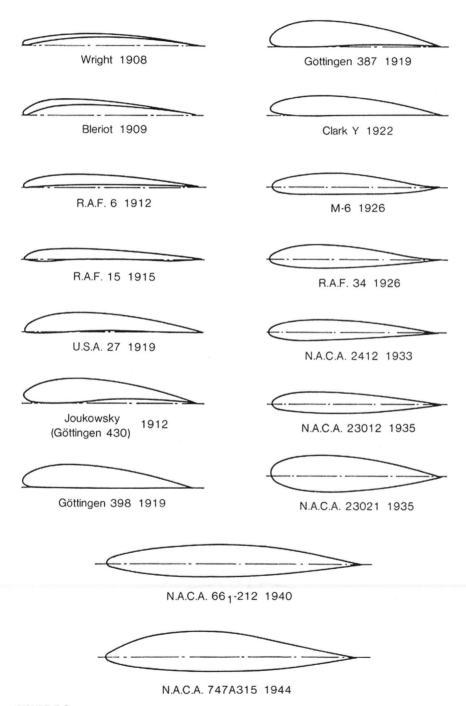

Wright 1908

Göttingen 387 1919

Bleriot 1909

Clark Y 1922

R.A.F. 6 1912

M-6 1926

R.A.F. 15 1915

R.A.F. 34 1926

U.S.A. 27 1919

N.A.C.A. 2412 1933

Joukowsky 1912
(Göttingen 430)

N.A.C.A. 23012 1935

Göttingen 398 1919

N.A.C.A. 23021 1935

N.A.C.A. 66_1-212 1940

N.A.C.A. 747A315 1944

FIGURE 5.6

Evolution of airfoil sections, over the period 1908 through 1944.

evolution of airfoil sections, over the period 1908 through 1944, is illustrated in Figure 5.6. The last two shapes shown (NACA 66_1-212 and NACA 747A315) are considered low-drag sections. The NACA 6-series airfoils were designed to maintain laminar flow over 60–70% of chord on both the upper and the lower surface. The objective was to reduce drag and increase the critical Mach number, that is, the speed at which compressibility effects become important. The NACA 7-series airfoils were designed to have a greater extent of laminar flow on the lower surface than on the upper surface. This was aimed at producing lower pitching moments at the expense of some reduction in critical Mach number. Note that these later laminar flow airfoils are thickest near the center of their chords. Among the classical airfoil sections, those of the NACA 6-series laminar flow airfoils are most appropriate for the design project because they have the best high-speed characteristics.

A full discussion of the systematic development of NACA airfoils, along with tabular listings of their coordinates and graphical presentation of their aerodynamic characteristics, is given in the book by Abbott and Von Doenhoff (1959) which is based on earlier work by Abbott et al. (1945). A computer program for generating the shapes of the NACA airfoils is presented by Ladson et al. (1996).

5.2.1 Airfoil sections

The NACA families airfoils are obtained by combining a mean line and a thickness distribution as illustrated in Figure 5.7. In the NACA 6-series of airfoils the mean camber lines (the locus of the centers of circles tangent to the upper and lower surfaces) are designed to have uniform chordwise pressure loading, $\Delta p = p_l - p_u$, up to a particular chordwise station $x/c = a$ and then to linearly decrease to zero at the trailing edge. The thickness distribution is designed to fix the chordwise (x/c) location of the minimum pressure for the symmetric form of the airfoil (straight camber line, $y = 0$) at the point of maximum thickness $(t/c)_{max}$.

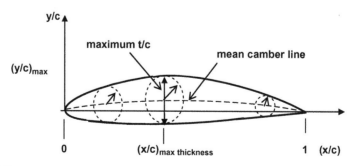

FIGURE 5.7

Airfoil showing the mean camber line, the maximum thickness to chord ratio and, its position along the chord.

The numbering system for the NACA 6-series may be explained using an example, say the NACA 65_3-218, $a=0.5$ airfoil. In this case

6:   denotes the NACA 6-series of sections

5:   denotes the chordwise position of the minimum value of pressure coefficient ($C_{p,min}$), measured in tenths of chord, for the basic symmetric thickness section at zero lift

3:   denotes the range of lift coefficients, in tenths, above and below the design lift coefficient $c_{L,des}$, for which a favorable pressure gradient exists on both the upper and lower surfaces and drag coefficient is lowest

2:   denotes the design lift coefficient multiplied by 10; here $c_{l,des}=0.2$

18: denotes the thickness in percent chord, here 18%; for values below 10% the thickness ratio would be entered with a prefix 0, as, for example, 08 for 8%

a: denotes the type of mean line used, in this case the mean line has a uniform load up to $x/c=a=0.5$. When no mean line designation is given it is understood that the uniform load mean line ($a=1$) has been used

Modifications to the basic airfoil numbering system sometime arise as a result of alterations to a basic airfoil shape. The most common is the replacement of the dash by a capital letter, such as in the NACA 64_1A212. The letter A denotes a change in the shape of the aft portion of the airfoil where the upper and lower surfaces remain essentially straight from about $x/c=0.8$ to the trailing edge. This modification was introduced to simplify manufacturing as well as to avoid a wing with a thin and sharp trailing edge which is prone to stress concentration and buckling. For explanations of other, less common, variations in the airfoil designations see Abbott and Von Doenhoff (1959).

Loftin (1948) points out that NACA 6-series airfoil sections with small thickness to chord ratios have relatively high critical Mach numbers but have the disadvantage of being impractically thin near the trailing edge so that substantial fabrication difficulties are encountered. To alleviate this problem the cusped trailing edge has been eliminated from a number of NACA 6-series basic thickness forms and replaced by essentially straight segments from approximately 80% chord to the trailing edge so as to form a finite trailing edge angle. These new sections have been designated the NACA 6A-series airfoil sections as previously indicated. Experimental results obtained at Reynolds numbers of 3–9 million based on chord length indicate that the section minimum drag and maximum lift characteristics of comparable NACA 6-series and 6A-series airfoil sections are essentially the same. The quarter-chord pitching-moment coefficients and angles of zero lift of NACA 6A-series airfoil sections are slightly more negative than those of corresponding NACA 6-series airfoil sections. The position of the aerodynamic center and the lift curve slope of smooth NACA 6A-series airfoil sections appear to be essentially independent of airfoil thickness ratio in contrast to the trends shown by NACA 6-series sections.

There was a resurgence of airfoil development in the 1960s and 1970s following a long hiatus in the two decades following the Second World War. NASA launched a concerted effort to develop new airfoil sections that would have improved performance at high subsonic Mach numbers and thereby improve the performance of the newly introduced turbojet airliners. In particular, designs were sought that would delay the drag divergence Mach number and maintain reasonable drag coefficients at the turbulent flow conditions typical of high-speed flight while retaining acceptable maximum lift and stall characteristics at the low speeds typical of landing. This research led to the so-called supercritical airfoil, one that has a distinctive shape compared to standard airfoils. A description of these developments is given by Harris (1990).

Airframe manufacturers have been actively engaged in this work, but their results are proprietary, and generally not publicly released. Several of the NASA supercritical airfoils are described by Harris (1990) and relevant NASA publications cited therein are generally available online through the NASA Technical Report Server. McCormick (1995) also discusses some supercritical airfoils. There is as yet no other single source which collects and summarizes a wide range of supercritical airfoil theories and results as do Abbott and Von Doenhoff (1959) for conventional airfoils. For these reasons, and for those of expediency and completeness of information supplied, it is suggested that NACA airfoils be used for the preliminary design. The use of advanced CFD codes for airplane design in industry, as described, for example, by Jameson (1989), has led to integrated design of complete three-dimensional wings, bypassing the approach presented in this book, which uses two-dimensional airfoil characteristics to fashion a three-dimensional wing. However, the latter approach is important in developing an understanding of the contribution of the different elements of a wing to its overall performance. Therefore in the preliminary design phase it is considered practical, efficient, and educationally sound to use this building block approach.

5.2.2 Airfoils at angle of attack

A great deal of theoretical and experimental work has been devoted to the development of airfoil sections. Theoretical airfoil design is hampered by the existence of viscous effects in the form of a "boundary layer" of low-energy air between the airfoil surface and the outer flow within which friction is important and outside of which friction is negligible. The viscous boundary layer has major effects on airfoil drag and maximum lift characteristics but only relatively minor effects on lift curve slope, angle of attack for zero lift, and section pitching-moment coefficient.

Because the boundary layer itself is influenced by surface roughness, surface curvature, pressure gradient, heat transfer between the surface and the boundary layer, and viscous interaction with the free stream it is apparent that no simple theoretical considerations can accurately predict all the airfoil characteristics. For these reasons, experimental data are always preferable to theoretical calculations. Airfoils have been optimized for many specific characteristics, including: high maximum lift, low drag at low lift coefficients, low drag at high lift coefficients, low pitching moments, low drag in the transonic region, and favorable lift characteristics beyond the critical Mach number. Optimization of an airfoil in one direction usually compromises it in another. Thus, low-drag airfoil designs often are prone to exhibiting poor high lift characteristics while high lift airfoil designs tend to show low critical Mach numbers.

Graphical representations of the experimental data gathered for many NACA airfoils by Abbott and Von Doenhoff (1959) are represented in Appendix A for selected airfoils. The data are mainly for smooth surface conditions and are shown for Reynolds numbers from 3×10^6 to 9×10^6 based on chord length. However, some data at a Reynolds number of 6×10^6 are also shown for airfoils where fine grit has been lightly deposited on the surface of the leading edge region for distances up to about 8% of the chord length in order to initiate turbulence in the boundary layer. These cases are denoted by the term "standard roughness." From these data the following airfoil characteristics for smooth surfaces have been collected:

1. Angle of attack at zero lift, α_0.
2. Moment coefficient at the quarter-chord point at zero lift, $c_{m,0}$.
3. Lift curve slope, $c_{l,\alpha}$.
4. Aerodynamic center location in percent chord, $a.c.$
5. Angle of attack for maximum lift coefficient, $\alpha_{c_{\ell,max}}$.
6. Maximum lift coefficient, $c_{\ell,max}$.
7. Angle of attack at which the lift curve deviates from linear variation, α^0.

Results for a number of smooth NACA 6-series airfoils at a Reynolds number $Re_c = 9 \times 10^6$ are presented in Table 5.1. From items 1, 3, 5, 6, and 7 an approximate section lift curve shape can be synthesized, as illustrated in Figure 5.8. It is apparent from the above that any generalized charts for airfoil section characteristics, including the ones in this book, must be used with caution. Tabulated experimental and theoretical data for other NACA airfoils are presented in Hoak et al. (1978).

The systematic compilation of aerodynamic data for various families of NACA airfoils presented by Abbott et al. (1945) and mentioned above covered a range of Reynolds numbers suitable for the aircraft of that period. However, soon afterwards, the development of larger and faster aircraft exposed the need for airfoil data at still higher Reynolds numbers. In response to this need Loftin and Bursnall (1948) carried out experiments on selected NACA airfoils at Reynolds numbers up to 25×10^6. The main conclusion of that research was that the airfoil lift curve slope was essentially unaffected by the increase in Reynolds number, remaining remarkably close to the theoretical value $a = 2\pi$ per radian or $a = 0.11$ per degree. On the other hand, the airfoil maximum lift coefficient $c_{l,max}$ for the relatively thin airfoils tested ($t/c \leq 12\%$) showed a constant value for Re_c up to about 6×10^6, then a small increase of up to 10% as Re_c approached 15×10^6, and finally a constant or slightly falling value up to $Re_c = 25 \times 10^6$, the maximum value tested. Relatively thick airfoils ($t/c > 18\%$) displayed different behavior with a slow but monotonic increase in $c_{l,max}$ up to the maximum Reynolds numbers tested. The effect of roughness was found to be minimal for both lift curve slope and maximum lift coefficient throughout the range of Reynolds number tested. The roughness is assumed to trip the boundary layer so that it is completely turbulent over the entire Reynolds number range and therefore is no longer sensitive to Reynolds number. Abbott and Von Doenhoff (1959) discuss the

Table 5.1 Basic Characteristics of Selected NACA Airfoils at a Reynolds number $Re_c = 9 \times 10^6$

Airfoil	α_0 (deg)	$c_{m,0}$	$c_{l\alpha}$	a.c.	$\alpha_{cl,max}$	$c_{l,max}$	$\alpha^0 w$
63-006	0	0.005	0.112	0.258	10.0	0.87	7.7
63-009	0	0	0.111	0.258	11.0	1.15	10.7
63_1-012	0	0	0.116	0.265	14.0	1.45	12.3
63_2-015	0	0	0.117	0.271	14.5	1.47	11.0
63_3-018	0	0	0.115	0.271	15.5	1.54	11.2
63_4-021	0	0	0.118	0.273	17.0	1.38	9.0
63-206	−1.9	−0.037	0.112	0.254	10.5	1.06	6.0
63-209	−1.4	−0.032	0.11	0.262	12.0	1.4	10.3
63-210	−1.2	−0.035	0.113	0.261	14.5	1.56	9.6
63_1-212	−2.0	−0.035	0.114	0.263	14.5	1.63	11.4
63_2-215	−1.0	−0.03	0.116	0.267	15.0	1.60	8.8
63_3-218	−1.4	−0.033	0.118	0.271	14.5	1.85	8.0
63_4-221	−1.5	−0.035	0.118	0.269	15.0	1.44	9.2
63_4-421	−2.2	−0.056	0.109	0.265	14.0	1.42	7.6
64-206	−1.0	−0.040	0.110	0.253	12.0	1.03	8.0
64-208	−1.2	−0.039	0.113	0.257	10.5	1.23	8.8
64-209	−1.5	−0.040	0.107	0.261	13.0	1.40	8.9
64-210	−1.6	−0.040	0.110	0.253	14.0	1.45	10.8
64_1-212	−1.3	−0.027	0.113	0.262	15.0	1.55	11.0
64_2-215	−1.6	−0.030	0.112	0.265	15.0	1.57	10.0
64_3-218	−1.3	−0.027	0.115	0.271	16.0	1.53	10.0
64_4-221	−1.2	−0.029	0.117	0.271	13.0	1.32	6.8
64_1-412	−2.6	−0.065	0.112	0.267	15.0	1.67	8.0
64_2-415	−2.8	−0.070	0.115	0.264	15.0	1.65	8.0
64_3-418	−2.9	−0.065	0.116	0.273	14.0	1.57	8.0
64_4-421	−2.8	−0.068	0.120	0.276	13.0	1.42	6.4
63A010	0	0.005	0.105	0.254	13.0	1.20	10.0
63A210	−1.5	−0.040	0.103	0.254	14.0	1.23	10.0
64A010	0	0	0.110	0.253	12.0	1.23	10.0
64A210	−1.5	−0.040	0.105	0.251	8.0	1.44	10.0

effects of Reynolds number for all the NACA airfoils originally presented by Abbott et al. (1945) in some detail, but only up to the value of 9×10^6. They do not deal with other tests at higher Reynolds number except to mention that an NACA 63-series 22% thick airfoil displays slow monotonic growth in $c_{l,max}$ for Re_c up to 26×10^6, in keeping with the behavior of thicker airfoils as described previously.

To achieve Reynolds numbers up to 26×10^6 NACA used the variable density wind tunnel in which high-density levels were achieved by operating at increased pressure levels. However, in the current era with very large airplanes operating at high subsonic Mach numbers the need for achieving in the laboratory still higher

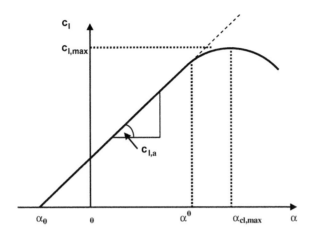

FIGURE 5.8

Schematic diagram showing the five elements of the airfoil lift curve.

Reynolds numbers, on the order of 50–100 million, led to the development of cryogenic wind tunnels. In these wind tunnels high-density levels are achieved by operating at low temperatures, which simultaneously results in lowered viscosity. NASA's National Transonic Facility (Wahls, 2001) and the European Wind Tunnel (Green and Quest, 2011) are large-scale facilities built expressly for supporting industrial design and development efforts. The experimental results are typically obtained for industrial airframe builders and represent a substantial investment in expense and intellectual property so that the data do not receive wide dissemination. Thus there isn't available a compilation of aerodynamic characteristics of airfoils at very high Reynolds number like that presented by Abbott and Von Doenhoff (1959) for NACA airfoils at Reynolds numbers up to 9 million.

5.2.3 Airfoil selection

The airfoil selected for the proposed design depends upon the cruising speed, which in turn is related to the powerplant chosen. If the selected powerplant is a turboprop the speed range for cruise will be in the range of 250–300 kts. Sweepback is unnecessary in this speed range so aspect ratios can be high ($\geq$10). The maximum airfoil thickness to chord ratio at the root is typically from 15% to 18% for such aircraft. A larger thickness ratio is generally chosen for the root airfoil and a smaller thickness ratio for the tip airfoil in order to provide a deep section at the root to reduce the bending stresses acting there as a result of the long wingspan of high-aspect-ratio aircraft. The tip chord thickness ratio is made smaller to provide an average value which optimizes $c_{l,\max}$, and typically lies in the range of 12–13%.

It is recommended that the airfoil selected be chosen both on the value of $c_{l,\max}$ and upon the post stall variation of c_l with angle of attack. An abrupt drop in section lift coefficient is to be avoided, and the airfoil with the smallest decrease in c_l for

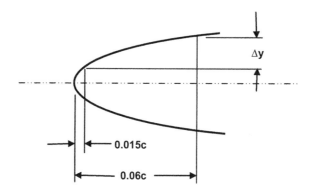

FIGURE 5.9

Definition sketch for the leading edge sharpness parameter Δy.

angles of attack above the stall is highly desirable, even at the expense of a smaller value of $c_{l,\max}$. In the 1930s and 1940s designers chose wings which used the NACA 2412 at the root and the 4412 at the tip.

The NACA 23012 or 23015 airfoils have higher maximum lift but these airfoils exhibit a large and abrupt loss in c_l beyond the stall. The NACA 6-series have smaller leading edge radii than the NACA 4-series and the NACA 5-series airfoils. The maximum thickness of the 4- and 5-digit airfoils is at 30% chord. The position of the maximum thickness of the 63, 64, and 65 series is located progressively aft. The 63 series airfoils might be considered for the turboprop aircraft. The NACA 63-215 is a suggested airfoil section because of its favorable stall characteristics. It is advisable to investigate those airfoil sections used by the competition (market survey aircraft) to aid in justifying the choice of airfoil.

Turbofan aircraft will typically cruise at high subsonic Mach number, typically in the range of $0.74 < M < 0.84$. The upper limit chosen is dependent upon the extent to which the drag rise associated with the transonic speed regime can be tolerated. The average wing thickness will generally be in the 10–12% range, the lower value associated with relatively unswept wings and the higher value with moderately swept $(\Lambda_{c/4} \sim 30^\circ)$ wings. The NACA 64-2xx or the 64-4xx series airfoils might prove satisfactory for these aircraft. Again it is suggested that the market survey be utilized to glean information on such matters. The maximum thickness-to-chord ratio at the root for commercial airliners generally lies in the range $1.5A \le (t/c)_r \le 1.9A$, where A is the wing aspect ratio. At the tip the range is $1.0A \le (t/c)_t \le 1.3A$. In both cases t/c is measured in percent.

In order to facilitate the airfoil selection process several airfoil characteristics are presented here. The stall characteristics of airfoils have been correlated by an airfoil leading edge sharpness parameter Δy which is shown in Figure 5.9. The value of Δy increases (linearly) with airfoil thickness ratio and depends upon the NACA airfoil family as given in Table 5.2. Similarly the leading edge radius (LER) for a number of selected airfoils in the NACA 6-series is presented in Table 5.3.

Table 5.2 Leading Edge Sharpness Parameter

NACA Airfoil Family	LE Sharpness Parameter $\Delta y/c$
63 Series	$0.221t/c$
64 Series	$0.205t/c$
65 Series	$0.192t/c$
66 Series	$0.183t/c$

Table 5.3 Leading Edge Radius (LER) for Selected NACA 6-Series Airfoils

t/c (%)	63 Series LER (%)	64 Series LER (%)	65 Series LER (%)	66 Series LER (%)
6	0.297	0.256	0.240	0.223
8	–	0.455	–	–
9	0.631	0.579	0.552	0.53
10	0.770	0.720	0.687	0.662
12	1.087	1.040	1.000	0.893
14	–	–	1.311	–
15	1.594	1.590	1.500	1.435
16	–	–	–	1.575
18	2.120	2.208	1.960	1.955

Table 5.4 Guidelines for Wing Configurations

Engine Type	Cruise Speed	$\Lambda_{c/4}$	λ	A
Turboprop	250–300 kts	$0°$	0.33–0.55	8–12
Turbojet	$0.75 < M < 0.85$	$25–35°$	0.2–0.3	7–10

Values of α_0, $c_{m,0}$, $c_{l,\alpha}$, x_{ac}/c, $\alpha_{cl,max}$, $c_{l,max}$, and α^0, for selected airfoils were listed in Table 5.1. The values of $c_{m,0}$ are observed to increase significantly with the design lift coefficient (the fourth digit in the 6-series airfoil) and these large negative values are to be avoided in the high-speed designs because they tend to put the aircraft in a dive and/or cause the wing to twist to negative angles of attack. A design lift coefficient of 0.2 is preferred to one of 0.4 for this reason. For airfoil selection purposes Abbott et al. (1945) and Abbott and Von Doenhoff (1959), as well as Hoak et al. (1978), may be consulted. An initial airfoil selection should be made before proceeding further and the value of $c_{l,max}$ corresponding to $M = 0.2$, $Re_c = 9 \times 10^6$, and smooth surface condition noted. On the basis of the market survey and the results of Chapter 4 one may also select the wing aspect ratio A, the taper ratio λ, and the sweepback angle of the quarter chord line $\Lambda_{c/4}$. Table 5.4 provides some suggested guidelines for the current stage of the design process. Table 5.5 contains information on wing characteristics for several operational airliners of different type.

Table 5.5 Wing Data for Several Operational Airliners[a]

Manufacturer	Aircraft	A	Sweep $\Lambda_{c/4}$ (deg)	Taper λ (deg)	MAC c_{MAC} (ft)	Root $(t/c)_r$	Tip $(t/c)_t$	Dihedral (deg)	Incidence (deg)
Bombardier	Q100	12.4	0.00	0.50	6.91	0.18	0.13	2.5[b]	
P.R. of China	Y-7	11.4	6.8[b]	0.37	8.15			-2.2[b]	3.00
Avions de T.R.	ATR72-500	12.0	3.1[b]	0.50	6.30	0.18	0.13	1.5[b]	2.00
Embraer	ERJ145LR	7.86	22.8	0.25	7.03			5.5	
Bombardier	CRJ200LR	8.28	24.5	0.20	7.00	0.13	0.10	2.3	3.40
Bombardier	CRJ700ER	7.87	26.0	0.25	6.60			2.0	
Embraer	E175	9.28	25.5	0.25	10.60			5.0	
Boeing	B737-700ER	10.3	25.0	0.22	13.50			6.0	
Airbus	A320-200	9.35	25.0	0.23	13.32	0.15	0.11	5.1	
Airbus	A310-300	8.79	28.0		19.10	0.21	0.11	4.0	
Boeing	B767-200ER	7.99	31.5	0.22	22.00	0.15	0.10	6.0	4.25
Boeing	B777-200	8.68	31.5	0.20	26.00			8.0	
Airbus	A340-300	10.1	30.0	0.25	26.90	0.15	0.11	5.5	
Boeing	B747-400	7.91	37.5	0.24	30.60	0.13	0.08	7.0	2.00
Airbus	A380-800	7.53	35.8	0.20	40.33	0.15	0.12	5.4	

[a] Data taken mainly from Jane's (2010) and manufacturers' specifications.
[b] Panels outboard of engines.

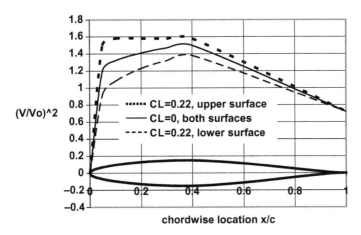

FIGURE 5.10

Theoretical velocity distribution on upper and lower surfaces of an NACA 64_2-015 symmetric airfoil for zero lift and moderate lift angles of attack.

5.2.4 Compressibility effects on airfoils

Consider a typical high-speed NACA airfoil, the 64_2-015, a symmetric section with 15% thickness. The theoretical inviscid surface velocity distribution, as given in Abbott and von Doenhoff (1959), is shown in Figure 5.10. Two distributions are illustrated, one for zero angle of attack where $c_l = 0$, and one for a moderate angle of attack where $c_l = 0.22$. Because the section is symmetric, the zero angle of attack case has exactly the same velocity distribution on both the upper and lower surfaces. As a result the pressure distributions are identical on both surfaces and therefore the net lift is zero. However, in the moderate angle of attack case the upper surface of the airfoil has a consistently higher velocity on the upper surface than on the lower surface, resulting in lower pressures on the upper surface than on the lower surface, thereby producing a net lift force. Note that the square of the velocity is plotted since this quantity is proportional to the pressure; the difference between the upper and lower surface curves is basically the net pressure force.

In the lifting case shown in Figure 5.10, the upper surface velocity V is approximately 26% greater than the free stream velocity V_0. Therefore as the free stream Mach number approaches 0.8, the velocity on the upper surface approaches the sonic value, i.e., $M = 1$. Thus the surface of the airfoil starts to feel compressibility effects before the free stream might suggest they are important. We may define the critical Mach number for an airfoil as that free stream Mach number at which the velocity at some point on the surface of the airfoil reaches the sonic value. For the airfoil considered the critical Mach number $M_{crit} = 0.71$ for $c_l = 0$ and $M_{crit} = 0.66$ for $c_l = 0.22$. Abbott et al. (1945) present graphs of M_{crit} as a function of airfoil lift coefficient for a wide variety of NACA airfoils; an extract of these data for NACA 6-series airfoils is presented in Appendix D. We will make use of this material to estimate drag in Chapter 9.

Research into means of delaying the onset of compressibility effects led to the development of the "supercritical" airfoil by Richard Whitcomb, a NASA researcher who also pioneered the "area rule" that prompted a "coke-bottle" shape for fuselages that reduced transonic drag (Whitcomb and Clark, 1965) that will be discussed in Chapter 9. Because speed is of importance in air transport, modern airliners are designed to cruise as close as possible to the local sonic speed without incurring an undue drag penalty. Though there is continuing interest in traveling even faster than sound, the generation of ground-level pressure disturbances ("sonic booms") limited the supersonic portions of flight of the Concorde supersonic transport to areas over the sea. As a consequence, supersonic transports are very specialized vehicles and their design is outside the scope of this book. As was just pointed out, flight at Mach numbers above 0.65 is likely to result in regions of supersonic flow developing over the wing. The deceleration of the supersonic flow to subsonic values over the aft sections of the wing produces shock waves which disturb the boundary layer flow there and can cause substantial flow separation with the concomitant penalty of increased drag. A schematic illustration of the flow field and pressure distributions over conventional and supercritical airfoils, as presented and discussed in detail by Harris (1990), is shown in Figure 5.11.

Therefore, to improve performance in the low-transonic-speed range, $0.7 < M < 1$, the region of locally supersonic flow over an advanced design airfoil must develop in a manner that ensures the terminating normal shock is weak, rather than the strong shock typical of conventional airfoils, like the NACA 6-series. Advanced airfoils include Whitcomb's supercritical airfoils (Whitcomb and Clark, 1965) and Pearcy's peaky airfoils (Pearcy, 1962), the generic shapes of which are shown in Figure 5.12. A comparison presented by Morrison (1976) of the typical chordwise pressure coefficient distribution over supercritical airfoils of the Pearcy "peaky" and Whitcomb type is shown in Figure 5.13.

The "peaky" airfoil is so called because it is designed to have a high suction peak near the rather slender nose of the airfoil. It is seen in Figure 5.13 that the negative of the pressure coefficient $C_p = \frac{p - p_0}{q_0}$ climbs rather rapidly and then settles to a relatively constant value over the middle of the chord. Note that it is common to display the negative of the pressure coefficient, $-C_p$, along the positive y-axis so the initial pressure actually drops rapidly. The critical value of $C_{p,crit}$, where the local Mach number on the airfoil $M = 1$ and $p = p^*$, for the peaky airfoil shown is $C_{p,crit} \sim 0.6$. Therefore, the peaky airfoil is shocking down from a relatively low locally supersonic Mach number resulting in a terminating normal shock that is weak, as desired. For the Whitcomb-type airfoil the critical value shown is $C_{p,crit} \sim 0.5$ and once again the airfoil is shocking down from a relatively low local Mach number keeping the resulting normal shock weak. In Figure 5.13 the Whitcomb-type airfoil has $c_l \sim 0.6$ at $M = 0.78$ and the peaky airfoil has $c_l \sim 0.4$ at $M = 0.75$.

Thus all supercritical airfoil designs are marked by controlling the supersonic flow region so as to produce weak terminal shocks, but Whitcomb's airfoil is seen to have substantial aft loading, as pointed out in Figure 5.13, due to the lower surface reflex curvature near the trailing edge. The Whitcomb-type supercritical airfoil

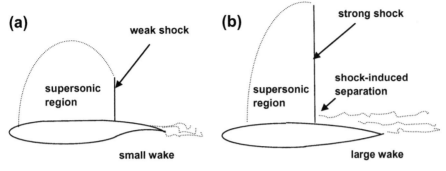

FIGURE 5.11

The flow field over (a) a supercritical airfoil and (b) a conventional airfoil; from Harris (1990).

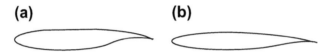

FIGURE 5.12

Typical shapes of the (a) Whitcomb type and (b) "peaky" type of supercritical airfoils.

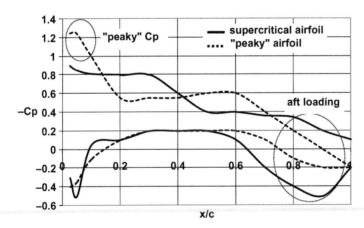

FIGURE 5.13

Comparison of the typical chordwise pressure coefficient distribution of supercritical airfoils of the Pearcy "peaky" and Whitcomb types.

combines the geometrical features of a rounder leading edge, flatter upper surface, and more reflexed trailing edge as compared to the "peaky" type of airfoil developed by Pearcy. These design differences reduce the leading edge pressure coefficient

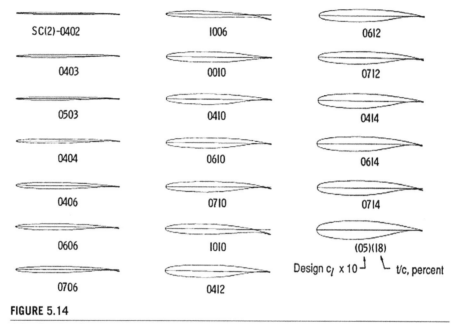

FIGURE 5.14

A family of NASA supercritical (SC) airfoils; from Harris (1990).

"peakiness," extend the near sonic flow further aft on the airfoil, and yield a more highly loaded aft portion of the airfoil. The result is a somewhat higher lift coefficient developed at a higher free stream drag divergence Mach number.

As mentioned previously, the supercritical airfoil has a flatter upper surface designed to provide a smoother deceleration of the supersonic flow so that a weaker shock wave is produced than on a conventional airfoil. A family of NASA SC-series supercritical airfoils is shown in Figure 5.14. Note that the airfoil shape, when flipped vertically, bears a resemblance to that of a conventional airfoil.

The extent of drag reduction possible is shown in Figure 5.15(a) for an 11% thick supercritical airfoil, and a 12% thick conventional NACA 64_1-212 low-drag airfoil. The major improvement provided by the supercritical airfoil design is found to be in delaying the onset of the drag divergence Mach number from $M_{DD}=0.7$ for the conventional airfoil to $M_{DD}=0.8$ for the supercritical airfoils. There are some additional benefits of the supercritical airfoil in that the lift is preserved and even augmented at the higher free stream Mach numbers possible. This is shown in Figure 5.15(b) where the normal force coefficient for the supercritical airfoil and the NACA 64_1-212 conventional airfoil is shown as a function of free stream Mach number.

However, there is an increase in pitching moment, not shown in Figure 5.15(b), that turns out to be not as much of a trim drag penalty for swept-wing aircraft fitted with supercritical airfoils because the optimum wing twist increases as the Mach number increases. This increased wing twist alleviates the penalty arising from increased pitching-moment coefficient.

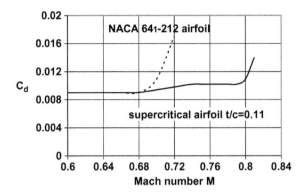

FIGURE 5.15(a)

Drag coefficients as a function of Mach number for a supercritical airfoil and a conventional airfoil; from Harris (1990).

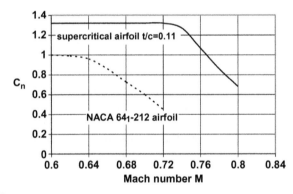

FIGURE 5.15(b)

The normal force coefficients for a supercritical and a conventional airfoil as a function of free stream Mach numbers; from Harris (1990).

The drag divergence Mach number is defined as that Mach number where the derivative of the drag coefficient with respect to Mach number has a particular value; NASA uses 10% as its criterion, i.e., $dc_d/dM = 0.10$. Thus the use of supercritical airfoils on the wings of modern airliners has provided substantial performance increases. So-called "peaky" airfoils, described previously, began to see operational use on several aircraft, including McDonnell Douglas DC-9 Series 30, DC-8 Series 63, and DC-10 Series 10. Then first-generation Whitcomb-type supercritical airfoils were introduced on airliners like the Airbus A300-600, A310-300, and the Boeing 767-200, while next-generation versions were used on the Bombardier CRJ 200 LR,

Table 5.6(a) Supercritical Airfoils Compiled by Schiktanz and Scholz (2011)

Airfoil	(t/c)max%	References
BAC 1	10	Johnson and Hill (1985)
CAST 7	11.8	AGARD (1979)
CAST 10-2/DOA 2	12.1	Dress et al. (1984)
Cessna EJ	11.5	Allison and Mineck (1996)
DFVLR R4	13.5	Jenkins, Johnson, Jr., Hill, et al. (1984)
NLR 7301	16.3	AGARD (1979)
NPL 9510	11	Jenkins (1983)
SC(2)-0012	12	Mineck and Lawing (1987)
SC(2)-0710	10	Harris (1975a)
SC(2)-0714	14	Harris (1975b); Harris et al. (1980)
SC(3)-0712(B)	12	Johnson et al. (1985)
SKF 1.1	12.07	AGARD (1979)

the Embraer EMB-145, and the Airbus A321-200 and A340-300. Wing designs are so valuable that aircraft manufacturers consider them proprietary and detailed information on them is not readily available. Schiktanz and Scholz (2011) present a compilation of wind tunnel test data on supercritical airfoils taken from publically available reports, as presented in Table 5.6(a). They noted that conventional NACA laminar flow airfoils showed good characteristics at supercritical speeds and included the NACA 65$_1$-213 studied by Plentovich et al. (1984) in their survey. The issue of the transonic drag reductions possible with supercritical airfoils will be treated in detail in Chapter 9.

5.2.5 Computational resources for airfoil analysis and design

As is evident from the material presented in the previous sections, there is a broad and diverse literature on airfoils and their characteristics. The approach taken in this book is to make expedient use of extant data and empirical methods based on theory and extensive experience in order to arrive at reasonably accurate design solutions for the many components of the complex system that is a modern commercial aircraft. The theoretical background on the aerodynamics of airfoils is described in Appendix C along with some of the techniques which may be used to calculate the flow field around airfoils. The availability of increasingly powerful personal computers has encouraged the development of a wide range of software applicable to various aspects of airplane design. In the case of airfoils alone, UIUC (2013) maintains a readily accessible library of almost 1600 airfoil designs and provides information on geometry and performance. They also provide links to several of the application codes that have proven to be successfully used by students, for example, XFOIL, developed by Drela (1989) and released under the GNU general public license. Public domain software provided by PDAS (2013) includes PABLO, an airfoil analysis program with boundary layer analysis. An online search will reveal a number of other codes for airfoil analysis and development.

For the present purpose of learning the basic principles of commercial aircraft design it seems prudent to determine the lifting characteristics of the design project airplane by simply using the reported characteristics of the NACA 6-series or SC-series airfoils. It is common for substantial time to be expended in learning how to use various software packages, and this effort often intrudes on the total time available for the entire project. As one gains more understanding of the aerodynamics involved and experience in assessing the results of various analyses it becomes expeditious to incorporate more complete theoretical tools.

5.3 Lifting characteristics of the wing

The maximum lift coefficient of the airplane $C_{L,\max}$ depends upon many factors. Only the most important of these will be considered here and they are listed below.

a. Airfoil maximum lift coefficient $c_{l,\max}$.
b. Wing aspect ratio A, taper ratio λ, and sweepback angle Λ.
c. Trailing edge flap design and deflection angle.
d. Leading edge flap design and deflection angle.

The methods that will be used to estimate $C_{L,\max}$ for the various configurations of a wing are taken mainly from the USAF Stability and Control DATCOM, Hoak et al. (1978) and are empirical in nature. There are other, more sophisticated, approaches based on different computational fluid dynamics (CFD) schemes which will be discussed in varying degrees of detail. Such scientifically richer methods will generally have been covered in the fluid dynamics analysis courses of an engineering degree program and may be implemented, if desired. However, it is important to develop some familiarity with empirical and approximate techniques rarely covered in academic courses. The rapid turnaround they provide is of great value in preliminary design situations in industry.

5.3.1 Determination of the wing lift curve slope

The airfoil characteristics described thus far are based on two-dimensional flow whereas wings have finite span and three-dimensional effects must be considered. Basic wing theory and analysis is presented in Appendix C. This background material on wings should be reviewed to complement the mainly empirical approaches presented here. The three-dimensional lift curve slope of conventional wings $C_{L\alpha}$ is given, per radian, by the following equation:

$$\frac{C_{L_\alpha}}{A} = \frac{2\pi}{2 + \left[\frac{A^2 \beta^2}{\kappa^2} \left(1 + \frac{\tan^2 \Lambda_{c/2}}{\beta^2} \right) + 4 \right]^{1/2}} \tag{5.18}$$

Thus $C_{L\alpha}$ is a function of wing aspect ratio, mid-chord sweep angle $\Lambda_{c/2}$, Mach number, and airfoil section (defined parallel to the free stream) lift curve slope. The factor κ in the figure is the ratio of the experimental two-dimensional (i.e., airfoil) lift curve slope (per radian) at the appropriate Mach number $(c_{la})_M$ to the theoretical value at that Mach number, $2\pi/\beta$, or $\kappa = (c_{la})_M/(2\pi\beta)$. Note that the theoretical (Prandtl-Glauert) correction for subsonic compressibility is $(c_{la})_M = c_{l\alpha}/\beta$, so in the absence of an experimental value for $(c_{la})_M$ the value for $\kappa = c_{l\alpha}/2\pi$, that is, the ratio of the actual low-speed airfoil lift curve slope to that of the airfoil in ideal incompressible flow will suffice. The section lift curve slope (per degree) is obtained from Table 5.1 or from, for example, Hoak et al. (1978), and β is the Prandtl-Glauert factor

$$\beta = \sqrt{1 - M^2} \tag{5.19}$$

The sweep-conversion formula, Equation (5.11), may be used to find the mid-chord sweep for any straight-tapered wing as follows:

$$\tan \Lambda_{1/2c} = \tan \Lambda_{LE} - \frac{2}{A} \left(\frac{1 - \lambda}{1 + \lambda} \right) \tag{5.20}$$

Recall that λ is the taper ratio, c_t/c_r. Writing Equation (5.11) to find the sweepback angle of the leading edge from that at any other constant percent chord line $(n = \%c/100)$, for trapezoidal wing planforms, yields

$$\tan \Lambda_{LE} = \tan \Lambda_{nc} + \frac{4n}{A} \left(\frac{1 - \lambda}{1 + \lambda} \right) \tag{5.21}$$

For example, if the quarter chord sweepback angle is known $(n = 1/4 = 0.25)$, the sweepback angle of the leading edge is easily determined. In a similar fashion, once the sweepback angle of the leading edge is known, the sweepback angle of any other constant percent chord line can be easily found:

$$\tan \Lambda_{nc} = \tan \Lambda_{LE} - \frac{4n}{A} \left(\frac{1 - \lambda}{1 + \lambda} \right) \tag{5.22}$$

5.3.2 Sample calculation of the wing lift curve slope

Consider the 64A010 airfoil, a symmetric section of thickness ratio $t/c = 10\%$ being used in a wing with an aspect ratio $A = 5$, a leading edge sweepback $\Lambda = 46.6°$, and a taper ratio $\lambda = 0.565$. To find the lift curve slope of this wing at a Mach number $M = 0.4$ we may use Equation (5.18). First we must determine the sweepback of the half-chord line, which may be found using Equation (5.22):

$$\tan \Lambda_{c/2} = \tan \Lambda_{LE} - \frac{2}{A}\left(\frac{1-\lambda}{1+\lambda}\right) = \tan(46.6) - \frac{2}{5}\left(\frac{1-0.565}{1+0.565}\right) = 0.946$$

We also require the value of $\kappa = (c_{l\alpha})_M / (2\pi/\beta)$. Because $(c_{l\alpha})_M$, the experimental value for $c_{l\alpha}$ at $M = 0.4$, is not provided here, we approximate it by $c_{l\alpha}/\beta$, where $c_{l\alpha} = 0.110/°$ (=6.303 per radian) is the low-speed value given in Table 5.1. Then Equation (5.18) yields

$$C_{L_\alpha} = \frac{2\pi A}{2 + \left[\frac{A^2 \beta^2}{\kappa^2}\left(1 + \frac{\tan^2 \Lambda_{c/2}}{\beta^2}\right) + 4\right]^{1/2}} = \frac{2\pi\,(5)}{2 + \sqrt{\frac{5^2\left(1-0.4^2\right)}{\left(\frac{6.303}{2\pi}\right)^2}\left(1 + \frac{0.946^2}{1-0.4^2}\right) + 4}}$$

$$= 3.48$$

This result, $C_{L\alpha} = 3.48$ per radian, may also be written as $C_{L\alpha} = 3.48/57.3 = 0.061\,\text{deg}^{-1}$. This is about 1.7% higher than the experimental result of Johnson and Shibata (1951) which is $C_{L\alpha} = 0.060\,\text{deg}^{-1}$. If we increase the Mach number to $M = 0.8$, then $C_{L\alpha} = 0.068\,\text{deg}^{-1}$, which is about 3.9% higher than the reported result of 0.0654. We see that the lift curve slope of the finite wing is substantially less than that of the airfoil of which it is comprised. However, note that doubling the Mach number from 0.4 to 0.8 increases the lift curve slope of the wing by almost 10%.

5.4 Determination of wing maximum lift in the cruise configuration

Section 4.1.3.4 of DATCOM (Hoak et al., 1978) presents methods of rapidly estimating the maximum lift and angle of attack for wings at subsonic, transonic, supersonic, and hypersonic speeds. The material pertinent to subsonic speed is used in the current design approach. At subsonic speeds a distinction is made between low- and high-aspect-ratio wings. The maximum lift of high-aspect-ratio wings at subsonic speeds is directly related to the maximum lift of the wing airfoil sections. Wing planform shape does influence the maximum lift obtainable, but its effect is distinctly subordinate to the influence of the section characteristics. For low-aspect-ratio wings, like fighter plane delta wings or the ogee wing of the Concorde SST, the maximum lift is primarily related to planform shape, while the airfoil section characteristics are secondary. Because commercial airliners are characterized by high-aspect-ratios only the portion of the DATCOM method pertinent to such wings is presented here. Other methods for calculating maximum lift, usually involving additional complexity but with increased accuracy are also covered in this section and may be used for a collective comparison.

5.4.1 Subsonic maximum lift of high-aspect-ratio wings

The maximum lift and stalling characteristics of high-aspect-ratio wings, are, to a first approximation, determined by section properties which, for selected NACA 6-series sections, have been presented in Table 5.1. See Section 4.1.1.4 of DATCOM (Hoak et al., 1978), for methods for dealing with non-standard airfoils. Obviously, three-dimensional effects arising from tip, taper, or sweepback effects have an influence on the stalling characteristics of a wing. As a result, the stall of a wing, even a simple unswept, untwisted wing using a constant airfoil sections, starts at a specific angle of attack at a particular point on the wing and then rapidly spreads across the span as the angle of attack increases further. Highly tapered or swept back wings tend to stall near the tips, while wings with little sweep or taper tend to stall near the root. As a first step in the process of accounting quantitatively for the existence of this effect in swept-wing design, it is necessary to examine the nature of the separation process which limits $c_{l,max}$ for the two-dimensional (airfoil) and three-dimensional (wing) cases.

On thick or highly cambered airfoils separation begins at the trailing edge and spreads upstream as the angle of attack increases, finally fixing $c_{l,max}$. The typical pressure distribution on the upper surface shows a sharp peak near the leading edge and an area of constant pressure over the aft portion where separation exists. On very thin airfoils the flow separates from the surface starting at the leading edge but then reattaches to the surface farther aft. This point of reattachment moves downstream as the angle of attack increases and finally fixes $c_{l,max}$ when it reaches the trailing edge. The pressure distribution on the upper surface shows a slight peak near the leading edge followed by relatively constant pressure region up to the point of reattachment, and then recovery to essentially free stream pressure. Intermediate-thickness airfoils with about 10-percent thickness and little camber typically have both types of separation simultaneously. The value of $c_{l,max}$ is fixed when the trailing edge separation nears or reaches the point of reattachment of the leading edge separation. The pressure distribution on the upper surface in this case shows little or no sharp peak near the leading edge and lack of recovery to the free stream value at the trailing edge.

On the basis of these distinctions and from examination of the chordwise pressure distributions just prior to stall of a given airfoil section in two- and three-dimensional flow, an insight can be had into the mechanism by which sweepback provides a degree of natural boundary layer control. Harper and Maki (1964) discuss the case of a wing swept back at $45°$. The two-dimensional pressure distributions for the airfoil used in the wing show evidence of both leading and trailing edge types of separation discussed above. This same type of separation pattern is found in the three-dimensional flow over the outboard region of the swept wing. However, on the inboard sections the separation pattern changes to the thin airfoil, leading edge type of separation. From this, it is concluded that as the root is approached from the tip, the spanwise flow due to sweep becomes increasingly effective in controlling the boundary layer resulting in suppression of the trailing edge type of separation of the swept wing.

The two major effects of wing sweep may be said to be:

- suppression of inboard stall, particularly at the trailing edge, through the natural boundary layer control just discussed,
- outboard movement of the peak of the span loading distribution, which increases as taper is increased.

These two factors combine to produce a stalling pattern which is unlike that commonly experienced by unswept wings.

It must be recognized that the maximum lift of the wing alone, as given in this section, may be substantially altered by interference effects. The addition of fuselages, nacelles, pylons, and other protuberances can change the aerodynamic characteristics of a given configuration near the stall. Interference effects of this type are discussed in Section 4.3.1.4 of Hoak et al. (1978).

5.4.2 DATCOM method for untwisted, constant-section wings

We will consider high-aspect-ratio wings, defined as those with an aspect ratio

$$A \geq \frac{4}{(C_1 + 1)\cos \Lambda_{LE}} \tag{5.23}$$

The quantity C_1 is a correlation factor that depends on the taper ratio λ; a graph of C_1 as a function of λ is shown in Figure 5.16.

An empirically derived method, based on experimental data, for predicting the subsonic maximum lift and the angle of attack for maximum lift of high-aspect-ratio, untwisted, constant-section (symmetrical or cambered) wings is given. The development follows the DATCOM method (Hoak et al., 1978). The equations and directions for using the charts are as follows:

$$C_{L,\max} = \left(\frac{C_{L,\max}}{c_{\ell,\max}}\right) c_{\ell,\max} + \Delta C_{L,\max} \tag{5.24}$$

$$\alpha_{C_{L,\max}} = \frac{C_{L,\max}}{C_{L_\alpha}} + \alpha_0 + \Delta\alpha_{C_{L.\max}} \tag{5.25}$$

The first term on the right-hand side of Equation (5.24) is the maximum lift coefficient at $M=0.2$ and the second term is the lift increment to be added for Mach numbers between 0.2 and 0.6. Here $C_{L,\max}/c_{\ell,\max}$ is obtained from Figure 5.17 and $c_{\ell,\max}$ is the section maximum lift coefficient at $M=0.2$ obtained from Table 5.1. The quantity $\Delta C_{L,\max}$ is a Mach number correction obtained from Figure 5.18. For cruise Mach numbers greater than 0.6, no general empirical correlation is readily available, so reasonable extrapolation to the cruise Mach number may be carried out. This is acceptable for preliminary design purposes because flight under normal conditions

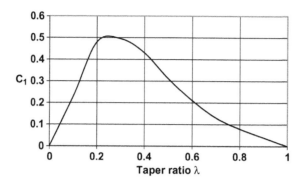

FIGURE 5.16

Variation of C_1 as a function of λ.

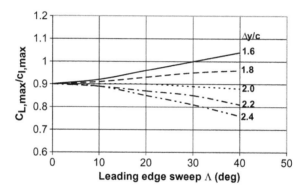

FIGURE 5.17

Variation of $C_{L,max}/c_{l,max}$ with leading edge sweep for different values of the airfoil leading edge sharpness parameter $\Delta y/c$ (Hoak et al., 1978).

will not involve maximum lift at the cruise speed. However, the more accurate methods for wing lift given in Appendix C and used subsequently in this chapter can accommodate the high subsonic cruise speeds of modern airliners. In Equation (5.24) C_{L_α} is the wing lift curve slope for the Mach number under consideration, obtained previously from Equation (5.18) and α_0 is the wing zero-lift angle of attack. Sharpes (1985) suggests that the zero-lift angle of attack for swept wings as calculated by DATCOM tends to overestimate the experimentally observed angles and should be replaced by an improved correlation given as follows:

$$\alpha_0 = \alpha_{0,\Lambda=0} \cos^2 \Lambda \tag{5.26}$$

The angle of attack for zero sweep $\alpha_{\Lambda=0}$ is equivalent to the airfoil zero-lift angle of attack, again for the Mach number under consideration, and may be

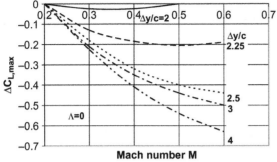

FIGURE 5.18(a)

Mach number correction for maximum wing lift for various values of the airfoil leading edge sharpness parameter $\Delta y/c$ and leading edge sweep $\Lambda = 0°$ (Hoak et al., 1978).

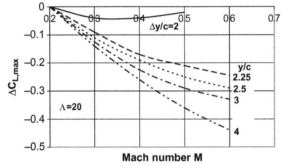

FIGURE 5.18(b)

Mach number correction for maximum wing lift for various values of the airfoil leading edge sharpness parameter $\Delta y/c$ and leading edge sweep $\Lambda = 20°$ (Hoak et al., 1978).

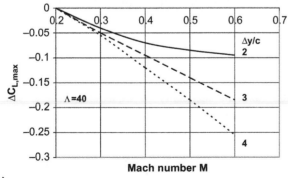

FIGURE 5.18(c)

Mach number correction for maximum wing lift for various values of the airfoil leading edge sharpness parameter $\Delta y/c$ and leading edge sweep $\Lambda = 40°$ (Hoak et al., 1978).

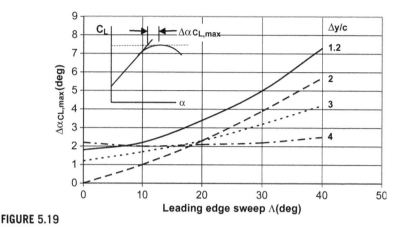

FIGURE 5.19

Angle of attack increment (defined in inset) for wing maximum lift in subsonic flight (Hoak et al., 1978).

obtained from Table 5.1. The angle of attack increment $\Delta\alpha_{C_{L\max}}$ is obtained from Figure 5.19. The leading edge parameter Δy, which does not explicitly appear in the equations, must be used in reading values from the charts. The value of Δy is expressed in percent chord and is obtained from Table 5.2. In calculating $\Delta\alpha_{C_{L\max}}$ the value of $C_{L\max}$ calculated from Equation (5.18) is used as the numerator of the first term of Equation (5.24).

5.4.3 Sample calculation for an untwisted, constant-section wing

Consider the case of a swept wing with an NACA 64A010 airfoil, a quarter-chord sweep $\Lambda_{c/4} = 45°$, taper ratio $\lambda = 0.568$, and aspect ratio $A = 5$ operating at $M = 0.4$ and a chord-based Reynolds number $\mathrm{Re}_c = 2$ million. The leading edge sweepback may be found by using Equation (5.21) as follows

$$\tan\Lambda_{LE} = \tan\Lambda_{c/4} + \frac{4n}{A}\left(\frac{1-\lambda}{1+\lambda}\right) = 1 + \frac{4\,(1/4)}{5}\left(\frac{0.465}{1.565}\right) = 1.059$$

Then $\Lambda_{LE} = 46.65°$ and the inequality of Equation (5.23) is satisfied because Figure 5.16 yields $C_1 = 0.24$, therefore

$$A = 5 \geq \frac{4}{(C_1 + 1)\cos\Lambda_{LE}} = \frac{4}{1.24\cos(46.65)} = 4.7$$

Thus the high-aspect-ratio requirement is met and the DATCOM approach of Section 5.4.2 may be applied. The maximum lift coefficient for the airfoil section may be found from Table 5.1 to be $c_{l,\max} = 1.23$. The leading edge thickness parameter for the NACA 64-series airfoils is taken from Table 5.2 to be $(\Delta y/c) = 0.205(t/c)$ and this may be used in Figure 5.17 (extrapolating out to $\Lambda_{LE} = 46.65°$) to estimate $C_{L,\max}/c_{l,\max} = 0.85$. Similar extrapolation using Figure 5.18 permits one to estimate $\Delta C_{L\max} = -0.53$. Then Equation (5.24) yields

$$C_{L,\max} = \left(\frac{C_{L,\max}}{c_{L,\max}}\right) c_{l,\max} + \Delta C_{L,\max} = (0.85)(1.23) - 0.053 = 0.99$$

This result is 4.2% higher than the experimental result of 0.95 for this case, as reported by Johnson and Shibata (1951). The angle of attack increment for maximum lift may be estimated from Figure 5.19 as $\Delta\alpha_{CL\max} = 7°$. Then the angle of attack for maximum lift is obtained from Equation (5.25) as

$$\alpha_{C_{L,\max}} = \frac{C_{L,\max}}{C_{L_\alpha}} + \alpha_0 + \Delta\alpha_{C_{L,\max}} = \frac{0.99}{0.061} + 0 + 7 = 23.2°$$

The zero-lift angle of attack for this symmetric airfoil is given in Table 5.1 as $\alpha_0 = 0$ and the lift curve slope was previously calculated as $C_{L\alpha} = 0.061$ in Section 5.3.1. This empirically calculated result of $23.3°$ is 3.3% lower than the experimental result of $24°$ for this case reported by Johnson and Shibata (1951).

5.4.4 Maximum lift of unswept twisted wings with varying airfoil sections

High-aspect-ratio wings are often slightly twisted along a spanwise axis and may have varying airfoil sections along the span in order to obtain favorable stalling characteristics. Abbott and von Doenhoff (1959) suggest that a reasonable estimate for the maximum lift of the wing, that is, the stall point, is given by location on the span where the local section lift coefficient of the wing is equal to the maximum lift coefficient of the airfoil used at that station. Though this estimate has no strong theoretical justification it is also given as the preferred method by DATCOM. Of course, this approach requires the availability of a span loading method which can supply the variation of the local lift coefficient with spanwise coordinate. Some simple lifting line and lifting surface methods for estimating the span loading are described in detail in Appendix C and various sample problems are addressed there. The simple approach for estimating the maximum lift of the wing may be described as follows:

1. Using any appropriate theoretical span loading method, as discussed in Appendix C, plot the calculated section lift coefficient c_l as a function of spanwise position $\eta = y/(b/2)$ and angle of attack α.
2. Plot the section lift coefficient $c_{l,\max}$ for the airfoil section(s) used on the given wing as a function of spanwise station for the appropriate Reynolds number and Mach number.

3. The angle of attack and spanwise position for maximum lift is approximated by the angle of attack at which the curves of steps 1 and 2 become tangent.

The integrated value of the curve of step 1 approximates the maximum lift coefficient of the wing. For example, the analysis in Appendix C for an unswept wing with aspect ratio $A = 7$ and taper ratio $\lambda = 0.5$ results in the spanwise distribution of local lift coefficient shown in Figure 5.20. Assuming that an NACA 63_2-215 airfoil is used throughout the span of the wing, the data of Abbott and Von Doenhoff (1959) show a two-dimensional maximum lift coefficient of $c_{l,max} = 1.6$ at $\alpha = 16°$ for a smooth finish airfoil in the Reynolds number range $6 \times 10^6 < Re_c < 9 \times 10^6$. Such an airfoil is appropriate for an unswept wing on a turboprop aircraft taking off at about 130 kts ($M = 0.2$) at sea level and flying at a cruise speed of 300 kts ($M = 0.5$) at 25,000 ft altitude. If the wing under consideration has a typical value for the mean aerodynamic chord of $c_{MAC} = 10$ ft, the root chord is $c_r = 12.86$ ft and the tip chord is $c_t = 6.43$ ft. Then at takeoff the Reynolds number varies linearly from 18 million at the root to 9 million at the tip. Similarly, in cruise the Reynolds number varies linearly from 20.6 million at the root to 10.3 million at the tip. Based on the discussion at the beginning of Section 5.2 we assume there is little change in $c_{l,max}$ from its value at $Re_c = 9$ million and therefore $c_{l,max}$ may be considered constant along the span. In Figure 5.20 this value of $c_{l,max} = 1.6$ is approximately tangent to the calculated local lift coefficient at a spanwise location $\eta \sim 0.6$ and the DATCOM method being employed suggests that stall will originate at or near that spanwise station. The lifting surface result for the wing lift coefficient is found to be $C_L = 1.50$ at $\alpha = 17.7°$ which is actually slightly larger than the airfoil stall angle.

If instead of a smooth finish the NACA 63_2-215 airfoil used throughout the span of the wing has the standard roughness finish, the two-dimensional maximum lift coefficient is given as $c_{l,max} = 1.25$ at $\alpha = 15°$ at a Reynolds number of 6×10^6. As mentioned at the beginning of Section 5.2 the maximum lift coefficient of an airfoil appears to be insensitive to Reynolds number if the boundary layer over the airfoil is turbulent everywhere beyond the immediate region of the leading edge. Therefore we expect the line $c_{l,max} = 1.25$ to be constant over the entire span and find that, once again, it is approximately tangent to the calculated local lift coefficient at a spanwise location $\eta \sim 0.6$. The DATCOM method then suggests that the wing stall will originate at or near that spanwise station. The lifting surface result for wing lift coefficient is $C_L = 1.185$ at $\alpha = 15°$ which is equal to that of the airfoil stall angle.

5.4.5 Reynolds number in flight

The Reynolds number based on the chord length may be written in terms of the Mach number V/V^* as follows:

$$Re_c = \frac{Vc}{\nu} = Mc\frac{V^*}{\nu} \tag{5.27}$$

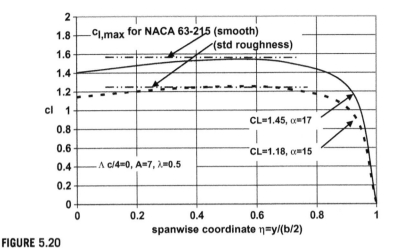

FIGURE 5.20

Spanwise distribution of the local lift coefficient for an unswept wing with $A=7$ and $\lambda=0.5$ at two angles of attack. Also shown is the maximum lift coefficient for an NACA 63_2-215 airfoil with smooth and standard roughness surface.

Using the information on the atmosphere presented in Appendix B, one may determine the ratio V^*/ν as a function of altitude and then show that $Re_c/Mc=7\times10^6\exp(-z/32{,}000)$, where z is the altitude in feet, represents a good fit to the atmospheric data. For typical commercial aircraft applications the Reynolds number per foot of chord length lies between 1.5 and 2 million per foot, as illustrated in Figure 5.21. Note that for the major operations of takeoff and cruise, the unit Reynolds numbers for turboprop and turbofan airliners fall in the range of 1.5–2 million per foot. The data given in Table 5.5 show that the mean aerodynamic chord for turboprop aircraft lies in the range $6.9\,\text{ft}<c_{MAC}<10.6\,\text{ft}$ and for turbofan airliners in the range of $13.3\,\text{ft}<c_{MAC}<26.9\,\text{ft}$, while for very large aircraft the range is $30.6\,\text{ft}<c_{MAC}<40.33\,\text{ft}$. Thus the actual Reynolds number based on mean aerodynamic chord is 10–20 million for turboprop airliners, 20–54 million for turbofan airliners, and 45–80 million for very large aircraft like the B747 and A380. Note that for highly tapered wings the outboard chords may be considerably smaller than the mean aerodynamic chord and will experience lower Reynolds numbers. For example, with a taper ratio $\lambda=0.4$ the tip chord $c_t\sim0.4c_{MAC}$. During low-speed operations like landing and takeoff, high lift devices such as flaps and slats are deployed. Because the characteristic lengths of these elements are considerably smaller than the local wing chord they will be operating at lower Reynolds number and therefore more susceptible to flow separation and stalling.

In the limited Reynolds number range achieved for the smooth NACA airfoils, $3\times10^6<Re_c<9\times10^6$, the maximum lift coefficient has its lowest value at 3×10^6 and then increases to a constant value for Reynolds numbers of 6×10^6 and 9×10^6.

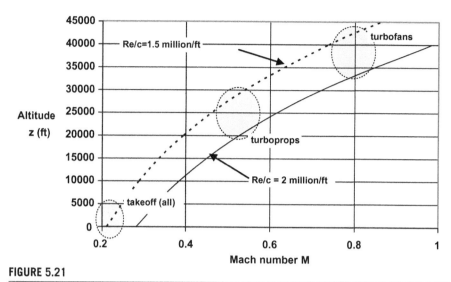

FIGURE 5.21

The variation of unit Reynolds number Re/c_{MAC} is shown for typical commercial aircraft applications.

The general trend is for $c_{l,max}$ to increase slowly, if at all, with increases in Reynolds number and for that increase to be more substantial for thicker sections. This is understandable because as the Reynolds number increases the boundary layer effects become relatively weaker allowing the flow to remain attached to the airfoil for longer distances along the airfoil surface. The lift of an airfoil depends primarily on keeping the flow attached to the airfoil while friction drag itself weakly influences the lift of an airfoil. However, little experimental data are available at higher Reynolds numbers, being limited to about 25 million in the traditional variable-density wind tunnels, but rising to as much as 100 million in cryogenic wind tunnels, as previously described in Section 5.2.2.

5.4.6 Maximum lift of swept and twisted wings with varying airfoil sections

As mentioned in the previous section, Abbott and Von Doenhoff (1959) suggest that a reasonable estimate for the spanwise location at which stall is initiated is given by that spanwise station where the local section lift coefficient $c_l(y)$ for the wing first becomes equal to the maximum lift coefficient of the airfoil used. The wing lift coefficient at this condition is considered the maximum lift of the wing $C_{L,max}$. The usefulness of this method is that the effects of airfoil section and wing planform may be treated independently and its success in providing reasonable estimates has been established in practice. However, as pointed out by Harper and Maki (1964), the measured maximum lift of swept wings is lower than that which would be expected based on the experience with unswept wings. They proposed that the same criterion

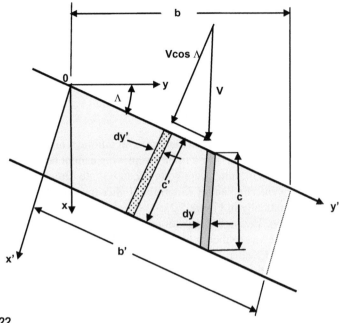

FIGURE 5.22

Infinite swept wing showing velocity components normal and tangent to the leading edge.

used for unswept wings can be correctly applied to sweptback wings if two conditions are satisfied: first, that an appropriate span loading analysis is used to compute the spanwise distribution of the section lift coefficient $c_l(y)$, and second, that the concepts of simple sweep theory be used in applying two-dimensional airfoil data.

Simple sweep theory states that the section characteristics of an infinite wing are invariant with yaw angle provided that these characteristics are defined along a line normal to the line of constant chord and that the appropriate reference velocity is the component of velocity along that line. The characteristics involved include not only the inviscid pressure coefficient distributions, but also the associated boundary layer characteristics, whether laminar or turbulent. For example, an infinite wing swept at some angle Λ may be considered as an unswept wing encountering a free stream velocity of $V\cos\Lambda$, as shown in Figure 5.22. The other component of the free stream velocity $V\sin\Lambda$ runs solely along the span and, in an inviscid flow, has no effect on the pressure field developed. The incremental lift of the wing is given by

$$dL = c_{l,\Lambda=0}\frac{1}{2}\rho\,(V\cos\Lambda)^2\,c'dy' = c_l\frac{1}{2}\rho V^2 cdy$$

For a given span segment of length $b = b'/\cos\Lambda$ the lift is constant and given by

$$L = c_{l,\Lambda=0}\frac{1}{2}\rho V^2\cos^2\Lambda\int_0^{b'} c'dy' = c_l\frac{1}{2}\rho V^2\int_0^b cdy$$

Then the lift coefficients are related by

$$c_{l,\Lambda=0} \cos^2 \Lambda \left(b'c'\right) = c_l (bc)$$

Because the planform area of the segment $S = b'c' = bc$, the lift coefficients normal to and along the leading edge are related by

$$c_l = c_{l,\Lambda=0} \cos^2 \Lambda \qquad (5.28)$$

This is usually called the simple sweep theory for an infinite wing at a given angle of sweep. Obviously the flow field over a finite swept wing cannot be directly related to that over a segment of an infinite swept wing because three-dimensional effects due to the presence of a centerline and a wingtip will alter the surface pressure field in regions shown schematically in Figure 5.23.

The invariance of the pressure distributions and the laminar boundary characteristics according to simple sweep theory have been demonstrated theoretically by Jones (1947a,b). Experimental evidence supports these results and suggests that turbulent boundary layer characteristics behave in a similar manner. Harper and Maki (1964) show that the method described for predicting maximum lift of swept wings of widely varying planform and profile geometries is consistently conservative by around 20%. This conservatism suggests that the spanwise flow over swept wings provides natural boundary layer control which permits local section maximum lift coefficients to reach higher values than could be achieved in purely two-dimensional flow.

Therefore, as in the case of the unswept wing, we use an applicable span loading method which supplies the variation of the local lift coefficient with spanwise

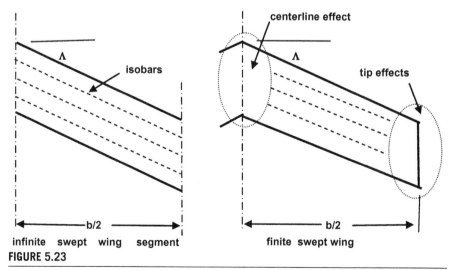

FIGURE 5.23

The isobars for an infinite swept wing and a finite swept wing showing the regions disturbed by three-dimensional effects.

coordinate. For swept wings a lifting surface method for estimating the span loading must be used and several are described in detail in Appendix C along with various sample problems. This approach is recommended in DATCOM for estimating the maximum lift of a swept and twisted wing and may be described as follows:

1. Using any appropriate theoretical span loading method, as discussed in Appendix C, plot the calculated section lift coefficient c_l as a function of spanwise position $\eta = y/(b/2)$ and angle of attack α.
2. Plot the maximum lift coefficient $c_{l,max}$ for the airfoil section(s) used on the given wing as a function of spanwise station for the appropriate Reynolds number and Mach number. Here the maximum lift coefficient is that appropriate to the streamwise airfoil section used in the wing. The maximum lift coefficient of the streamwise section is approximated by $c_{l,max} = (c_{l,\Lambda=0})_{max}\cos^2\Lambda$ where $(c_{l,\Lambda=0})_{max}$ is the maximum lift coefficient of the airfoil section normal to the leading edge. Simple sweep theory requires that the Reynolds and Mach numbers to be used are those for which the streamwise velocity V is replaced by $V\cos\Lambda$ and the streamwise chord length c by $c\cos\Lambda$.
3. The angle of attack and spanwise position for maximum lift is assumed to be determined by the angle of attack at which the curves of steps 1 and 2 first become tangent.

Turning our attention to a turbofan airliner with a swept ($\Lambda_{c/4} = 30°$) and twisted wing ($\Omega = 5°$) of aspect ratio $A = 8$ and taper ratio 0.25 we would select a relatively thin airfoil section. The NACA 64A210 airfoil is a possible choice and, like the thicker NACA 63_2-215 airfoil used in the example for the unswept wing for a turboprop airliner, it also has $c_{l,max} = 1.44$ for a smooth surface finish in the Reynolds number range of 6–9 million, as reported by Abbott and Von Doenhoff (1959). Such a wing would have a typical value for the mean aerodynamic chord in the range 13.3 ft $< c_{mac} <$ 26.9 ft. The tip chord $c_t = \lambda c_r$ and the root chord is given by

$$c_r = \frac{3c_{mac}(1+\lambda)}{2\left(1+\lambda+\lambda^2\right)} \qquad (5.29)$$

Thus $c_r = 1.43c_{mac}$ so that typical root chord for such an airliner would lie in the range 19.0 ft $< c_r <$ 39 ft while the tip chord would be in the range of 5 ft $< c_t <$ 10 ft. For the mid-range case of $c_r = 30$ ft and $c_t = 7.5$ ft and using the approximation $Re_c/Mc = 7 \times 10^6\exp(-z/32,000)$, the Reynolds number at takeoff ($M = 0.20$) would vary linearly from 10.5 million at the tip to 42 million at the root. In cruise ($M = 0.8$) the Reynolds number varies linearly from 14.1 million at the tip to 56.3 million at the root. Based on the discussion at the beginning of Section 5.2 we assume there is little change in $c_{l,max}$ from its value at $Re_c = 9$ million and therefore $c_{l,max}$ may be considered constant along the span.

Applying the simple sweep approximation we set $c_{l,max} = c_{l,max,0}(\cos30°)^2$. We also consider that the appropriate Reynolds number in this approximation is based on $c(\cos\Lambda_{c/4})$ and the appropriate Mach number is $M(\cos\Lambda_{c/4})$. Then, in takeoff, the Reynolds number varies from 7.9 million at the tip to 31.5 million at the root. In cruise the Reynolds number varies linearly from 10.6 million at the tip to 42.2 million at the root. In the spanwise load distribution shown in Figure 5.24 the sweep-corrected value of $c_{l,max} = 1.44(0.866)^2 = 1.08$ is seen to be approximately tangent to the calculated local lift coefficient at a spanwise location $\eta \sim 0.6$ and the DATCOM method being employed suggests that stall will originate at or near that spanwise station. The lifting surface result for the wing lift coefficient is found to be $C_L = 0.996$ at $\alpha = 12.7°$.

5.4.7 A simple modified lifting line theory for $C_{L,max}$

Phillips and Alley (2007) present a method for estimating the maximum wing lift that is based on the classical lifting line theory discussed in Appendix C. Their approach involves providing correction factors for the effects of twist and sweepback on the span loading of unswept, untwisted, tapered wings. The correction factors are developed with the aid of CFD panel method computations for a variety of conventional wing planforms. Curves for the correction factors are presented for a number of specific taper and aspect ratios along with the general method for calculating these effects for other wing geometry. If we confine our attention to taper ratios in the range $0.25 < \lambda < 0.6$, small twist angles $0 < \Omega < 5°$, and aspect ratios in the range $8 < A < 12$, which are typical of commercial airliners, we may use the results presented by Phillips and Alley (2007) to develop the following approximations:

1. For the range of aspect and taper ratios considered here the ratio of wing lift coefficient to maximum section lift coefficient may be fitted, within an error of about 1%, by the following simple equation:

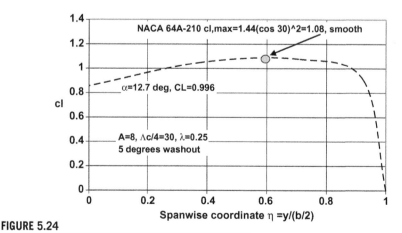

FIGURE 5.24

Spanwise distribution of the local lift coefficient for a swept wing with $A=8$, $\lambda=0.25$, and $5°$ washout at an angle of attack $\alpha = 12.7°$. Also shown is the maximum lift coefficient for an NACA 64A210 airfoil with smooth surface corrected for sweep.

$$\frac{C_L}{c_{l,\max}} = \left[0.952 - 0.45\,(\lambda - 0.5)^2\right]\left(\frac{A}{12}\right)^{0.03} \tag{5.30}$$

2. The sweep correction is given by Phillips and Alley as

$$K_\Lambda = 1 + K_{\Lambda 1}\Lambda - K_{\Lambda 2}\Lambda^{1.2} \tag{5.31}$$

The coefficients in this equation may be approximated by the following simple relations:

$$K_{\Lambda 1} = 0.15 + 18.5\frac{\lambda - 0.4}{A} \tag{5.32}$$

$$K_{\Lambda 2} = 0.55 + 12\frac{\lambda - 0.275}{A} \tag{5.33}$$

This gives errors in $K_{L\Lambda}$ less than 10% except where $K_{\Lambda 1} \ll 1$, but in that case it has little effect on the final result for $K_{L\Lambda}$.

3. The twist correction curves given by Phillips and Alley (2007) cover a wide range of specific wing parameters. For wing planforms of the type considered here, we may use the curves presented to provide simple estimates for the twist correction factor $K_{L\Omega}$, which appears in the term $K_{L\Omega}C_{L,\alpha}\Omega/c_{l,\max}$:

   a. For turboprop airliners with $10 < A < 12$ and $\lambda \sim 0.5$ we may estimate $C_{L,\alpha} \sim 5$ per radian, $\Omega \sim 0.1$ radian, and $c_{l,\max} \sim 1.6$ so that $C_{L,\alpha}\Omega/c_{l,\max} \sim 0.31$, corresponding to $K_{L\Omega} \sim 0.1$.

   b. For turbofans with $8 < A < 10$ and $\lambda \sim 0.25$ we may estimate $C_{L,\alpha} \sim 4.5$ per radian, $\Omega \sim 0.1$ radian, and $c_{l,\max} \sim 1.6$ so that $C_{L,\alpha}\Omega/c_{l,\max} \sim 0.28$, which corresponds to $K_{L\Omega} \sim -0.2$.

The twist correction term has a magnitude lying in the range of $0.056 < K_{L\Omega}\,C_{L,\alpha}\Omega/c_{l,\max} < 0.031$, with the low end of the range corresponding to low taper ratios (λ around 0.25) and the high end of the range corresponding to moderate taper ratios (λ around 0.5).

Phillips and Alley also introduce a stall factor given by

$$K_{LS} = 1 + (0.0042A - 0.068)\left(1 + 2.3\frac{C_{L\alpha}\Omega}{c_{l,\max}}\right) \tag{5.34}$$

For typical values of the parameters in the second parentheses, a change of Ω from $0°$ to $5°$ results in a reduction of K_{AS} by between 1% and 2%.

To estimate the maximum lift coefficient of a wing these quantities may be combined according to the following relation given by Phillips and Alley (2007):

$$C_{L,\max} = \left(\frac{C_L}{c_{l,\max}}\right)_{\Lambda=0,\Omega=0} K_{LS}K_{L\Lambda}c_{l,\max}\left(1 - \frac{K_{L\Omega}C_{L,\alpha}\Omega}{c_{l,\max}}\right) \tag{5.35}$$

According to the order of magnitude estimate provided in item (3) above, the term in parentheses lies in the range of 1–1.033 for moderate taper ratios and to 1–0.967

for low taper ratios. Of course, as the washout is reduced the term in parentheses approaches unity.

5.4.8 Comparison of span loading and the modified lifting line methods

Applying this approach to the unswept ($\Lambda_{c/4}=0$), untwisted ($\Omega=0$) wing treated in the previous section we find from Equation (5.30) for $A=7$ and $\lambda=0.5$ that $C_L/c_{l,\max}=0.937$. Because the wing is unswept we find, from Equation (5.31), that $K_{LA}=1$. The wing is untwisted ($\Omega=0$) and therefore Equation (5.34) yields the stall factor $K_{LS}=0.961$. From Equation (5.35) we estimate that the maximum lift coefficient is $C_{L,\max}=0.9c_{l,\max}$. For the NACA 63_2-215 airfoil at high Re_c the section maximum lift coefficient is $c_{l,\max}=1.6$ and therefore the wing maximum lift coefficient is $C_{L,\max}=1.44$ at $\alpha=17°$. This compares well with the DATCOM technique applied to this wing in Section 5.4.4 where $C_{L,\max}$ was found to be 1.49 at an angle of attack of $17.6°$.

Phillips and Alley (2007) note that the stall correction factor is based on CFD studies alone and only at one Reynolds number, 3 million, which is a relatively low value for practical wings. They suggest that this correction should be applied with discretion, and probably not applied at all in preliminary studies. In this case the use of $K_{LS}=1$ leads to $C_{L,\max}=1.50$, a value much closer to that obtained by the span loading method.

Turning our attention to the swept ($\Lambda_{c/4}=30°$) and twisted wing ($\Omega=5°$) with an aspect ratio $A=8$ and taper ratio 0.25, we find from Equation (5.30) that $C_L/c_{l,\max}=0.913$. From Equations (5.32) and (5.33) we find $K_{A1}=-0.197$ and $K_{A2}=0.513$ so that Equation (5.31) yields $K_{LA}=0.66$. The lift curve slope for the wing may be found from Equation (5.18) to be $C_{L,\alpha}=4.51$ per radian. Using the curves presented by Phillips and Alley (2007) we estimate $K_{L\Omega}=-0.2$ and therefore $K_{L\Omega}C_{L,\alpha}\Omega/c_{l,\max}=-0.0787/c_{l,\max}$. The NACA 64A210 airfoil is used for the swept wing and, like the thicker NACA 63_2-215 airfoil, it also has $c_{l,\max}=1.44$. Then the term $K_{L\Omega}C_{L,\alpha}\Omega/c_{l,\max}=-0.055$, while Equation (5.34) yields a stall factor $K_{LS}=0.969$, resulting in a maximum lift coefficient from Equation (5.35) as follows:

$$C_{L,\max} = (0.913)(0.969)(0.660)(1.44)[1 - (-0.055)] = 0.888$$

This result is 11.9% lower than the value of 0.996 obtained using the span loading method. Note that if we select $K_{LS}=1$ we find $C_{L,\max}=0.916$, a value 8% lower than that obtained using the span loading method. Once again it seems appropriate to avoid using the stall factor K_{LS} until its accuracy is more fully corroborated by comparison with experiments carried out over a broad range of Reynolds numbers.

It is worth recalling that Harper and Maki (1964) point out that the simple sweep approximation provides a conservative estimate for the maximum lift coefficient of the wing. Therefore the approach of Phillips and Alley (2007) also seems to yield a conservative estimate for stall. As an example of this, take the case of a swept ($\Lambda_{c/4} = 35°$) but untwisted wing ($\Omega = 0$) with an aspect ratio $A = 6$ and taper ratio 0.5 studied by Koven and Graham (1948) for which Equation (5.30) yields $C_L/c_{l,max} = 0.94$. From Equations (5.32) and (5.33) we find $K_{A1} = 0.46$ and $K_{A2} = 1$ so that Equation (5.31) yields $K_{L\Lambda} = 0.73$. Because there is no twist $K_{L\Omega} = 0$ and therefore $K_{L\Omega}C_{L,\alpha}\Omega/c_{l,max} = 0$. The NACA 64_1-212 airfoil normal to the quarter-chord line is used for this swept wing and has $c_{l,max} = 1.55$. Equation (5.34) yields a stall factor $K_{LS} = 0.957$, resulting in a maximum lift coefficient from Equation (5.35) as follows:

$$C_{L,max} = (0.94)(0.957)(0.73)(1.55)[1 - (0)] = 1.02$$

The experimental result for this wing is $C_{L,max} = 1.27$ at $19°$ angle of attack so that the simple modified lifting line technique yields a conservative result which is about 20% lower than the experimental value. On the other hand, if we apply Diederich's simple span loading method, as described in Appendix C, along with the DATCOM approach for estimating wing maximum lift coefficient, we arrive at the results shown in Figure 5.25, which suggests $C_{L,max} = 1.35$ at $18.6°$ angle of attack. Here the estimated $C_{L,max}$ is 6.3% higher than the experimental value while the estimated angle of attack is 2% lower than the observed value. The only difference in this estimate is that because the airfoil used is normal to the quarter-chord line, the simple sweep correction is not applied to the section lift coefficient. The Reynolds numbers ranged from 2 million to 9.35 million and the section lift coefficient is quite insensitive to values above 3 million.

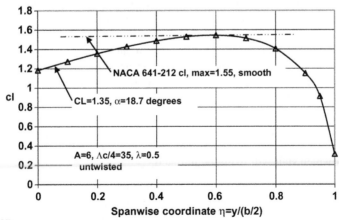

FIGURE 5.25

Spanwise distribution of the local lift coefficient for a swept wing with $A = 6$, $\Lambda = 35°$, and $\lambda = 0.5$, and no twist at an angle of attack $\alpha = 18.7°$. Also shown is the maximum lift coefficient for the smooth NACA 64_1-212 airfoil section which is set normal to the quarter-chord line.

5.4.9 The pressure difference rule for $C_{L,max}$

Panel methods for predicting the surface pressure over a real wing are described in Appendix C. Such computational methods provide a distribution of pressure, chordwise and spanwise, as detailed as the paneling used to approximate the wing's surface. Valarezo and Chin (1994) use a panel method to develop the inviscid flow over a wing or wing body, with or without high lift devices, and then introduce an empirical relation for the airfoil characteristics to estimate the maximum lift of a wing. The method is based on the experimental observation that, under maximum lift conditions, the difference between $C_{p,peak}$, the peak suction pressure coefficient of an airfoil section of the wing, and $C_{p,t.e.}$, the pressure coefficient at the trailing edge, is defined solely by the Mach number and Reynolds number of the flow over the airfoil and not the details of the airfoil's shape. Because Valarezo and Chin (1994) are mainly interested in takeoff characteristics the Mach number range they deal with is limited to $0.15 < M < 0.25$. They point out that Smith (1975) suggested two empirical criteria for defining the point at which the maximum lift is reached: first, the case where the upper surface velocity becomes sonic, $C_p = C_p^*$ and second, where $M^2 C_{p,peak} = -1$. The former suggestion, which for isentropic flow yields $C_p = -13$ at $M = 0.223$, has been often used as a rule of thumb for wings with no leading edge devices. The latter suggestion is based on data obtained at $M > 0.4$ and leads to negative pressure coefficients at low M which are below those observed in experiments.

The results of the method of Valarezo and Chin (1994), which they call the "pressure difference" rule, are shown to be quite accurate and its application has found a following in the airplane design community. The method is applied as follows:

1. Compute the flow field for the wing or wing-body combination using a panel method, assuring adequate definition of the flow at the leading and trailing edges, and plot the computed pressure difference $C_{p,peak} - C_{p,t.e.}$ as a function of spanwise distance for several values of the wing lift coefficient.
2. Plot the absolute value pressure difference quantity $C_{p,peak} - C_{p,t.e.}$ as a function of spanwise coordinate for the Mach number and Reynolds number based on local chord of interest. Curves representing their empirical results for the pressure difference $C_{p,peak} - C_{p,t.e.} = f(M, Re_c)$ are given in Figure 2 of their paper. Interestingly, $13 < (C_{p,peak} - C_{p,t.e.}) < 14$ for $Re_c > 10$ million.
3. The spanwise position at which the curve of step 2 is tangent to one of the curves of step 1 defines the maximum lift coefficient and the point along the span at which stall will likely be initiated.

Note that this method is very similar to the method suggested by Abbott and Von Doenhoff as discussed previously. Consider that the maximum section lift coefficient may be computed from the following equation involving $C_{p,l}$ and $C_{p,u}$, the pressure coefficients on the lower and upper surfaces of the wing, respectively:

$$c_{l,max} = \left\{ C_{p,peak} \left[\int_0^1 \left(\frac{C_{p,l}}{C_{p,peak}} \right) \frac{dx}{c} - \int_0^1 \left(\frac{C_{p,u}}{C_{p,peak}} \right) \frac{dx}{c} \right] \right\}_{max}$$

Valarezo and Chin (1994) show data that indicate the scaled pressure coefficient on the upper surface of the airfoil is essentially a unique function of the normalized distance along the chord and therefore the second integral in the equation above is simply a number. No data are shown for the scaled pressure coefficient on the lower surface, but making the assumption that it also exhibits similarity with respect to the normalized chord, we may expect that

$$c_{l,\max} \sim \left(C_{p,peak}\right)_{\max}$$

The pressure difference may be written as

$$\left(C_{p,peak} - C_{p,t.e.}\right)_{\max} = \left(C_{p,peak}\right)_{\max} \left[1 - \left(\frac{C_{p,t.e.}}{C_{p,peak}}\right)\right]_{\max}$$

The scaled data presented by Valarezo and Chin (1994) indicate that, at maximum lift, the ratio of the pressure coefficient at the trailing edge to that at the suction peak is a number independent of Reynolds number so that

$$\left(C_{p,peak} - C_{p,t.e.}\right)_{\max} \sim \left(C_{p,peak}\right)_{\max}$$

This result suggests that the use of the section maximum lift coefficient to aid in determining the maximum wing lift as suggested by Abbott and Von Doenhoff (1959) is analogous to the pressure difference rule introduced by Valarezo and Chin (1994) and should be acceptable for preliminary design work when panel method results are not available.

Valarezo and Chin (1994) carry out a panel method solution for the clean wing tested by Lovell (1977) which has $A=8.35$, $\Lambda_{c/4}=28°$, and $\lambda=0.35$. Using their pressure difference method they predict $C_{L,\max}=1.04$ at $\alpha=12°$ which agrees well with the wind tunnel data. The spanwise distribution of section lift coefficient for this wing, according to Diederich's (1952) method described in Appendix C, is shown in Figure 5.26. The maximum section lift coefficient for stall as predicted by the DATCOM method of Section 5.4.6 is $c_{l,\max}=1.2$. The airfoil used by Lovell is 10.6% thick (streamwise) and though the shape is specified no specific airfoil data are given, only wing data. This airfoil would be 12% thick normal to the quarter-chord, for which we may assume a range of $1.5<c_{l,\max,\Lambda=0}<1.6$. Applying simple sweep theory suggests that the airfoil $c_{l\max}$ is in the range of 1.17–1.25. Figure 5.26 shows that this range is consistent with a prediction of stall. Thus we may expect a reasonable prediction of $C_{L,\max}$ using the simpler prediction methods if a more accurate panel method is unavailable.

We may also examine the prediction of $C_{L,\max}$ for this wing according to the simplified lifting line method presented previously in Section 5.4.7. Here Equation (5.30) yields $C_L/c_{l,\max}=0.931$. From Equations (5.32) and (5.33) we find $K_{A1}=0.0392$ and $K_{A2}=0.658$ so that Equation (5.31) yields $K_{LA}=0.74$. Because there is no twist $K_{L\Omega}=0$ and therefore $K_{L\Omega}C_{L,\alpha}\Omega/c_{l,\max}=0$. The airfoil normal to the quarter-chord line for this swept wing has been assumed to have $1.5<c_{l,\max}<1.6$. Equation

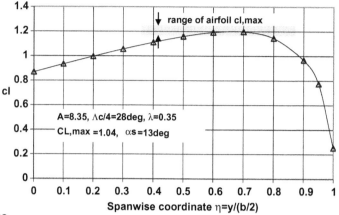

FIGURE 5.26

Spanwise distribution of the local lift coefficient for a swept wing with $A=8.35$, $\Lambda_{c/4}=28°$, $\lambda=0.35$, and no twist at an angle of attack $\alpha=13°$. Also shown is the estimated range of maximum lift coefficient for the smooth 10.6% thick (streamwise) RAE airfoil section used on the wing tested by Lovell (1977).

(5.34) yields a stall factor $K_{LS}=0.967$, resulting in a maximum lift coefficient from Equation (5.35) as follows:

$$C_{L,\max} = (0.931)(0.967)(0.74)(c_{l,\max})[1 - (0)] = 0.666c_{l,\max}$$

Using the assumed values for $c_{l,\max}$ yields $1<C_{L,\max}<1.067$. The panel method result for this wing is $C_{L,\max}=1.04$ at $13°$ angle of attack so that the simple modified lifting line technique yields a result about equivalent to that obtained with the pressure difference rule.

5.5 High lift devices

Airfoils that provide high lift to drag ratios necessary for efficient high-speed cruise are limited to values of $c_{l,\max}<1.4$. As pointed out in Chapter 4, the operational values of C_L for a wing in takeoff and landing are limited to values less than $C_{L,\max}/1.44$ to provide a stall margin for safety. Therefore, for an airfoil under takeoff or landing conditions the maximum c_l achievable is $c_l\sim1$. In takeoff, for example, the ratio of lift coefficient at takeoff to that at landing is given by

$$\frac{C_{L,to}}{C_{L,cr}} = \frac{L_{to}}{q_{to}S}\frac{q_{cr}S}{L_{cr}} = \frac{(W/S)_{to}}{(W/S)_{cr}}\frac{\rho_{cr}}{\rho_{to}}\left(\frac{V_{cr}}{V_{to}}\right)^2 \approx \sigma_{cr}\left(\frac{V_{cr}}{V_{to}}\right)^2 \tag{5.36}$$

In Equation (5.36) we assumed that the wing loading in takeoff and at the start of cruise are approximately equal ($W_4/W_3=0.98$ from Table 2.3) and that takeoff occurs near sea level. For a typical cruise altitude around 33,000 ft (10 km) the atmospheric density ratio $\sigma \sim 1/3$. Then with a typical cruise to takeoff velocity ratio $V_{cr}/V_{to} \sim 3$, we find that $C_{L,to}/C_{L,cr} \sim 3$. The lift coefficient in cruise is

$$C_{L,cr} = \frac{(W/S)_{cr}}{q_{cr}} = \frac{(W/S)_{cr}}{\frac{1}{2}\gamma\, p_{SL}\, \delta\, M_{cr}^2}$$

The dynamic pressure under typical cruise conditions, where the atmospheric pressure ratio $\delta \sim 1/4$, is $q_{cr} \sim p_{SL}/10 \sim 200\,\text{lb/ft}^2$ (10 kPa). With typical wing loading values of around $125\,\text{lb/ft}^2$ (6000 N/m²) we see that $C_{L,cr} \sim 0.6$. This result suggests that the takeoff value of lift coefficient is $C_{L,to} \sim 1.8$. Taking the stall margin safety factor into account shows that the maximum lift coefficient in takeoff required is $C_{L,\max,to} \sim 2.6$, which is far more than an efficient high-speed airfoil can supply. As a consequence, some means must be found for enhancing the lift of an airfoil at low speed without affecting its aerodynamic efficiency at high speed. The most straightforward approach is to consider a variable geometry airfoil.

The lift is proportional to the circulation Γ that can be developed by the airfoil and this is primarily a function of turning the flow efficiently by means of a highly curved camber line. In order to generate the full circulation possible for a given airfoil the Kutta condition, i.e., the condition that the flow leaves the trailing edge of the airfoil smoothly, must be met. Separation of the low-momentum viscous boundary layer over the airfoil limits the degree to which this condition is satisfied. Thus, to achieve an effective high lift airfoil we must be able to increase the camber of the airfoil while simultaneously maintaining attachment of the boundary layer. A multi-element airfoil, comprised of movable elements, each of which is generating its own circulation and thereby contributing to an enhanced lift coefficient for the system, is the most cost-effective practical approach currently available.

Flow over airfoils with various types of moving elements in the vicinity of the trailing edge is illustrated in Figure 5.27 and those with various types of moving elements in the vicinity of the leading edge are shown in Figure 5.28. The airfoil elements are schematically indicated in these figures in order to emphasize the point that movable elements comprise the airfoil. These elements are stowed for high-speed flight so that the airfoil would appear as a straight line. Obviously, there are costs associated with the mechanisms required to deploy and stow the various elements shown. The tradeoff between airfoil performance and weight and cost penalties must be considered carefully.

The aerodynamic effects of the elements were described in detail by Smith (1975) and have been summarized by van Dam (2002) as follows:

1. The wake flow from an upstream element reduces the suction peak on the following element thereby reducing pressure recovery demands and delaying separation on that following element. This is called the slat effect.
2. Circulation produced by a downstream element increases the loading on the previous element, increasing its lift as well as increasing its pressure recovery demands. This is the circulation effect.
3. High-velocity flow on the upper surface of the downstream element permits the flow to leave the upstream element at higher speed, reducing the pressure recovery requirements on the forward element. This is the "dumping" effect.

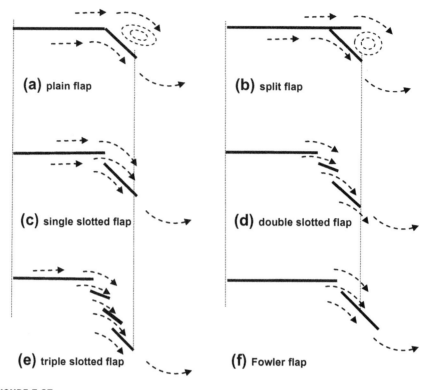

FIGURE 5.27

Schematic representation of typical trailing edge flap systems illustrating the nature of the flow guidance provided by the movable elements of each system.

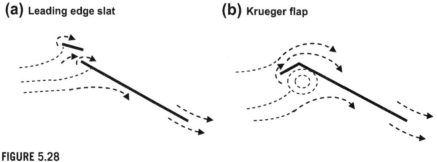

FIGURE 5.28

Schematic diagrams of two leading edge flap systems illustrating the nature of the flow guidance provided by the movable elements of each system.

4. Off-surface pressure recovery is more efficient than pressure recovery on a wall where pressure equilibrates less easily than in a free wake.
5. Fresh boundary layers appear on each element and the thinner boundary layers can withstand adverse pressure gradient better than thicker ones. Sufficient gap width ensures that the boundary layer grows independently on each element.

5.5.1 Airfoil with trailing edge flaps

The deployment of a trailing edge flap changes the camber of the airfoil and thereby increases the maximum lift coefficient and decreases the zero-lift angle of attack as depicted schematically in Figure 5.29.

A schematic diagram of a double-slotted flap was shown in Figure 5.27(d). The practical embodiment of such a flap on an airfoil is presented, for example, in Abbott and Von Doenhoff (1959) and is shown in Figure 5.30. The mechanism for driving the flap is not shown but the travel of the individual elements should be clear. In the stowed position, Figure 5.30(a), the airfoil shape is suitable for efficient high-speed flight. In the fully deployed position, Figure 5.30(b), the two slots separated by a small airfoil

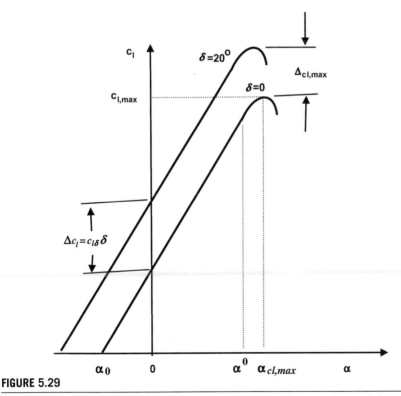

FIGURE 5.29

The generic change in the lift curve of an airfoil when a trailing edge flap is deflected.

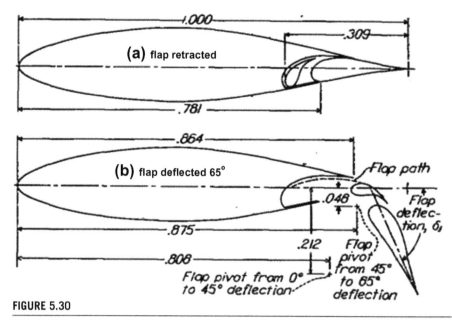

FIGURE 5.30

Airfoil with a double-slotted flap shown in the (a) retracted position and (b) fully deflected position, $\delta_f = 65°$.

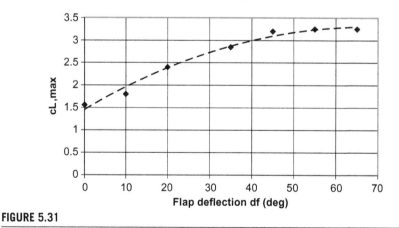

FIGURE 5.31

The maximum lift coefficient as a function of flap deflection for the double-slotted flap shown in Figure 5.20.

element are readily seen. Note that the travel of the flap is such that the projected elemental wing area dS is not materially changed. The increase in maximum lift coefficient for that airfoil is shown as a function of flap deflection angle in Figure 5.31.

FIGURE 5.32

Upper-surface view showing triple-slotted flap and spoilers on Boeing 737 airplane [NASA].

FIGURE 5.33

Lower-surface view of triple-slotted flap on Boeing 737 airplane. [NASA].

Though the maximum lift coefficient may be increased further by using triple-slotted flaps, such as those employed on the Boeing 737 shown in Figures 5.32 and 5.33, and on the Boeing 747, experience has shown that the added weight and complexity of such flaps are not completely cost-effective in airline operations. As a consequence, most trailing edge flap systems on jet transports are of the double-slotted type.

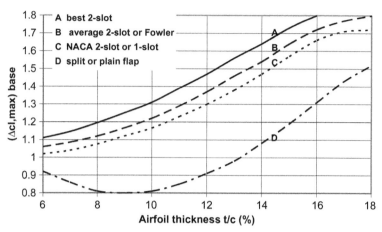

FIGURE 5.34

Base maximum lift increments for 25% chord trailing edge flaps of various types at the reference flap angle. Curve A is for best 2-slot flaps with NACA airfoils, B is for 2-slot flaps with NACA airfoils or Fowler flaps with any airfoil, C is for NACA 2-slot flaps with NACA 6-series airfoils or NACA 1-slot flaps with any airfoil, and D is for split and plain flaps with any airfoil.

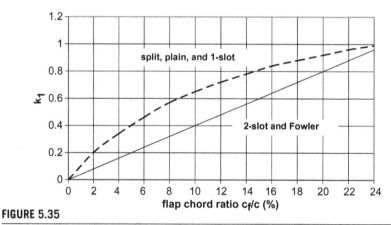

FIGURE 5.35

Correction factor for trailing edge flap chord to airfoil-chord ratios, c_f/c, other than 0.25.

5.5.2 DATCOM method for trailing edge flaps

An empirical method for predicting airfoil maximum lift increments for plain, split, and slotted flaps is presented in DATCOM and will be followed here. The maximum lift increment provided to an airfoil by the deflection of a trailing edge flap is given by

$$\Delta c_{\ell \, max} = k_1 k_2 k_3 \left(\Delta c_{\ell \, max}\right)_{base} \tag{5.37}$$

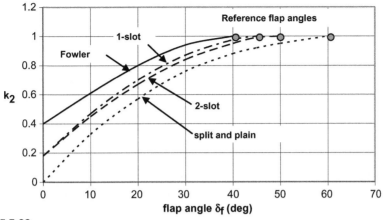

FIGURE 5.36

Flap angle correction factor. The reference flap angle for each type of flap is shown as a solid symbol at $k_2 = 1$.

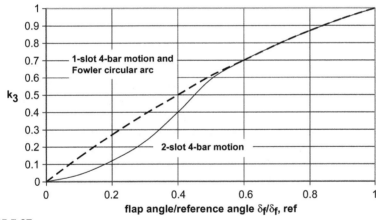

FIGURE 5.37

Flap motion correction factor.

Here $(\Delta c_{lmax})_{base}$ is the section maximum lift increment for 25 percent-chord flaps at the reference flap-deflection angle and is shown in Figure 5.34 for different flap systems. The quantity k_1 is a factor accounting for flap-chord-to-airfoil-chord ratios, c_f/c, other than 0.25 and is shown in Figure 5.35. The quantity k_2 is a factor accounting for flap deflections other than the reference value and is shown in Figure 5.36. Finally, k_3 is a factor accounting for flap motion as a function of flap deflection and is shown in Figure 5.37. A sample calculation for the maximum lift increment for an airfoil with a trailing edge flap according to the DATCOM method is presented in the next section.

5.5.3 **Sample calculation of airfoil with a trailing edge flap**

Assume that one selects a smooth NACA 65-210 airfoil operating at a Reynolds number $Re_c = 6 \times 10^6$. Consider the use of a double-slotted flap on this airfoil and a deflection through $\delta_f = 40°$. The chord length of the flap is $c_f/c = 0.312$. First it is necessary to compute $(\Delta c_{l,max})_{base}$ using Figure 5.34. Selecting an NACA double-slotted flap with the chosen NACA 6-series airfoil requires the use of curve C. The NACA 65-210 airfoil is 10% thick, as indicated by the last two digits in the airfoil designation, and entering this thickness on curve C in Figure 5.34 yields $(\Delta c_{l,max})_{base} = 1.165$. The required flap chord ratio $c_f/c = 0.312$ is off the scale shown in Figure 5.35 so the value of k_1 must be extrapolated. For a double-slotted flap the k_1 curve is linear and is given by $k_1 = 0.04(c_f/c)$ where c_f/c is in percent. Therefore the value of $k_1 = 1.25$. Then k_2 is found, from Figure 5.36 for a double-slotted flap at $\delta_f = 40°$, to be $k_2 = 0.95$. To find k_3 the ratio of the actual flap angle selected to the reference flap angle, $\delta_f/\delta_{f,ref}$, for a double-slotted flap must be found. Following the curve for a double-slotted flap in Figure 5.36 up to $k_2 = 1$ locates the flap reference angle as $\delta_{f,ref} = 50°$. Then $\delta_f/\delta_{f,ref} = 0.8$ is entered in Figure 5.29 to find $k_3 = 0.87$. Using Equation (5.37) we find the increment in $c_{l,max}$ for the airfoil to be

$$\Delta c_{l,max} = k_1 k_2 k_3 \left(\Delta c_{l,max}\right)_{base} = (1.25)(0.95)(0.87)(1.165) = 1.20$$

This result is about 10% less than the experimental value reported by Cahill (1947).

We may also consider the trailing edge flap for a wing in the takeoff configuration where the flap deflection is much smaller than the maximum, $\delta_f = 15.6°$. The wing has an aspect ratio $A = 5.1$, a taper ratio $\lambda = 0.383$, and a quarter-chord sweep angle $\Lambda_{c/4} = 46°$. The wing uses an NACA 64-210 airfoil normal to the leading edge with a resulting thickness to chord ratio $t/c = 0.072$ in the streamwise direction. The wing has a single-slotted flap with a flap to airfoil-chord ratio $c_f/c = 0.258$ with a rated area $S_f/S = 0.378$ and the Reynolds number based on chord $Re_{,c} = 6.0 \times 10^6$. We first use Figure 5.34 to find $(\Delta c_{l,max})_{base} = 1.045$, then enter Figure 5.35 to find $k_1 = 1.010$. Figure 5.36 yields $k_2 = 0.605$ and we calculate (flap angle)/(reference flap angle) $= 15.6/45 = 0.347$. Figure 5.37 provides $k_3 = 0.445$. We may compute the section maximum lift coefficient from Equation (5.37) as follows:

$$(\Delta c_{l,max}) = k_1 k_2 k_3 (\Delta c_{l,max})_{base} = (1.010)(0.605)(0.445)(1.045) = 0.284$$

The sweepback correction $K_\Lambda = 0.730$ and therefore the total maximum lift coefficient is

$$\Delta C_{L,max} = \Delta c_{l,max}(S_{Wf}/S)K_\Lambda = (0.284)(0.378)(0.730) = 0.0784$$

This result is of course based on the reference wing planform area S and compares well with a laboratory test value of 0.075.

5.5.4 **Airfoil with leading edge slats or flaps**

A method has been developed for predicting the stall of comparatively thin airfoils with leading edge flaps or slats. Modern airliners generally are equipped with slats since they are relatively simple and provide lift augmentation with little drag penalty. A schematic illustration of the operation of a typical leading edge slat appears in Figure 5.38. The change in the lift curve produced by the addition of a leading edge slat is shown in Figure 5.39. Leading edge slat installations on a Boeing 737 and an Airbus A310 are shown in Figures 5.40 and 5.41, respectively. Leading edge flaps in use on commercial transport are Krueger flaps (see, for example Torenbeek, 1982) and the approach presented is not applicable to them. The DATCOM estimation method, which is presented here, is based on the assumption that the flapped and unflapped airfoils stall when the respective pressure distributions about the noses are the same. The method gives best results for slat deflections less than 20° and slat-chord-to-airfoil-chord ratios less than 0.20.

Leading edge flaps are of two types: a drooping leading edge, which is a simple flap hinged downstream of the leading edge, and an extensible leading edge that basically increases the chord. The former is complicated in application and, though well studied, has found application only on the furthest inboard station on the Airbus A380 and not on any other operating commercial jet transports. One type of leading edge flap is the Krueger flap whose simplicity makes it a commonly used leading edge high lift device. Several manifestations are schematically illustrated in Figure 5.42.

The feature differentiating leading edge flaps from slats is that the former have their trailing portions coincident with the airfoil surface so that there is no slot formed as there is in a leading edge slat system. These simple embodiments of a leading edge flap may be made more elaborate by having a position which produces a slot, thereby emulating a slat. This variation culminates in the highly developed

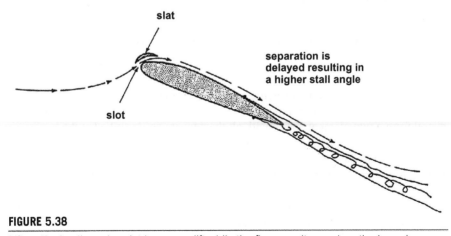

FIGURE 5.38

Lift on the leading edge slat increases lift while the flow over it energizes the boundary layer over the main airfoil delaying separation.

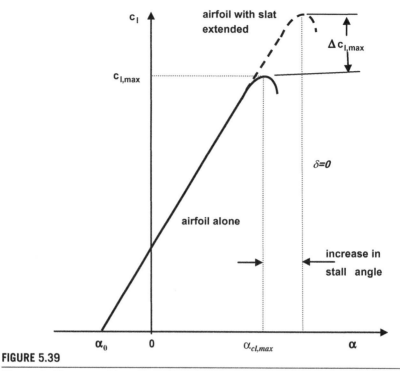

FIGURE 5.39

Change in lift curve achieved by using a leading edge slat.

variable camber Krueger (VCK) flap used by Boeing. In the VCK flap a linkage system permits the flap to be extended and cambered in various configurations in order to achieve optimal lift production in several flap settings.

5.5.5 DATCOM method for leading edge slats and flaps

Though leading edge flaps, that is, hinged nose sections on airfoils, have been the subject of much study they are not used on any commercial aircraft. The leading edge slat is the most common leading edge device in wide use. The leading edge slat is analogous to the slotted trailing edge flap in that the slat and the remainder of the airfoil form a two-element airfoil with all the advantages described at the beginning of this section. As explained previously, the leading edge slat increases the maximum lift coefficient of the airfoil along with the stall angle. As a result, they are commonly used, particularly in landing, though they are also useful on takeoff because the lift increment they develop comes with little drag penalty.

The DATCOM method for leading edge flaps and slats proposes that the maximum lift increment for leading edge flaps or slats may be approximated by the following empirical relation:

$$\Delta c_{l,\max} = \left(\frac{\partial c_l}{\partial \delta}\right)_{\max} \eta_{\max} \eta_\delta \delta_s \frac{c'}{c} \tag{5.38}$$

FIGURE 5.40

Leading edge slats on a Boeing 737 airplane as seen from below.

FIGURE 5.41

Leading edge slats and trailing edge flaps on Airbus A310-300. The fairings for the trailing edge flap drive mechanisms are in the deflected position.

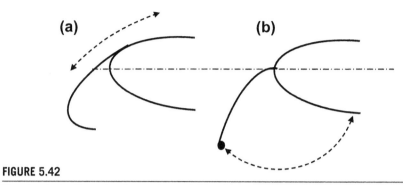

FIGURE 5.42

Two simple leading edge flap systems: (a) leading edge flap, (b) simple Krueger flap.

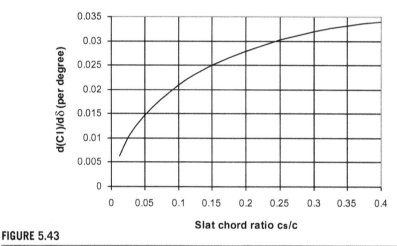

FIGURE 5.43

Rate of change of airfoil lift coefficient with slat deflection (per degree).

The first term in Equation (5.38) is the theoretical lift effectiveness which gives the rate of change of the lift coefficient with change in deflection angle; it is shown in Figure 5.43 as a function of the leading edge flap or slat chord to airfoil-chord ratio c_s/c. The second term in Equation (5.38), η_{max}, is an empirical factor which accounts for the effects of airfoil leading edge radius and maximum thickness. A graph of this factor is presented in Figure 5.44; the discontinuity in the curve for slats is said to be due to a lack of data in the region of the discontinuity. An *ad hoc* correction is proposed there which provides more accurate results as will be shown in the sample problems to follow.

The third term, η_δ, is another empirical factor which corrects for flap or slat deflections different from the optimum flap angle. This parameter is shown in Figure 5.45 as a function of the flap or slot deflection angle δ_s. The deflection angle is defined in Figure 5.46. To understand the angle δ_s one may first imagine drawing a chord line on the slat-airfoil combination when the slat is stowed.

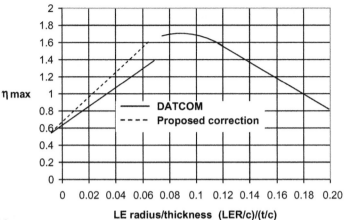

FIGURE 5.44

Correction factor for leading edge radius and airfoil thickness ratio.

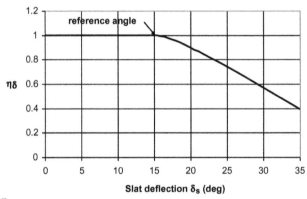

FIGURE 5.45

Slat deflection correction factor as a function of deflection angle.

Then when the slat is deflected, the segment of the chord line that was drawn on the slat in the stowed position has now rotated through the deflection angle δ_s. This is the standard used in the DATCOM method and is not necessarily used throughout the literature as a definition of slat deflection angle. The ratio c'/c accounts for the apparent increase in chord length when the slat is deflected and a slot is formed between the two airfoil elements; these dimensions are illustrated in Figure 5.46.

5.5.6 Sample calculations of airfoil with a leading edge slat

Consider a smooth NACA 64A010 airfoil operating at a Reynolds number $Re_c = 6 \times 10^6$. The NACA 64-series airfoil has a leading edge radius of 0.72% chord, according to Table 5.3. However, the letter A in the airfoil designation denotes a

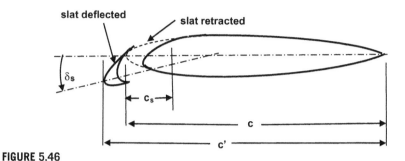

FIGURE 5.46

Geometry of the leading edge as used in the DATCOM method.

modified thickness distribution and experiments performed by Hayter and Kelly (1953) indicate a leading edge radius $LER=0.687\%$ chord. The airfoil is 10% thick, the slat deflection $\delta_s=25.6°$, the slat chord to airfoil-chord ratio is $c_s/c=0.17$, and the ratio $c'/c=1$. From Figure 5.43 we estimate $\left(\frac{\partial c_l}{\partial\delta}\right)_{max}=0.026$ and from Figure 5.44 we enter $(LER/c)/(t/c)=0.687\%/10\%=0.0687$ and find $\eta_{max}=1.32$. Entering Figure 5.45 with $\delta_s=25.6°$ yields $\eta_\delta=0.75$. Then

$$\Delta c_{l,max} = \left(\frac{\partial c_l}{\partial\delta}\right)_{max}\eta_{max}\eta_\delta\delta_s\frac{c'}{c} = \left(0.026\frac{1}{deg}\right)(1.32)\,(0.74)\,(25.6\ deg)\,(1.092)$$

$$= 0.71$$

This result obtained is conservative, about 15% less than the result of 0.84 reported by Hayter and Kelly (1953). However, using the proposed correction in Figure 5.44 leads to $\eta_{max}=1.59$ which changes the answer for the maximum lift increment to $\Delta c_{l,max}=0.855$ which is now 1.8% larger than the measured value.

Gottlieb (1949) studied the effect of leading edge slats on two smooth airfoils: a NACA 64_1-212 airfoil and a NACA 65A109 airfoil, both operating at a Reynolds number based on airfoil-chord $Re_c=6\times10^6$. The NACA 64-series airfoil has a leading edge radius of 0.72% chord, according to Table 5.3. The airfoil is 12% thick, the slat deflection $\delta_s=14.3°$, the slat chord to airfoil-chord ratio $c_s/c=0.14$, and the ratio $c'/c=1.1$. From Figure 5.43 we estimate $\left(\frac{\partial c_l}{\partial\delta}\right)_{max}=0.024$ per degree and from Figure 5.44 we enter $(LER/c)/(t/c)=0.72\%/12\%=0.06$ and find $\eta_{max}=1.2$. Entering Figure 5.45 with $\delta_s=14.3°$ yields $\eta_\delta=1$. Then

$$\Delta c_{l,max} = \left(\frac{\partial c_l}{\partial\delta}\right)_{max}\eta_{max}\eta_\delta\delta_s\frac{c'}{c} = \left(0.024\frac{1}{deg}\right)(1.2)\,(1.0)\,(14.3\ deg)\,(1.1)$$

$$= 0.45$$

This result obtained is quite conservative, 25% less than the result of 0.6 reported by Gottlieb (1949). However, using the proposed correction in Figure 5.44 we obtain $\eta_{max}=1.47$ and the new result for the lift increment is $\Delta c_{l,max}=0.55$ which is about 8% smaller than the reported value. The 65A109 airfoil is 9% thick, the slat deflection

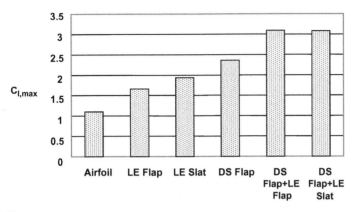

FIGURE 5.47

Maximum lift coefficients for an NACA 64A010 airfoil alone and with a leading edge flap (LE Flap), a leading edge slat (LE Slat), a double-slotted flap (DS Flap), a combination of double-slotted flap and leading edge flap (DS Flap + LE Flap), and a combination of double-slotted flap and leading edge slat (DS Flap + LE Slat).

$\delta_s = 24.3°$, the slat chord to airfoil-chord ratio $c_s/c = 0.14$, and the ratio $c'/c = 1.09$. From Figure 5.43 we estimate $\left(\frac{\partial c_l}{\partial \delta}\right)_{max} = 0.024$ per degree and from Figure 5.44 we enter $(LER/c)/(t/c) = 0.55\%/9\% = 0.061$ and find $\eta_{max} = 1.22$. Entering Figure 5.45 with $\delta_s = 24.3°$ yields $\eta_\delta = 0.76$. Then

$$\Delta c_{l,max} = \left(\frac{\partial c_l}{\partial \delta}\right)_{max} \eta_{max} \eta_\delta \delta_s \frac{c'}{c} = \left(0.024 \frac{1}{\deg}\right)(1.22)(0.76)(24.3 \deg)(1.09)$$
$$= 0.59$$

This result obtained is also conservative, about 15% less than the result of 0.69 reported by Gottlieb (1949). However, using the proposed correction in Figure 5.44 we obtain $\eta_{max} = 1.47$ and the new result for the lift increment is $\Delta c_{l,max} = 0.71$ which is about 3% larger than the reported value.

5.5.7 Combining leading and trailing edge devices on airfoils

Leading and trailing edge high lift devices have been shown to provide improved lifting performance over that possible with the airfoil alone. It follows that using them in combination should lead to better performance than that achievable with either leading edge or trailing edge devices alone. Two-dimensional experimental results for an NACA 64A010 airfoil section equipped with various combinations of a leading edge slat, leading edge flap, split flap, and double-slotted flap, were reported by Hayter and Kelly (1953). Wind tunnel experiments were performed at $M = 0.18$, and Re_c from 2 million to 7 million with and without leading and trailing edge devices. The maximum lift coefficients obtained for these tests are shown in Figure 5.47.

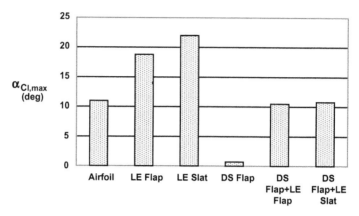

FIGURE 5.48

Angle of attack for the maximum lift coefficients for an NACA 64A010 airfoil alone and with a leading edge flap (LE Flap), a leading edge slat (LE Slat), a double-slotted flap (DS Flap), a combination of double-slotted flap and leading edge flap (DS Flap + LE Flap), and a combination of double-slotted flap and leading edge slat (DS Flap + LE Slat).

The effect of the high lift devices on the angle of attack for maximum lift coefficient was also presented by Hayter and Kelly (1953) and is illustrated in Figure 5.48. When used alone, the leading edge devices dramatically increase the angle of attack capability of the airfoil. However, the angle of attack for maximum lift coefficient when the double-slotted flap is used alone is equally dramatically decreased from that

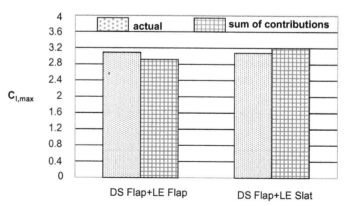

FIGURE 5.49

Measured maximum lift coefficients for an NACA 64A010 airfoil with a combination of double-slotted flap and leading edge flap (DS Flap + LE Flap) and a combination of double-slotted flap and leading edge slat (DS Flap + LE Slat) showing measured values as well as those obtained by summing the contributions of each component.

for the airfoil alone. The use of either leading edge device in conjunction with the double-slotted flaps nearly restores the original angle of attack capability of the airfoil.

Using the results presented in Figure 5.47 we may show that the maximum lift coefficient for the combination of trailing and leading edge devices is approximately equal to the sum of the airfoil contribution, the leading edge device increment, and the trailing edge flap increment. In other words, the contribution of the high lift devices was additive. This result for an airfoil is shown in Figure 5.49. It must be noted that the results reported were for optimal positions and configuration of the high lift devices.

5.6 Determination of $C_{L,max}$ for the wing in takeoff and landing configurations

The DATCOM method for estimating the increment of ΔC_{Lmax} for the wing with flaps and slats deflected is used here. These devices have different drag penalties associated with their operation and each will be addressed in Chapter 7. A compilation of leading and trailing edge devices currently in use is presented in Table 5.6(b). The information included is drawn largely from Rudolph (1996) and manufacturers' literature, among other sources.

5.6.1 Application of high lift devices on wings

We have discussed the determination of the lift of a three-dimensional wing in Section 5.4. The discussion of leading and trailing edge high lift devices has thus far been confined to airfoils, that is, two-dimensional wings. However, Furlong and McHugh (1953) point out and Hoak et al. (1978) repeat that the lift increments due to leading and trailing edge devices are not, in general, additive. When both are deployed the resulting lift increment is not necessarily the sum of the lift increments produced by each when working alone. To appreciate this fact we turn to the early experimental work compiled by Furlong and McHugh (1953) which illustrates some of the features of leading and trailing edge high lift devices used independently and in concert on three-dimensional swept wings.

One wing considered by Furlong and McHugh (1953) is reasonably typical of commercial jet transports, having an aspect ratio $A=6$, quarter-chord sweep $\Lambda_{c/4}=35°$, and taper ratio $\lambda=0.5$. A sketch of the wing is shown in Figure 5.50. Wind tunnel experiments were performed at $M=0.18$, and $Re_c=9.35 \times 10^6$ with and without leading and trailing edge devices. The wing was constructed using the NACA 64$_1$-212 airfoil aligned with the free stream direction.

We compare the performance of the airfoil to that of each high lift device alone and then to that of a combination of both leading and trailing edge devices for the case in which high lift devices have a spanwise extent of 50%. The leading edge devices include a flap or a slat, while the trailing edge device is a double-slotted flap. The maximum lift coefficient achieved by the wing of Figure 5.50 using various

Table 5.6(b) High Lift Systems on Commercial Jet Transports

Airplane	Trailing Edge Devices	Leading Edge Devices
B707	Fixed vane/main double-slotted	Simple Krueger
B727	Triple-slotted	3-position slats, Inboard FBNK
B737	Triple-slotted	3-position slats, Inboard FBNK
B747	Triple-slotted	VC Krueger
B757	Main/aft double-slotted	3-position slats
B767	Main/aft double-slotted inboard, Single-slotted outboard Drooped inboard aileron	3-position slats
B777	Main/aft double-slotted inboard Single-slotted outboard Drooped and slotted inboard aileron	3-position slats,
B787-8	Single-slotted in- and outboard Flaperon with droop function	3 position slats, Inboard S/S Krueger
DC-8	Fixed vane/main double-slotted	3-position slats
DC-9-10	Fixed vane/main double-slotted	none
DC-9-30/50	Fixed vane/main double-slotted	3-position slats
DC-10	Articulating vane/main double-slotted	3-position slats
MD-80	Fixed vane/main double-slotted	3-position slats
MD-11	Articulating vane/main double-slotted Drooped inboard aileron	3-position slats
A300B	Main/aft double-slotted	3-position slats, Inboard FBNK
A300-600	Single-slotted Drooped inboard aileron	3-position slats, Inboard FBNK
A310	Articulating vane/main inboard, Single-slotted outboard Drooped inboard aileron	3-position slats

Table 5.6(b) Continued

Airplane	Trailing Edge Devices	Leading Edge Devices
A320	Single-slotted	3-position slats
A321	Main/aft double-slotted	3-position slats
A330	Single-slotted Drooped outboard aileron	3-position slats
A340	Single-slotted Drooped outboard aileron	3-position slats
A380	Trailing edge flaps	Inboard slat and drooped nose Outboard short chord slats Fuselage side fixed leading edge

Notes: FBNK = folding, bull-nose Krueger, S/S Krueger = slat and seal Krueger.

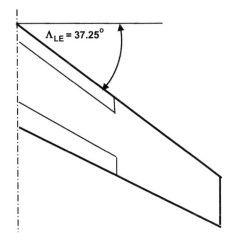

$\Lambda_{LE} = 37.25°$

FIGURE 5.50

Planform of a wing with an NACA 64_1-212 airfoil, $A=6$, $\Lambda_{LE}=37.25°$, and $\lambda=0.5$ showing leading and trailing edge high lift devices.

combinations of high lift devices is shown in Figure 5.51. It is seen that the two leading edge devices give about the same increment in maximum lift coefficient over that produced by the airfoil alone—about a 10% improvement. The double-slotted flap alone provides a 50% improvement over the airfoil alone. However, when either of the leading edge devices is used in conjunction with the double-slotted flap there is essentially no change, or even a slight decrease in maximum lift coefficient. At the same time, it is also shown in Figure 5.51 that if the spanwise extent of the leading edge flap is increased to 65% the use of both the leading edge flap and the (50% span)

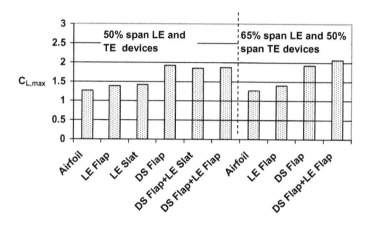

FIGURE 5.51

Maximum lift coefficients for the wing of Figure 5.38 with different combinations of high lift devices. The notation is as follows: NACA 64_1-212 airfoil alone and with a leading edge flap (LE Flap), a leading edge slat (LE Slat), a double-slotted flap (DS Flap), a combination of double-slotted flap and leading edge flap (DS Flap + LE Flap), and a combination of double-slotted flap and leading edge slat (DS Flap + LE Slat).

double-slotted flaps leads to an improvement over that with the flap alone. The extent to which the notion of the effects of each high lift coefficient contribution being additive is illustrated in Figure 5.52.

Here we see that the increase in spanwise extent of the leading edge flap to 65% from 50% resulted in a situation where the sum of the individual contributions of the high lift devices gives a good approximation to the actual value for the components in combination.

It must be noted that the results discussed in some detail here are intended to demonstrate that the effects of leading and trailing edges on wing performance are not necessarily additive. However, the widespread use of these devices in commercial jet transports indicates that better combined performance can be achieved than this set of wind tunnel experiments might suggest. To achieve synergetic performance between the leading and trailing edge devices careful attention must be paid to the integration of these elements, as explained by Smith (1975). Reasons beyond lift performance alone help explain the widespread use of combined leading and trailing edge high lift devices. An illustration of the implementation of elaborate high lift systems in practice is shown in Figure 5.53. There a Boeing 747-400 is depicted in the landing configuration. The triple-slotted trailing edge flaps and the variable camber Krueger flaps on the leading edges are clearly visible. The inboard Krueger flaps seal to the upper surface while the outboard ones have a clearly visible slot.

Leading edge devices may be safely deployed at higher speeds than can the trailing edge flaps. Thus, in operations during the final stages of descent, where flight

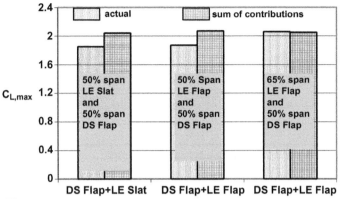

FIGURE 5.52

Comparison of measured $C_{L,max}$ for the wing of Figure 5.38 with different combinations of high lift devices with the result of simply summing the individual contributions for each component. Notation is as given in Figure 5.39.

speeds are much lower than in cruise but substantially higher than in landing, leading edge flaps or slats may be extended to improve the lifting capability of the wing without incurring very large drag penalties. Besides a lift coefficient increase, using the leading edge flap or slot alone also produces a smaller pitch-up moment than

FIGURE 5.53

A Boeing 747-400 in the landing configuration showing the triple-slotted trailing edge flaps and the variable camber Krueger flaps on the leading edges. The inboard Krueger flaps seal to the upper surface while the outboard ones have a clearly visible slot and therefore act as slats.

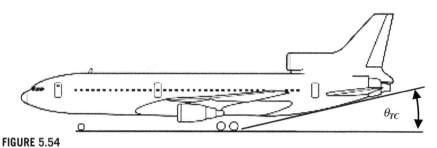

FIGURE 5.54

Maximum angle of rotation θ_{TC} before a tail strike is incurred.

does a trailing edge flap alone. This improves the handling qualities and reduces trim requirements. It is clear from Figure 5.48 that the use of a leading edge device alone increases the angle of attack for maximum lift coefficient while a trailing edge flap reduces it. Properly combining the use of leading and trailing edge high lift devices provides the benefit of increased lift at higher angles of attack than can be achieved using just the trailing edge flap alone. This is important in avoiding a tail strike, a situation where there is a rotation of the aircraft beyond the tail cone flare angle θ_{TC}, as shown in Figure 5.54. The higher maximum lift coefficients possible with a combination of leading and trailing edge devices generally result in reduced pitching-moment variation with angle of attack. This essentially neutral longitudinal motion response of the wing during terminal operations improves handling properties and reduces trim requirements. The fact that leading edge systems are relatively simple and easy to maintain means that the incremental cost of providing them is usually low enough to be attractive.

5.6.2 Determination of $\Delta C_{L,\max}$ for the wing due to flaps

The increment in maximum lift coefficient for the wing due to the trailing edge flap deflection is given in the DATCOM method by the following equation:

$$\Delta C_{L,\max,f} = \Delta c_{l,\max} \frac{S_{w,f}}{S} K_\Lambda \tag{5.39}$$

Here $\Delta c_{l,\max}$ is the increment in lift coefficient due to flap deflection as defined in Equation (5.37). The quantity $S_{w,f}/S$ is the ratio of wing area affected by the trailing edge flap deflection (including both port and starboard wings) to the total wing area, as shown in Figure 5.55. The wing area affected by the flap may be written as

$$S_{w,f} = \left(\frac{b}{2}\right) c_r \left[2 - (1 - \lambda)(\eta_i - \eta_o)\right](\eta_i - \eta_o) \tag{5.40}$$

In Equation (5.39) the quantity $\lambda = c_t / c_r$ is the taper ratio, i.e., the ratio of the tip chord to the root chord, and b is the wingspan. The non-dimensional inboard

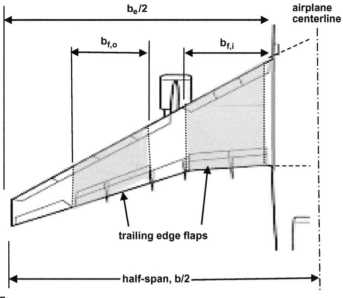

FIGURE 5.55

Plan view of wing where shaded area, including both the starboard and port wing, is the rated area for the flaps, that is, the portion of the wing planform area affected by the flaps.

location of a flap is $\eta_i = y_i/(b/2)$, while the outboard location is $\eta_o = y_o/(b/2)$. If the flaps do not extend continuously along the trailing edge, as is the case in Figure 5.55, then the affected area for each may be calculated independently and added together. The wing planform area S is given by Equations (5.5) and (5.12) for straight-tapered wings and cranked wings, respectively. The correction factor for sweepback is

$$K_\Lambda = \left(1 - 0.08\cos^2 \Lambda_{c/4}\right)\cos^{3/4} \Lambda_{c/4} \tag{5.41}$$

It should be noted that the flap deflection angles and all dimensions are measured in planes parallel or perpendicular to the plane of symmetry. Then the lift coefficient for the wing with flaps deflected is

$$C_{L,\max,w} = C_{L,\max} + \Delta C_{L,\max,f} \tag{5.42}$$

5.6.3 Determination of $\Delta C_{L,\max}$ for the wing due to slats

For preliminary design purposes the contribution to the maximum lift coefficient of the wing due to leading edge slat deflection, we assume an approach based on that used in estimating the contribution of the trailing edge flap and on the approach presented previously in Section 5.5.5. Then, in analogy with Equation (5.39), the airfoil results are extended to the wing as given by

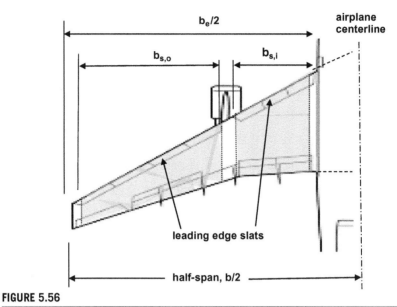

FIGURE 5.56

Plan view of wing where shaded area, including both the starboard and port wing, is the rated area for the slats, that is, the portion of the wing planform area affected by the slats.

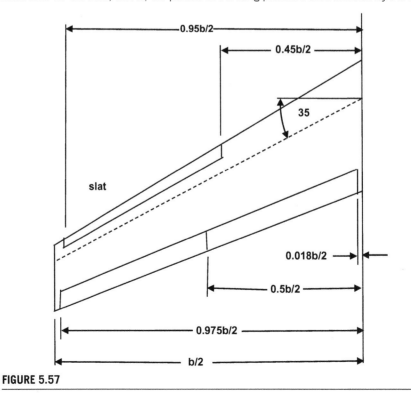

FIGURE 5.57

Wing with trailing edge flaps and leading edge slats.

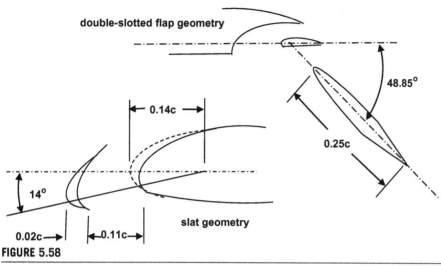

double-slotted flap geometry

48.85°

0.14c

0.25c

14°

slat geometry

0.02c

0.11c

FIGURE 5.58

Details of high lift devices for the wing in Figure 5.46.

$$\Delta C_{L,\max,s} = \Delta c_{l,\max,s} \frac{S_{w,s}}{S} K_\Lambda \tag{5.43}$$

The definitions of Equations (5.39), (5.40) and (5.41) hold, where now $S_{w,s}$ is interpreted in terms of the wing area affected by the operation of the slat or flap, as depicted in Figure 5.56. Then the maximum lift coefficient of the wing with a leading edge slat deflection is

$$C_{L,\max,w} = C_{L,\max} + \Delta C_{L,\max,s} \tag{5.44}$$

Because the careful integration of leading and trailing edge devices can lead to increments in maximum lift coefficient greater than either one alone we will assume that the estimates of Equations (5.39) and (5.43) may be added so that the combined lift coefficient for the wing is

$$C_{L,\max,w} = C_{L,\max} + \Delta C_{L,\max,f} + \Delta C_{L\max,s} \tag{5.45}$$

At least two values of the total maximum lift coefficient of the wing $C_{L,\max,w}$ should be calculated; one for takeoff and the other for landing.

5.6.4 Sample calculation for flaps and slats

Consider the wing investigated by Koven and Graham (1948) and illustrated in Figures 5.57 and 5.58. The wing has an aspect ratio $A=6$, a taper ratio $\lambda=0.5$, and a leading edge sweep angle $\Lambda_{c/4}=37.25°$. The wing uses an NACA 64$_1$-212 airfoil normal to the free stream with a thickness to chord ratio $t/c=0.098$ in the streamwise direction. The wing has a double-slotted flap with a flap to airfoil-chord ratio $c_f/c=0.25$ and the rated area may be calculated to be $S_{w,f}/S=0.951$ for the full-span flap and 0.573 for the half-span flap. The Reynolds number based on chord $Re_{,c} = 6.0 \times 10^6$ and the full flap deflection is $\delta_f = 48.85°$.

We first use Figure 5.34 to find $(\Delta c_{l,max})_{base} = 1.31$, then enter Figure 5.35 to find $k_1 = 1$. Figure 5.36 yields $k_2 \sim 1$ and for the flap angle of 48.85° Figure 5.37 provides $k_3 = 1$. We may compute the section maximum lift coefficient from Equation (5.37) as follows:

$$(\Delta c_{l,max}) = k_1 k_2 k_3 (\Delta c_{l,max})_{base} = (1)(1)(1)(1.31) = 1.31$$

The quarter-chord sweepback angle is $\Lambda_{c/4} = 35.17°$ and the sweepback correction $K_\Lambda = 0.815$ and therefore the total maximum lift coefficient is

$$\Delta C_{L,max} = \Delta c_{l,max}(S_{w,f}/S)K_\Lambda = (1.31)(0.951)(0.815) = 1.015$$

The wing maximum lift coefficient with no high lift devices deployed is 1.27 so that with the full-span flaps employed the maximum lift coefficient becomes

$$C_{L,max} = 1.27 + 1.015 = 2.28$$

The experimental value with full-span flaps is $C_{L,max} = 2.32$ so the error is -1.5%. For the case of half-span flaps the increment in maximum lift coefficient is

$$\Delta C_{L,max} = \Delta c_{l,max}(S_{w,f}/S)K_\Lambda = (1.31)(0.573)(0.815) = 0.612$$

The wing maximum lift coefficient with no high lift devices is 1.27 so that with the half-span flaps deployed the maximum lift coefficient becomes

$$C_{L,max} = 1.27 + 0.612 = 1.88$$

The experimental value with full-span flaps is $C_{L,max} = 1.92$ so the error is -2%.

The wing also has slats and the rated area may be calculated to be $S_{w,s}/S = 0.42$.

The nose flap to chord ratio is 0.14 and from Figure 5.43 we find $\left(\frac{\partial c_l}{\partial \delta}\right)_{max} = 0.024$.

From Table 5.3 the leading edge radius $LER = 1.04\%$ chord so that the ratio $(LER/c)/(t/c)$ for this airfoil is 0.087 and from the curve in Figure 5.36 $\eta_{max} = 1.7$. For 14° deflection Figure 5.37 suggests $\eta_\delta = 1$ and from Figure 5.47 we calculate $c'/c = 1.11$. Using these values in Equation (5.38) yields

$$\Delta c_{l,max} = \left(\frac{\partial c_l}{\partial \delta}\right)_{max} \eta_{max} \eta_\delta \delta_s \frac{c'}{c} = (0.024)\,(1.7)\,(1)\,(14)\,(1.11) = 0.634$$

Then the increment in maximum lift for the wing with slats deployed is

$$\Delta C_{L,max,s} = \Delta c_{l,max,s} \frac{S_{w,s}}{S} K_\Lambda = (0.634)\,(0.42)\,(0.815) = 0.217$$

The wing maximum lift coefficient with no high lift devices is 1.27 so that with the half-span slats deployed the maximum lift coefficient becomes

$$C_{L,max} = 1.27 + 0.217 = 1.49$$

The experimental value with half-span slats is $C_{L,max} = 1.42$ so the error is $+4.9\%$.

Although the estimates presented are quite accurate for the flaps alone or the slats alone, the same is not true if the increments for half-span flaps and half-span slats are simply added. In that case the maximum lift coefficient for the wing would be

$$C_{L,max} = C_L + \Delta C_{L,max,flaps} + \Delta C_{L,max,slats} = 1.27 + 0.612 + 0.217 = 2.10$$

The experimental value for this case is reported as $C_{L,max} = 1.85$, an error of $+13.5\%$. Indeed, the experimental value for the maximum lift coefficient with only the half-span flaps deployed is better than that observed with both flaps and slats deployed. The slats had little effect on maximum lift coefficient or maximum angle of attack when used in combination with the flaps. It was also observed, as expected, that the use of the slats alone increased the maximum angle of attack from $19°$ for the airfoil alone to $26°$, while the use of the half-span flaps decreased the maximum angle of attack to $14.3°$. However, when both are deployed the maximum angle of attack remains the same as with the flaps alone. This provides another demonstration that the incorporation of both leading and trailing edge high lift devices on a wing must be carried out carefully because the effects of each are not necessarily additive.

5.7 Development and layout of the preliminary wing design

The weight of the aircraft at all stages of the mission, as well as the selected cruise Mach number and altitude, is known from the results in Chapter 2. The fuselage dimensions, including the length l_f, the diameter d_f, and the tail cone flare angle θ_{TC}, were determined in Chapter 3. The required wing area S, along with the estimated values of lift coefficient in takeoff and landing is known from the results of Chapter 4. Using this information and the methods presented in the current chapter a preliminary layout of the wing design may be carried out.

5.7.1 Aspect ratio

The jet transport database suggests that there is no reliable simple correlation between the aspect ratio and previously calculated characteristics of the proposed aircraft. For example, consider the relationship between the aspect ratio A and the takeoff weight W_{to} for 49 commercial turbofan and turboprop transport aircraft as shown in Figure 5.59. There is no clear correlation, but it is apparent that turboprops with unswept wings employ the highest values of A. Turbofan-powered swept wing aircraft with fewer than 200 passengers ($N_p < 200$) show a clear upward trend from $A = 8$ to $A = 10$, while turbofans with $N_p > 200$ cover a much broader range of the design space, but with a somewhat downward trend as the takeoff weight grows.

As another example, we may consider the variation of aspect ratio as a function of a combined variable which is also known for the design aircraft, the takeoff

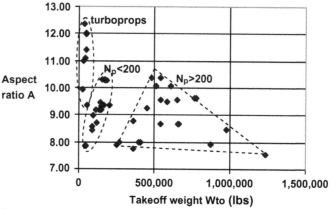

FIGURE 5.59

Variation of the aspect ratio with takeoff weight for 49 commercial turbofan and turboprop transport aircraft with different passenger loads N_p.

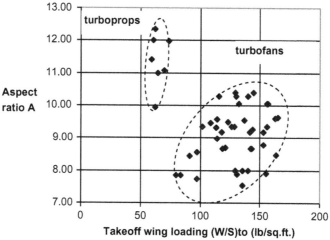

FIGURE 5.60

Variation of the aspect ratio with takeoff wing loading for 49 commercial turbofan and turboprop transport aircraft.

wing loading $(W/S)_{to}$, as shown in Figure 5.60. Again, there are no clear-cut trends although the turboprops and turbofans are more clearly separated. The turbofan-powered aircraft show an upward trend in aspect ratio as the wing loading increases, but the range $7.5 < A < 10.5$ is covered in the wing loading range $100 \, \mathrm{lb/ft^2} < (W/S)_{to} < 150 \, \mathrm{lb/ft^2}$. Thus, the choice of aspect ratio is well bounded, but without a straightforward means of choosing a value. However, the span loading $(W/b^2)_{to}$ of the aircraft

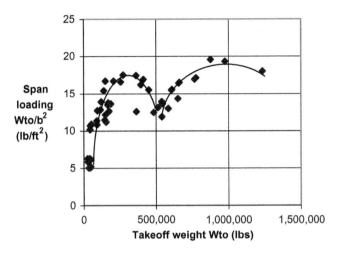

FIGURE 5.61

Span loading as a function of takeoff weight for 49 commercial turbofan and turboprop transport aircraft. Faired curve is added to aid in depicting trends.

in the database, which is shown as a function of takeoff weight in Figure 5.61, shows a rather unusual trend. This parameter is related to induced drag and one may use this figure to determine a reasonable span loading from which an estimate may be made of the span of the design aircraft. From that result and the known wing area a provisional aspect ratio $A = b^2/S$ may be determined. The result of this approach for selecting an initial value of aspect ratio should be weighed in light of the market survey aircraft values and modified accordingly, if necessary.

5.7.2 Taper ratio and the root chord

Assuming a straight leading edge, Equation (5.5) may be used to determine the root chord c_r once a taper ratio λ is selected. De Young and Harper (1948) provide a theoretical estimate for the variation of the taper ratio as a function of quarter-chord sweep angle for wings with an approximately elliptic span loading independent of aspect ratio, and this is shown as a curve in Figure 5.62. Also shown on that plot is the region occupied by current jet transport configurations. As one moves to the right of the curve in Figure 5.62 the loading moves outboard as the aspect ratio increases.

The taper ratio may be selected using the market survey values as a guide, which will generally show $0.25 < \lambda < 0.35$. Then the tip chord c_t is determined from the definition of the taper ratio, $\lambda = c_t/c_r$. The wing sweepback angle remains to be selected and this may be varied while keeping constant all other parameters chosen thus far. This is shown in Figure 5.63 for a wing with aspect ratio $A = 7$ and taper ratio $\lambda = 0.33$ but two different leading edge sweep angles, $25°$ and $35°$, which are typical for jet transports. It should be clear from Figure 5.56 that the larger the sweepback, the

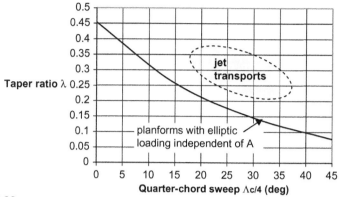

FIGURE 5.62

Variation of the taper ratio with quarter-chord sweep angle for wings with an approximately elliptic span loading. The shaded region contains values typical of current jet transports.

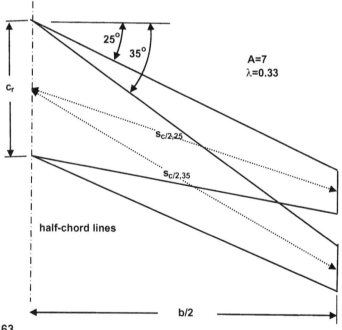

FIGURE 5.63

The length of the locus of half-chord points along a wing of constant A and λ for different leading edge sweepback angles.

larger the cantilever length of the 50% chord line $s_{c/2,\Lambda}=b(2\cos\Lambda_{c/2})$. This suggests that wing root bending moment that must be resisted by the structure will also grow with increasing sweepback.

Here the thickness of the root chord t_r resists the bending moment caused by the lift distribution of the wing. Consider the simple case of the bending stress in a cantilever beam of length $s_{c/2} = b/(2\cos\Lambda c_{/2})$ and hollow rectangular cross-section with width c_r, depth t_r, and uniform thickness t. If the half-span wing lift acts at the spanwise center of pressure location $y_{c.p.}$, and equals half the aircraft weight W, then the bending stress is given by

$$\sigma_b = \frac{\left(\frac{1}{2}L\frac{y_{c.p.}}{\cos\Lambda_{c/2}}\right)t_r}{I} \approx \frac{\frac{W}{4}\left(\frac{b}{\cos\Lambda_{c/2}}\right)\eta_{c.p.}t_r}{\frac{1}{2}c_r t_r^2 t\left[1+\frac{1}{3}\left(\frac{t}{c}\right)\right]} \approx \frac{W\eta_{c.p.}}{2c_r t_r\left(\frac{t}{t_r}\right)}\left(\frac{b}{t_r\cos\Lambda_{c/2}}\right)$$

(5.46)

We may normalize the bending stress by the ratio of supported weight to approximate structural cross-sectional area as follows:

$$\frac{\sigma_b}{\left(\frac{W}{t_r c_r}\right)} = \frac{1}{\left(\frac{t}{t_r}\right)}\frac{1}{2}\left(\frac{b}{t_r\cos\Lambda_{c/2}}\right)$$

(5.47)

Thus we see that $b/(2t_r\cos\Lambda_{c/2})$ logically appears as a basic structural parameter for the wing design. Indeed, Torenbeek (1982) shows data indicating that for jet transports, over a wide range of weights, the parameter

$$\frac{1}{2}\frac{b}{t_r\cos\Lambda_{c/2}} = \frac{1}{2}\frac{\left(\frac{b}{c_r}\right)}{\left(\frac{t}{c}\right)_r\cos\Lambda_{c/2}} = \frac{A(1+\lambda)}{4\left(\frac{t}{c}\right)_r\cos\Lambda_{c/2}} \approx 20\pm2$$

(5.48)

Now, with a good estimate for the root thickness ratio we can identify an equally reasonable value for the wing sweepback. We realize that for a given value of A and λ, a lower sweepback permits a thinner wing and this would save wing weight. On the other hand, a greater sweepback reduces the component of Mach number normal to the leading edge, $M_n = M\cos\Lambda_{LE}$, and this reduces compressibility drag. Striking an appropriate balance between conflicting requirements such as this is typical of the tradeoffs surfacing in all aspects of airplane design.

5.7.3 Wing relative thickness and airfoil selection

With the planform of the wing determined it is necessary to select an airfoil with appropriate shape and thickness appropriate to the cruise Mach number M_{cr}, cruise altitude z_{cr}, and sweep angle $\Lambda_{c/4}$ that have already been selected. Scholz and Ciornei (2005) reviewed and evaluated a dozen theoretical and empirical approaches for estimating the thickness to chord ratio of a commercial jet transport wing at high subsonic speeds in terms of M, $\Lambda_{c/4}$, C_L, and airfoil type. They compared the results obtained for all approaches with the measured relative thickness for 29 jet transport wings using the following definition for relative thickness suggested by Jenkinson, Simpkin, and Rhodes (1999):

$$\left(\frac{t}{c}\right)_{rel} = \frac{1}{4}\left[\left(\frac{t}{c}\right)_r + 3\left(\frac{t}{c}\right)_t\right] \tag{5.49}$$

Scholz and Ciornei (2005) found that the two most accurate results were obtained by regression analysis and by the theoretical approach proposed by Torenbeek (1982). Inasmuch as the regression equation is more readily implemented it is used here in the form

$$\left(\frac{t}{c}\right)_{rel} = 0.127\frac{k_a^{0.556} C_L^{0.065} \left(\cos \Lambda_{c/4}\right)^{0.573}}{M_{DD}^{0.204}} \tag{5.50}$$

Here $k_a = 0.921$ for conventional airfoils, 0.928 for peaky airfoils, 0.932 for modern supercritical airfoils, and 1.012 for older supercritical airfoils. The quantity M_{DD} is the drag divergence Mach number which is defined several ways in practice, but represents the Mach number at which the wing drag increases rapidly. For example, M_{DD} is often chosen to be the Mach number at which the drag coefficient increases a specific amount or where the rate of change of drag coefficient dC_D/dM reaches a certain value. For preliminary design purposes we may take $M_{DD} = M_{cr}$. The lift coefficient in cruise follows from aircraft weight, wing area, and altitude according to the usual lift relation

$$C_{L,cr} = \left(\frac{W}{qS}\right)_{cr} = \frac{2}{\gamma\, p_{SL}\delta\, M_{cr}^2}\left(\frac{W}{S}\right)_{cr} \tag{5.51}$$

The weight of the aircraft decreases as fuel is used during the cruise so taking its value at the start of cruise, that is, W_4 from the results of Chapter 2, is the conservative choice. With these assumptions we may write the relative thickness as follows:

$$\left(\frac{t}{c}\right)_{rel} = 0.127\frac{k_a^{0.556} \left(\cos \Lambda_{c/4}\right)^{0.573}}{M_{cr}^{0.334} \left(0.7 p_{SL}\delta\right)^{0.065}}\left(\frac{W}{S}\right)_4^{0.065} \tag{5.52}$$

Consider the case of an aircraft having a conventional NACA 6-series airfoil and a wing loading of $(W/S)_4 = 125\,\text{lb/ft}^2$ (6 kPa) flying at an altitude of $z = 36{,}000\,\text{ft}$ (11 km). The relative wing thickness is shown as a function of Mach number and quarter-chord sweepback angle in Figure 5.56. We see that the relative thickness for a jet transport with typical value of quarter-chord sweep lies between 11% and 12% for Mach numbers between 0.75 and 0.85.

The definition for relative thickness in Equation (5.49) biases the results to the tip thickness ratio so as to permit a thicker root section. Note that if an airfoil with a given value of t/c, say 12%, is used throughout the wing, the actual thickness of the wing will decrease toward the tip for a taper ratio different from unity, but the thickness ratio remains constant. Using a single airfoil throughout in the preliminary

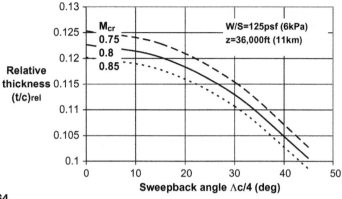

FIGURE 5.64

Relative thickness of a conventional airfoil as a function of quarter-chord sweepback for various Mach numbers at a wing loading of 125 lb/ft² and an altitude of 36,000 ft.

design phase simplifies the process; more complex configurations are best left to refinements of the design study. Therefore it is suggested that the relative thickness design approach leading to results like those in Figure 5.64 be employed and that a single airfoil type be chosen. A conventional airfoil that has a high critical Mach number and good low-speed characteristics may be found among the NACA 64-series airfoils. The details of the drag performance of various airfoils will be discussed in Chapter 9.

5.7.4 Wing-mounted engines and a cranked trailing edge

If a cranked trailing edge at $y = y_b$ is desired, like that shown in Figure 5.3, then Equation (5.12) is used to determine the root chord. At this stage of the design, the market survey aircraft data should be used to help choose the location of the break in the trailing edge, y_b. Typically the kink is placed just inboard of the engine location for wing-mounted engines. An estimate of the location of the engine may be developed from controllability considerations in emergency situations where one engine is shut down. We consider the specific case of a twin-engine aircraft; the approach may be readily extended to a four-engine airplane. The imbalance of thrust due to the shutdown of an engine on one side of the aircraft will induce a yawing moment N which must be countered by sufficient control authority from the rudder on the vertical tail. The restoring moment generated by the vertical tail is the product of the lift (side) force L_v produced by the vertical tail acting through the moment arm l_v measured from the vertical tail aerodynamic center to the aircraft center of gravity as follows:

$$L_v l_v = C_{L,v} q S_v l_v \qquad (5.53)$$

The yawing moment N may be estimated by noting that the available thrust F_{avail} from the operating engine located at a spanwise station y_e must be equal to the drag of the aircraft D at a new speed commanded by the pilot as follows:

$$N = C_D q S y_e \qquad (5.54)$$

Thus, for equilibrium flight, the yawing moment caused by thrust loss on one side of the aircraft must be balanced by a restoring moment provided by the vertical tail. Then, equating Equations (5.53) and (5.54) yields

$$\frac{y_e}{c_{MAC}} = \frac{C_{L,v}}{C_D} \frac{S_v l_v}{S c_{MAC}} = \frac{C_{L,v}}{C_D} V_v \qquad (5.55)$$

Equation (5.55) introduces the volumetric coefficient of the vertical tail V_v which is discussed in more detail in the following chapter on tail design. We may rewrite this equation to show that an estimate for the spanwise location of an engine, in terms of the mean aerodynamic chord of the wing, is

$$\frac{y_e}{c_{MAC}} \approx \frac{C_{L,v}}{C_D} V_v = \frac{C_{L,v}}{C_L} \left(\frac{L}{D}\right) V_v \qquad (5.56)$$

The volume coefficient for the vertical tail of jet transports lies in the range $0.6 < V_v < 0.8$, while the lift to drag ratio with one engine shut down will be about half the cruise value, that is, in the range $6 < (L/D) < 9$. Because the lift coefficients of the tail and the airplane will be roughly proportional to their aspect ratios, a reasonable spanwise engine location is about $1.5c_{MAC}$. A more accurate result is developed in Chapter 6 and the engine location on the market survey aircraft may also be used to provide guidance at this stage in the design.

Engines mounted on the aft fuselage have much smaller yawing moments when an engine must be shut down because the effective moment arm for the thrust force is much smaller than for wing-mounted engines. This is an attractive safety feature of fuselage-mounted engines. In addition, because of their naturally greater ground clearance they better avoid foreign object damage due to ingestion by the engine inlet. However, the maintenance, repair, and overhaul expense for fuselage-mounted engines militates against their adoption, except for smaller jet transports, like regional jets, where there is insufficient ground clearance for wing-mounted turbofan engines. Interestingly, the Honda Jet has engines mounted above the wing to minimize ground clearance problems.

5.7.5 Placement of high lift devices

The location, type, chord size, and spanwise extent of the high lift devices is largely determined by meeting the lift coefficient requirements for takeoff and landing, as calculated in Chapter 4. Though the leading edge can support slats or flaps along

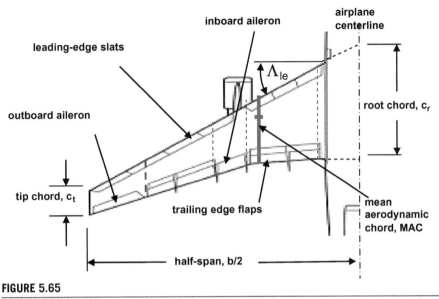

FIGURE 5.65

Typical wing layout showing high lift and control surfaces.

Table 5.7 Typical Flap Settings for Different Configurations		
Flight Configuration	**Flap Deflection δ_{LE} Leading Edge (deg)**	**Flap Deflection δ_{TE} Trailing Edge (deg)**
Takeoff	0–20	10–25
Landing	10–20	30–50
Cruise	0	0

most of the exposed span, there are constraints at engine pylon junctions for wing-mounted engines and at the wing root junctions. Trailing edge flaps have to share trailing edge space with low-speed (outboard) and possibly high-speed (inboard) ailerons as well as being clear of hot jet exhaust from wing-mounted engines. It is suggested that reasonable values of the trailing edge and leading edge flap layouts be chosen based on Figure 5.65 and on the market survey aircraft. For typical deflections the guidelines given in Table 5.7 should be considered.

5.7.6 Wingtip treatments

As described in the discussion of flow over finite wings in Appendix C, the trailing vortex sheet induces a swirling component into the flow field in the vicinity of the wing. At the same time it was demonstrated that the distribution of lift across the span influences the induced drag arising as a result of the trailing vortex sheet. The

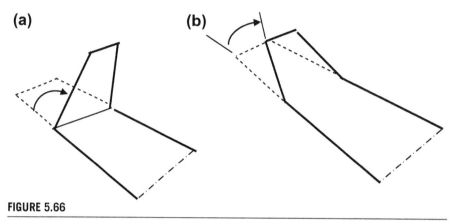

FIGURE 5.66

Two popular wingtip treatments: (a) winglet and (b) raked tip.

downwash at the lifting line tilts the velocity vector down by an amount equal to the induced angle of attack, $\tan^{-1}(w/V)$. This in turn tilts the lift vector away from the normal to the free stream velocity producing a force component in the drag direction, and that component is the induced drag. Increasing the aspect ratio factor of the wing reduces the induced drag since the induced drag coefficient is

$$c_{D,i} = \frac{c_L^2}{\pi e A}$$

The aspect ratio is $A = b^2/S$ so the induced drag is proportional to the square of the span loading

$$c_{D,i} \sim \left(\frac{c_L}{b}\right)^2$$

Although this effect is beneficial it has drawbacks because increasing the span for the same lift effectively makes the wing more slender in planform leading to increased bending moment at the root of the wing. The increased bending moment must be countered by a strengthened structure which typically involves additional weight. By the same token, increased span can be a limiting factor in ground operations of commercial aircraft. Thus, the desire to improve the efficiency of aircraft by reducing wing drag has inherent limitations that must be considered.

The current rapid rise in fuel costs has made the search for drag reduction techniques even more important. Because induced drag is a large fraction of the total drag for subsonic aircraft, on the order of 35%, modifications of the spanwise lift distributions which reduce induced drag are desirable. Two wingtip treatments have received substantial attention from airplane manufacturers and operators: winglets and raked wingtips. These two types of wingtips are shown schematically in Figure 5.66. The winglet case presents a non-planar wing because the tips of the wing are not in the plane of the major portion of the wing. The raked wingtip presents a case of leading and trailing edges which are not straight although the wing remains planar.

A number of different wingtip treatments aimed at improving aerodynamic performance are discussed in NRC (2007). The wide range of alterations to the outboard tenth of wings of operational aircraft illustrates the variety of design choices possible. However, the underlying concept is based on increasing the developed length of the wing, and this may be done in several ways:

- Increasing the span horizontally. Constraints include hangar, gate, taxiway, and runway width, increased structural weight of the wing, greater bending loads, and increased wetted area.
- Adding vertical, or canted, extensions at the wingtip above and/or below the plane of the wing. Constraints include increased wetted area and corner flow effects at junctions, increased weight and bending moments.

Kroo (2001) points out that various tip devices may lead to incremental but important gains in aircraft performance but there are challenges in integrating such features with the total aircraft system. In any case, the increased weight and cost of larger span wings leads to diminishing returns as span is increased. This, combined with the geometric issues noted above, determines the optimal span. Commercial experience with winglet retrofits on the Boeing 737, 757, and 767 aircraft indicates a fuel saving of up to around 5% for longer ranges.

5.7.7 **Winglets**

The flow around the tips of a finite wing involves a whirling motion caused by the difference in pressure between the upper and lower surfaces of the wing, as illustrated schematically in Figure 5.67. Therefore there is an inboard flow on the upper surface and an outboard flow on the lower surface. The swirling flow downstream of the trailing edge of the wing organizes, that is, "rolls up" itself into two concentrated trailing vortices after traveling a downstream distance on the order of the wingspan. It seems clear that preventing this motion would increase the total lift produced and a simple solution would appear to be the placement of endplates at each wingtip to block the cross-flow. However, this intuitive solution mainly increases drag because of the increased surface area needed, thereby eroding any benefit in improving lift. A more sophisticated exploitation of the three-dimensional flow field itself can indeed reduce the induced drag produced by the wing downwash.

In much the same way that a sailboat may be sailed against the wind by appropriate pointing of the sail, the placement of a winglet in such a fashion as to actually produce a negative drag force, that is, a thrust, would be a distinct advantage. A view of the flow looking down on the upper surface of the wing with a winglet mounted vertically on the tip is shown in Figure 5.68. The resultant force in the flight direction produced by the three-dimensional flow on the winglet is given by

$$F_{x,winglet} = L \left(\sin \beta - \frac{\cos \beta}{L/D} \right)$$

(a)

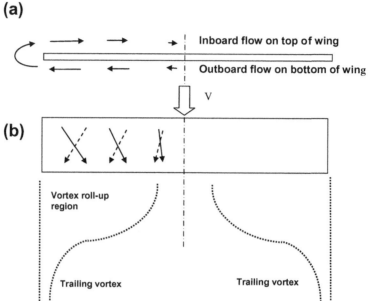

Inboard flow on top of wing

Outboard flow on bottom of wing

V

(b)

Vortex roll-up
region

Trailing vortex

Trailing vortex

FIGURE 5.67

Schematic illustration of the flow over a finite lifting wing: (a) front view facing the leading edge showing cross-flow velocity components due to the higher pressures on the bottom wing surface and lower pressures on the top wing surface; (b) top view looking down on the wing illustrating the velocity on the top surface (solid arrows) and on the bottom surface (dashed arrows).

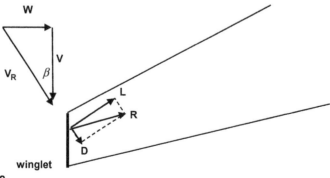

W

V

V_R β

L

R

D

winglet

FIGURE 5.68

View looking down on the upper surface of the port wing showing the resultant velocity V_R producing a resultant force R on the vertically mounted winglet, here represented as a simple flat plate. The magnitude of the angle β is exaggerated for clarity.

This force is a thrust force so that the drag of the wing is reduced. The actual design of the winglet is rather complicated and tends to have superior performance only in the vicinity of the design point chosen, for example, the cruise condition. Discussion of winglets may be found in textbooks on aerodynamics, for example McCormick (1995) and Bertin and Smith (1998). Early research on winglets applied to subsonic jet transports is reported by Flechner et al. (1976) and later studies are described by Lazos and Visser (2006). An interesting general discussion of the subject is presented by Jones (1979).

The important design variables for the winglet are

- the leading edge sweep angle of the winglet,
- the cant angle measured between the vertical and the plane of the winglet,
- the toe-in angle measured between the flight direction and the winglet root.

Installing a winglet doesn't add the additional wing root bending moment that would be developed if the span was merely increased by the height of the winglet and this is a benefit. On the other hand, there are local forces and moments at the winglet junction with the wing and these tend to add some weight to the wing. Note that the action of the winglet is tied to properly exploiting the induced whirling flow generated by the finite wing so a winglet may be employed above the wing, below the wing, or both. The performance of a winglet is difficult to generalize in the preliminary design process and consideration of winglets is typically left to later detail design stages. In general, the induced drag reduction will not be equal to that provided by simply extending the wingspan by an amount equal to the winglet lengths, but perhaps by only half that amount. However, in recognition of their fairly widespread use they may be employed if desired, using market survey information as a guide. It is suggested that they be assumed to act aerodynamically as an increase in span equal to half the total length of the winglets. Furthermore, when the refined weight estimation procedures of Chapter 8 are carried out the weight of the wing with winglets should be considered to be equal to the weight of the wing with a span which includes the total length of the winglets.

5.7.8 Raked wingtips

Boeing has been prominent in the application of raked wingtips to their latest aircraft rather than winglets. The efforts at modifying the wingtips are aimed at altering the spanwise loading in order to achieve the desired reduction in induced drag. The idea may be broadly understood as taking a wing of given span and then increasing the span to achieve the corresponding reduction in the induced drag coefficient. This extended wingtip may be bent up to fashion a winglet or swept back further to form a raked wingtip. The details of the flow resulting from such changes are important to determining whether worthwhile improvements can be achieved. Of course, in addition to any performance improvements which may accrue to the wingtip treatment the other effects such as structural loading, weight implications, off-design performance, and the like must be assessed.

Rudolph (1991) patented a raked wingtip described as a modification to the main wing of a conventional airliner made by adding a highly tapered wingtip extension member to it. This addition increases the wingspan and aspect ratio with a relatively small increase in wetted area. Adding this extension and suitably changing the camber of the extension provides a means of tailoring the spanwise lift distribution into an almost elliptic shape so as to produce the least induced drag (as discussed in Appendix C) while incurring a minimal friction drag penalty.

A detailed computational and experimental study of raked wingtips is presented by Vijgen et al. (1989). They point out that the trailing vortex sheet behind a swept wing deforms rapidly and is not well described by classical wing theory as described in Appendix C. Nonlinear effects and non-planar attributes of the trailing vortex sheet must be treated by more sophisticated lifting surface methods. The performance improvements accruing to raked wingtips, or to winglets, are difficult to quantify in a general manner. Therefore, in the preliminary design stage, raked wingtips may be treated in the same empirical fashion as recommended for winglets. More accurate evaluation of their efficacy must be deferred to later stages of detailed design.

5.7.9 **Wing dihedral and incidence**

The configuration of the wing has been determined and now the disposition of the wing with respect to the fuselage is sought. The location of the wing along the longitudinal axis of the fuselage relative to the center of gravity of the complete aircraft is determined by stability and control requirements in pitch. However, the position of the center of gravity of the complete aircraft depends on where the wing is placed. Therefore setting the wing on the fuselage at this stage must be considered to be provisional. Locating the center of gravity position requires a fairly detailed knowledge of the weights and positions of all the major aircraft components. A refined weight estimate for the proposed aircraft along with the appropriate positioning of the wing is presented in Chapter 8.

Stability and control requirements in roll suggest that a certain amount of dihedral angle is desirable, as illustrated in Figure 5.69. This aspect of the airplane's design is discussed in Chapter 6. An idea of the magnitude of the dihedral angle ϕ for typical airliners is presented in Table 5.5. Note that jet transports are low- or mid-wing aircraft and generally have dihedral angles of about 5–$8°$, while turboprop transports have no dihedral inboard of the engines and small values of dihedral, ϕ about $2°$, on the section of the wing outboard of the engines, or even a small negative dihedral angle, generally called anhedral. Horizontal tail surfaces may also exhibit small values of dihedral or anhedral.

In order to ensure passenger comfort the cabin floor should remain essentially horizontal during cruising flight. Therefore the incidence angle i between a fuselage reference line, like the cabin floor, and the chord of the aircraft should be equal to the design angle of attack at the cruise condition, as illustrated in Figure 5.70. Typical values, as suggested by the database in Table 5.5, lie in the range of $2°<i<5°$. During

FIGURE 5.69

The dihedral angle ϕ is illustrated in this front view of an airliner.

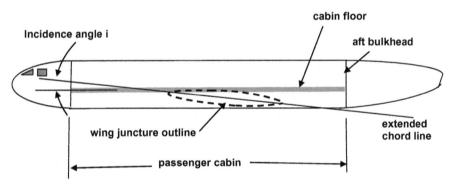

FIGURE 5.70

A typical elevation view of a fuselage illustrating the wing incidence angle i as measured between a fuselage reference line, taken here as the passenger cabin floor.

ground maneuvers the cabin floor is maintained level by the design of the landing gear. Obviously, during rotation and the subsequent climb the cabin floor will be tilted at the flight path angle. Such inclinations, like those due to banked turns, are generally small and acceptable to the passengers as routine. Because the lift required during cruise is not yet known in detail, the incidence angle, like the axial location of the wing, must remain a provisional value until the required lift is determined by the methods in Chapter 10.

5.8 Design summary

The features of the wing designed for the proposed aircraft may now be collected for use in the further development of the design. The geometrical configuration of the

Table 5.8 Summary of Estimated Values of $C_{L,\max}$

Airfoil Choice:		NACA 64-212 (root)
		NACA 64-209 (tip)
Aspect Ratio:		$A = 9.25$
Taper Ratio:		$\lambda = 0.30$
Flap Type, c_f/c:		Fowler, 23%
Spanwise Extent η_i and η_o:		16.4% and 76.0%
Rated Area:		$S_{wf}/S = 0.53$

Configuration	$\delta_{flap, \text{Trailing Edge}}$ (deg)	$\delta_{flap, \text{Leading Edge}}$ (deg)	$C_{L,\max}$ Low Speed M=0.2	$C_{L,\max}$ Cruise Speed M=0.8
Cruise	0	0	1.23	0.99
Takeoff	25	0	1.64	–
Landing	40	20	1.92	–

Note: Numbers in this table are merely representative.

wing of the proposed aircraft is now in hand, as well as the lifting characteristics in the takeoff, cruise, and landing configurations. These characteristics may be calculated by the methods described in this chapter in a relatively simple and rapid manner with little recourse to the computer outside of a spreadsheet. If desired, and if time and resources permit, the student may wish to explore CFD approaches for wing analysis. For example, the AVL code by Drela (2013) for the aerodynamic and flight-dynamic analysis of rigid aircraft of arbitrary configuration employs an extended vortex lattice model for the lifting surfaces, together with a slender-body model for fuselages and nacelles. It is available under the Gnu General Public License. The Computerized Environment for Aircraft Synthesis and Integrated Optimization Methods CEASIOM (2013) provides another available code that integrates computational tools for conceptual design.

The format illustrated in Table 5.8 is suggested for the presentation of the results for the maximum lift coefficients in the three basic configurations. Note that the numerical values shown in the sample table are merely representative of a typical design. It is expected that the maximum lift coefficients in takeoff and landing are appropriate to the requirements developed in selecting the engines in Chapter 4.

The maximum lift coefficient for the cruise condition in Table 5.8, in combination with the lift curve slope of the wing determined by the methods in Section 5.3, the zero-lift angle of attack as given by the chosen airfoil section, and the angle of attack for maximum lift found using the methods of Section 5.4, provide a means for constructing a plot of C_L as a function of α. These data will be used with the drag characteristics determined in Chapter 9 to compute the lift to drag ratio and the ceiling altitude of the aircraft in the cruise configuration. It is useful to calculate additional values of $C_{L\max}$ at different Mach numbers so as to develop a plot of $C_{L\max}$

Table 5.9 Wing Characteristics of Several Airliners. Other Features and Identification of Manufacturers are Given in Table 5.5

Aircraft	Gross Weight (lb) W_g	Aspect Ratio $A/b^2/S$	Sweep (deg) $\Lambda_{c/4}$	Taper Ratio λ	MAC (ft) c_{mac}	Root Thick. $(t/c)_r$	Tip Thick. $(t/c)_t$	Engine Loc. y_e (ft)	Engine loc. y_e/c_{mac}
Q100	36,300	12.4	0.00	0.50	6.91	0.18	0.13	13.3	1.92
Y-7	48,000	11.4	6.80[a]	0.37	8.15			12.7	1.56
ATR72-500	48,500	12.0	3.10[a]	0.50	6.30	0.18	0.13	13.1	2.08
ERJ145LR	46,275	7.86	22.8	0.25	7.03			6.60	0.94
CRJ200LR	53,000	8.28	24.5	0.20	7.00	0.13	0.10	7.34	1.05
CRJ700ER	75,250	7.87	26.0	0.25	6.60			7.85	1.19
E175	85,500	9.28	25.5	0.25	10.6			12.8	1.21
B737-700ER	154,500	10.3	25.0	0.22	13.5			16.0	1.17
A320-200	169,800	9.35	25.0	0.23	13.3	0.15	0.11	18.3	1.37
A310-300	361,600	8.79	28.0	0.20	19.1	0.21	0.11	25.3	1.32
B767-200ER	395,000	7.99	31.5	0.22	22.0	0.15	0.10	26.0	1.18
B777-200	545,000	8.68	31.5	0.20	26.0			31.55	1.21
A340-300	609,580	10.1	30.0	0.25	22.0	0.15	0.11	60.3[b]	2.74
B747-400	875,000	7.91	37.5	0.24	30.6	0.13	0.08	68.3[b]	2.23
A380-800	1,234,600	7.53	35.8	0.20	40.3	0.15	0.12	84.3[b]	2.09

[a] Outboard panel of wing.
[b] Outboard engine.

as a function of Mach number in the range $0.2 < M < M_{cr}$ so that the minimum flying speed of the aircraft at any altitude and Mach number may be calculated for the performance section of the report described in Chapter 10.

A scaled and dimensioned layout of the planform of the wing showing the location of engines, flaps, and other control devices is presented in Figure 5.50. These dimensions, along with the fuselage drawings and the tables prepared in this chapter, will be used in the design of the tail surfaces, which is discussed in the next chapter. The other features of the wing design should be presented in the manner illustrated by Table 5.9 which presents data for the wings of a representative range of commercial airliners for reference.

5.9 Nomenclature

A	aspect ratio
a	lift curve slope
b	wingspan
c	chord length
c'	chord length with slat extended
c_d	airfoil drag coefficient
c_f	flap chord
c_l	airfoil lift coefficient
$c_{l\alpha}$	airfoil lift curve slope
C_L	wing lift coefficient
$C_{L\alpha}$	wing lift curve slope
C_p	pressure coefficient
D	drag
d_f	wing diameter
e	span efficiency factor
F	thrust
g	acceleration of gravity
i	wing incidence angle
k	constant
k_a	constant in Equation (5.50)
L	lift
l_w	wing length
l_v	vertical tail moment arm
M	Mach number
N	yawing moment
N_p	number of passengers
p	pressure
q	dynamic pressure
R	range
Re	Reynolds number
S	wing planform area

S_v	area of vertical tail
$S_{w,f}$	wing planform area affected by flaps
$S_{w,s}$	wing planform area affected by slats
$s_{c/2}$	length of wing along the half-chord from root to tip
t	thickness
V	velocity
$V*$	local sound speed
V_h	horizontal tail volume coefficient
V_v	vertical tail volume coefficient
v	local velocity on airfoil
W	weight
x	chordwise distance from wing leading edge
y	spanwise distance from centerline
y_b	spanwise distance to break in LE or TE sweepback
y_e	spanwise distance from centerline to engine centerline
α	angle of attack
α^0	angle of attack where a departs from linearity
β	Prandtl-Meyer factor $(1-M^2)^{1/2}$ or sideslip angle
δ	atmospheric pressure ratio or angular deflection
ϕ	dihedral angle
γ	ratio of specific heats
η	$y/(b/2)$
η_{max}	empirical factor for leading edge radius, Equation (5.38)
η_δ	empirical factor for slat deflection, Equation (5.38)
κ	$c_{l\alpha}/2\pi$
Λ	wing sweepback angle
λ	wing taper ratio
ρ	density
σ	atmospheric density ratio
σ_b	wing root bending stress
θ_{TC}	tail cone strike angle

5.9.1 Subscripts

base	reference condition
cr	cruise
crit	critical
DD	drag divergence
f	flap
i	induced or inboard
LE	leading edge
l	lower surface
MAC	mean aerodynamic chord
max	maximum
min	minimum

o	outboard
p	pressure
r	root
ref	reference condition
rel	relative
s	slat
SL	sea level
TE	trailing edge
t	tip
to	takeoff
u	upper surface
0	zero-lift

References

Abbott, I.H. et al., 1945. Summary of Airfoil Data, NACA Technical Report No 824. Available from: <http://ntrs.nasa.gov/archive/nasa/casi.ntrs.nasa.gov/19930090976_1993090976.pdf>.

Abbott, I.H., Von Doenhoff, A.E., 1959. Theory of Wing Sections. Dover, NY.

AGARD, 1979. Experimental Data Base for Computer Program Assessment: Report of the Fluid Dynamics Panel Working Group 04, AGARD-AR-138.

Allison, D.O., Mineck, R.E., 1996. Assessment of Dual-Point Drag Reduction for an Executive-Jet Modified Airfoil Section, NASA TP-3579.

Bertin, J.J., Smith, L.M., 1998. Aerodynamics for Engineers, third ed. Prentice Hall, New Jersey.

Cahill, J.F., 1947. Two-Dimensional Wind-Tunnel Investigation of Four Types of High-Lift Flaps on an NACA 65–210 Airfoil Section, NACA TN 1191.

CEASIOM, 2013. Computerized Environment for Aircraft Synthesis and Integrated Optimization Methods. Available from: <www.ceasiom.com>.

De Young, J., Harper, C.W., 1948. Theoretical Symmetric Span Loading at Subsonic Speeds for Wings Having Arbitrary Planforms, NACA Report 921.

Diederich, F.W., 1952. A Simple Approximate Method for Calculating Spanwise Lift Distributions and Aerodynamic Influence Coefficients at Subsonic Speeds. NACA TN 2751.

Drela, M., 1989. XFOIL: an analysis and design system for low Reynolds number airfoils. In: Mueller, T.J. (Ed.), Lecture Notes in Engineering: Low Reynolds Number Aerodynamics, vol. 54. Springer-Verlag, New York. Available from: <http://web.mit.edu/drela/Public/web/xfoil/>

Drela, M., 2013. Available from: <http://web.mit.edu/drela/Public/web/avl/>.

Dress, D.A., Stanewsky, E., McGuire, P. et al., 1984. High Reynolds Number Tests of the CAST 10–2/DOA 2 Airfoil in the Langley 0.3-Meter Transonic Cryogenic Tunnel – Phase II, NASA TM-86273.

Flechner, S.G., Jacobs, P.F., Whitcomb, R.T., 1976. A High Subsonic Speed Wind Tunnel Investigation of Winglets on a Representative Second Generation Jet Transport Wing, NASA Technical Note TN-8264.

Furlong, G.C., McHugh, J.G., 1953. A Summary and Analysis of the Low-Speed Longitudinal Characteristics of Swept Wings at High Reynolds Number, NACA Report 1339

Gottlieb, S.M., 1949. Two-Dimensional Wind-Tunnel Investigation of Two NACA 6-Series Airfoils with Leading-Edge Slats, NACA RM L8K22.

Green, J., Quest, J., 2011. A short history of the European transonic wind tunnel ETW. Progress in Aerospace Sciences 47 (5), 319–368.

Harper, C.W., Maki, R.L., 1964. A Review of the Stall Characteristics of Swept Wings, NASA TN D-2373.

Harris, C.D., 1990. NASA Supercritical Airfoils, NASA Technical Paper 2969. Available from: <http://hdl.handle.net/2002/13874>.

Harris, C.D., 1975a. Aerodynamic Characteristics of the 10-Percent-Thick NASA Supercritical Airfoil 33 Designed for a Normal-Force Coefficient of 0.7, NASA TM X-72711.

Harris, C.D., 1975b. Aerodynamic Characteristics of a 14-Percent-Thick NASA Supercritical Airfoil Designed for a Normal-Force Coefficient of 0.7, NASA TM X-72712.

Harris, C.D., McGhee, R.J., Allison, D.O., 1980. Low Speed Aerodynamic Characteristics of a 14-Percent-Thick NASA Phase 2 Supercritical Airfoil Designed for a Lift Coefficient of 0.7, NASA TM -81912.

Hoak, D.E. et al., 1978. USAF Stability and Control DATCOM, Flight Control Division, Air Force Flight Dynamics Laboratory, Wright-Patterson AFB.

Hayter, N.L.F., Kelly, J.A., 1953. Lift and Pitching Moment at Low Speeds of the NACA 64A010 Airfoil Section Equipped with Various Combinations of a Leading Edge Slat, Leading Edge Flap, Split Flap and Double Slotted Flap, NACA TN 3007.

Jameson, A., 1989. Computational Aerodynamics for Aircraft Design. Science 245, 361–371.

Jenkins, R.V., 1983. Reynolds Number Tests of an NPL 9510 Airfoil in the Langley 0.3-Meter Transonic Tunnel, NASA TM-85663.

Jenkins, R.V., Johnson, Jr., W.G., Hill, A.S., et al., 1984. Data from Tests of a R4 Airfoil in the Langley 0.3-Meter Transonic Cryogenic Tunnel. NASA TM-85739.

Jenkinson, L.R., Simpkin, P., Rhodes, D., 1999. Civil Jet Aircraft Design, AIAA, Reston, VA.

Johnson Jr., W.G., Hill, A.S., 1985. Pressure Distributions From High Reynolds Number Tests of a Boeing BAC I Airfoil in the 0.3-Meter Transonic Cryogenic Tunnel, NASA TM-87600.

Johnson Jr., W.G., Hill, A.S., Eichmann, O., 1985. High Reynolds Number Tests of a NASA SC(3)-0712(B) Airfoil in the Langley 0.3-Meter Transonic Cryogenic Tunnel, NASA TM-86371.

Johnson, B.H., Shibata, H.H., 1951. Characteristics Throughout the Subsonic Speed Range of a Plane Wing and a Cambered and Twisted Wing, Both Having 45° of Sweepback, NACA RM A51D27.

Jones, R.T., 1947a. Wing Plan Forms for High-Speed Flight, NACA TR863.

Jones, R.T., 1947b. Effects of Sweepback on Boundary Layer and Separation, NACA TN 1402.

Jones, R.T., 1979. Minimizing Induced Drag. Soaring. pp. 26–29.

Koven, W., Graham, R.R., 1948. Wind Tunnel Investigation of High-Lift and Stall Control Devices Over A 37o Sweptback Wing of Aspect Ratio 6 at High Reynolds Numbers, NACA RM No L8D29.

Kroo, I., 2001. Drag due to lift: concepts for prediction and reduction. Annual Reviews of Fluid Mechanics 33, 587–617.

Ladson, C.L., 1996. Computer Program to Obtain Ordinates for NACA Airfoils, NASA Technical Memorandum 4741. Available from: <www.pdas.com>.

Lazos, B.S., Visser, K.D., 2006. Aerodynamic Comparison of Hyper-Elliptic Cambered Span (HECS) Wings with Conventional Configurations, AIAA 2006–3469, 25th AIAA Applied Aerodynamics Conference, San Francisco.

Loftin, L.K., 1948. Theoretical and Experimental Data for a Number of NACA 6A-Series Airfoil Sections NACA, Report 903.

Loftin, L.K., Bursnall, W.J., 1948. The Effects of Variations in Reynolds Number between 3.0x106 And 25.0x106 upon the Aerodynamic Characteristics of a Number of 6-Series Airfoil Sections, NACA Technical Note No 1773.

Lovell, D.A., 1977. A Wind-Tunnel Investigation of the Effects of Flap Span and Deflection Angle, Wing Planform, and a Body on the High-Lift Performance of a 28o Swept Wing, RAE C.P. No 1372, Farnborough, UK.

McCormick, B.H., 1995. Aerodynamics, Aeronautics, and Flight Mechanics. Wiley, NY.

Mineck, R.E., Lawing, P.L., 1987. High Reynolds Number Tests of the NASA SC(2)-0012 Airfoil in the 0.3-Meter Transonic Cryogenic Tunnel, NASA TM-89102.

Morrison Jr., W.D., 1976. Advanced Airfoil Design: Empirically Based Transonic Aircraft Drag Buildup Technique NASA CR-137928.

NRC, 2007. Assessment of Wingtip Modifications to Increase the Fuel Efficiency of Air Force Aircraft, Air Force Studies Board, National Research Council, National Academy of Sciences, Washington, DC. Available from: <http://www.nap.edu/catalog/11839.html>.

PDAS, 2013. Public Domain Aeronautical Software. Available from: <www.pdas.com/index.html>.

Pearcy, H.H., 1962. The aerodynamic design of section shapes for swept wings. In: Advances in the Aeronautical Sciences, , vol. 3. Pergamon Press, Oxford.

Phillips, W.F., Alley, N.R., 2007. Predicting maximum lift coefficient for twisted wings using lifting line theory. Journal of Aircraft 44 (3), 898–910.

Plentovich, E.B., Ladson, C.L., Hill, A.S., 1984. Tests of a NACA 651–213 Airfoil in the NASA Langley 0.3-Meter Transonic Cryogenic Tunnel, NASA TM-85732.

Rudolph, P.K.C., 1991. High Taper Wingtip Extension. US patent, No. 5,039,032.

Rudolph, P.K.C., 1996. High Lift Systems on Commercial Subsonic Airliners, NASA CR 4746.

Schiktanz, D., Scholz, D., 2011. Survey of Experimental Data of Selected Supercritical Airfoils, Airport2030-TN-Supercritical Airfoils. Available from: <http://www.fzt.haw-hamburg.de/pers/Scholz/Airport2030/Airport2030TN_Supercritical_Airfoils_11-12-21.pdf>.

Scholz, D., Ciornei, S., 2005. Mach number, Relative Thickness, Sweep and Lift Coefficient of the Wing – An Empirical Investigation of Parameters and Equations, DGLR Paper No. 2005-122, in Jahrbuch 2005, Brandt, P. (Ed.), DGLR, Bonn, Germany.

Sharpes, D.E., 1985. Validation of USAF Stability and Control Datcom Methodologies for Straight-Tapered Swept-forward Wings, AFWAL-TR-84-3084, Flight Dynamics Laboratory Air Force Wright Aeronautical Laboratories.

Smith, A.M.O., 1975. High lift aerodynamics, 37th Wright brothers lecture. Journal of Aircraft 12 (6), 501–530.

Whitcomb, R.T., Clark, L.R., 1965. An Airfoil Shape for Efficient Flight at Supercritical Mach Numbers, NASA TM X-1109.

Torenbeek, E., 1982. Synthesis of Subsonic Airplane Design. Kluwer Academic Publishers, Dordrecht, The Netherlands.

UIUC, 2013. Airfoil Data Site. Available from: <http://www.ae.illinois.edu/m-selig/ads.html>.

Valarezo, W.O., Chin, V.D., 1994. Method for the prediction of wing maximum lift. Journal of Aircraft 31 (1), 103–109.

Van Dam, C.P., 2002. The aerodynamic design of multi-element high-lift systems for transport airplanes. Progress in Aerospace Science 38 (2), 101–144.

Vijgen, P.M.H.W., Van Dam, C.P., Holmes, B., 1989. Sheared wing-tip aerodynamics. Journal of Aircraft 26 (3), 207–213.

Wahls, R.A., 2001. The National Transonic Facility: A Research Retrospective, AIAA-2001-0754, 39th AIAA Aerospace Sciences Meeting and Exhibit.

Tail Design

6

CHAPTER OUTLINE

6.1 Preliminary tail design

At the present stage of the design process all the characteristics of the wing and fuselage have been determined with only the size and configuration of the horizontal and vertical tails remaining unknown. The tail surfaces are primarily trim, stability, and control appendages and methods for estimating the horizontal and vertical tail sizes based on stability considerations are presented. The two most important parameters are the tail planform areas and the moment arms through which the lifting forces generated by the tail surfaces act. Therefore, initial estimates of the size, shape, and

Commercial Airplane Design Principles. http://dx.doi.org/10.1016/B978-0-12-419953-8.00006-1

location of the horizontal and vertical tails are needed to carry out the suggested design methods by which those estimates may be refined. Here is another situation where the detailed market survey data are valuable. It is important to determine the following wing and tail surface and location characteristics from the market survey information:

$c_{MAC,w}$ mean aerodynamic chord of the wing

$c_{MAC,v}$ mean aerodynamic chord of the vertical tail

$c_{MAC,h}$ mean aerodynamic chord of the horizontal tail

S_v the area of the vertical tail as defined in Figure 6.1

l_v the distance from the aerodynamic center of the vertical tail to the aerodynamic center of the wing as in Figure 6.1

S_h the area of the horizontal tail as defined in Figure 6.2

l_h the distance from the aerodynamic center of the horizontal tail to the aerodynamic center of the wing as in Figure 6.2

Here S is the wing area and c_{MAC} is the mean aerodynamic chord (MAC) of the wing. We assume that the aerodynamic centers of the wing and tail surfaces are located at the quarter-chord points of their respective mean aerodynamic chords. The properties listed above will be used to form the characteristic parameters for the tail surfaces and help guide the design process. It will become apparent in subsequent sections that the appropriate parameters for the horizontal and vertical tails are given by

$$V_h = \frac{S_h l_h}{S c_{MAC}}$$ (6.1)

$$V_v = \frac{S_v l_v}{S b}$$ (6.2)

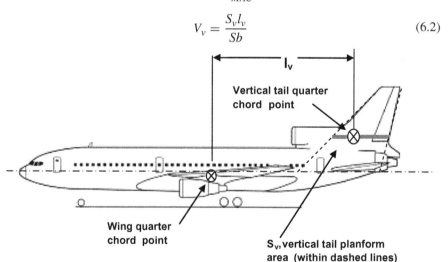

FIGURE 6.1

Dimensions for use in estimating the vertical tail parameters.

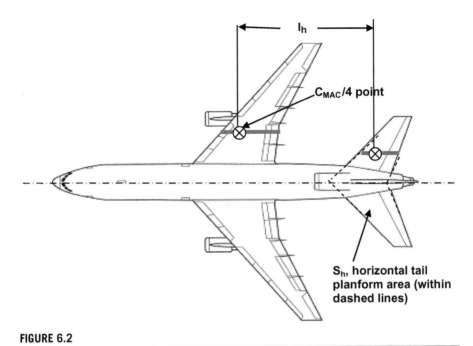

FIGURE 6.2

Dimensions for use in estimating the horizontal tail parameters.

These ratios are called the volume coefficients of the horizontal and vertical tails, respectively, because their numerators and denominators have the units of volume. The forces developed by the tail surfaces are proportional to their planform areas and the moments those forces produce are proportional to the distances through which they act. It should be clear that for a given wing, that is, a given S, b, and c_{MAC}, larger volume coefficients indicate more effective tail surfaces.

6.1.1 Tail surface characteristics

The horizontal tail surface area is obtained by reasonable extrapolation of the leading and trailing edges in to the fuselage centerline, as is done for the wing. Because the vertical tail is not a symmetric figure its area is defined somewhat more arbitrarily, but generally it involves extrapolation of leading and trailing edges in to the fuselage centerline. It should be noted that sometimes the said extrapolation is extended only to the plane of the horizontal tail, so care must be taken in collecting data on tail areas. Detailed dimensional data for tail surfaces of market survey aircraft are best obtained by working from scaled three-view drawings so that consistency for all data is maintained. Similarly, the root and tip chords of the tail surfaces should be collected and the respective taper ratios and aspect ratios calculated. In addition, the sweepback angles of the tail surfaces should be noted. Then the mean aerodynamic chords of all surfaces may be determined using the definitions provided in Section 5.1. The aerodynamic centers for the various surfaces may be extracted, or at least estimated, from the market survey aircraft as being situated at the quarter-chord point of their mean aerodynamic chords. It should be obvious from the above that scaling dimensions from three-view drawings will be necessary. Such drawings may often be found in the websites of the manufacturers under the heading of technical data for airport operations.

Turbofan-powered airliners that cruise at high subsonic Mach numbers ($0.7 < M < 0.9$) have horizontal and vertical tail surfaces that are swept back at angles greater than that of the wing in order to make their effective moment arms as long as possible and to maintain their critical Mach numbers higher than that of the wing. Airfoils for the tail surfaces have thickness ratios smaller than that of the wing as an additional aid in keeping the critical Mach number of the tail surfaces higher than that of the wing. The thinner sections are practical because they save weight and because of the smaller aerodynamic loads they experience. Turboprop-powered airliners cruise at lower Mach numbers ($0.5 < M < 0.6$) and have minimal compressibility problems so the wing and horizontal tail generally have little or no sweepback. On the other hand, the vertical tails are usually swept back in order to increase the vertical tail moment arm. The airfoils used on both the horizontal and vertical tail surfaces are generally symmetrical sections so as to produce the same force magnitude for a given angle of attack in the positive or negative direction.

Some of the features of the horizontal and vertical tail surfaces for a representative range of airliners are collected in Tables 6.1 and 6.2. The aircraft are listed in ascending weight order; the first three are regional turboprops, the next four are regional turbofans, the next three are narrow-body jetliners, the next three wide-body jetliners, and the last two jumbo jets. These tables show that horizontal tail area ratios are typically in the range of $0.2 < S_h/S < 0.35$, with an average of around 0.25. The vertical tail area ratios tend to be somewhat smaller, lying in the range of $0.15 < S_v/S < 0.30$, with an average of around 0.21.

Aspect ratios of horizontal and vertical tail surfaces are shown as a function of takeoff weight in Figure 6.3. Note that horizontal tail aspect ratios lie in the range $3.5 < A_h < 5.5$, with an average of 4.5, while vertical tail aspect ratios are considerably smaller, lying in the range $1 < A_v < 2.5$, with an average of 1.6. This difference arises because the horizontal tail continually supplies a positive or negative lifting force and therefore benefits from a higher aspect ratio which reduces its lift-induced drag. On the other hand, the vertical tail is essentially unloaded in non-maneuvering flight and therefore incurs no induced drag penalty. A reduced aspect ratio may then be employed resulting in a smaller vertical tail span and therefore a stiffer structure. The sweepback angles of the horizontal and vertical tails are shown in Figure 6.4 and illustrate that the vertical tails for the turbofan aircraft have sweepback angles about $5° - 10°$ larger than the horizontal tails, with both larger than those of their respective wings. The turboprop aircraft tend to employ large sweepback on the vertical tail and little or none on their horizontal tail.

The taper ratios for the horizontal and vertical tail surfaces are shown in Figure 6.5 where it is seen that they tend to be the same for both. It is also clear that taper ratios are smaller for the heavier jetliners and have greater variability for the lighter aircraft.

Table 6.1 Horizontal Tail Properties for a Range of Airliners[a]

Aircraft	Gross Weight W_g (lb)	Wing Area S (sq.ft.)	S_h/S	Horizontal Tail Span b_h (ft)	A_h	Taper λ_h	Sweep Λ_h (deg)
Dash 8 Q100	36,300	585	0.28	26.8	4.43	0.80	6.5
XAC MA60	48,050	810	0.22	31.0	5.49	0.44	20.0
ATR72-500	48,500	657	0.20	24.2	4.37	0.56	8.0
ERJ145LR	46,275	551	0.24	24.8	4.68	0.59	20.0
CRJ200LR	53,000	587	0.19	20.5	3.72	0.50	34.5
CRJ700ER	75,250	739	0.24	28.0	4.36	0.40	33.0
E175	85,517	783	0.32	32.8	4.29	0.50	35.0
B737-700ER	154,500	1341	0.33	47.1	5.04	0.33	36.0
A320-200	169,800	1320	0.26	40.9	4.84	0.31	32.5
A310-300	361,600	2360	0.30	53.2	3.99	0.44	37.0
B767-200ER	395,000	3050	0.28	61.1	4.42	0.30	36.0
B777-200	545,000	4605	0.25	70.6	4.40	0.33	39.5
A340-300	609,580	3890	0.20	63.5	5.12	0.40	32.5
B747-400	875,000	5650	0.26	72.8	3.57	0.24	42.5
A380-800	1,234,600	9104	0.24	99.7	4.45	0.37	37.5

[a] *Manufacturers: Airbus (A320, A310, A340, A380), Avions de Transport Regional (ATR72), Boeing (B737, B747, B767, B777), Bombardier (CRJ200, CRJ700, Dash8), Embraer (ERJ145, E175), Xian Aircraft Industry Co. (XAC MA60).*

6.1.2 **Preliminary tail sizing**

To initiate the design process for the tail surfaces a first approximation to the size of the horizontal and vertical tail surfaces must be made. The tail surfaces are located at the aft end of the fuselage so that the moment arms $l_{h,ac}$ and $l_{v,ac}$ for the horizontal and vertical tails ultimately depend upon the location of the wing on the fuselage. In the subsequent sections we will show that stability requirements require that the wing be located within some precise range of distances from the airplane's center of gravity. However, the position of the center of gravity depends heavily upon the location of the wing. In Chapter 8 we will carry out a refined weight analysis and then proceed to present a method for locating the wing properly on the fuselage. At this point it is sufficient to rely on past practice as demonstrated by the market survey aircraft.

The approximations $l_h = 0.49l_f$ and $l_v = 0.45l_f$ are fairly representative of actual values for a range of commercial airliners. Because the fuselage length l_f of the design aircraft has been determined, as were the span and mean aerodynamic chord of the wing, approximate values for the moment arms l_h and l_v may be determined. One may use these approximations for the tail moment arms to define approximate volume coefficients given by

$$V_h' = \frac{0.49 S_h l_f}{S c_{MAC}} \tag{6.3}$$

Table 6.2 Vertical Tail Properties for a Range of Airliners[a]

Aircraft	Gross Weight W_g (lb)	Wing Area S (sq.ft.)	S_v/S	Vertical Tail Height b_v	A_v	Taper λ_v	Sweep Λ_v (deg)
Dash 8 Q100	36,300	585	0.28	14.0	1.20	0.67	32.0
XAC MA60	48,050	810	0.18	15.2	1.57	0.40	27.0
ATR72-500	48,500	657	0.26	15.8	1.48	0.50	38.5
ERJ145LR	46,275	551	0.18	11.4	1.29	0.64	36.0
CRJ200LR	53,000	587	0.19	11.4	1.16	0.69	44.5
CRJ700ER	75,250	739	0.17	12.0	1.16	0.71	40.0
E175	85,517	783	0.29	20.5	1.84	0.30	40.0
B737-700ER	154,500	1341	0.22	26.8	2.45	0.20	37.0
A320-200	169,800	1320	0.24	23.8	1.77	0.30	39.0
A310-300	361,600	2360	0.26	32.3	1.70	0.34	44.0
B767-200ER	395,000	3050	0.21	33.5	1.77	0.33	45.0
B777-200	545,000	4605	0.18	40.7	2.03	0.25	45.0
A340-300	609,580	3890	0.16	33.6	1.81	0.34	44.0
B747-400	875,000	5650	0.19	38.3	1.39	0.31	49.5
A380-800	1,234,600	9104	0.17	52.6	1.76	0.37	44.0

[a] *Manufacturers: Airbus (A320, A310, A340, A380), Avions de Transport Regional (ATR72), Boeing (B737, B747, B767, B777), Bombardier (CRJ200, CRJ700, Dash8), Embraer (ERJ145, E175), Xian Aircraft Industry Co. (XAC MA60).*

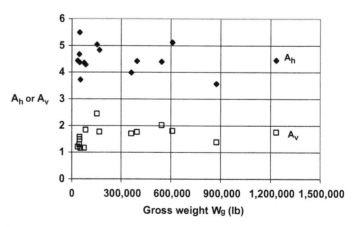

FIGURE 6.3

Variation of the aspect ratio of the horizontal tail (solid symbols) and of the vertical tail (open symbols) with gross weight for a range of representative airliners.

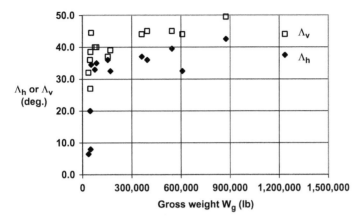

FIGURE 6.4

Variation of the leading edge sweepback of the horizontal tail (solid symbols) and of the vertical tail (open symbols) with gross weight for a range of representative airliners.

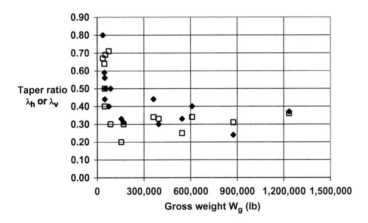

FIGURE 6.5

Variation of the taper ratio of the horizontal tail (solid symbols) and of the vertical tail (open symbols) with gross weight for a range of representative airliners.

$$V_v' = \frac{0.45 S_v l_f}{Sb} \qquad (6.4)$$

The range of values for the approximate horizontal and vertical tail volume coefficients V_h' and V_v' for the representative airliners of Tables 6.1 and 6.2 is illustrated in Figures 6.6 and 6.7. Note that there is considerable variability for the smaller aircraft and much less for the larger aircraft.

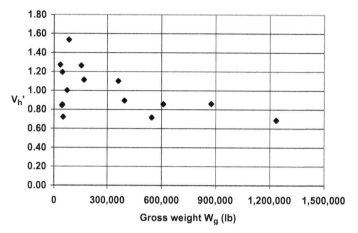

FIGURE 6.6

The approximate horizontal tail volume coefficient V_h' is shown as a function of gross weight for the airliners of Table 6.1.

Selecting target values for V_h' and V_v' using both the market survey aircraft information and Figures 6.6 and 6.7 as a guide permits determination of reasonable values for the tail areas S_h and S_v. These values are important for laying out the preliminary shape of the tail surfaces in terms of span, sweepback angle, taper ratio, and airfoil section along the lines suggested previously in this section. With the configuration for the tail surfaces now selected they may be located on the fuselage previously designed. Using the aerodynamic centers for the wing and tail surfaces along

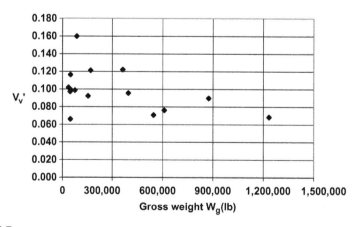

FIGURE 6.7

The approximate vertical tail volume coefficient V_v' is shown as a function of gross weight for the airliners of Table 6.2.

with their approximate moment arms the wing now may be provisionally located on the fuselage. A three-view drawing of the entire aircraft, excepting the landing gear configuration, may now be drawn. The overall configuration of the aircraft provides the dimensional information necessary for carrying out the refined design of the tail surfaces discussed in subsequent sections.

Additional refinement of the tail design, taking into account some basic stability and control requirements, is described in the subsequent sections of this chapter. The inclusion of additional constraints on the design permits the final size, shape, and location of the tail surfaces to be developed through a series of iterations aimed at satisfying these constraints. If time considerations preclude further calculated refinements to the tail configuration developed thus far, then the remaining material in the chapter may be used to provide some ad hoc refinements to finalize the tail design.

6.2 Refined horizontal tail design

In the simplest analysis we may separate the aircraft into two parts: the wing-fuselage-vertical tail and the horizontal tail. We are interested in the longitudinal characteristics of the aircraft, that is, the pitching motion of the aircraft in its plane of symmetry, that is, about the y, or spanwise axis, as shown in Figure 6.8. To facilitate analysis we will consider the various components of the aircraft to act independently and will account only for the most important features of any mutual interference between them. The center of gravity location indicated is that of the complete aircraft.

6.2.1 Equilibrium conditions

The aircraft, as shown in Figure 6.8, is in equilibrium in straight and level flight. This requires that there be no net force or moment acting on it. The equilibrium of forces in the vertical direction is

$$L_w + L_h = W \tag{6.5}$$

As discussed in Section 5.7, the contribution of the fuselage to the lift is neglected with respect to that produced by the wing. The moment acting around the center of gravity of the wing-body-tail M_{cg} is given by

$$M_{cg} = L_w \left(x_{cg} - x_{ac}\right) + M_{ac,w} - L_h l_h + M_{ac,h} + M_f \tag{6.6}$$

The lift and moment about the aerodynamic center of the wing are denoted by L_w and $M_{ac,w}$, respectively, the lift and moment about the aerodynamic center of the tail are denoted by L_h and $M_{ac,h}$ while the moment contribution of the fuselage about the center of gravity is shown as M_f. In Equation (6.6) the moment of the horizontal tail about its aerodynamic center is neglected with respect to the moment produced by the lift of the tail acting through the moment arm l_h. Note that moments about the spanwise, or y-coordinate, are taken as positive in the nose-up direction. Dividing

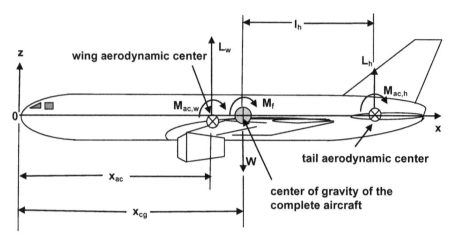

FIGURE 6.8

Forces acting in the symmetry plane: L_w, the lift of the wing, L_t, the lift of the horizontal tail, and W, the weight of the aircraft. The moments of the wing, $M_{ac,w}$, the tail, $M_{ac,h}$, and the fuselage, M_f, are also shown; they are considered positive in the direction shown.

Equation (6.5) by qS and Equation (6.6) by qSc_{MAC} permits the equations to be cast in coefficient form as follows:

$$C_{L,w} + C_{L,h}\left(\frac{q_h}{q}\right)\frac{S_h}{S} = \frac{W}{qS} = C_L \tag{6.7}$$

$$C_m = C_{L,w}\left(\frac{x_{cg}}{c_{MAC}} - \frac{x_{ac,w}}{c_{MAC}}\right) + C_{m,w} - C_{L,h}\left(\frac{q_h}{q}\right)\frac{S_h l_h}{Sc_{MAC}} + C_{m,f} \tag{6.8}$$

Equation (6.8) may be rearranged and written as

$$C_m = C_{m,w} + \left(\frac{x_{cg}}{c_{MAC}} - \frac{x_{ac,w}}{c_{MAC}}\right)C_{L,w} - C_{L,h}\eta_h V_h + C_{m,f} \tag{6.9}$$

In Equation (6.9) we note the appearance of the horizontal tail volume coefficient V_h of Equation (6.1) and the introduction of the horizontal tail efficiency η_h which is given by

$$\eta_h = \frac{q_h}{q} \approx \left(\frac{V_{d,h}}{V}\right)^2 \tag{6.10}$$

The tail efficiency η_h accounts for the possibility that the horizontal tail will see a somewhat reduced velocity $V_{d,h}$ due to the downwash generated by the lifting wing, particularly at higher angles of attack of the wing. The influence of induced downwash on the velocity and the angle of attack seen by the horizontal tail will be discussed subsequently. The horizontal tail volume coefficient V_h expresses the relative strength of the moment the horizontal tail can produce.

The first trim condition, that the total lift equals the aircraft weight, as expressed in Equation (6.7), may be written as

$$C_{L,h} = (C_L - C_{L,w}) \frac{l_h}{c_{MAC} \eta_h V_h}$$

Note that the lift of the horizontal tail is negative if the wing produces a lift greater than the aircraft weight. This can be the case when the aircraft needs to not only be trimmed but also be statically stable, as described in the next section.

The pitching moment coefficient of the wing, which appears in Equation (6.9), is shown in Appendix C to be given by

$$C_{m,w} = C_{m0} + C_L \left(\frac{x_{cg} - x_{cp}}{c_{MAC}} \right) = C_{m0} + C_L \Delta \bar{x}_{cp}$$

Here x_{cp} denotes the location of the center of pressure of the wing, where the lift could be considered to act without producing a pitching moment. For brevity we have introduced

$$\Delta \bar{x}_{cp} = \frac{x_{cg} - x_{cp}}{c_{MAC}}$$

The second trim condition, that the moment about the center of gravity equals zero, transforms Equation (6.9) to

$$C_m = 0 = C_{m0} + C_{L,w} \Delta \bar{x}_{cp} + C_{L,w} \Delta \bar{x}_{ac} - (C_L - C_{L,w}) \frac{l_h}{c_{MAC}} + C_{m,f}$$

Here, for consistency, we have introduced

$$\Delta \bar{x}_{ac} = \frac{x_{cg} - x_{ac}}{c_{MAC}}$$

Then the lift coefficient required from the wing to trim the aircraft is

$$C_{L,w} = \frac{C_L \frac{l_h}{c_{MAC}} - C_{m0} - C_{m,f}}{\left(\Delta \bar{x}_{cp} + \Delta \bar{x}_{cg} + \frac{l_h}{c_{MAC}} \right)}$$

The displacements of the aerodynamic center and the center of pressure from the center of gravity, $\Delta \bar{x}_{cp}$ and $\Delta \bar{x}_{ac}$, respectively, are much smaller than the normalized tail moment arm l_h/c_{MAC} so we may approximate the trimmed lift coefficient of the wing as follows:

$$C_{L,w} \approx C_L - (C_{m0} + C_{m,f}) \frac{c_{MAC}}{l_h}$$

For a given weight and flight condition the total lift coefficient C_L is fixed and therefore, the sign of C_{m0} and $C_{m,f}$ influences the lift required of the wing. If the sum of the two is negative the lift required from the wing is increased which in turn increases the induced drag. In general the zero-lift moment coefficient of the wing is the determining factor. As pointed out in Chapter 5, C_{m0} for untwisted wings is generally negative, and more so for those using supercritical airfoils. Twisted wings tend to have less negative and often even positive values for C_{m0} and thereby pay less of a trim drag penalty.

Thus there is a trim penalty which is primarily associated with the zero-lift moment coefficient of the wing and the fuselage pitching moment coefficient. We have tacitly assumed that the drag and thrust pass through the center of gravity of the aircraft and therefore make no contribution to the pitching moment. The mounting of the engines and the drag associated with a large vertical T-tail, for example, may require assessment as the design progresses. Similarly, propeller slipstream effects can influence the longitudinal stability; see, for example Wolowicz and Yancey (1972).

6.2.2 Trim and longitudinal static stability

The aircraft can, in general, be trimmed, that is, put into an equilibrium state where the combined lift of the wing and the tail balances the weight while the moment about the center of gravity is zero. But the question remains as to whether the equilibrium so achieved is statically stable. The aircraft is said to be statically stable if the response of the aircraft to a disturbance in angle of attack is to tend to return to the original equilibrium position. Thus, if flying in equilibrium at one angle of attack and a disturbance increases the angle of attack, the moment produced at this angle of attack must act to reduce it, that is, tend back toward the original equilibrium state. Conversely, if a disturbance decreases the angle of attack, the moment at the new, lower, angle of attack must serve to increase that angle. In other words, the rate of change of the moment about the center of gravity must be negative for static stability: an increase in α should reduce $C_{m,cg}$ and a decrease in α should increase $C_{m,cg}$.

The general trend of the moment coefficient for an airplane with and without a horizontal tail is illustrated in Figure 6.9 along with the trim points where the net moment, as well as the net force, is zero. However if we displace the aircraft from these trim points by a small positive angle of attack, that is with the nose rising, the aircraft with a tail will have a negative pitching moment which will tend to drive the nose back down to the trim point. Thus the aircraft with a horizontal tail is statically stable in pitch. Conversely, the aircraft without a horizontal tail will experience a positive pitching moment which will tend to drive the aircraft further from equilibrium. In general, a tailless aircraft is unstable in pitch.

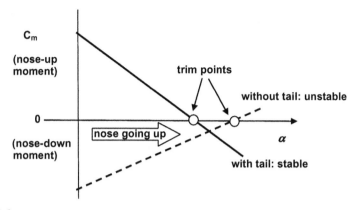

FIGURE 6.9

Moment coefficient of an aircraft about the aerodynamic center of the wing-body-vertical tail assembly with and without a horizontal tail.

Taking the derivative of Equation (6.9) with respect to α yields

$$\frac{\partial C_{m,cg}}{\partial \alpha} = \left(\frac{x_{cg}}{c_{MAC}} - \frac{x_{ac}}{c_{MAC}} \right) \left(\frac{\partial C_{L,w}}{\partial \alpha} \right) - \frac{\partial C_{L,h}}{\partial \alpha} \eta_h V_h + \frac{\partial C_{m,f}}{\partial \alpha} \quad (6.11)$$

The moment of the wing makes no contribution to Equation (6.11) because the moment coefficient of the wing about its aerodynamic center is independent of angle of attack. The first term on the right-hand side of Equation (6.11) will be negative and contribute to longitudinal static stability if $x_{ac} > x_{cg}$, that is, if the center of gravity of the aircraft is forward of the aerodynamic center of the wing. This situation requires a tail down force ($L_h < 0$) so that the wing lift must be greater than the weight $L_w > W$ to trim the aircraft. Increasing the wing lift increases the induced drag and this increase is called trim drag and is considered a penalty. The contribution of the fuselage is destabilizing because as the angle of attack increases the moment produced is positive, or nose-up. However, the fuselage contribution is generally small and therefore should not upset the overall stability of the airplane.

6.2.3 The stick-fixed neutral point

It is possible now to find the aerodynamic center of the aircraft, or, as it's usually called, the neutral point, by setting the derivative of the moment coefficient with respect to angle of attack in Equation (6.11) equal to zero. In this case we assume the control stick is fixed so that the horizontal tail acts without any deflection of the control surface, that is, without deflecting the elevator. The center of gravity location which satisfies this condition is called the neutral point, denoted by x_n. The resulting expression is

$$\frac{x_n}{c_{MAC}} = \frac{x_{ac}}{c_{MAC}} + \frac{\frac{\partial C_{L,h}}{\partial \alpha}}{\frac{\partial C_{L,w}}{\partial \alpha}} \eta_h V_h - \frac{\frac{\partial C_{m,f}}{\partial \alpha}}{\frac{\partial C_{L,w}}{\partial \alpha}} \quad (6.12)$$

The coefficients of wing lift and horizontal tail lift may be expressed, respectively, as follows:

$$C_{L,w} = a_w \left(\alpha - \alpha_0 \right) \tag{6.13}$$

$$C_{L,h} = a_h \left(\alpha - \varepsilon + i_h \right) \tag{6.14}$$

Because horizontal tail airfoils are generally symmetric sections we may set $\alpha_0 = 0$ in Equation (6.13). On the other hand, the wing lift-induced downwash angle ε has been introduced in Equation (6.14) to account for a reduced angle of attack seen by the horizontal tail, as shown in Figure 6.10. The term i_h represents the geometric incidence angle of the horizontal tail with respect to the chord line of the wing. Many jet transports have the capability to adjust the geometric incidence of the horizontal tail. Then, in terms of the lift curve slopes of the aircraft and the tail, a and a_h, respectively, and the downwash angle ε, Equation (6.12) becomes

$$\frac{x_n}{c_{MAC}} = \frac{x_{ac}}{c_{MAC}} + \frac{a_h}{a_w} \left(1 + \frac{\partial \varepsilon}{\partial \alpha} \right) \eta_h V_h - \frac{1}{a_w} \frac{\partial C_{m,f}}{\partial \alpha} \tag{6.15}$$

We may estimate how far the neutral point of the airplane is from the aerodynamic center of the wing by examining the second and third terms on the right-hand side of Equation (6.15). The second term may be written as

$$\frac{a_h}{a_w} \left(1 - \frac{\partial \varepsilon}{\partial \alpha} \right) \eta_h V_h \approx \left(\frac{3}{4} \right) \left(\frac{1}{2} \right) (1) \left(\frac{1}{4} \frac{b}{2 c_{MAC}} \right)$$

$$\approx \left(\frac{3}{4} \right) \left(\frac{1}{2} \right) \left(\frac{1}{4} \frac{3A}{8 \left[1 - \frac{\lambda}{(1+\lambda)^2} \right]} \right) \approx \left(\frac{3}{8} \right) \tag{6.16}$$

Here we have assumed that $l_h \sim b/2$, used the relations for aspect ratio and mean aerodynamic chord from Section 5.1, and the definitions of wing lift curve slope from Section 5.3. It is further assumed that for jet transports $A \sim 8$ and $\lambda \sim 1/3$ in order to arrive at the final result in Equation (6.16).

Analysis of the contribution of the fuselage is complicated by the fact that at angle of attack the flow over the forward portions is reasonably well described by potential theory, but the aft portions are strongly influenced by cross-flow separation and related viscous effects. The nature of the flow over a typical fuselage shape at angle of attack is shown in Figure 6.11. The notional pressure distribution over the fuselage shape shown in Figure 6.11 suggests that the load on the fuselage is essentially a pure couple. This is the reason that the fuselage contribution to the lift is not shown in Figure 6.8 and was neglected in Equation (6.5).

The classical result for a symmetric body of revolution obtained from inviscid small perturbation theory is

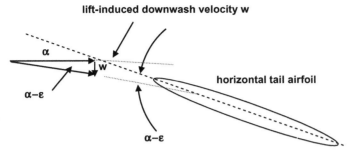

FIGURE 6.10

Velocity field at the horizontal tail showing the reduction in angle of attack due to the downwash velocity.

$$\frac{\partial C_{m,f}}{\partial \alpha} = 2\frac{\upsilon_f}{Sc_{MAC}} \approx 2\frac{\pi \frac{d_f^2}{4}l_f\left(1 - 2\frac{d_f}{l_f}\right)}{Sc_{MAC}} = \left[\frac{\pi}{2}\left(1 - 2\frac{d_f}{l_f}\right)\right]\frac{d_f^2 l_f}{Sc_{MAC}} \quad (6.17)$$

The quantity υ_f denotes the volume of the fuselage which is approximated using a relation suggested by Torenbeek (1982). Typical values of fineness ratio for commercial jet transports are in the range $8 < l_f/d_f < 11$ and therefore the coefficient in square brackets in Equation (6.17) lies between 1.2 and 1.3.

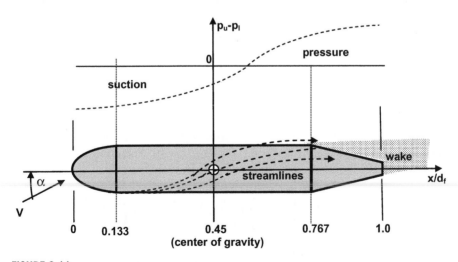

FIGURE 6.11

A typical fuselage shape at angle of attack α showing surface streamlines and separation wake. A notional variation of pressure difference between upper (p_u) and lower (p_l) surface is also depicted.

It is fairly common to see a similar form which was proposed by Gilruth and White (1941):

$$\frac{\partial C_{m_f}}{\partial \alpha} = K_f \frac{d_f^2 l_f}{S c_{MAC}} \tag{6.18}$$

In Equation (6.18) however, the coefficient K_f depends upon the location on the fuselage of the quarter-chord point of the wing in percent l_f, rather than on the fuselage fineness ratio l_f/d_f as in Equation (6.17). A curve fit to Gilruth's data is

$$K_f \approx 0.14 \exp \left(\frac{x_{c/4}}{l_f} \right)^5 - 0.045 \tag{6.19}$$

This gives agreement to within ±3% of Gilruth's data for quarter-chord wing locations $x_{c/4}$ from 20% to 60% of the overall fuselage length l_f. For typical jet transports where the wing position is between 40% and 45% l_f the coefficient K_f lies in the range of $1 < K_f < 1.3$. This compares well with the constant given by potential flow theory in Equation (6.17).

Hopkins (1951) proposed an approach which uses potential flow theory forward and cross-flow drag considerations aft. A semi-empirical relation based on a number of experimental results determined the axial location at which the calculation would switch. The approach gives reasonably good results but typically underestimates the pitching moment obtained in experiments. Among the experiments, which dealt primarily with airship shapes, were two which closely approximated the configuration of typical jet transport fuselages, having fineness ratios of 7.9 and 10.1. In addition, these two models had moments measured about locations that also approximate those of jet transports, $0.456l_f$ and $0.485l_f$, respectively. The experimental results for these two cases were in the range

$$1.29 \frac{\upsilon_f}{S c_{MAC}} < \frac{\partial C_{m_f}}{\partial \alpha} < 1.44 \frac{\upsilon_f}{S c_{MAC}}$$

Potential flow results for these two models overestimated the stability derivative of the fuselage by 25–30%. The experimental results put the coefficient K_f in Equation 14 in the range $0.76 < K_f < 0.9$ for $8 < l_f/d_f < 10$. Therefore both Equations (6.17) and (6.18) should yield similar results. Then we may write

$$\frac{d_f^2 l_f}{S c_{MAC}} = \left(\frac{d_f}{b} \right)^2 \left(\frac{b^2}{S} \right) \left(\frac{l_f}{c_{MAC}} \right) \approx \left(\frac{d_f}{b} \right)^2 A \left(\frac{1.09b}{c_{MAC}} \right)$$

$$\approx \left(\frac{d_f}{b} \right)^2 A (1.09 \times 0.89A) \approx \left(\frac{d_f A}{b} \right)^2 \tag{6.20}$$

Here, we made the same estimates for the typical parameter values for jet transports that were used in developing Equation (6.16). Now Equation (6.15) has the following estimated form:

$$\frac{x_n}{c_{MAC}} \approx \frac{x_{ac}}{c_{MAC}} + \frac{3}{8} - \frac{1}{a_w}K_f\left(A\frac{d_f}{b}\right)^2 \approx \frac{x_{ac}}{c_{MAC}} + \frac{3}{8} - \frac{1}{5}\left(8\frac{1}{8}\right)^2 \approx \frac{x_{ac}}{c_{MAC}} + 0.175$$

$$(6.21)$$

In Equation (6.21) K_f is approximately equal to unity so the second and third terms on the right-hand side are small and therefore their difference is smaller still. Thus the location of the neutral point of a conventional jet transport is fairly close to that of the aerodynamic center of the wing. We shall see that the center of gravity will lie between these two points and therefore the moment arm l_h is approximately equal to $l_{h,ac}$.

6.2.4 The stick-fixed static margin

In the development thus far, the controls have been held fixed, that is, no elevator deflection is employed. We found the neutral point and estimated that it is approximately coincident with the aerodynamic center of the wing. The derivative of $C_{m,cg}$ with respect to angle of attack in Equation (6.11) may be written as

$$\frac{\partial C_{m,cg}}{\partial \alpha} = \left(\frac{x_{cg}}{c_{MAC}} - \frac{x_{ac}}{c_{MAC}}\right)a_w - a_h\left(1 - \frac{\partial \varepsilon}{\partial \alpha}\right)\eta_h V_h + \frac{\partial C_{m,f}}{\partial \alpha} \quad (6.22)$$

When the center of gravity is at the neutral point Equation (6.22) becomes

$$\frac{\partial C_{m,cg}}{\partial \alpha} = 0 = \left(\frac{x_n}{c_{MAC}} - \frac{x_{ac}}{c_{MAC}}\right)a_w - a_h\left(1 - \frac{\partial \varepsilon}{\partial \alpha}\right)\eta_h V_h + \frac{\partial C_{m,f}}{\partial \alpha} \quad (6.23)$$

Subtracting Equation (6.23) from Equation (6.22) yields

$$\frac{\partial C_{m,cg}}{\partial \alpha} = \left(\frac{x_{cg}}{c_{MAC}} - \frac{x_n}{c_{MAC}}\right)a_w = (h - h_n)a_w \quad (6.24)$$

The quantity h is the normalized distance to the aircraft center of gravity measured from the leading edge of the mean aerodynamic chord, while h_n is the normalized distance to the neutral point measured from the leading edge of the mean aerodynamic chord. The quantity $h - h_n$ is called the static margin. It expresses how far the center of gravity of the airplane is forward of the neutral point, expressed as a fraction of the mean aerodynamic chord. The larger $h - h_n$, the more stable the aircraft and the less maneuverable it is. However, with greater stability the pilot workload decreases because fewer control inputs are required to keep a particular course. The airplane is said to be stiffer as the static margin increases. Typical commercial aircraft have a static margin of around 5–10%. When the static margin is zero the airplane is neutrally stable, while if the static margin is negative the aircraft is statically unstable. Modern fighter aircraft employ relaxed static stability (RSS) in order to achieve higher maneuverability, but this requires a flight control system which senses motions and uses redundant computers to provide stabilizing control inputs thus relieving the pilot of a heavy and continuous workload.

6.2.5 Estimate of horizontal tail area based on a stability requirement

The horizontal tail of an aircraft, in conjunction with other aerodynamic components, power plant, and weight characteristics, determines the longitudinal stability and control characteristics of an aircraft. In the following analysis for determining horizontal tail area only one of several stability requirements is enforced. Thus the horizontal tail area so found may be insufficient to meet other stability and control requirements. Here we present a procedure which determines the horizontal tail area S_h needed to produce a controls-fixed neutral point at a position aft of the center of gravity (CG). The aft CG position is determined after the refined weight analysis in Chapter 8 and should be placed at about $30\% c_{MAC}$. The preliminary determination of the horizontal tail volume coefficient of Equation (6.3) was used to define an initial layout of the proposed aircraft and this layout will ultimately change somewhat once the center of gravity location is found using the methods described in Chapter 8.

The refined analysis presented below includes only the most important contributions of the airplane components to the stability of the whole. Effects of the vertical position of both the horizontal tail and the CG are neglected except in the determination of the downwash derivative $\partial \varepsilon / \partial \alpha$ at the tail. The effect of power will be handled with an empirical shift of the neutral point. Other characteristics will be found for the power-off case. In addition, aerodynamic interference effects are neglected. More detailed analyses may be found in Etkin and Reid (1995) and Stengel (2004), among others.

The basic moment equation relating center of gravity position, neutral point, and moment coefficient is Equation (6.24) which may also be written as

$$h - \left(h_{no} + \Delta h_{np}\right) = \frac{C_{m_\alpha}}{a_w} \tag{6.25}$$

The quantity h is the CG location, h_{no} is the power-off neutral point location, and Δh_{np} is the shift in neutral point location due to power (typically around 0.03); all locations are given in fractions of the mean aerodynamic chord c_{MAC}. The power-off, controls-fixed longitudinal stability derivative is denoted by C_{m_α} while a_w is the wing lift curve slope C_{L,w_α}. Then with the center of gravity at its most rearward point, say $0.3c_{MAC}$, the quantity $h_{no} + \Delta h_{np} = 0.3$ and the stability would be neutral. Any further rearward movement of the center of gravity would cause the aircraft to become statically unstable. The contribution of the wing, fuselage, and tail to the stability derivative may be written as follows:

$$C_{m_\alpha} = \frac{\partial C_m}{\partial \alpha} = \left(C_{m_\alpha}\right)_w + \left(C_{m_\alpha}\right)_f + \left(C_{m_\alpha}\right)_t \tag{6.26}$$

Note that the effect of nacelles could be included in the fuselage derivative term if more detail is desired. The wing contribution alone is given by

$$\left(C_{m_\alpha}\right) = a_w \left(h - h_{ac}\right) \tag{6.27}$$

Here the quantity h_{ac} is the location of the aerodynamic center of the wing given as a fraction of mean aerodynamic chord. The longitudinal stability derivative of the horizontal tail is

$$C_{m_\alpha} = -a_h \eta_h V_h \left(1 - \frac{\partial \varepsilon}{\partial \alpha}\right) \tag{6.28}$$

The tail efficiency η_h may be taken to be about 95% unless the tail is completely out of the wing and fuselage wake, where it can be taken as 100%. Using Equations (6.25)–(6.27) we may determine the horizontal tail area to be given by

$$S_h = \frac{c_{MAC} S}{l_h a_h \eta_h} \frac{\left(C_{m_\alpha}\right)_f - a_w \left(h_{ac} - h_{no} - \Delta h_{np}\right)}{1 - \frac{\partial \varepsilon}{\partial \alpha}} \tag{6.29}$$

Now it is necessary to describe how the various terms in Equation (6.29) may be evaluated. The lift curve slope of the wing $a_w = \left(\frac{\partial C_L}{\partial \alpha}\right)_w = \left(C_{L_\alpha}\right)_w$ and this may be found using Equation (5.18), which is presented in units of per radian. As discussed previously, a reasonable estimate for the fuselage contribution may be found from slender-body theory, and is given in Equation (6.17).

To find $\frac{\partial \varepsilon}{\partial \alpha}$, the downwash derivative at the tail, a simplified vortex theory is used so as to permit analytic solutions to the equations involved. It accounts for the following wing characteristics: aspect ratio, sweepback angle, taper ratio, and dihedral angle. Both the vertical and longitudinal positions of the horizontal tail are also included. The simplified analysis typically yields good estimates for conventional transport aircraft configurations, but may be inadequate for unusual tail assemblies.

We assume that the vortex sheet from the wing trailing edge is completely rolled up into two trailing tip vortices but partial roll-up is accounted for in the span of the tip vortices as shown in the flight photograph in Figure 6.12. A detailed discussion of wing theory which describes the trailing vortex system appears in Appendix C.

The maximum value of the downwash variation due to fully rolled up tip vortices at large downstream distances is given by

$$\left(\frac{\partial \varepsilon}{\partial \alpha}\right)_{tip,\infty} = \frac{3.24 \cos \Lambda_{c/4}}{A + 2 \cos \Lambda_{c/4}} \tag{6.30}$$

This value occurs in the plane normal to the plane of symmetry of the aircraft-trailing vortex system. For typical aft tail locations the contribution to the downwash due to the bound wing vortex must also be taken into account. The contribution of the wing is then given by

$$\frac{\left(\frac{\partial \varepsilon}{\partial \alpha}\right)_{wing,max}}{1 - \left(\frac{\partial \varepsilon}{\partial \alpha}\right)_{tip,max}} = \frac{\left(\frac{2x_0}{b}\right) \sqrt{1 + \left(\frac{2x_0}{b}\right)^2}}{\left[\left(\frac{2x_0}{b}\right) + \left(\frac{2l_{h,2}}{b}\right)\right] \sqrt{1 + \left(\frac{2x_0}{b} + \frac{2l_{h,2}}{b}\right)^2}} \tag{6.31}$$

FIGURE 6.12

Roll-up of wingtip vortices on a B727 aircraft. *Courtesy NASA Dryden Flight Research Center.*

Here the quantity $l_{h,2}$ is the distance (taken parallel to the longitudinal axis) from the aerodynamic center of the horizontal tail to the trailing edge of the wing root chord, as shown in Figure 6.13 and the term x_0 may be found from the following equation:

$$A\left(\frac{2x_0}{b}\right)\sqrt{1+\left(\frac{2x_0}{b}\right)^2} = 1 \tag{6.32}$$

Then the total maximum downwash derivative on the centerline of the plane of the tip vortices is

$$\left(\frac{\partial\varepsilon}{\partial\alpha}\right)_{max} = \left(\frac{\partial\varepsilon}{\partial\alpha}\right)_{tip,\infty} + \left(\frac{\partial\varepsilon}{\partial\alpha}\right)_{wing,max} \tag{6.33}$$

The result given by Equation (6.33) would be the maximum value of the tail downwash derivative when the plane of the horizontal tail is coincident with the plane of the trailing vortex system. For other vertical positions y_h above the extended root chord line of the wing, the magnitude of the tail downwash derivative is smaller and is given by

$$\left(\frac{\partial\varepsilon}{\partial\alpha}\right)_{tail} = \frac{\left(\frac{\partial\varepsilon}{\partial\alpha}\right)_{max}}{1+\left(\frac{2y'}{b_v}\right)^2} \tag{6.34}$$

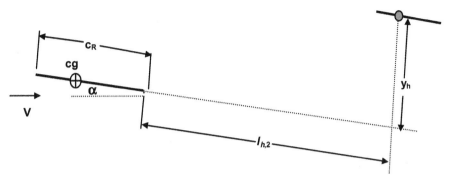

FIGURE 6.13

Elevation view of the symmetry plane of the aircraft showing the definition of the horizontal tail height y_h and the distance $l_{h,2}$.

The correction factor for tail heights above the extended root chord line involves the height of the quarter-chord point of the tail measured from the plane of the trailing vortices y' which is given by

$$y' = y_h - l_{h,3}\left[\alpha - \frac{0.41C_L}{\pi A}\right] - \frac{b}{2}\tan\gamma \tag{6.35}$$

Here $l_{h,3}$ is the longitudinal distance from the quarter-chord line of the wingtip to the quarter-chord point of the tail mean aerodynamic chord and γ is the dihedral angle of the wing as shown in Figure 6.14. The dihedral angle is related to stability in rolling motion of the aircraft and will be discussed subsequently. All the angles in Equation (6.35) are measured in radians and the other quantities are depicted in Figure 6.15. The second factor in the correction term of Equation (6.31) is the tip vortex span b_v which is given by

$$b_v = b - (b - b_{v,ru})\sqrt{\frac{\left(\frac{2l_{h,3}}{b}\right)}{\left(\frac{2l_{h,ru}}{b}\right)}} \tag{6.36}$$

For a wing with taper ratio λ, at far downstream distances where the tip vortices are completely rolled up, the span between them is given by

$$\frac{b_{v,ru}}{b} = 0.78 + 0.10\,(\lambda - 0.4) + 0.003\Lambda_{c/4} \tag{6.37}$$

Once again, the angle $\Lambda_{c/4}$ in Equation (6.37) is measured in radians. Note that the tip vortices move closer to each other in the far field and the distance between them is about 75% of the wingspan. The distance from the wingtip quarter-chord point $c_t/4$ to full roll-up of the tip vortices is given by

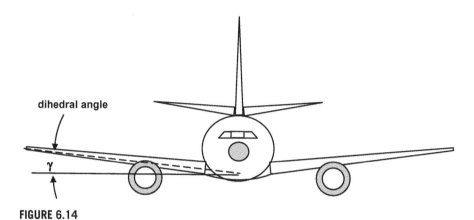

FIGURE 6.14

Front view of an aircraft showing the dihedral angle γ.

$$\frac{2l_{h,ru}}{b} = 0.56\frac{A}{C_L} \tag{6.38}$$

All the previous information may be used to calculate the various terms in Equation (6.29) from which the horizontal tail area may be determined.

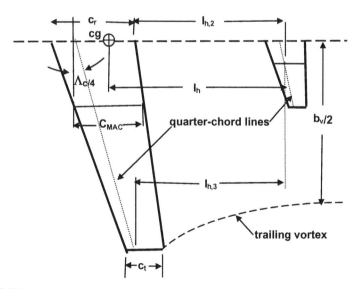

FIGURE 6.15

Plan view of wing and tail showing the various dimensions used in the analysis.

6.3 **Refined vertical tail design**

The vertical tail provides stability in sideslip, or weathervane stability, by developing a yawing moment which moves the aircraft nose back to the original heading after it has been displaced to one side by a small angle β, as indicated in Figure 6.16. In the coordinate frame of the aircraft the free stream velocity V is equal and opposite in sense to the velocity of the center of gravity of the aircraft. In this situation the vertical tail produces a side force Y_v and a moment $Y_v l_v$ acting at the aerodynamic center of the vertical tail. The wing-body combination produces a yawing moment N_{wb} about the center of gravity due to the asymmetrical nature of the sideslip flow. Note that the yawing moments are considered positive in the clockwise direction shown.

The asymmetry of the flow field suggests that it may have some influence on the effectiveness of the vertical tail as indicated by the shaded areas in Figure 6.17 which denote the regions of interference of the wake of the wing-body combination with the vertical tail. Note that the sideslip may occur while the aircraft is at some angle of pitch, as it would be in takeoff or landing, and this expands the interference region in the vertical direction, affecting more of the vertical tail. The fuselage itself is destabilizing in sideslip, just as it was in pitch, while the wings, which experience different drag levels, provide some reduction in the instability of the wing-fuselage combination. A second feature of the asymmetry of the flow field is the generation of a rolling moment by the side force of the vertical tail which must be counteracted by the asymmetric deflection of the ailerons, as depicted in Figure 6.18.

6.3.1 **Equilibrium conditions**

Taking moments about the center of gravity for a small angular displacement β, as depicted in Figure 6.16, results in the following:

$$N_{cg} = N_{wb} + Y_v l_v + N_v \tag{6.39}$$

In coefficient form Equation (6.39) becomes

$$C_{n,cg} = C_{n,wb} + C_{Y,v} \frac{q_v}{q} \frac{S_v l_v}{Sb} + C_{n,v} \tag{6.40}$$

Note that the convention in stability and control analyses is to normalize the yawing and rolling moments by the product qSb, unlike the pitching moment normalization which uses qSc_{mac}. Then, taking the derivative of the moment about the center of gravity with respect to the angle of sideslip yields

$$\frac{\partial C_{n,cg}}{\partial \beta} = \frac{\partial C_{n,wb}}{\partial \beta} + \frac{\partial C_{Y,v}}{\partial \beta} \frac{q_v}{q} \frac{S_v l_v}{Sb} + \frac{\partial C_{n,v}}{\partial \beta} \tag{6.41}$$

For stability the change of the yawing moment with respect to angular displacement must be positive, that is, the yawing moment must increase as the deflection β increases. The contribution of the wing alone to the yawing moment coefficient of

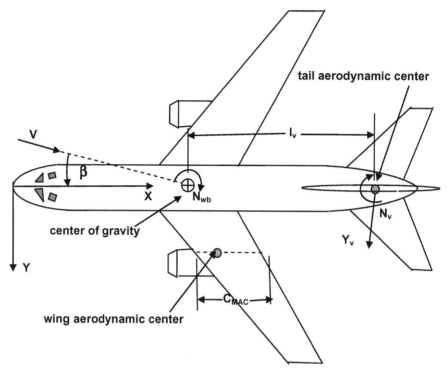

FIGURE 6.16

Plan view of aircraft showing displacement of nose away from the original flight direction by an angle β and the restoring moment due to the vertical tail side force Y_v acting through a distance to the center of gravity. Moments are considered positive in the direction shown.

the wing-body combination is small compared to that of the fuselage alone because the thin wing is moving in its own plane and therefore we assume that $C_{n,wb} \approx C_{n,f}$. The moment coefficient of the vertical tail about its aerodynamic center is independent of β so that $\frac{\partial C_{n,v}}{\partial \beta} = 0$ and Equation (6.40) becomes

$$\frac{\partial C_{n,cg}}{\partial \beta} = \frac{\partial C_{n,f}}{\partial \beta} + \frac{\partial C_{Y,v}}{\partial \beta} \eta_v V_v \qquad (6.42)$$

In Equation (6.42) we introduce the vertical tail efficiency η_v which is given by

$$\eta_v = \frac{q_v}{q} \approx \left(\frac{V_{s,v}}{V}\right)^2 \qquad (6.43)$$

The vertical tail volume coefficient V_v, given in Equation (6.2), expresses the relative strength of the moment the vertical tail can produce and is repeated here for convenience:

$$V_v = \frac{S_v l_v}{Sb}$$

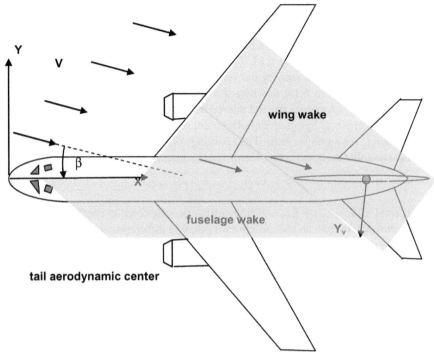

FIGURE 6.17

Wakes produced by the wing and fuselage due to the asymmetrical flow in sideslip are shown indicating how they may affect the response of the vertical tail.

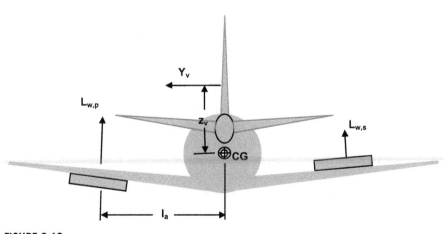

FIGURE 6.18

Aircraft shown looking forward from the aft end showing sideslip-induced side force producing a rolling moment and differential aileron deflection employed to counter that moment ($L_{w,p} > L_{w,s}$).

The vertical tail efficiency η_v accounts for the possibility that the vertical tail will see a somewhat reduced velocity $V_{s,v}$ due to the sidewash generated by the fuselage and the wings, particularly at higher angles of attack, as discussed at the start of this section and illustrated in Figure 6.17. The influence of induced sidewash on the velocity and the angle of attack seen by the vertical tail will be discussed subsequently.

As shown in Figure 6.18, downward aileron deflection on the port (left) wing increases the lift ($L_{w,p}$) while upward aileron deflection decreases the lift ($L_{w,s}$) on the starboard (right) wing. Symmetrical placement of the ailerons makes the aileron lift moment arm l_a the same for both. The resulting rolling moment equilibrium may be expressed as follows:

$$Y_v z_v - \left(L_{w,p} - L_{w,s}\right) l_a = 0$$

The total lift $L = L_{w,p} + L_{w,s} = W$ and putting the side force in coefficient form and dividing through by qSb and accounting for possible differences in dynamic pressure seen by the port and starboard wings $q_{w,p}$ and $q_{w,s}$ yields

$$C_{Y,v}\eta_v V_v \frac{z_v}{l_v} - \left(C_{L,s}\frac{q_{w,s}}{q} - C_{L,p}\frac{q_{w,p}}{q}\right)\frac{l_a}{b} = 0 \tag{6.44}$$

The vertical tail sizing approach presented in the following section requires the aircraft to be able to provide sufficient yawing moment to restore the aircraft to its original heading after a sideslip. Equation (6.44) can provide information on the aileron effectiveness only after the vertical tail has been sized. Therefore aileron sizing will be determined by the resulting vertical tail sizing. The allowable aileron span and associated moment arm l_a are limited by the flap span and location required for landing. To make full use of the wingspan, jet transports often carry outboard ailerons for roll control at low speeds where aerodynamic loads are relatively low and inboard ailerons for roll control at high speeds where aerodynamic loads are high.

6.3.2 Trim and lateral static stability

The aircraft will tend to remain headed in the desired direction in the face of disturbances, like wind shifts, that produce small sideslip angles if the vertical tail is large enough to ensure that the moment about the center of gravity is zero. From a stability point of view, the right-hand side of Equation (6.42) must be positive, so that the vertical tail contribution must be larger than the fuselage contribution, including engine nacelles. With β increasing, as shown in Figure 6.16, we see that the moment produced by the vertical tail is positive. The fuselage response is essentially the same as for pitching motion, so using Equations (6.18) and (6.20) we obtain

$$\frac{\partial C_{n,f}}{\partial \beta} = -k_2 \left(A\frac{d_f}{b}\right)^2 \tag{6.45}$$

From Figure 6.16 we note that the stability derivative for the fuselage in sideslip requires a negative sign because the moment coefficient produced by the fuselage, which corresponds to a negative yawing moment, tends to increase the angle of sideslip, rather than reduce it. The vertical tail, like the horizontal tail, is in a disturbed flow field as shown in Figure 6.17. The sidewash felt by the vertical tail is analogous in effect to the downwash experienced by the horizontal tail, and the angle of attack of the vertical tail is not simply the undisturbed sideslip angle β, but the sum of β and a small sidewash angle. This is usually represented as $\beta \left(1 + \frac{\partial \sigma}{\partial \beta}\right)$. Thus, the variation of the side force coefficient of the vertical tail with sideslip angle may be written as

$$\frac{\partial C_{Y,v}}{\partial \beta} = C_{Y_\beta} = a_v \left(1 + \frac{\partial \sigma}{\partial \beta}\right) \tag{6.46}$$

Equation (6.46) is similar in form to the stability derivative for the horizontal tail given in Equation (6.28). The downwash on the horizontal tail was amenable to analytic treatment but the asymmetric sidewash is more complex so that we will have to treat the sideslip problem with a more accurate empirical approach. Assuming there is no control surface deflection, that is, the rudder is not used, Equation (6.42) may be rewritten as

$$\frac{\partial C_{n,cg}}{\partial \beta} = -k_2 \left(A \frac{d_f}{b}\right)^2 + a_v \eta_v V_v \left(1 + \frac{\partial \sigma}{\partial \beta}\right) \tag{6.47}$$

Solving Equation (6.47) for the area of the vertical tail yields

$$S_v = \frac{c_{MAC} S}{l_v a_v \eta_v \left(1 + \frac{\partial \sigma}{\partial \beta}\right)} \left[k_2 \left(A \frac{d_f}{b}\right)^2 + \frac{\partial C_{n,cg}}{\partial \beta}\right] \tag{6.48}$$

Using some estimates of typical characteristics of jet transports in Equation (6.48) suggests that

$$S_v = \left(\frac{1}{3}\right) \left[1 + \frac{\partial C_{n,cg}}{\partial \beta}\right] \left(\frac{2}{A}\right) S \approx \frac{1}{12} \left[1 + \frac{\partial C_{n,cg}}{\partial \beta}\right] S \tag{6.49}$$

Torenbeek (1982) indicates that for jet transports the stability derivative, in our notation, lies in the range

$$0.09A < \frac{\partial C_{n,cg}}{\partial \beta} < 0.22A$$

For the smaller value of the applied moment about the center of gravity, 0.1, the vertical tail area would need to be about $0.14S$, which is near the low end of the values for jet transports. However, for the larger value of the moment about the center of gravity, 0.25, the vertical tail area would need to be about $0.23S$, which is near the high end of the values for jet transports.

Although an airplane with a vertical tail is stable to small disturbances, an important criterion for the vertical tail size is its ability to handle the one-engine-out condition. This situation is depicted in Figure 6.19. This is especially important for aircraft with wing-mounted engines, as was discussed briefly in Chapter 5 with respect to locating the engines on the wing. The vertical tail must provide sufficient turning force to balance the yaw moment due to the asymmetric thrust produced when one engine is inoperative. For this situation Equation (6.40), with the assumptions made thus far, may be written as

$$C_{n,cg} = -k_2\beta \left(A\frac{d_f}{b}\right)^2 + a_v\beta\left[k\left(1 + \frac{\partial\sigma}{\partial\beta}\right)\right]\eta_v V_v + C_{n,1eo} \qquad (6.50)$$

Consider the port engine to be shut down so that only the starboard engine is producing thrust at the level F_{1eo}, as shown in Figure 6.19. This thrust, commanded by the pilot from the $n_e - 1$ operating engine(s), defines a new equilibrium speed where the total drag of the airplane, including the drag of the inoperative engine, matches the new total thrust level. The moment coefficient produced by one engine being inoperative, as appears in Equation (6.50), may be written as

$$C_{n,1eo} = \frac{-F_{eo}y_{eo}}{qSc_{MAC}} = \frac{-\left(\frac{D}{n_e-1}\right)y_{eo}}{qSc_{MAC}} = \frac{-C_D}{n_e - 1}\frac{y_{eo}}{c_{MAC}} \qquad (6.51)$$

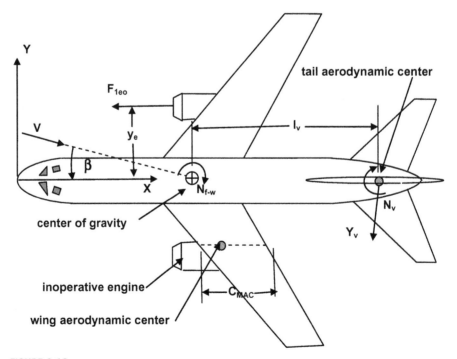

FIGURE 6.19

Plan view of aircraft showing displacement of nose away from the original flight direction by an angle β because of the asymmshut down.

Here y_{eo} denotes the spanwise position of the inoperative engine and F_{eo} represents the new thrust level produced by the engine at the same spanwise position on the other wing. In the case of one engine out on an aircraft with four wing-mounted engines the approach is the same because two engines will still be operating symmetrically. This is also true if one wishes to consider the more stressing case of two engines out on the same wing, as long as one combines the thrust of the two engines as if they were one. The total thrust is equal to the drag D developed at the new speed in the one-engine-out configuration. Setting the moment about the center of gravity in Equation (6.50) equal to zero, inserting the result from Equation (6.51) for the asymmetric thrust, and solving for the required vertical tail area yields

$$S_v = \frac{1}{a_v \eta_v k \beta \left(1 + \frac{\partial \sigma}{\partial \beta}\right)} \left[\beta k_2 \left(A \frac{d_f}{b}\right)^2 + \frac{C_D}{(n_e - 1)} \frac{y_e}{c_{mac}}\right] \left(\frac{c_{MAC}}{l_v}\right) S \quad (6.52)$$

An evaluation of Equation (6.52) for typical jet transport parameters yields the results shown in Figure 6.20, which also includes data for some modern jet transport aircraft. In this analysis a low-speed stressing case is assumed: takeoff configuration with an engine failure and a corresponding drag coefficient of 0.06, while the sideslip angle is taken as 0.1 radians. The two calculated curves shown have only one difference: the rudder power characterized by the angle of zero-lift shift Δ_v. It is clear that the vertical tail itself may have insufficient control authority to keep the aircraft stable without a significantly larger surface area. However, deflecting the rudder provides additional side force so as to keep the vertical tail surface area at lower values. It can be seen that current aircraft can maintain a margin of safety with reasonable vertical tail size because of the rudder effectiveness.

6.3.3 **Horizontal and vertical tail placement**

In Equation (6.48) it is clear that the lift curve slope of the vertical tail a_v directly influences the required vertical tail surface area. From the study of wings in Chapter 5 it is equally clear that the lift curve slope is directly dependent on the aspect ratio of the wing. Considering the vertical tail as a wing we are struck first by the rather small aspect ratio of that surface taken as an isolated body. The aspect ratio in this case is simply defined as $A_v = b_v^2/S_v$ where these quantities are defined in Figure 6.21. As indicated in Figure 6.3, vertical tail aspect ratios for jet transports are typically in the range $1 < A_v < 2$.

The DATCOM method described by Hoak et al. (1978) for estimating the side force coefficient C_Y clearly indicates that the lift curve slope is enhanced by the presence of a fuselage at the base of the vertical tail and that a T-tail also improves the lift curve slope. This is the so-called endplate effect where a surface bordering the vertical tail helps reduce three-dimensional tip effects which reduce the lift curve slope. One may think of the difference between testing a finite wing in a wind tunnel as opposed to a wing that completely spans the tunnel.

Three basic tail configurations are shown in Figure 6.22: (a) the low tail, common to most large jet transports ($W_{to} > 100{,}000\,\text{lb}$), (b) the cruciform tail, which is rarely

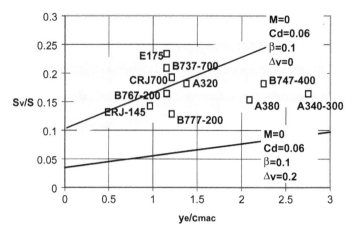

FIGURE 6.20

Variation of the vertical tail area with the spanwise location of the engines. Shaded area indicates range for current aircraft.

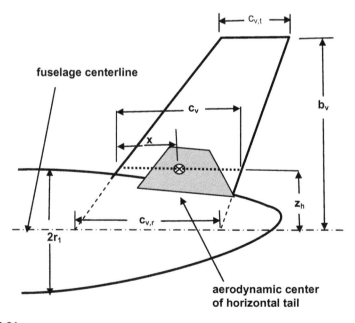

FIGURE 6.21

Shown here are the parameters of the fuselage wing which contribute to the sideslip derivative $C_{Y\beta}$. The shaded area denotes the horizontal tail.

seen on jet transports, and (c) the T-tail which is common on smaller jet transports ($W_{to}<100,000\,lb$), particularly those with fuselage-mounted engines. The cruciform tail is rarely encountered because it provides the least enhancement of the natural lift curve slope of the vertical tail and it is structurally complex. The T-tail is common on aircraft with fuselage-mounted engines because the jet exhaust would interfere catastrophically with a low tail setting and because it provides a substantial increase in the effective aspect ratio of the vertical tail alone, developing an A_{eff} up to $2.8A_v$. The low set tail is structurally sound and provides much the same improvement in lift curve slope as the T-tail, up to $2A_v$, while avoiding the complications of a cantilevered structure highly loaded at its tip.

An empirical equation for the vertical tail side force stability derivative is given in DATCOM as

$$\frac{\partial C_{Y,v}}{\partial \beta} = -k \left(\frac{\partial C_L}{\partial \alpha}\right)_v \left(1 - \frac{\partial \sigma}{\partial \beta}\right) \frac{q_v}{q} \frac{S_v}{S} \tag{6.53}$$

The quantity k is a coefficient that is a function of the ratio $b_v/2r_1$ where b_v is the vertical tail span and $2r_1$ is the depth of the fuselage in the vicinity of the vertical tail. The definition of both these parameters is rather arbitrary but b_v is often taken as the distance from the fuselage centerline measured vertically to the tip of the tail and $2r_1$ is often taken as the diameter of the fuselage near the junction of the leading edge of the vertical tail with the fuselage. The magnitude of the ratio $b_v/2r_1$ is indicative of the endplate effect of the fuselage on the vertical tail. In the range $0.5<b_v/2r_1<3.5$ the fuselage has a beneficial effect on the vertical tail to the extent that its effective aspect ratio $A_{v,eff}$ is from 20% to 60% greater than the isolated tail value, defined as $A_v=b_v^2/S_v$. At larger values of the ratio there is little or no benefit

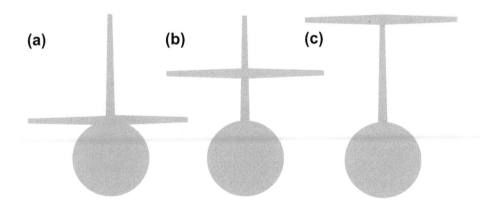

(a)   (b)   (c)

FIGURE 6.22

Three possible tail configurations for transport aircraft. (a) Low horizontal tail, (b) cruciform tail and (c) T-tail.

of the presence of the fuselage. The factor k in Equation (6.53) has the following behavior:

$$\frac{b_v}{2r_1} < 2 : k = 0.75$$

$$2 \le \frac{b_v}{2r_1} \le 3.5 : k = 0.167\left(2.5 + \frac{b_v}{2r_1}\right) \tag{6.54}$$

$$\frac{b_v}{2r_1} > 3.5 : k = 1$$

Typical commercial transports have vertical tails that fall in the middle range $2 < b_v/2r_1 < 3$ with a representative value being about 2.4.

The combined effect of sidewash angle and local dynamic pressure is given by DATCOM as follows:

$$\left(1 + \frac{\partial\sigma}{\partial\beta}\right)\eta_v = 0.724 + \frac{3.06}{1 + \cos\Lambda_{c/4}}\frac{S_v}{S} + 0.4\frac{z_{w,c/4}}{d_f} + 0.009A \tag{6.55}$$

The quantity $z_{w,c/4}$ is the perpendicular distance from the wing root quarter-chord point to the fuselage centerline. The remaining quantity in Equation (6.53) is the lift curve slope of the vertical tail, considering it as a wing. The lift curve slope for a wing, given in Equation (5.18), depends upon the aspect ratio of the wing and we have just indicated that endplate effects due to the presence of the fuselage can alter the geometric value of the aspect ratio of the vertical tail. The position of the horizontal tail in the x, z plane can also contribute to endplate effects. In addition, the relative size of the vertical and horizontal tail influences the effective aspect ratio. In order to account for these effects DATCOM presents three graphs that permit estimation of the effective aspect ratio of the vertical tail.

The effective aspect ratio of the vertical tail is given by

$$A_{v,eff} = A_v\left(\frac{A_{v,f}}{A_v}\right)\left\{1 + K_h\left[\left(\frac{A_{v,fh}}{A_v}\right)\left(\frac{A_{v,f}}{A_v}\right)^{-1} - 1\right]\right\} \tag{6.56}$$

The term $A_{v,f}/A_v$ is the ratio of the aspect ratio of the vertical tail in the presence of a fuselage to that of an isolated vertical tail panel which is considered to be the extension of the vertical tail leading and trailing edges to the fuselage centerline as illustrated in Figure 6.21. The quantity $A_{v,f}/A_v$ is shown in Figure 6.23 as a function of the span to local body diameter $b_v/2r_1$; the taper ratio of the vertical tail $\lambda_v = c_{v,r}/c_{v,t}$ appears as a parameter. It is clear that the effective aspect ratio for a vertical tail can be from 20% to 60% larger than its geometric aspect ratio. Indeed, for commercial airliners where $b_v/2r_1$ is around 2, the effective aspect ratio $A_{v,f}/A_v \sim 1.6$. However, this effect does not carry over to the case of conventional fuselage-wing combinations. Because the wingspan is much larger than the local fuselage diameter, the endplate effect of the fuselage on the wing is negligible.

The coefficient K_h in Equation (6.56), which accounts for the relative size of the horizontal and vertical tails, is illustrated in Figure 6.24. For commercial airliners the ratio of horizontal to vertical tail area is around 1.2 so that the coefficient $K_h \sim 1$. The term $A_{v,fh}/A_v$ is the ratio of the aspect ratio of the vertical tail in the presence of both a fuselage and a horizontal tail to that of the isolated vertical tail panel. This quantity is shown in Figure 6.25 as a function of z_h/b_v, where z_h is the location of the horizontal tail normal to the fuselage centerline as shown in Figure 6.21. The parameter x/c_v appearing in Figure 6.25 is the ratio of the longitudinal distance from the leading edge of the vertical tail to the aerodynamic center of horizontal tail normalized by the chord of the vertical tail measured at z_h. For aircraft with low horizontal tails z_h/b_v is relatively small, and from Figure 6.25 we see that $1 < A_{v,fh}/A_v < 1.3$. For T-tails where $z_h/b_v = 1$ we see that $A_{v,fh}/A_v = 1.7$. For the low tail case we would find moderate improvements of about 10–20% in the effective aspect ratio of the vertical tail $A_{v,eff}$. For the T-tail the improvement would rise from 50% to 70%. The improvement for the T-tail comes at some cost because the aerodynamically loaded horizontal tail would place substantial loads on the vertical tail and it would require strengthening that would entail added weight. The decision to use a T-tail is usually predicated on other factors, for example, to ensure that the horizontal tail won't be immersed in the jet exhaust from turbofan engines mounted on the aft fuselage.

6.3.4 Example calculation of vertical tail stability derivative

A wind tunnel model consisting of a fuselage, wing, horizontal tail, and vertical tail has wing area $S = 576\,\text{in.}^2$, quarter-chord sweepback $\Lambda_{c/4} = 30°$, and aspect ratio

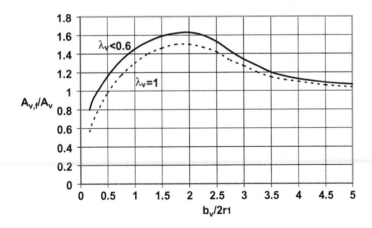

FIGURE 6.23

$A_{v,f}/A_v$, the ratio of the aspect ratio of the vertical tail in the presence of a fuselage to that of the isolated vertical tail, is shown as a function of $b_v/2r_1$. The taper ratio of the vertical tail λ_v appears as a parameter.

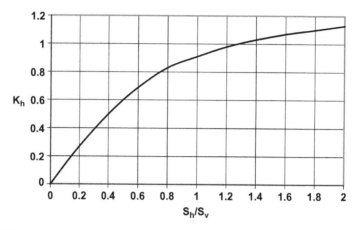

FIGURE 6.24

The coefficient K_h of Equation (6.51) is shown as a function of the ratio of the areas of the horizontal and vertical tails.

$A=6$. The wing root quarter-chord point is located on the body centerline so that $z_{w,c/4}=0$. The vertical tail uses an NACA 63-006 airfoil and the effect of adding this vertical tail to the fuselage-wing-horizontal tail combination is sought. The details of the tail assembly are illustrated in Figure 6.26.

From Figure 6.26 the ratio $b_v/2r_1=2.72$ and $\lambda_v=0.16$. Entering Figure 6.23 with these values suggests $A_{v,f}/A_v=1.47$. Again using the information in Figure 6.26

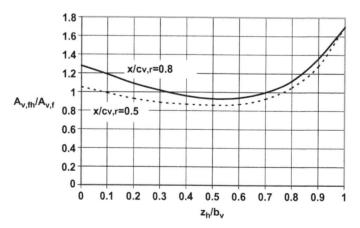

FIGURE 6.25

$A_{v,fh}/A_{v,f}$, the ratio of the aspect ratio of the vertical tail in the presence of both a fuselage and a horizontal tail to that of the vertical tail in the presence of a fuselage alone is shown as a function of z_h/b_v, where z_h is the location of the horizontal tail normal to the fuselage centerline as shown in Figure 6.21.

we may calculate $S_v = 153.7\,\text{ft}^2$ and $A_v = b_v^2/S_v = 1.51$, as well as $S_h = 121.5\,\text{in.}^2$, $A_h = 4.14$, and $\lambda_h = 0.5$. Using Equation (5.6) for the mean aerodynamic chord leads to $c_{MAC,v} = 11.86$ and $c_{MAC,h} = 5.62$. The aerodynamic centers for the horizontal and vertical tails are shown at their respective quarter-chord points in Figure 6.26.

The ratio $S_h/S_v = 0.79$ and using Figure 6.24 we select $K_h = 0.82$. From Figure 6.26 we know that $x/c_{v,r} = 0.8$ and that $z_h/b_v = 0$. Then, using this information in Figure 6.25 yields the approximate value $A_{v,fh}/A_{v,f} = 1.28$. Then Equation (6.56) may be solved for the effective aspect ratio of the vertical tail as $A_{v,eff} = 2.71$. Using this result for the effective aspect ratio of the vertical tail in Equation (5.18) for the lift curve slope gives

$$a_v = \left(\frac{\partial C_L}{\partial \alpha}\right)_v = \frac{2\pi A_{v,eff}}{2 + \sqrt{\frac{A_{v,eff}^2}{\kappa^2}\left(\beta^2 + \tan^2 \Lambda_{c/2}\right) + 4}}$$

The experimental value for the lift curve slope of the vertical tail, which has an NACA 63-006 airfoil, is given as $a = 0.112$ per degree or 6.42 per radian in Table 5.1. The corresponding value of $\kappa = a/2\pi = 1.021$. The sweepback of the mid-chord of the vertical tail is shown in Figure 6.26 to be $\Lambda_{c/2} = 41.9°$ so that for incompressible flow ($M = 0$) the lift curve slope is found to be $a_v = 2.76$ per radian, while for $M = 0.8$ the lift curve slope is $a_v = 2.93$ per radian.

The ratio of vertical tail area to the given wing area is $S_v/S = 0.267$ and $z_{w,c/4} = 0$ so that Equation (6.55) may be solved to yield

$$\left(1 + \frac{\partial \sigma}{\partial \beta}\right)\eta_v = 1.22\eta_v$$

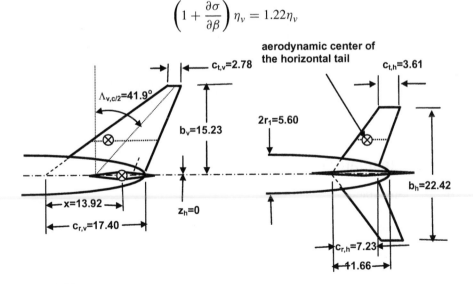

FIGURE 6.26

Tail configurations for the sample calculation of the effect of adding the vertical tail to the fuselage-wing-horizontal tail combination. All dimensions are given in inches.

We may use this information in Equation (6.53) to find the change in side force coefficient due to the addition of the vertical tail to the wing-fuselage-horizontal tail combination. The constant k in that equation may be found from Equation (6.54) to be $k=0.87$ so that

$$\left(\frac{\partial C_{Y,v}}{\partial \beta}\right)_{f,h} = -0.87a_v\,(0.267)\,\eta_v = -0.232a_v\eta_v$$

Then assuming $\eta_v=0.95$ we find $\left(\frac{\partial C_Y}{\partial \beta}\right)_{v,f,h} = -0.61$ per radian for $M=0$ and -0.69 per radian for $M=0.8$.

6.4 Design summary

At this stage of the design a preliminary configuration of the design aircraft may be completed with specific dimensions, except for the landing gear, which will be treated in the next chapter. The fuselage dimensions are known from Chapter 3 and the wing dimensions and characteristics are known from Chapters 4 and 5. The horizontal and vertical tail sizes and characteristics required for basic static stability are determined either by using the simple empirical approach of Section 6.1 or, preferably, by using the more refined analyses leading to Equations (6.29) and (6.52). The sizing of all these components is necessary for carrying out the refined weight estimate described in Chapter 8. Indeed, the final placement of the wing on the fuselage which defines the location of the center of gravity for the aircraft will be accomplished in Chapter 8 after the refined weight analysis is carried out. However, in order to carry out the tail sizing a nominal center of gravity location is required. Therefore, at this point it is sufficient to follow the suggestion of Torenbeek (1982) and take the approximate center of gravity to be about 42–45% of the fuselage length for aircraft with wing-mounted engines and about 47% of the fuselage-length for aircraft with aft-fuselage-mounted engines.

6.5 Nomenclature

A	aspect ratio
a	lift curve slope
b	wingspan
b_v	span of trailing vortices
C_D	drag coefficient
CG	center of gravity
C_L	lift coefficient
C_m	pitching moment coefficient
C_n	yawing moment coefficient

c_{MAC}	mean aerodynamic chord
C_r	side force coefficient
d	diameter
h	normalized center of gravity location, x_{cg}/c_{MAC}
h_n	normalized neutral point location, x_n/c_{MAC}
h_{no}	normalized power-off neutral point location
i	incidence angle
K_f	fuselage coefficient, Equation (6.19)
K_h	horizontal tail coefficient, see Equation (6.56)
k	constant
L	lift
l	length
l_a	moment arm for lift increment due to aileron deflection, see Figure 6.13
l_h	longitudinal distance between aerodynamic centers of the wing and horizontal tail
$l_{h,2}$	see Figure 6.13
$l_{h,3}$	see Figure 6.15
l_v	longitudinal distance between aerodynamic centers of the wing and vertical tail
M	pitching moment
N	yawing moment
n_e	number of engines
p	pressure
q	dynamic pressure
r_1	local body radius, see Figure 6.21
S	projected wing area
S_h	projected area of horizontal tail
S_v	projected area of vertical tail
V	free stream velocity
V_h	horizontal tail volume coefficient $= S_h l_h / S c_{MAC}$
V_v	vertical tail volume coefficient $= S_v l_v / S b$
W	weight
x	longitudinal distance from aircraft nose
x_0	Equation (6.30)
Y	side force
y	spanwise distance from centerline
y'	Equation (6.35)
y_h	see Figure 6.13
z_h	see Figure 6.21
z_v	moment arm for vertical tail side force, see Figure 6.18
α	angle of attack
β	sideslip angle
Δh_{np}	shift of neutral point due to power application
ε	downwash angle

γ	wing dihedral angle
λ	taper ratio
η_h	horizontal tail efficiency q_h/q
η_v	vertical tail efficiency q_v/q
Λ	sweepback angle
υ	volume
ς	sidewash angle

6.5.1 Subscripts

ac	aerodynamic center
cg	center of gravity
cp	center of pressure
$c/4$	quarter-chord
d	downwash
eff	effective
f	fuselage
g	gross
h	horizontal tail
n	neutral point
p	port (left) side of aircraft
s	starboard (right) side of aircraft
ru	vortex roll-up
v	vertical tail
w	wing
wb	wing-body combination
α	derivative with respect to angle of attack α
0	zero lift
eo	one engine out

References

Etkin, B., Reid, L.D., 1995. Dynamics of Flight: Stability and Control. Wiley, NY.

Gilruth, R.R., White, M.D., 1941. Analysis and Prediction of Longitudinal Stability of Airplanes, NACA Report 711.

Hoak, D.E., et al., 1978. USAF Stability and Control DATCOM, Flight Control Division, Air Force Flight Dynamics Laboratory, Wright-Patterson AFB.

Hopkins, E.J., 1951. A Semiempirical Method for Calculating the Pitching Moment of Bodies of Revolution at Low Mach Numbers, NACA RM A51C14.

Stengel, R., 2004. Flight Dynamics. Princeton University Press, Princeton, NJ.

Torenbeek, E., 1982. Synthesis of Subsonic Airplane Design. Kluwer Academic Publishers, Dordrecht, The Netherlands.

Wolowicz, C.H., Yancey, R.B., 1972. Longitudinal Aerodynamic Characteristics of Light, Twin-Engine, Propeller-Driven Airplanes, NASA TN D-6800.

Landing Gear Design

CHAPTER OUTLINE

7.1 Introduction

The landing gear serves to support the aircraft while standing on the ground, to provide a stable platform during the takeoff roll, and to accommodate the stress of landing and braking. In addition, it must be capable of retracting sufficiently to be stowed

Commercial Airplane Design Principles. http://dx.doi.org/10.1016/B978-0-12-419953-8.00007-3

aboard the aircraft within an aerodynamically streamlined contour. A detailed development of landing gear theory and practice is given by Curry (1988).

7.2 General characteristics of commercial jet transport landing gear

Jet transports use tricycle-type landing gear with nose gear aligned with the fuselage centerline and main gear symmetrically disposed with respect to the aircraft centerline and located aft of the aircraft center of gravity, as shown schematically in Figures 7.1 and 7.2. The basic requirement is that the landing gear safely support the aircraft weight under all static loading conditions, provide adequate stability and maneuverability during takeoff acceleration, properly retract and extend from locations within the aircraft contour, and accommodate the dynamic loads incurred in transitioning from flight to ground operation.

7.2.1 Quasi-static loads on landing gear

When the aircraft is stationary or taxiing at constant speed such that accelerations are very small, force equilibrium requires that

$$F_{NG} + F_{MG} = W_{oe} \qquad (7.1)$$

The symmetric disposition of the main gear about the fuselage centerline ensures that the forces on the port and starboard components of the main gear are equal, with each supporting a load equal to $F_{MG}/2$. The farthest aft center of gravity location, which occurs in the operating empty weight configuration, produces the greatest static load on the main gear. Taking moments about the center of gravity in the operating empty weight configuration, and enforcing the condition that there be no net moment acting yields

$$F_{NG} \left(x_{oe} - x_{NG} \right) = F_{MG} \left(x_{MG} - x_{oe} \right) \qquad (7.2)$$

From Equations (7.1) and (7.2) we find the static loads on the nose and main gear to be as follows:

$$F_{NG} = W_{oe} \frac{\left(x_{MG} - x_{oe} \right)}{l_{wb}} \qquad (7.3)$$

$$F_{MG} = W_{oe} \frac{\left(x_{oe} - x_{NG} \right)}{l_{wb}} \qquad (7.4)$$

Commercial aircraft typically have $x_{oe}/l_f \sim 0.45$, $x_{NG}/l_f \sim 0.10$, and $x_{MG}/l_f \sim 0.50$ so that $l_{wb} = x_{MG} - x_{NG} \sim 0.40 l_f$. It may be expected then that about 85–90% of the static load is taken by the main gear and the remaining 10–15% by the nose gear. The loading of the payload, that is, passengers and cargo, onto the aircraft must be carried out in such a way that the center of gravity is kept forward of x_{oe}.

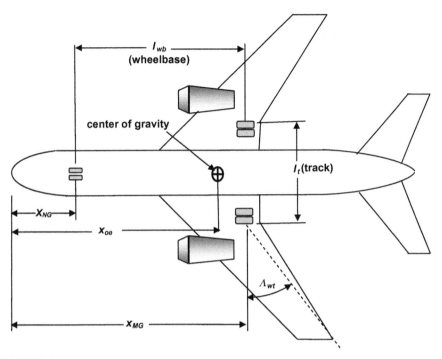

FIGURE 7.1

General layout of landing gear for commercial jet transports as seen from below.

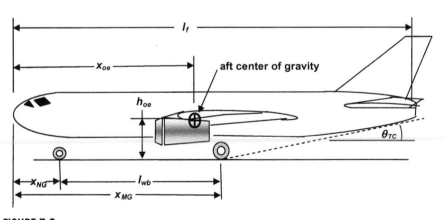

FIGURE 7.2

Elevation view of the general landing gear layout for a commercial jet transport.

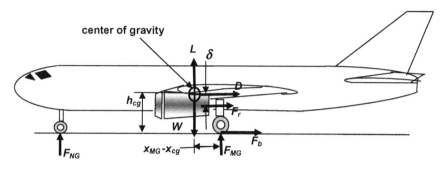

FIGURE 7.3

Forces acting on the aircraft center of gravity during braking.

7.2.2 Dynamic loads in landing

In a nominal landing the aircraft has a nose up attitude so that the main gear touch first, activating the lift dumpers, or spoilers, on the wings thereby reducing the lift. The aircraft nose is rotated down to contact the runway surface and when all wheels are firmly on the ground, reverse thrust and wheel braking are initiated. The effect of wheel braking serves to put a high load on the nose gear, as can be appreciated by first considering the schematic diagram of forces in Figure 7.3. The aircraft has a deceleration parallel to the ground but essentially no angular acceleration about its axis because of the constraint of the ground plane.

Because there is no angular acceleration about the center of gravity there can be no unbalanced moment about that point. Therefore we can consider the moments acting around the center of gravity of the aircraft as a static problem with the following result:

$$F_{NG}\left(x_{cg} - x_{NG}\right) - F_{MG}\left(x_{MG} - x_{cg}\right) - F_b h_{cg} - F_r \delta = 0 \qquad (7.5)$$

Equation (7.5) may be written as

$$F_{NG} = \frac{F_{MG}\left(x_{MG} - x_{cg}\right) + F_b h_{cg} + F_r \delta}{l_{wb} - \left(x_{MG} - x_{cg}\right)} \qquad (7.6)$$

Because there is also no vertical acceleration we require that

$$F_{NG} + F_{MG} = W - L \qquad (7.7)$$

Using Equation (7.7) transforms Equation (7.6) into

$$F_{NG} = (W - L)\frac{x_{MG} - x_{cg}}{l_{wb}} + F_b \frac{h_{cg}}{l_{wb}} + F_r \frac{\delta}{l_{wb}} \qquad (7.8)$$

The first term in Equation (7.8) is of the form of the static load on the nose gear as given previously in Equation (7.3), less the alleviating effect of lift. As shown in Equation (7.3) the furthest aft center of gravity location, x_{oe}, provides the least static load on the nose gear. However, the second term in Equation (7.8) represents the added load due to deceleration forcing the nose down and increasing the load on the nose gear. Thus braking poses the most stressing case for the load on the nose gear. The third term in Equation (7.8) accounts for the effect of thrust reversal and is shown as providing a nose down moment for wing-mounted engines. For fuselage-mounted engines, which are generally located somewhat above the airplane's center of gravity, the sign of the third term in Equation (7.8) should be reversed.

We see that the maximum load on the nose gear occurs during braking when L and F_r are small. The lift is dumped upon touchdown and drops to low levels once the nose gear contacts the runway. A discussion of the use of thrust reversers is given in Chapter 10, but for the present analysis one may consider the worst case, that is, $F_r = 0$. Commercial airline operators typically call for cut-off of the thrust reversers at a speed of 60–80 knots, according to Yetter (1995). If we characterize the braking force as $F_b = (W/g)a$, where a is the average deceleration, then the maximum force on the nose gear is

$$F_{NG,\max} = W \left[\frac{(x_{MG} - x_{cg})}{l_{wb}} + \left(\frac{a}{g}\right) \frac{h_{cg}}{l_{wb}} \right] \qquad (7.9)$$

The important parameters here are the deceleration level, which is typically in the range of $0.33 < (a/g) < 0.45$ for dry concrete runways, and the ratio of center of gravity height to wheelbase, which is generally in the range $0.16 < h_{cg}/l_{wb} < 0.23$. The first term in the square brackets in Equation (7.9), using the approximate dimensions in Section 7.2.1, is around 0.15, while the second term can range from 0.05 to 0.1. Thus the effect of braking is to increase the load on the nose gear by a factor of 33–67% over the static load. Federal Air Regulations (FAR) require nose wheel tires to withstand a load under braking 50% greater than the static load, as will be discussed in a subsequent section.

The ratios h_{cg}/l_{wb} and l_{wb}/l_f are given for a number of commercial aircraft in Table 7.1. The dimensions are estimated from drawings or published data from, for example, Jackson (2010). A graphical representation of the data appears in Figure 7.4. It is clear that the variation of these ratios over a wide range of aircraft sizes is rather slight and the average value of h_{cg}/l_{wb} is 0.2 and that for l_{wb}/l_f is 0.42.

Impact loads occur during touchdown. The main gear touch first and the sink rate of the aircraft must be low enough to avoid high loads on the main gear which may lead to a blowout of one or more tires. The desired sink rate, or vertical velocity, is generally kept to less than 3 ft/s for commercial aircraft. The Space Shuttle Orbiter was designed for 3 ft/s, while naval carrier aircraft have allowable sink rates of 15 ft/s or more. As described in Chapter 10, the approach flight path angle $\gamma \sim 3°$ so the sink rate $w \sim V_a \tan\gamma \sim 0.05 V_a$. In Chapter 4 approach equivalent airspeeds for large transport aircraft were shown to be about 150 kts (173 mph = 253 ft/s) so that the corresponding sink rate is $w \sim 12.7$ ft/s, a value too high for typical commercial aircraft operations. A sink rate of even 4 or 5 ft/s is considered very bumpy and uncomfortable for passengers. The flare maneuver prior to touchdown reduces the flight path

Table 7.1 Landing Gear Characteristics of Operational Airliners

Aircraft	h_{oe}/l_{wb}	l_{wb} (ft)	l_f (ft)	l_{wb}/l_f	W_{to} (lbs)	l_t (ft)	$l_t/2l_{wb}$	b (degrees)	h_{oe}/l_t
B737-700	0.21	41	105	0.39	133,000	18.75	0.227	12.8	0.463
B737-800	0.18	57	125	0.46	155,500	18.75	0.164	9.3	0.549
B737-900	0.16	56	133	0.42	164,000	18.75	0.166	9.4	0.481
B747-400	0.2	84[a]	225	0.37	800,000	36.08[b]	0.215	12.1	0.466
B767-200	0.23	65	155	0.42	345,000	30.5	0.236	13.3	0.487
B767-300	0.2	75	176	0.42	380,000	30.5	0.204	11.5	0.490
B777-200	0.2	95	209	0.45	545,000	36	0.190	10.7	0.527
B777-300	0.17	103	240	0.43	660,000	36	0.176	10.0	0.484
B787-8	0.19	75	183	0.41	484,000	32.16	0.215	12.2	0.441
B787-9	0.19	85	204	0.42	540,500	32.16	0.190	10.8	0.500
A320	0.23	42	123	0.34	162,040	24.92	0.300	16.7	0.383
A330-300	0.21	84	209	0.40	507,060	34.42	0.205	11.6	0.512
A340-300	0.2	83	209	0.40	606,275	34.42	0.207	11.7	0.484
A350-900	0.2	94	219	0.43	584,225	34.77	0.185	10.5	0.541
A380	0.19	115[a]	231	0.50	1,234,590	40.85[b]	0.178	10.1	0.535
E175	0.21	37	102	0.36	82,673	13.5	0.182	10.3	0.577
CRJ700	0.17	45	97	0.46	72,750	18.76	0.209	11.8	0.406
Average	0.20			0.42			0.203	11.5	0.490

[a] Nose gear to mean value between main gear bogies.
[b] Track of outboard bogies.

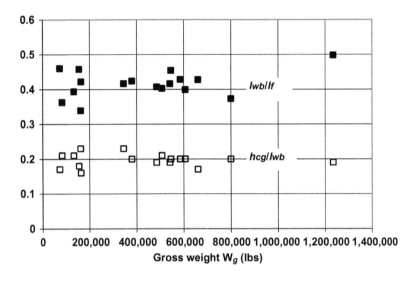

FIGURE 7.4

Estimated ratios of h_{cg}/l_{wb} (open symbols) and l_{wb}/l_f (closed symbols) are shown for 17 operational jet transports as a function of aircraft gross weight.

angle to zero and in the process reduces the sink rate to a much more comfortable 2 or 3 ft/s. Lift is still being produced at the instant of touchdown and while the aircraft settles the lift dumpers, or spoilers, have been deployed and the lift rapidly diminishes causing the aircraft to settle more rapidly on the main gear.

The main gear struts are essentially shock absorbers, often called oleo-pneumatic struts, which operate on the principle of compressing a gas in the strut by means of a piston inside the strut to store the energy of the touchdown. Oil carried between the gas and the piston is then forced by the high-pressure gas through internal orifices to extend the duration of impact and thereby distribute the impact load over a finite time period, lessening the effect of impact. Details of the operation of the oleo-pneumatic shock strut will be discussed subsequently in Section 7.4. In addition, the nose gear is exposed to impact loads when downward rotation of the nose is complete and the nose gear wheels contact the runway surface. The horizontal tail provides control over the rotation rate.

7.2.3 Location of the main gear

The touchdown of the aircraft is in a nose-up attitude with the main gear contacting the runway first. A schematic diagram of the landing gear at this point is shown in Figure 7.5. To preserve clarity the gear legs are shown attached to a plane which is tilted at the maximum touchdown pitch angle θ_{td} and the center of gravity is shown at its most aft and highest position. This is the worst-case situation for which we must place the main gear so that the maximum pitch angle is not exceeded. The limits on the touchdown pitch angle will be discussed in a subsequent section.

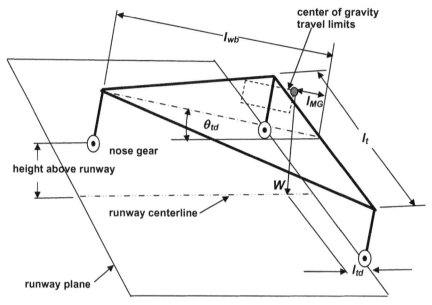

FIGURE 7.5

Schematic of aircraft landing gear geometry at the point of main gear touchdown where the pitch angle is θ_{td} and the center of gravity is at the highest and most aft point.

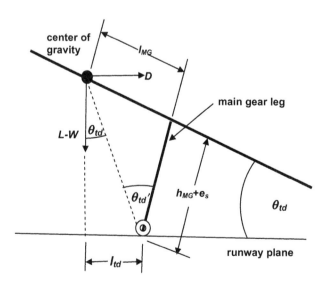

FIGURE 7.6

Detailed view of the main gear touchdown shown in Figure 5.

A detailed view of the main gear in its extended position just at the point of contact with the runway surface is shown in Figure 7.6. The height of the main gear measured between the ground and the aircraft center of gravity under essentially static conditions is h_{MG} and when the aircraft is well off the runway the oleo strut relaxes and extends a further distance, e_s. Note that as l_{MG} decreases, the angle θ'_{td} approaches θ_{td} and the main gear wheel contact point would be directly beneath the center of gravity. In order to ensure stability it is appropriate to require that

$$l_{MG} > \left(h_{cg} + e_s\right) \tan \theta_{td} \tag{7.10}$$

This requirement sets the minimum distance of the main gear from the farthest aft and highest center of gravity location and the least aft position of the main gear is thereby specified. The center of gravity location is influenced by the weight and placement of the landing gear, so detailed positioning of the landing gear is part of the weight and balance analysis which will be carried out in Chapter 8.

7.2.4 Location of the nose gear

After the concern expressed in the previous section regarding a backward turnover of the aircraft, it is apparent that preventing a possible sideways turnover must also be considered. Once the aircraft is on the ground the main forces acting are those due to braking deceleration, to centrifugal acceleration in turning, and cross wind drag. A schematic diagram of the footprint the landing gear makes on the ground surface is shown in Figure 7.7.

Sideways turnover may occur along the line joining the nose gear footprint and either main gear footprint as a result of the inertia forces that may arise, shown in Figure 7.7 as a longitudinal component $a_x W/g$ and a spanwise component $a_y W/g$, where a denotes the acceleration and g the gravitational acceleration. Because these forces are acting at the center of gravity they produce a moment about the turnover line $h_{cg} F_n$, where F_n is the component of the resultant force normal to the turnover line. The weight W of the aircraft resists the turnover through the moment arm l_{to} so that the moments balance where

$$\frac{W}{F_n} = \frac{h_{cg}}{l_{to}} = \tan \psi \tag{7.11}$$

The quantity ψ is called the turnover angle and is typically considered to be limited to a maximum of 60° although a value of 55–57° may be considered a better choice. The turnover distance

$$l_{to} = l_{NG} \sin \beta \tag{7.12}$$

The angle β in Equation (7.12) is defined by $\tan\beta = l_t/2l_{wb}$. The maximum acceleration load on the aircraft in landing is the braking deceleration, which may be as much as 0.5 g. Therefore, we may take the worst case, one where $a_x = a_y = 0.5$ so that $F_n = 0.707W\cos\beta$ and

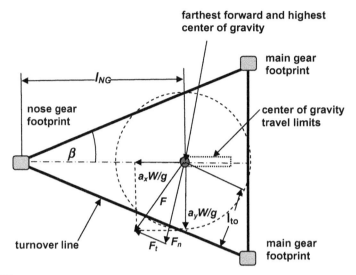

farthest forward and highest center of gravity

main gear footprint

l_{NG}

nose gear footprint

center of gravity travel limits

β

$a_x W/g$

F

$a_y W/g$

l_{to}

turnover line

F_t F_n

main gear footprint

FIGURE 7.7

Schematic diagram of landing gear footprint showing inertia forces acting at the highest, farthest forward center of gravity location.

$$\tan \psi = \frac{W}{0.707W \cos \beta} = \frac{1.414}{\cos \beta} \tag{7.13}$$

We may estimate $\beta = 11.5°$ from the average value of $l_t/2l_{wb} = 0.203$ given in Table 7.1. Thus, Equation (7.13) yields $\psi = 55.3°$, which is approximately the value suggested as a rule of thumb at the start of the discussion of the turnover condition. Using a value for ψ and knowing the highest and farthest forward location of the center of gravity, Equation (7.11) defines the distance l_{to}. The axial location of the main gear l_{MG} is fixed by Equation (7.10) for the farthest aft and highest center of gravity location so that the track l_t can be determined for any nose gear position l_{NG}, as illustrated in Figure 7.8. The static load on the nose gear, discussed in Section 7.2.1, should be greater than about $0.09W_g$ to ensure adequate controllability during the ground roll and is typically around $0.15W_{oe}$, where W_{oe} is the empty operating weight.

The previous discussion assumes that the main gear is rigid, but the shock absorbing nature of the struts permits some degree of roll and that case is illustrated (with an exaggerated roll angle for clarity) in Figure 7.9. Here we see that for a given turnover line the roll-induced movement of the center of gravity toward the turnover line reduces the moment arm for turnover while also reducing the moment arm for the restoring effect of the aircraft weight W. It is clear that to maintain the same line of action of the resultant force $F_{n,roll} < F_n$ so that the case with roll is more susceptible to turnover. Additional detail on the case with roll is given by Torenbeek (1982).

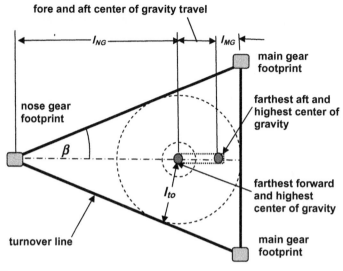

FIGURE 7.8

General configuration of the landing gear footprint.

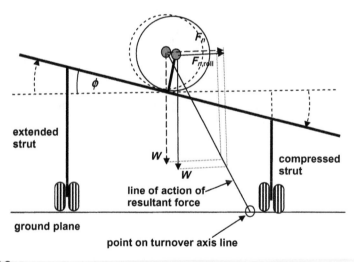

FIGURE 7.9

Schematic diagram of force structure with and without a roll angle looking at the aircraft from the front. The roll angle is exaggerated for clarity.

7.2.5 Ground clearance in takeoff and landing

During flight operations close to the ground, adequate clearance between aircraft components and the ground must be maintained. Therefore, limits on pitch and bank

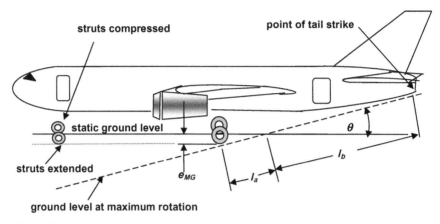

FIGURE 7.10

Elevation view of aircraft showing aft fuselage clearance requirements.

angles must be prescribed and it is preferable that ground clearance limits are outside the performance limits required for safe aircraft performance. In Chapter 3 the need for a kick-up of the aft underside of the fuselage is necessary to avoid tail strikes when operating at high angles of attack typical of landing and takeoff. The geometry to be considered is shown in Figure 7.10. It is desirable for the clearance angle θ to be larger than the maximum angle of attack α_{max} of the aircraft during the takeoff or landing operation. In takeoff, as the aircraft nose is rotated up at a rate $d\theta/dt$, which is typically around 3 or 4 deg/s, the aft fuselage rotates downward as the aircraft begins to rise above the runway. As the airplane climbs the rear of the fuselage is still moving down toward the runway so that the clearance angle may be estimated by

$$\theta \approx \alpha_{max} + \frac{d\theta}{dt}\Delta t \tag{7.14}$$

Here the characteristic time interval Δt is estimated by Torenbeek (1982) to be given by

$$\Delta t = \frac{2l_a}{V_{off}} + \sqrt{\frac{l_b C_{L,off}}{ga}} \tag{7.15}$$

The subscript *off* in Equation (7.11) refers to the condition when the landing gear struts are fully extended and a denotes the lift curve slope for the aircraft. Torenbeek (1982) presents an approach for calculating the terms in Equation (7.11). However, at this point in the design process we may estimate a global value for Δt by first rewriting Equation (7.15) as

$$\Delta t = \frac{l_b}{V_{off}}\left[2\left(\frac{l_a}{l_b}\right) + \sqrt{\frac{2}{\rho a g l_b}\left(\frac{W}{S}\right)_{to}}\right] \tag{7.16}$$

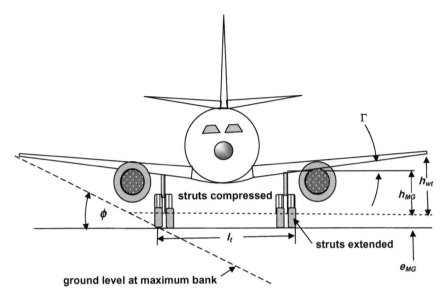

FIGURE 7.11

Front view of an aircraft showing wingtip clearance requirements.

For jet transports the lift curve slope a is around 4.5 per radian, the takeoff wing loading $(W/S)_{to}$ is around 125 lb/ft^2, and at sea level $\rho g = 0.076$ lb/ft^3. Furthermore, l_a/l_b is typically around 15% so that

$$\Delta t \approx \frac{l_b}{V_{to}} \left[0.3 + \sqrt{\frac{730}{l_b}} \right] \approx \frac{100}{V_{to}} [0.3 + 2.7] = \frac{300}{V_{to}} \qquad (7.17)$$

In Equation (7.17) we have assumed that $l_b \sim l_f/2 \sim 100$ ft so that the characteristic time is about 1.5 s and $\theta = \alpha_{max} + (4 \deg/s)(1.5 s) = \alpha_{max} + 6°$. For aircraft with the usual high lift devices α_{max} is usually around 8° or 9° so θ should be about 15°.

In the same manner the bank angle limit during takeoff may be estimated. The geometry of this situation is shown in Figure 7.11. Banking of the aircraft due to, for example, crosswinds might cause the wingtip to strike the ground. Though not apparent from the geometry shown in Figure 7.10, it may also be the case that before the wingtip strikes the ground the nacelle lip might do so. The bank angle must be limited so that neither event occurs. The maximum bank angle allowable is

$$\tan \phi = \frac{2 \left(h_{wt} + e_{MG} \right)}{(b - l_t)} - \tan \theta \tan \Lambda_{wt} = \frac{2 \left(h_{MG} + e_{MG} \right)}{(b - l_t)} + \tan \Gamma - \tan \theta \tan \Lambda_{wt}$$

$$(7.18)$$

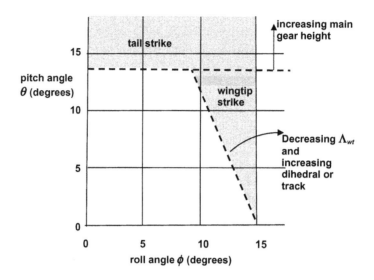

FIGURE 7.12

Envelope of pitch and roll limits for given aircraft design.

Here we account for the furthest aft point on the wingtip, defined by Λ_{wt} and illustrated in Figure 7.1, touching the ground at any time during the rotation through the pitch angle θ shown in Figure 7.10.

For a given aircraft configuration the allowable bank angle has the form $\tan\phi = c_1 - c_2\tan\theta$ and one may construct an envelope for allowable bank angle as a function of pitch angle, as shown in Figure 7.12. As mentioned in the discussion following Equation (7.17) the largest value of θ for jet transports is about 15° and since the expansion for small θ is $\tan\theta = \theta + \theta^3/3! + \cdots \sim \theta$. The roll angle ϕ is similarly small so that we may write $\phi = c_1 - c_2\theta$. Note that the envelope for fuselage ground clearance may be increased by increasing the main gear height h_{MG} and the fuselage tail cone kick-up angle θ_{TC}. The ground clearance envelope for wingtip clearance may be increased by widening the track of the main gear l_t, increasing the dihedral, and/or decreasing the wing sweepback. All these factors are influenced by other aspects of the airplane design goals of weight and performance so they cannot be freely altered. It is more likely that they will be set by other issues and the ground clearance limitation will be only marginally alterable. It should be noted that in designs using four wing-mounted engines, the ground clearance for the outboard nacelles will be the limiting factor rather than the wingtips. These limits may be determined using the approach developed for the wingtips.

7.2.6 Operational considerations for positioning the landing gear

When the aircraft is in motion on the ground, the nose gear must be capable of providing adequate control of the aircraft heading in spite of possible runway surface

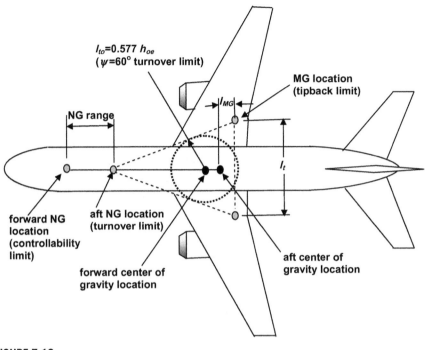

FIGURE 7.13

Positioning the landing gear in the planform view.

irregularities. This requirement favors placing the nose gear far forward because lengthening the wheelbase l_{wb} increases directional stability and reduces the possibility of turnover, as discussed in Section 7.2.4. On the other hand, at the same time there must be sufficient load on the nose gear to ensure maintenance of contact and the ability to provide appropriate steering forces but the additional loads due to braking would increase, as shown in Section 7.2.2. This requirement would seem to favor positioning the nose wheel to a more aft position but not beyond the point at which turnover becomes a critical consideration. Thus there is a range of nose gear locations that takes these factors into account, as illustrated in Figure 7.13, and a suitable point in this range must be selected.

The main gear must be positioned aft of the rearmost center of gravity to ensure that the fuselage does not tip backwards and strike the ground. Moving the main gear far aft would ensure that tipping is not a problem. However, in takeoff the aircraft must be rotated in exactly the same manner and this is achieved by downward force exerted by the horizontal tail. This force acts through a moment arm connecting the aerodynamic center of the horizontal tail and the main gear contact point with the runway surface. If the main gear is moved farther aft, the moment arm just described will be shortened and the moment available to rotate the aircraft to the takeoff attitude reduced. Once again there is a tradeoff, now in the main gear location, that takes the competing requirements into consideration. The basic requirement that the main gear be set at l_{MG}, as shown in Figure 7.13, was developed in Section 7.2.3.

The disposition of the landing gear in Figure 7.13 defines the wheelbase and track of the landing gear and the remaining global factor is the height of the landing gear. This dimension is primarily a function of ground clearance considerations as discussed in Section 7.2.5. Generally, the designer would select the shortest landing gear that satisfies all the clearance problems since that would be the lightest weight system possible. However, there are additional constraints on the landing gear height, For example, with the selected track and the shortest landing gear, will there be sufficient space for complete retraction of the landing gear? For example, Flottau and Norris (2012) report that the lengthened nose gear required for the B737MAX, an increased range version of the B737, resulted in the designers considering a bulge, or chin fairing, on the nose to accommodate the increased length. In a similar fashion, if a future stretched version of the design aircraft is desired the landing may not be able to tolerate the same degree of pitch rotation, θ. It is also possible that the available length for the shock absorber portion of the landing gear strut may be insufficient.

7.2.7 Maneuverability in ground operations

The aircraft must be able to execute turns in order to proceed from the gate to the runway and this will require a prescribed level of maneuverability because the layouts of runways and taxiways are existing features of airports. The basic maneuver is a circular turn, which takes place by steering the nose wheel about one of the main gear legs as shown in Figure 7.14.

The turning circle shown in Figure 7.14 assumes that the steering angle χ achievable by the nose wheel is given by $90° - \beta$ where $\tan\beta = l_t/2l_{wb}$. As mentioned in Section 7.2.4, the average value of β is $11.5°$ which suggests a steering angle of $78.5°$. Most jet transports have steering angles around in the range $65° < \chi < 75°$ and so the minimum turning radius is less than the track of the aircraft. Therefore to execute a complete $180°$ turn the runway must be wider than twice the track of the aircraft. The turn radius achievable as a function of nose wheel steering angle is schematically illustrated in Figure 7.15. The locus of centers of rotation is the main gear axis and the path of the nose wheel and main gear are indicated. To execute a $180°$ turn the diameter of the turn is larger than the track and is given by

$$d_{180} = l_t + 2l_{wb} \tan (90 - \chi) \tag{7.19}$$

The use of nose wheel steering alone results in some degree of scuffing of the tires and newer, larger aircraft, like the B747, B777, and the A380 have steering capability built into the main gear as well as in the nose wheel. The elaborate nature of the A380 main gear, which has 20 wheels, in fuselage and wing mountings, is shown in Figure 7.16. This reduces scrubbing wear on the tires and improved maneuverability, but at the expense of extra weight. Equation (7.19) should be sufficient for preliminary design studies.

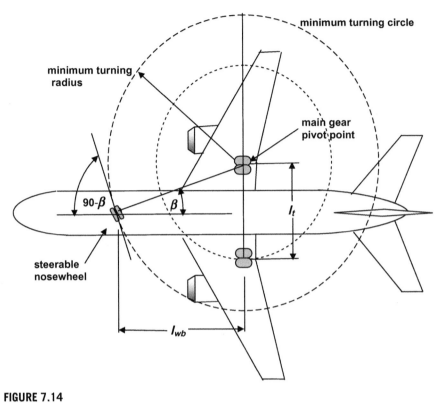

FIGURE 7.14

Minimum turning radius of a jet transport showing rotation about one main gear leg using steerable nose wheel.

7.2.8 Powered wheel-drive systems

In the attempt to curtail fuel consumption, serious attention is being paid to mounting electric motors on the landing gear to drive the wheels during taxiing operations. Such systems can reduce noise, exhaust emissions, and fuel usage by permitting the turbofans to be idling during ground movements. Wall (2012) reports that current testing of electric wheel-drive systems includes those driving the main gear wheels and those driving the nose gear wheels alone, with power coming from the onboard auxiliary power unit (APU). Because weight is such an important driver, two approaches are under consideration: a push back only system with a 3 mph speed capability and a full taxiing system with 30 mph capability. The tradeoff is between the weight of the electric wheel-drive system and the weight of fuel that can be offloaded because of fuel savings. Minimum fuel burn for a B737 taxiing on one engine is reported to be about 15 lb/min and crews typically add 30 min of fuel (about 450 lbs) for contingencies. Powering the electric wheel-drive with the APU cuts fuel

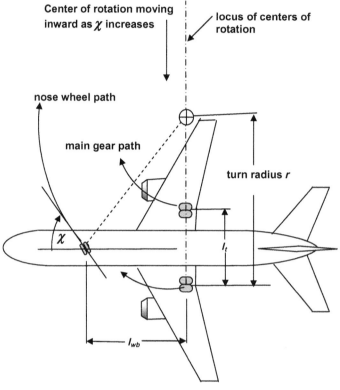

FIGURE 7.15

Turning radius *r* of a jet transport showing the path of the nose wheel and the main gear leg and the steering angle χ.

burn during taxiing by up to 85% and permits the loading of only 60 lbs of contingency fuel yet retaining the same degree of operational flexibility. Manufacturers of electric wheel-drive systems project savings of up to 130 tons of fuel per year for the average A320 or B737 operator.

7.3 Aircraft tires and wheels

Aircraft operating conditions require a wide variety of tire sizes and constructions. The modern aircraft tire is a complex composite structure designed to carry heavy loads at high rolling speeds while being as light in weight and small in size as is economically practical. By weight, an aircraft tire is approximately 50% rubber, 45% fabric, and 5% steel and different types of nylon and rubber compounds are combined in a tire so as to best achieve the operational extremes required. Their

FIGURE 7.16

The Airbus A380 in landing illustrates a complex main gear system.

composite construction is classified as either of bias-type or radial-type, the former using plies of material arranged at angles of 30° and 60° and the latter using plies arranged at 90°.

Tires must be able to take the loads imposed by the stressful conditions of landing and takeoff as well as maintain appropriate "flotation" on the airport runway surfaces. Flotation refers to the footprint of the tire on the runway surface; the larger the footprint, the lower the load per unit area on the runway surface. The load capacity of a tire depends directly upon the inflation pressure, the number of plies used in the construction, and on the size of the tire and the wheel. The load rating of the tire is the maximum permissible load carried by the tire at rest. The maximum load under braking is generally designed by the tire manufacturer to be 1.5 times the maximum static load in order to comply with the FAA regulations which are discussed subsequently in Section 7.3.1.

Consider, for example, specifying an H40 × 14.5-19 tire with a ply rating of 22. The first number indicates the nominal tire diameter d (40 in.), the next is the nominal tire width w (14.5 in.), and the last is the wheel diameter (19 in.); there are sometimes additional notations providing additional detail, such as the "H" in the example cited, which refers to a higher flotation, or greater deflection and therefore larger footprint than usual for that size tire. The ply rating refers to the load the tire is capable of supporting, and isn't necessarily equal to the actual number of plies used in the lay-up of the tire carcass. The general relationship between the number of plies and the rated

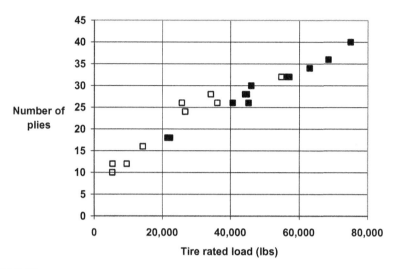

FIGURE 7.17

Relationship between the number of plies and the tire load rating as given by the data in Table 7.2. Closed symbols denote main gear tires and open symbols denote nose gear tires.

load is illustrated in Figure 7.17. One may make a reasonable estimate of the tire weight, in pounds, in terms of the tire diameter d_{ti}, the tire width b_{ti}, and the number of plies N_{plies} by calculating the following relation:

$$W_{ti} = \frac{d_{ti}b_{ti}N_{plies}}{107} \tag{7.20}$$

This correlation, which is in reasonable agreement with actual tire weights, is based on the idea that the tire weight is proportional to the volume of tire material used in its construction. Using this estimate one may get an appreciation for the relationship between load-carrying capability and weight as shown in Figure 7.18.

Detailed engineering data are available from tire manufacturers, e.g., Goodyear (2002) and Michelin (2012). For the example tire quoted above, the 22 ply rating indicates a maximum load-carrying capacity of 30,100 lb. If the tire is of radial construction the letter (R) appears in place of the hyphen (-) in the tire designation. Bias ply construction offers good high-speed performance and high load capacity with versatility for different runway types and a wide range of operating conditions. Radial tires have greater dimensional stability which contributes to long tire life. Radial construction generally provides more landings and reduced weight compared to radial ply tires. Typical tires used on some current airliners and their load ratings are given in Table 7.2.

The wheel diameter is also related to tire rated load, as shown in Figure 7.19. An estimate for the wheel diameter, in inches, may be made from the following relation:

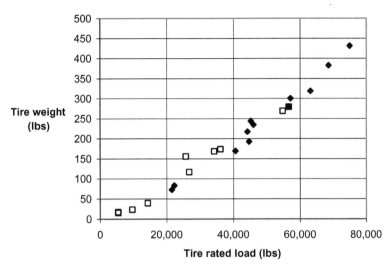

FIGURE 7.18

Approximate tire weight calculated using Equation (7.20) as a function of tire load rating data in Table 7.2. Closed symbols denote main gear tires and open symbols denote nose gear tires.

$d_w \sim 1.4$(tire rated load)$^{1/4}$, where the tire rated load is in pounds. Data on wheel weights are not readily available in the open literature, but an approximation based on the tire information may be written as follows:

$$W_w \approx 0.1\pi \left[d_w b_{ti} + \frac{d_w^2}{4} \right] \qquad (7.21)$$

This estimate assumes that the wheel weight depends upon the sum of the wheel web area, which is proportional to the square of the wheel diameter d_w, and the area of the wheel barrel, which is proportional to the product of the wheel circumference πd_w and the tire width b_{ti}. The results given by Equation (7.21) for the aircraft tires in Table 7.2 are shown in Figure 7.20. The few data shown for reported wheel weight seem consistent with the estimates from Equation (7.21).

7.3.1 Load requirements

A part 25.733 specifies that "when a landing gear axle is fitted with more than one wheel and tire assembly, such as dual or dual-tandem, each wheel must be fitted with a suitable tire of proper fit with a speed rating approved by the Administrator that is not exceeded under critical conditions, and with a load rating approved by the Administrator that is not exceeded by the loads on each main wheel tire, corresponding

Table 7.2 Typical Tires used on some Current Airliners, from Goodyear (2002) and Michelin (2012)

Aircraft	Main Gear Tire Size/ply	Rated Load (lbs)	Nose Gear Tire Size/ply	Rated Load (lbs)
B737-700	H43.5 × 16-20/26	40,600	27 × 7.75R15/12	9650
B737-800	H44.5 × 16.5-21/28	44,700	27 × 7.75R15/12	9650
B737-900	H44.5 × 16.5-21/28	44,700	27 × 7.75R15/12	9650
B747-400	H49 × 19-22/32	56,600	H49 × 19-22/32	56,600
B767-200	H46 × 18-20/28	44,200	H37 × 14-15/24	26,700
B767-300	H46 × 18-20/28	44,200	H37 × 14-15/24	26,700
B777-200	50 × 20R22/26	45,200	42 × 17R18/26	36,100
B777-300	50 × 20R22/32	57,100	42 × 17R18/26	36,100
B787-8	50 × 20R22/34	63,050	40 × 16R16/26	
B787-9	50 × 20R22/34	63,050	40 × 16R16/26	
A320	46 × 17R20/32	46,000	30 × 8.8R15/16	14,340
A330-300	54 × 21-23/36	68,500	1050 × 395R16/28	34,200
A340-300	54 × 21-23/36	68,500	1050 × 395R16/28	34,200
A350-900				
A380	1400 × 530R23/40	68,500	1270 × 455R22/32	54,800
E175	H38 × 13-18/18	22,250	24 × 7.7-5.5/10	5400
CRJ700	H36 × 12-18/18	21,525	20.5 × 6.75-10DT/12	5450

[a] Metric notation given in mm except for wheel diameter, which remains in inches.

[b] This is an older Type VII designation where wheel diameter is not given.

[c] Notation DT refers to a deflector tire which has a chine molded into the shoulder of the tire to deflect runway splash away from rear-mounted engines.

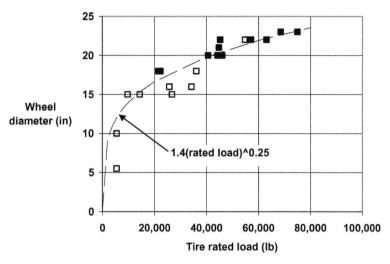

FIGURE 7.19

Wheel diameters as a function of tire load rating data as given in Table 7.2. Closed symbols denote main gear tires and open symbols denote nose gear tires. The dashed line represents a curve faired through the data.

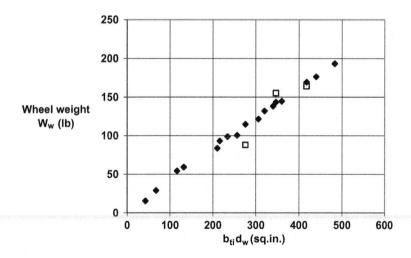

FIGURE 7.20

Wheel weight for the tires in Table 7.2 as a function of the product of tire width and wheel diameter as estimated by Equation (7.21) is shown by the closed symbols. The open symbols denote published data for main gear wheels.

to the most critical combination of airplane weight (up to the maximum weight) and center of gravity position, when multiplied by a factor of 1.07." Thus, for example, an H40 × 14.5-19 tire with a ply rating of 22, which has a maximum load of 30,100 lb, can be used to carry a maximum load of 28,131 lb, that is, 30,100 lb/1.07.

The applicable ground reactions for nose wheel tires, as given by FAA part 25.733, are as follows:

1. The static ground reaction for the tire corresponding to the most critical combination of airplane weight (up to maximum gross weight) and center of gravity position with a force of 1.0 g acting downward at the center of gravity. This load may not exceed the load rating of the tire.
2. The ground reaction of the tire corresponding to the most critical combination of airplane weight (up to maximum landing weight) and center of gravity position combined with forces of 1.0 g downward and 0.31 g forward acting at the center of gravity. The reactions in this case must be distributed to the nose and main wheels by the principles of statics with a drag reaction equal to 0.31 times the vertical load at each wheel with brakes capable of producing this ground reaction. This nose tire load may not exceed 1.5 times the load rating of the tire.
3. The ground reaction of the tire corresponding to the most critical combination of airplane weight (up to maximum gross weight) and center of gravity position combined with forces of 1.0 g downward and 0.20 g forward acting at the center of gravity. The reactions in this case must be distributed to the nose and main wheels by the principles of statics with a drag reaction equal to 0.20 times the vertical load at each wheel with brakes capable of producing this ground reaction. This nose tire load may not exceed 1.5 times the load rating of the tire.

FAA part 25.733 goes on to state that the tires must be mounted on wheels large enough to carry suitably sized brakes. In addition, for an airplane with a maximum certificated takeoff weight of more than 75,000 lbs, tires mounted on braked wheels must be inflated with dry nitrogen or other gases shown to be inert so that the gas mixture in the tire does not contain oxygen in excess of 5% by volume, unless it can be shown that the tire liner material will not produce a volatile gas when heated or that means are provided to prevent tire temperatures from reaching unsafe levels.

FAA part 25.733 also requires that each tire installed in a retractable landing gear system must, at the maximum size of the tire type expected in service, have a clearance to surrounding structure and systems that is adequate to prevent unintended contact between the tire and any part of the structure or systems. Because tires must fit within the aircraft over the life of the tire, provision must be made for their "growth" as use causes the tire carcass to stretch over time. One first calculates the grown tire envelope and then one may compute the radial clearance for a stationary and a spinning tire as well as the lateral clearance. The stowing of the main gear in

FIGURE 7.21

A Boeing B737 version which stows the main gear in the fuselage but leaves the wheels exposed.

the fuselage is clearly visualized in Figure 7.21 which shows a Boeing B737 model that leaves the tires and wheels exposed.

7.3.2 Landing gear configurations

The relatively low load-carrying capacity of tires of a size practical for retracting into an airliner fuselage, as indicated by Table 7.2, requires the use of multiple tire and wheel assemblies. At the same time, considerations of weight, volume, and reliability favor the use of the least number of landing gear supports, or struts, typically three: the nose gear strut and the two main gear struts. Each strut of the smaller airliners carries dual wheels, such as may be seen in the photograph of a Boeing B737-300 in Figure 7.22.

In order to support the load, heavier aircraft can require more than two wheels on each main gear strut, such as the four wheels in dual tandem arrangement on each main gear strut shown on the Airbus A330-300 in Figure 7.23. A wheel arrangement involving more than two wheels on a single strut is called a bogie. Larger and even heavier aircraft must resort to additional wheels as illustrated in Figure 7.24, where a Boeing 777-200ER is shown with a six-wheel bogie on each main gear strut, although the nose gear maintains a dual-wheel configuration.

The heaviest aircraft, like the Boeing B747 and the Airbus A380, employ an additional pair of main gear struts with multiple wheel bogies on each. The A380 shown

FIGURE 7.22

Boeing B737-300 in landing configuration illustrating dual wheels on the main and nose landing gear struts.

FIGURE 7.23

An Airbus A330-300 in landing configuration showing four wheels in a dual tandem arrangement (or bogie) on each main gear strut and a dual wheel assembly on the nose gear strut.

in Figure 7.16 has two four-wheel bogies and two six-wheel bogies as well as the usual two-wheel nose gear. Table 7.2 indicates that the tires and wheels used on the main gear are essentially the same size as those used on the nose gear for these jumbo jets, unlike the smaller airliners for which the nose wheels and tires are usually considerably smaller than those of the main gear.

FIGURE 7.24

A Boeing B777-200ER in landing configuration showing four wheels in a dual tandem arrangement (or bogie) on each main gear strut and a dual wheel assembly on the nose gear strut.

7.4 Shock absorbing landing gear struts

The dynamic loads in landing must be absorbed by the landing gear struts and tires and thereby transmitted to the airframe. The situation at the start and the end of the touchdown process is described schematically in Figure 7.25. A simplified oleo-pneumatic shock-absorbing strut is shown for clarity. The oleo-pneumatic strut combines gas compression, the pressure drop and turbulence attending the forcing of hydraulic oil through a small orifice, and the deflection of the tire in order to absorb the energy of the dropping process.

A schematic diagram of an oleo-pneumatic strut is shown in Figure 7.26. The cylinder is attached to the airframe and the outer wall of the hollow cylinder supports a bearing over which the piston can slide. The cylinder also supports an inner perforated cylindrical tube that terminates in an orifice plate. The piston is also a hollow cylinder that slides along the inner surface of the cylinder between the cylinder bearing and the orifice plate and has a bearing at its upper end. As the piston slides up from its extended position it forces oil through the orifice into the gas chamber of the cylinder thereby compressing the gas. The size of the orifice effectively determines the rate of filling of the cylinder and therefore the degree of shock absorption. Some pistons carry a cylindrical rod of variable cross-sectional area concentric with the cylinder axis. The rod, called a metering rod, changes the flow area of the orifice during the stroke to provide a desired shock absorption characteristic.

The total of the kinetic and potential energy to be absorbed during the touchdown is given by

$$\Delta E = \frac{1}{2} \frac{W}{g} w^2 + (W - L)\left(S + S_{ti}\right)$$

Here the velocity is reduced to 0 from initial sink rate w while the aircraft height drops by the amount $S + S_{ti}$. Although some lift may be acting during touchdown,

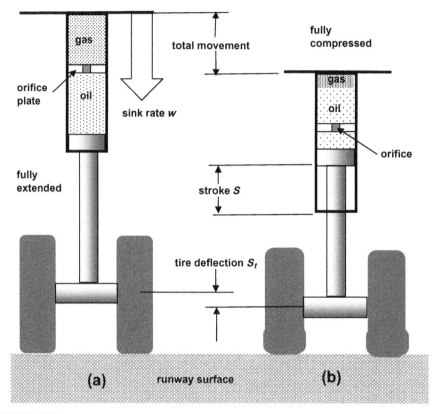

FIGURE 7.25

(a) The start of the touchdown process where the gear is fully extended and (b) the end of the touchdown process where the gear is fully compressed. A simplified shock-absorbing strut is shown here for clarity.

it ultimately reduces to zero as the lift dumpers, or spoilers are deployed and as the forward speed of the aircraft decreases. There is work done on the gas and hydraulic fluid in the strut and on the gas in the tire and on the deflection of its carcass. These contributions may be expressed as the product of a force, considered here to be the product of a load factor n and the weight W, and the distance through which the force travels modified by an efficiency factor for the process. For the strut the distance traveled is the stroke S and the process efficiency is η, while for the tire the distance traveled is the tire deflection S_{ti} and the process efficiency is η_{ti}. Such an expression may be written as follows:

$$\Delta W_k = \eta S n W + \eta_{ti} S_{ti} n W \tag{7.22}$$

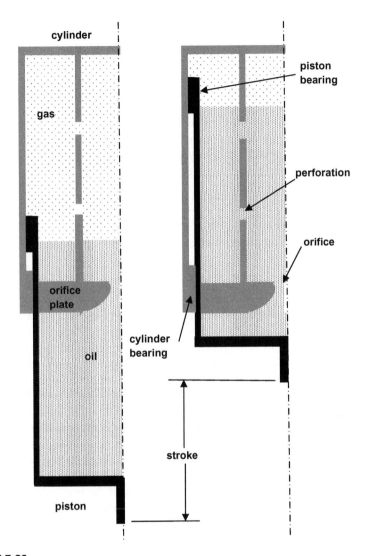

FIGURE 7.26

Schematic cross-sectional diagram of a typical oleo-pneumatic strut is shown in the extended position on the left and in the compressed position on the right.

Equating the change in energy to the work done yields

$$\frac{1}{2}\frac{w^2}{g} + \left(1 - \frac{L}{W}\right)(S + S_{ti}) = \eta Sn + \eta_{ti}nS_{ti} \qquad (7.23)$$

The stroke required is obtained from Equation (7.23) as

$$S = \frac{w^2}{2g\left(n\eta - 1 + \frac{L}{W}\right)} - S_{ti}\left(\frac{n\eta_{ti} - 1 + \frac{L}{W}}{n\eta - 1 + \frac{L}{W}}\right) \tag{7.24}$$

The FAR25.723 requires demonstration that the landing gear system of a commercial jet transport will not fail at a sink rate of $w = 12$ ft/s at the design landing weight. Note that this is approximately the vertical speed in approach, as discussed in Section 7.2.2. Using this value for sink rate and typical values for the efficiencies $\eta = 0.8$ and $\eta_{ti} = 0.5$ the equation for the strut stroke becomes

$$S = \frac{2.80 - 0.625\left[n - 2\left(1 - \frac{L}{W}\right)\right]S_{ti}}{n - 1.25\left(1 - \frac{L}{W}\right)} \tag{7.25}$$

We may estimate the tire deflection to be

$$S_{ti} = \frac{1}{3}\left(\frac{d_{ti}}{2} - \frac{d_w}{2}\right) \tag{7.26}$$

Here d_{ti} and d_w are the nominal tire and wheel diameters, respectively, as quoted by the manufacturer. Daugherty (2003) points out that statically supporting the rated load of a bias ply tire at its rated pressure will cause a deflection to approximately 35% of its available deflection before it bottoms out on the wheel, while a radial tire's deflection is more approximately 24–33% of the available deflection. If actual tire deflection data are unavailable a reasonable approximation is given by Equation (7.35) and with typical tires and wheels $0.08 < S_{ti}/d_{ti} < 0.09$. The effect of load factor, lift to weight ratio, and tire deflection on the stroke is illustrated in Figure 7.27 for the representative case where $L/W = 0.5$ and the least stressing case, $L/W = 1$. Nominal tire and wheel diameters of 44.5 in. and 21 in., respectively, were selected as representative.

According to Figure 7.27, the deflection response of the tires, which may have diameters in the range of 20–50 in. (see Table 7.2), decreases the stroke required of the shock strut by 10–15%. Note that the stroke doesn't depend upon the weight of the aircraft, but mainly on the sink rate and the load factor considered. For example, a sink rate of 5 ft/s and a load factor of 1.5 would require approximately the same stroke as the case shown for a sink rate of 12 ft/s and a load factor $n = 3$. The difference is that the force developed in the strut $F = nW = 1.5W$ rather than $3W$. The effective pressure in the strut $p = nW/A$ so that accepting a higher load factor requires a higher pressure in the strut or a larger piston area and therefore greater weight.

7.4.1 Sizing the landing gear struts

In normal ground handling operations, where the aircraft is nominally at rest at the static loading due to the weight of the aircraft, the pressure of the compressed gas in a strut would be about 1500 psia. Dry nitrogen is readily available commercially in

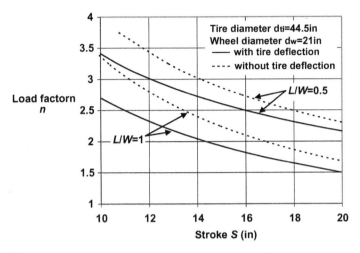

FIGURE 7.27

Variation of the stroke required as a function of the load factor n for the cases where $L/W=0.5$ and $L/W=1$ and a sink rate of 12 ft/s. The effect of tire deflection is based on representative tire and wheel diameters of 44.5 in. and 21 in., respectively.

standard cylinders at 2000 psia so that charging the gas side of a strut is easily carried out. The ratio of the static load pressure to the pressure in the fully extended strut is usually set at about 4. Thus, in the fully extended position the gas pressure would drop by a factor of 4, to about 375 psia. The gas compression may be considered to be a polytropic process with exponent k that is nearly isothermal ($k=1.1$) and may be expressed as follows:

$$p_2 = p_1 \left(\frac{\upsilon_1}{\upsilon_2}\right)^{1.1} = p_1 \left(\frac{1}{1-\frac{x_2}{l}}\right)^{1.1} \tag{7.27}$$

If we denote the hydraulic piston diameter as A_h, then $\upsilon_1 = A_h l$ is the volume occupied by the gas when the strut is fully extended and $\upsilon_2 = A_h(l-x_2)$ is the volume occupied by the gas when the aircraft is at rest with a static load on the landing gear. These situations are shown schematically in Figure 7.28a and c. Using the nominal pressure levels described previously, $p_1 = 375$ psia and $p_2 = 1500$ psia, the ratio $x_2/l = 0.716$. With the airplane at rest the force acting on the piston in the strut is $F_s = p_2 A_h$. For a given aircraft weight and center of gravity location the force on each main gear strut and on the nose wheel strut may be calculated. The maximum static main gear strut load occurs at maximum gross weight and farthest aft center of gravity. Using the static load pressure, say $p_2 = 1500$ psia, the required piston area is $A_h = F_{s,\max}/p_2$.

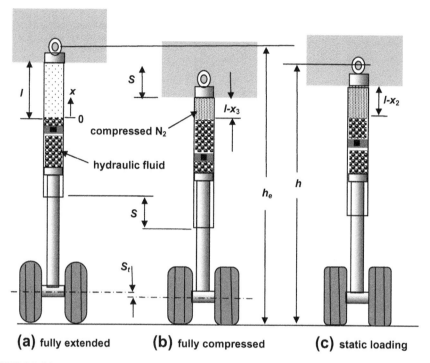

(a) fully extended   **(b)** fully compressed   **(c)** static loading

FIGURE 7.28

Position of the landing gear strut under (a) fully extended just at the point of touchdown, (b) fully compressed as a result of the impact at the sink rate, and (c) static loading at the gate or in slow taxiing.

In the most stressing touchdown case, that is, when $L=0$, the internal force supported by the strut is nF_s and therefore the pressure $p_3=np_2$ and, for a polytropic process with exponent k, we have

$$p_3 = p_2 \left(\frac{\upsilon_2}{\upsilon_3}\right)^k = p_2 \left(\frac{1 - \frac{x_2}{l}}{1 - \frac{x_3}{l}}\right)^k = np_2 \qquad (7.28)$$

Now the volume of gas $\upsilon_3 = A_h l(1 - x_3/l)$ and may be found from Equation (7.28) to be

$$\upsilon_3 = A_h L \left(1 - \frac{x_3}{l}\right) = A_h L \left(1 - \frac{x_2}{l}\right) n^{-\frac{1}{k}} \qquad (7.29)$$

With $n=3$ the pressure $p_3=4500\,\text{psia}$ and, using $k=1.1$, the final volume is $\upsilon_3=0.264Al$ and $x_3/l=0.895$. The remaining volume is available for additional compression of the gas and acts as a safety margin. In the same spirit the calculated maximum stroke is generally increased by a small amount, about an inch, to be available for overstressed situations.

FIGURE 7.29

Aircraft mechanics replacing a main landing gear tire on a U.S. Navy Lockheed P-3C Orion aircraft (W_{to} = 140,00 lb). The relative size of the tire, wheel, main gear strut, axle, and integral brake unit is well illustrated.

One may then use Equation (7.25) to determine the stroke required for the load factor of $n = 3$. Figure 7.27 shows that for this load factor and the most stressing case of $w = 12$ ft/s and $L/W = 0.5$ the stroke required is approximately 12 in. Adding on an extra inch as a safety margin yields a maximum stroke of 13 in. The height of the gear in the fully compressed static condition, h in Figure 7.28c, is set by the ground clearance requirements while the height of the fully extended gear h_e is slightly less than $h + S$; the exact value can be obtained when the details of the piston motion have been collected. The outside diameter of the strut may be approximated as being 30% larger than the piston diameter so $d_s = 1.46 A_h^{1/2}$.

An appreciation for the relative size of the tire, wheel, main gear strut, axle, and integral brake unit for a large aircraft is well illustrated in Figure 7.29. The airplane, a Lockheed Martin PC-3 Orion patrol and reconnaissance aircraft, is of substantial size, having a takeoff weight of around 140,000 lb.

7.4.2 Dynamic landing gear analysis

Milwitzky and Cook (1953) carry out a detailed analysis of landing gear behavior. They consider a two-degree of freedom system where the upper mass is the aircraft and the oleo-pneumatic cylinder while the lower mass is the oleo-pneumatic piston, strut, axle, and wheel. The hydraulic fluid effectively acts as a damper and the compressed gas as a spring. Tire deflection behavior is accounted for and time histories of the motion are presented for various conditions. After evaluating the effects of the parameters involved they arrive at a simplified set of non-dimensional equations which give quite reasonable results compared to the original equations. In particular, they present a solution for the normalized maximum stroke as a function of the normalized vertical velocity that may be approximated by

$$S_{n,\max} = 1.74 w_n^{0.58} \tag{7.30}$$

Here the normalized maximum stroke of the landing gear piston is given by

$$S_{n,\max} = \frac{\rho_h A_h^3}{2 \left(c_d A_o\right)^2 \cos \phi} \frac{g}{W} S = A' \frac{g}{W} S \tag{7.31}$$

The quantities ρ_h, A_h, and A_o denote the hydraulic fluid density, the hydraulic piston area, and the orifice area, respectively. The hydraulic fluid used in landing gear shock struts is a mineral-based fluid with a specific gravity of about 0.88. The area of the piston may be determined by the methods of the previous section. The orifice diameter is the most important independent parameter affecting the shock-absorbing capability of the landing gear strut. The discharge coefficient of the orifice is c_d and for a rounded entry orifice it is reasonable to take $c_d = 0.9$. The quantity ϕ is the angle the landing gear strut makes with the vertical and for commercial jet transports ϕ is small enough to permit the approximation $\cos \phi \sim 1$. The quantity W is the weight carried by the strut. Therefore everything but the orifice area is known for the coefficient of the strut stroke. Using the strut stroke obtained in the previous section permits us to write the normalized stroke solely as a function of the piston orifice diameter, that is, $S_n = S_n(A_o)$.

The non-dimensional vertical velocity is defined as

$$w_n = \sqrt{\frac{A'^2 g}{\tau W}} w \tag{7.32}$$

The quantity τ is a tire constant equal to the slope of the vertical force versus the tire deflection. Making a linear approximation, we may put $\tau \approx W/S_{ti}$ so that the non-dimensional vertical velocity becomes

$$w_n \approx \frac{A' g}{W} \sqrt{\frac{S_{ti}}{g}} w \tag{7.33}$$

The quantity W/gA' has the units of length and A' has the units of mass/length. Because the tires are already selected on the basis of the load-carrying capacity required of the strut, we may write the normalized vertical velocity as a function of the piston orifice diameter alone, that is, $w_n = w_n(A_o)$. We may then use this result and $S_n = S_n(A_o)$ in Equation (7.30) to find the required orifice area A_o.

7.4.3 Landing gear strut design example

Assume that the design airplane has a gross weight of 155,000 lb, similar to that of the Boeing B737-800 listed in Table 7.1 Using the midpoints of the weight distribution approximation of Section 7.2.1, we may calculate the load on the main gear to be about 88% of the total weight and that on the nose gear to be about 12% of the weight. Assuming one main strut on each side of the aircraft, each must support 68,200 lb, while the nose gear strut must support 18,600 lb. The maximum static main gear strut load occurs at maximum gross weight and farthest aft center of gravity. Using the static load pressure, say $p_2 = 1500$ psia, the required piston area is $A_h = F_{s,\max}/p_2$. Thus the area of the piston for the main gear is $A_{h,MG} = 45.4$ in.^{2} and for the nose gear $A_{h,NG} = 12.4$ in.^{2}. The corresponding piston diameters are $d_{h,MG} = 7.60$ in. and $d_{h,NG} = 3.97$ in. The outside diameter of the strut may be approximated as being 30% larger than the piston diameter so $d_{s,MG} = 9.49$ in. and $d_{s,NG} = 5.16$ in.

Assuming a load factor of $n = 3$ one may then use Equation (7.25) to determine the stroke required. For this load factor and the most stressing case of $w = 12$ ft/s, $S_{ti} = 0$, and $L = 0$, the stroke required is approximately 19 in. Adding on an extra inch as a safety margin yields a maximum stroke of $S_{\max} = 20$ in. The height of the gear in the static condition, h in Figure 7.28c, is set by the ground clearance requirements while the height of the fully extended gear h_e is slightly less than $h + S$; the exact value can be obtained when the details of the piston motion have been collected. At this stage of the design we do not yet know the exact location of either the landing gear or the center of gravity of the airplane. However, we may start by using estimates given in Table 7.1 and choose the value given for the B737-800, that is, $h_{oe}/l_{wb} = 0.18$ and $l_{wb} = 57$ ft, resulting in a center of gravity location $h_{oe} = 9.8$ ft. Based on three-view drawings and photographs of the aircraft listed in Table 7.1 we may estimate the static height of both the main and nose gear to be approximately $2 < h/d_{ti} < 3$.

The dynamics of the landing gear operation were discussed in Section 7.4.2 and we may first determine the normalized stroke given in Equation (7.31). Using the assumptions described in Section 7.4.2 we may write the product Ag in Equation (7.31) as follows:

$$Ag = \frac{\rho_h g A_h^3}{2\left(c_d A_o\right)^2 \cos\phi} \approx \frac{0.88\,(\rho g)_{water}\, A_h^3}{2\left(0.9 A_o\right)^2 (1)} = \left(33.9\frac{\text{lb}}{\text{ft}^3}\right)\frac{A_h^3}{A_o^2}$$

The normalized maximum stroke in Equation (7.31) may then be written as

$$S_{n,\max} = \frac{Ag}{W}S = \left(33.9\frac{\text{lb}}{\text{ft}^3}\right)\frac{A_h^3}{A_o^2}\frac{1}{W}S$$

In the same manner we may write the normalized velocity of Equation (7.33) as follows:

$$
w_n = \frac{Ag}{W}\sqrt{\frac{S_{ti}}{g}}w = \left(33.9\frac{\text{lb}}{\text{ft}^3}\right)\frac{A_h^3}{A_o^2}\frac{1}{W}\sqrt{\left(0.0028\frac{\text{s}^2}{\text{ft}}\right)d_{ti}}\left(12\frac{\text{ft}}{\text{s}}\right)
$$
$$
= \left(21.5\frac{\text{lb}}{\text{ft}^{5/2}}\right)\frac{A_h^3\sqrt{d_{ti}}}{A_o^2 W}
$$

Substituting for $S_{n,\text{max}}$ and w_n in Equation (7.30) and solving for the effective orifice area yields

$$
A_o = 4.12\sqrt{\frac{A_h^3}{W}\frac{S^{1.19}}{d_{ti}^{0.345}}}
$$

The numerical coefficients shown assume lengths in ft, areas in ft^2, and weights in lb. We determined that for a load $W=F_{s,\text{max}}=68,200\,\text{lb}$ on a main gear strut the area of the piston for that strut is $A_{h,MG}=45.4\,\text{in.}^2=0.315\,\text{ft}^2$. Similarly, with load $W=F_{s,\text{max}}=18,600\,\text{lb}$ on the nose gear strut the area of the piston for that strut is $A_{h,NG}=12.4\,\text{in.}^2=0.086\,\text{ft}^2$. We also found, from our previous calculation using Equation (7.25) that for a sink rate of $w_o=12\,\text{ft/s}$ and a load factor $n=3$ the stroke required was 20 in. $=1.67\,\text{ft}$. This leads to the result that $A_{o,MG}=0.0051/d_{ti}^{0.345}$ and $A_{o,NG}=0.0014/d_{ti}^{0.345}$. The tire sizes are not yet known, though they will be considered in a subsequent section. We may use the tire sizes for the B737-800 from Table 7.2 to finish the calculation of the oleo strut orifice sizes: main gear $d_{ti,MG}=44.5\,\text{in.}=3.71\,\text{ft}$ and nose gear $d_{ti,NG}=27\,\text{in.}=2.25\,\text{ft}$. Thus, $A_{o,MG}=0.005$ $1/d_{ti}^{0.345}=0.0032\,\text{ft}^2=0.467\,\text{in.}^2$ and the diameter of the orifice is $d_{o,MG}=0.77\,\text{in.}$ Finally, $A_{o,NG}=0.0014/d_{ti}^{0.345}=0.00106\,\text{ft}^2=0.152\,\text{in.}^2$ and the diameter of the orifice is $d_{oNG}=0.44\,\text{in.}$

The landing gear struts are typically made of high-strength steel and the associated structure may be roughly approximated by considering the weight of a solid piece of steel of length $h\sim2d_{ti}$ and the corresponding calculated diameter. In this case each of the main gear strut assemblies would have $h=2d_{ti,MG}=89\,\text{in.}$ and $d_{s,MG}=9.49\,\text{in.}$ so that $W_{MG}=3.64\,\text{ft}^3(489\,\text{lb/ft}^3)=1780\,\text{lb}$. The nose gear $h=2d_{ti,NG}=54\,\text{in.}$ and $d_{s,NG}=5.16\,\text{in.}$ so that $W_{NG}=0.648\,\text{ft}^3(489\,\text{lb/ft}^3)=317\,\text{lb}$. The total weight of the strut assemblies is therefore about 3,880 lb. As efforts to reduce weight escalate there is an increasing interest in expanding the use of titanium as well as composite materials in landing gear components.

A schematic diagram of one of the main gear assemblies developed in this section, drawn approximately to scale, is shown in Figure 7.30. A similar figure would be drawn for the nose gear. During retraction of the main gear the strut pivots about the trunnion and is drawn into the fuselage whereupon the landing gear doors are closed. Further details about landing gear retraction are described in Section 7.6.

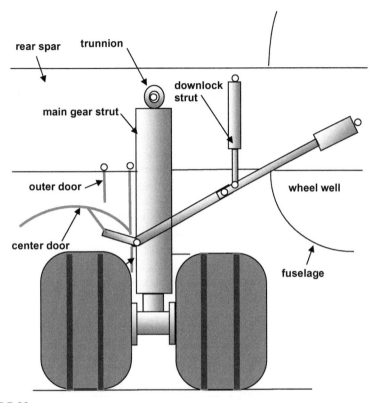

FIGURE 7.30

A schematic diagram of one of the main gear assemblies for the design example is shown, drawn approximately to scale.

7.5 Landing gear brake systems

The major portion of the deceleration of the aircraft is borne by the wheel brakes. Yetter (1995) indicates that airlines typically use thrust reversers on all landings, but their greatest contribution is in landings on wet or icy runways. Aerodynamic drag also accounts for some deceleration, but its magnitude drops off as the square of the velocity, which is already at a low level in terminal operations. Therefore, it is the wheel brakes that bear the brunt of the burden of stopping the aircraft safely in a reasonable distance. The brakes are applied once the aircraft is on the ground and they transform the kinetic energy E_k of the aircraft into heat Q which must be safely dissipated. The most stressing case is that of a refused takeoff where the aircraft is at about its maximum weight, $W_{to} \simeq W_g$, and at the critical velocity, typically 90% of the takeoff velocity. If instead of the critical velocity we take the more conservative case of $V = V_{to}$, the kinetic energy of the aircraft is

$$E_k = \frac{1}{2}mV_{to}^2 = \frac{1}{2}m\left(1.2V_s\right)^2 = 0.72\frac{W_{to}}{g}\left(\frac{W}{S}\right)_{to}\frac{2}{\rho_{s.l.}\sigma C_{L,\max,to}} \qquad (7.34)$$

This may also be written as

$$E_k = 18.82\frac{W_{to}}{\sigma C_{L,\max,to}}\left(\frac{W}{S}\right)_{to} \qquad (7.35)$$

A nominal case for a jet transport taking off at sea level with a wing loading of 125 lb/ft^2 and maximum lift coefficient of 2.5 has a kinetic energy of about $E_k = 941 W_{to}$ ft-lb or about $E_k = 1.21 W_{to}$ Btu. The equivalent of this energy in heat Q must be dissipated by the brakes during the time of the braking process without excessive temperature buildup in the brakes, wheels, and tires.

The simplest heat transfer situation is that in which the thermal conductivity κ of the material is very high, that is, the thermally diffusivity $\kappa/\rho c_p \gg 1$, such that heat passes so rapidly through the solid material that the temperature is essentially spatially uniform but varying with time. Assuming that an amount of heat Q is generated within the mass of the brake unit over the braking time interval Δt_b, the thermal conduction equation reduces to

$$mc_p\Delta T \approx Q \qquad (7.36)$$

Setting $Q = E_k = (W/g)(V^2/2)$ in Equation (7.36) leads to the following result for the bulk temperature of the brake material:

$$T = T(0) + \frac{Q}{\rho g v c_p} = T(0) + \frac{1.21W}{\rho g v c_p} \qquad (7.37)$$

The total weight of brake material available in the brake assembly needed to absorb all the heat and rise to a final allowable brake temperature T_b is

$$W_b = \rho g v = \frac{1.21W}{c_p\left[T_b - T(0)\right]} \qquad (7.38)$$

Characteristics of some brake materials are shown in Table 7.3. From the standpoint of brake assembly weight W_b, materials with the highest heat capacity c_p are to be favored, and this suggests the use of carbon or beryllium. However, because the brakes must be contained within the confines of the wheel, the volume of brake material required is also of importance. In that case, materials with the highest heat capacity per unit volume, ρc_p, are attractive, and that favors the use of steel. From the standpoint of high temperature capability all those mentioned in Table 7.3 are acceptable. However, steel brakes are the least expensive, while beryllium is expensive, as well as being a toxic material. Carbon brake material has become more popular in recent years as costs have been reduced. One must also consider the wear characteristics as part of the economic issue and carbon brakes tend to last longer.

Table 7.3 Properties of Several Brake Materials

Material	ρg (lb/ft^3)	c_p/g (Btu/lb-R)	$\rho g c_p$ (Btu/ft^3-R)	κ (Btu/hr-ft-R)	κc_p (ft^2/s)	T_{max} (R)
Steel	489	0.13	63.6	8.5	0.134	2500
Copper	559	0.092	51.4	230	4.47	1060
Carbon	105	0.31	32.7	100	3.06	4000
Beryllium	115	0.45	51.8	99	1.9	1460

As may be seen in Table 7.3, carbon brake material is low in weight and high in thermal conductivity and specific heat. In addition, the wear characteristics of carbon brakes provide more landings between overhauls than do steel brakes and this makes the life cycle costs of carbon brakes comparable to those of steel brakes. They are basic equipment on new aircraft like the Boeing B787 and may be retro-fitted to older aircraft. Steel brake wear is proportional to the kinetic energy absorbed by the brakes. Therefore, steel brake life is extended by light brake applications with time for cooling between each application. On the other hand, carbon brake wear mainly depends on the number of brake applications so that fewer, but firmer, brake applications help reduce wear.

Equation (7.38) shows that the higher the brake temperature is permitted to be, the lower the brake weight. However, there are other issues to be considered, particularly safety, reliability, and economics, and satisfying them favors lower brake temperatures. A realistic engineering compromise must be met to allow for a design of practical weight. The deciding factor is the allowable brake temperature, which may be permitted to be close to the maximum allowable temperature for the brake material or the surrounding structure, wheels, and tires, whichever is lower. In the case of steel brakes a temperature around 2300R (1840F or 1000 °C), which is close to the maximum given in Table 7.3, can be acceptable for a one-time emergency application like a rejected takeoff at maximum gross weight. A relatively low temperature of around 1400R (940F or 500 °C) should be acceptable for regular service. An intermediate temperature of about 1900R (1440F or 780 °C) is likely to safely accommodate several high deceleration stops at the maximum landing weight. Using these guidelines, a plot of the weight of steel brake assemblies is shown in Figure 7.31 as a function of the kinetic energy that must be dissipated as heat. Also indicated on the plot are the intersection points of three levels of kinetic energy with the three brake temperature curves mentioned previously. The intersection points show three different required weights and one would have to choose the heaviest among them, that is, the 325 lb assembly, to assure safety. The kinetic energy levels shown in Figure 7.31 are those associated with different braking conditions for a B737-800 steel heat sink brake assembly weighing 363.4 lb described in Honeywell (2008).

The use of carbon brake material leads to the results shown in Figure 7.32. It is clear that for the same conditions the weight of the brake assembly is about 150 lb, less than half that of the steel brakes described previously. It would seem that carbon

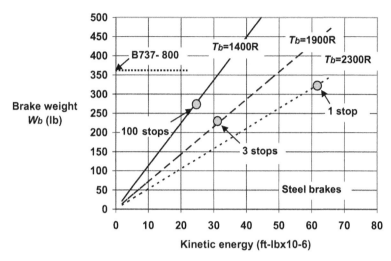

FIGURE 7.31

Weight of steel brake assemblies required is shown as a function of the kinetic energy that must be dissipated as heat for three levels of brake temperature T_b. Also shown are predicted values for three levels of usage.

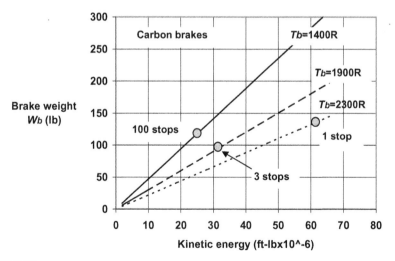

FIGURE 7.32

Weight of carbon brake assemblies required is shown as a function of the kinetic energy that must be dissipated as heat for three levels of brake temperature T_b. Also shown are predicted values for three levels of usage.

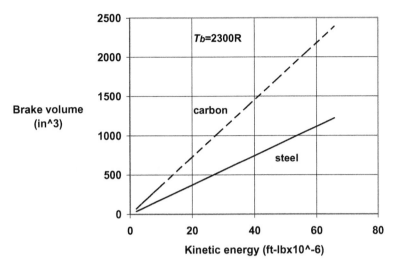

FIGURE 7.33

The volume of brake material, steel or carbon, required to keep the brake temperature to a maximum of 2300R (1840F) or 1278K (1005°C), is shown as a function of kinetic energy that must be dissipated.

would be the material of choice, but the volume occupied by the weight of carbon heat sink material is about twice that needed for steel brake assemblies as shown in Figure 7.33, which assumes the most stressing case of a rejected takeoff, that is, $T_b = 2300R$.

It is apparent that carbon brake material requires more space than steel brake material for the same performance. This volume of material is distributed among the individual wheels and is generally confined within a circular cylindrical space of diameter less than that of the wheel diameter d_w and a length l_b so that appropriate account of the space available is required. A schematic diagram of the brake assembly nested within the wheel drum is given in Figure 7.34. One part of a typical brake is comprised of a series of rotor disks attached to a wheel shaft on which they rotate, together with the wheel. The other part is a set of stator disks which are attached to the brake housing which, in turn, is attached to the landing gear strut by a backing plate. A piston assembly forces the rotor and stator disk to rub against one another to dissipate the kinetic energy of the rotating wheel. The photograph in Figure 7.29 shows the compact brake assembly clearly. More detail regarding the configuration of a brake assembly is given by Curry (1988).

The volume available for the compact brake system is essentially that within the wheel drum and therefore is proportional to the square of the wheel diameter. An approximate relation for the data presented by Curry (1988) for the available cross-sectional area is

$$A_{avail} = 0.36 d_w^{1.98} \tag{7.39}$$

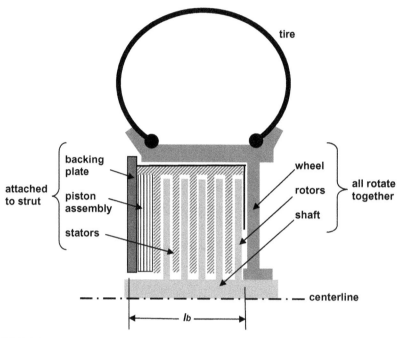

FIGURE 7.34

Schematic cross-sectional diagram of brake assembly nested in wheel drum showing the nominal brake length l_b.

The area A_{avail} and wheel diameter d_w given in Equation (7.39) are in square inches and inches, respectively. Considering the brake assembly to be an annular cylinder with cross-sectional area A_{avail} and length l_b the brake material volume is $v_b = A_{avail}l_b$. Using Equation (7.39) we may find the total length of the brake assembly, l_b, which is illustrated in Figure 7.34, required for a given wheel as follows:

$$l_b = \frac{v_b}{A_{avail}} = 2.78\frac{v_b}{d_w^{1.98}} \tag{7.40}$$

To gain an idea of the dimensions of the brake assembly, consider the case of a rejected takeoff where about 60 million ft-lb of kinetic energy must be dissipated. For a steel brake Figure 7.31 suggests about 1150 in.[3] of brake heat sink material is required. For a typical wheel of a B737, which is about 20 in. diameter, Equation (7.40) suggests that the length of the brake l_b should be about 8.85 in. This is reasonably consistent with an actual B737 brake, as described in Honeywell (2008). The volumetric heat capacity of carbon, as illustrated in Table 7.3, is only 51.4% that of steel so that carbon brakes require 94.5% more volume for equivalent thermal performance and the required volume, from Figure 7.33, is about 2200 in.[3] Because the available cross-sectional area of the brake is dependent only on the wheel diameter, the length of the multiple disk carbon brake assembly would be 17.2 in. for the B737-like case being examined. Another advantage of carbon brakes is the higher temperature capability, which may be employed to reduce weight and therefore required volume. Of course, final dimensions are dependent on detail design issues, but it should be clear that the substantial weight savings accruing to carbon brakes come at a price in volume as well as initial cost.

7.5.1 Tire, wheel, and brake selection example

Assume that the design airplane has a gross weight of 155,000 lb, similar to that of the Boeing B737-800 listed in Table 7.1. Using the midpoints of the weight distribution approximation of Section 7.2.1, we may calculate the load on the main gear to be about 88% of the total weight and that on the nose gear to be about 12% of the weight. Assuming one main strut on each side of the aircraft, each must support 68,200 lb, while the nose gear strut must support 18,600 lb. We consider each strut to carry at least two wheels so that the load on each main gear tire is 34,100 lb and on each nose gear tire 9300 lb. According to the FAA requirements discussed in Section 7.3.1, the main gear tires load rating should be at least 107% of the maximum load to be carried. Thus the tire rating for the main gear should be at least 36,500 lb. Examining Figure 7.17a we see that the main gear tires should have a ply rating of at least 24. Specifying a greater number of plies must be done with caution because Figure 7.17b shows that the weight of the tire increases essentially linearly with the rated load, and therefore the number of plies. Based on the data shown in Figure 17.18 the wheel diameter should be around 25 in. Then we may consult a tire handbook for a tire that is approximately of the size $d_{ti} \times b_{ti}$-20/24 with a rated load of at least 36,500 lb. Goodyear (2002) shows an H44.5 × 16.5-20/24 which has a rated load of 36,200 lb, which is 300 lb shy of the required value. Moving up to an H44.5 × 16.5-20/26 tire provides a rated load of 39,600 lb and would weigh, according to Equation (7.20), 178 lb. The wheel diameter is 20 in. and the width of the wheel between the flanges which support the tire is 10.5 in. This dimension will be important for housing the necessary brake size which will be determined subsequently. According to Equation (7.21) this wheel would weigh 135 lb. There are other combinations that will meet the load rating requirement, but at this stage the selected tire and wheel will be adequate. Note that Table 7.2 shows that the B737-800 typically uses an H44.5 × 16.58-21/28 tire, which has a rated load of 44,700 lb.

The FAA load requirements on nose gear tires were outlined in Section 7.3.1. The nose gear static load is 18,600 lb so that with two wheels each must support 9300 lb. FAR25.731 requires that this load not exceed the rating of the tire. From the dynamic load requirements discussed in Section 7.3.1 we know that the nose gear load under braking must not exceed 1.5 times the load rating of the tires. At this stage of the design we do not know the exact location of either the landing gear or the center of gravity of the airplane. Therefore we may start by using the estimates given in Section 7.2.1 and the data in Table 7.1 and find that $(x_{MG} - x_{oe})/l_{wb} \sim 1/8$ and $h_{oe}/l_{wb} = 1/5$.

First, for the case of a maximum ramp, or gross weight of 155,000 lb Equation (7.9) yields

$$F_{NG} = (155,000 \text{ lb}) \left[\frac{1}{8} + \left(\frac{a}{g} \right) \frac{1}{5} \right] = 3875 \left[5 + 8 \left(\frac{a}{g} \right) \right]$$

In the case of maximum landing weight, FAR25.731 requires $a/g = 0.21$, and therefore $F_{NG} = 25,900$ lb. Thus each tire is subjected to a dynamic load of 12,950 lb and this may not exceed 1.5 times the tire rated load. Therefore the load rating of the tire must exceed 8,633 lb. Second, we consider the case of maximum landing weight, which may be approximated using the correlation shown in Chapter 4, Figure 4.7:

$$\frac{W_{l,max}}{W_{to}} \approx 1 - 0.003 W_{to}^{1/3} = 0.839$$

Assuming, as usual, that gross weight and takeoff weight are approximately equal, we find the maximum landing weight for the design aircraft to be about 130,000 lb and therefore Equation (7.9) becomes

$$F_{NG} = (130,000 \text{ lb}) \left[\frac{1}{8} + \left(\frac{a}{g} \right) \frac{1}{5} \right] = 3250 \left[5 + 8 \left(\frac{a}{g} \right) \right]$$

In the case of maximum landing weight, FAR25.731 requires $a/g = 0.31$, for which $F_{NG} = 24,310$ lb so that each tire is subjected to 12,160 lb and therefore the load rating of the tire must exceed 8100 lb.

Recall that under the static loading condition the nose wheel tire load rating must be greater than 9300 lb and this is greater than the mandates under dynamic loading conditions. Therefore we may seek nose gear tires that are rated above 9300 lb. From Figure 7.17 we see that a tire with a 12 ply rating, or better, is suggested. Using the correlation in Figure 7.18 a wheel diameter of $d_w = 1.4(9300)^{0.25} = 13.75$ in. is calculated. We are therefore searching for a tire of approximate size $d_{ti} \times w_{ti} - 14/12$ that has a rated load greater than 9300 lb. Goodyear (2002) shows a 26.5×8.0-13/12 tire with a rated load of 9475 lb and a weight, according to Equation (7.20), of 24 lb. The 13 in. wheel has a width between flanges of 6.5 in. and according to Equation (7.21) this wheel would weigh 46 lb. Note that Table 7.2 shows the B737-800 using a nose wheel tire of size $27 \times 7.75R15/12$ with a rated load of 9650 lb. This is a radial ply tire which would also satisfy the requirement.

In considering the brake requirements we note that the nose wheels generally carry no brakes because their application would increase the structural load on the nose gear strut. Furthermore they only support about 12% of the weight of the aircraft and thus handle the same small fraction of the kinetic energy dissipation. Therefore, the main gear brakes are sized to accommodate the entire braking requirement. Solving Equation (7.39) for the available cross-sectional brake area for one 20 in. wheel yields $A_{avail} = 135.6$ in.^{2} For a rejected takeoff speed of about 190 kts and a maximum weight of 155,000 lb the kinetic energy to be dissipated is about

Table 7.4 Design Example Landing Gear Components

	Designation	Number	Unit Weight	Total Weight
Nose tire	26.5×8.0-13/12	2	24	48
Nose wheel	13 × 6.5	2	46	92
Nose strut	5.14 × 54	1	317	317
Main tire	H44.5 × 16.5-20/26	4	178	712
Main wheel	20 × 10.5	4	135	540
Main brake	18 × 9	4	345	1380
Main strut	9.49 × 89	2	1780	3560
Total				6650

248 million ft-lb. Using Equation (7.38) for a maximum brake temperature of 2300R yields the required total brake weight for steel to be $W_b = 1380$ lb and therefore each of the four wheel brakes weighs 345 lb. Each brake requires a volume of about $0.706 \, \text{ft}^3 = 1220 \, \text{in.}^3$. The required brake unit length is therefore $l_b = v/A_{avail} = 1070 \, \text{in.}^3/135.6 \, \text{in.}^2 = 9.00$ in. The width between the flanges of a 20 in. wheel is about 10.5 in., which suggests that there is sufficient room for a steel brake system to be at least partially shrouded by those wheels. A carbon brake unit for one main gear wheel weighs less than half that of a steel brake unit, about 176 lb as opposed to 345 lb. Carbon brakes, on the other hand, require about twice as much brake volume and therefore the brake unit length would be approximately 17.5 in., unless higher temperatures can be safely used.

The results for the landing gear components designed in Section 7.4.3 and the brake system designed in this section are collected in Table 7.4. The landing gear weight as calculated here is about 4.3% of the maximum gross weight, which is consistent with typical commercial airliner data, as will be discussed in Chapter 8.

7.6 Landing gear retraction

The landing gear struts transmit their load to the airframe through the trunnion mounted on the rear spar of the wing and pivot about the trunnion for stowage within the wing and fuselage, as shown in Figure 7.35. Landing retraction mechanisms are often quite different in configuration, as can be seen in Figures 7.16 and 7.21–7.24, but are usually based on the four-bar linkage of kinematics. The schematic diagram in Figure 7.35 is presented to provide a general idea of the landing gear arrangement in the down and stowed positions.

Clearance must be allowanced between the tires and nearby components or housings. The maximum overall tire dimensions given in the manufacturer's tire data tables must include a growth allowance due to use. In the case of a rotating tire allowance must be made for the diameter increase due to centrifugal force.

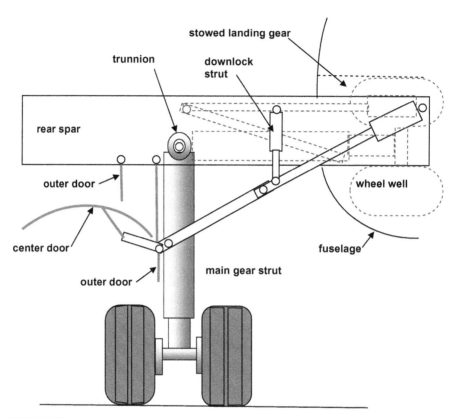

FIGURE 7.35

A schematic diagram of the general arrangement of landing gear components in the down and in the stowed positions (dashed lines).

Clearances between the tire and adjacent parts of the aircraft are determined by first computing the new tire envelope due to service according to the manufacturer's specifications. The new envelope is denoted by the dashed line called the used inflated tire in Figure 7.36. This expanded outer envelope of the tire is usually called the "grown" tire. The lateral clearance denoted by C_w may be found from the equation

$$C_w = 0.019w_g + 0.23 \tag{7.41}$$

This side clearance is given in inches, as is the width of the grown tire, w_g. The radial clearance depends upon the speed of the aircraft and the size of the tire and is shown in Figure 7.37 for rolling speeds up to 250 mph.

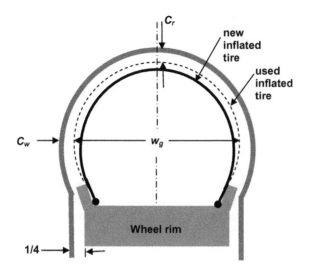

FIGURE 7.36

Side and radial clearance between tire and adjacent surfaces.

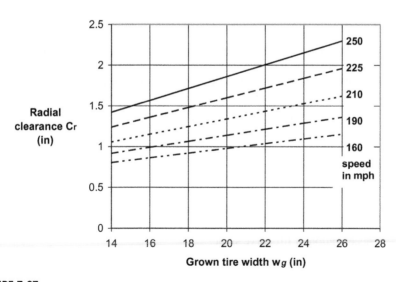

FIGURE 7.37

Radial clearances for grown tires as a function of aircraft ground speed.

7.7 Design summary

Using the configuration drawings and weight information for the design aircraft the landing gear layout and specifications can be estimated. The exact location of the wing on the fuselage will be determined after a refined weight estimate and center of gravity location is carried out as described in Chapter 8. However, using an estimated longitudinal center of gravity location of $x_{cg}/l_f = 0.44$ for aircraft with wing-mounted engines and $x_{cg}/l_f = 0.42$ for aircraft with fuselage-mounted engines one may use the methods of Section 7.2 to find the quasi-static loads which must be supported by the main and nose gear. Continuing with the methods of that section and the aircraft database of Table 7.1 one may find the dynamic load on the nose gear, approximately locate the main and nose gear so as to avoid the turnover condition, and determine the required ground clearance in pitch and roll for takeoff and landing. The provisional position of the main and nose gear may be shown on a planform view of the design aircraft, as illustrated in Figure 7.13, and the maneuverability in ground operations may be demonstrated.

Having determined the loads on the landing gear, the methods of Section 7.3 are used to specify the tires and wheels appropriate for the design aircraft. Then the landing gear struts may be sized as well as the necessary braking system. When estimating landing gear height, note that the extended height h_e of the main gear must be less than half the track so that the gear may be stowed within the fuselage. This is best illustrated by Figure 7.19 where it is clear that there is a noticeable distance between the stowed tires. The total weight of the landing gear system can then be developed and presented as in Table 7.4.

Knowing the layout and dimensions of the landing gear, provisions for retracting the landing gear into the fuselage may be determined. A scaled diagram, like that in Figure 7.35, can show the vertical location of the wing with respect to the fuselage, the location of the trunnion pivot on the rear spar of the wing, and the wheel well sufficient in size to accept the wheels upon retraction.

7.8 Nomenclature

A	orifice area
A'	coefficient defined in Equation (7.31)
A_h	hydraulic piston cross-sectional area
a	acceleration or lift curve slope
b	wingspan or width
C_L	lift coefficient
C_w	lateral tire clearance
C_r	radial tire clearance
c_d	orifice discharge coefficient
c_p	specific heat
D	drag
d	diameter

E_k	energy
e	extension
F	force
g	acceleration of gravity
h	height
k	polytropic exponent
L	lift
ℓ	length
l_a	see Figure 7.10
l_b	see Figure 7.10
m	mass
N	number
n	load factor
p	pressure
Q	heat
r	radius
S	stroke or wing planform area
S_n	normalized piston stroke, Equation (7.31)
T	temperature
t	time
V	velocity
W	weight
W_k	work
w	vertical velocity or tire width
w_n	normalized vertical velocity, Equation (7.32)
w_g	width of grown tire
x	longitudinal distance
α	angle of attack
β	angle defined by $\arctan(l_t/2l_{wb})$
δ	vertical distance between CG and line of action of F_{rev}
Γ	wing dihedral angle
γ	flight path angle
κ	thermal conductivity
ρ	density
η	strut compression efficiency
η_t	tire compression efficiency
Λ	sweepback angle
ϕ	roll angle
θ	pitch angle
σ	stress or atmospheric density ratio
ψ	turnover angle, Equation (7.11)
τ	tire constant, see Equation (7.32)
χ	steering angle
υ	volume

7.8.1 Subscripts

a	approach
b	brake
cg	center of gravity
f	fuselage
MG	main gear
max	maximum
n	normal
NG	nose gear
$plies$	tire plies
oe	operating empty
off	tires just off the ground
r	reverse
s	strut
st	static conditions
t	track
td	touchdown
ti	tire
to	turnover or takeoff
w	wheel
wb	wheelbase
wt	wingtip
180	180° turn

References

Curry, N.S., 1988. Aircraft Landing Gear Design: Principles and Practices. American Institute of Aeronautics and Astronautics, Reston, VA.

Daugherty, R.H., 2003. A Study of the Mechanical Properties of Modern Radial Aircraft Tires, NASA TM-2003-212415.

Flottau, J., Norris, G., 2012. In Sync. Aviation Week & Space Technology, July 16, pp. 26–27.

Goodyear, 2002. Goodyear Aircraft Tire Data Book, Goodyear Aviation Tires, Akron, OH, <www.goodyearaviation.com>.

Honeywell, 2008. Boeing 737NG Wheel and Cerametalix Brake, Honeywell Aerospace C61–0810-000-000.

Jackson, P. (Ed.), 2010. Jane's All the World's Aircraft 2009–2010. IHS (Global) Ltd., Coulsdon, Surrey, UK.

Michelin, 2012. Michelin Aircraft Tires, <www.airmichelin.com>.

Milwitzky, B., Cook, F.E. 1953. Analysis of Landing Gear Behavior, NACA Report 1154.

Torenbeek, E., 1982. Synthesis of Subsonic Aircraft Design. Kluwer Academic Press, Dordrecht, The Netherlands.

Wall, R., 2012. Taxi Electric. Aviation Week & Space Technology, February 27, pp. 47–48.

Yetter, J.A., 1995. Why Do Airlines Want and Use Thrust Reversers? NASA Technical Memorandum 109158.

Refined Weight and Balance Estimate

Commercial Airplane Design Principles. http://dx.doi.org/10.1016/B978-0-12-419953-8.00008-5

8.1 Process for refining the weight estimate

The aircraft preliminary design process executed thus far resulted in an airplane with a distinct configuration and specific dimensions. The procedures involved were based on a first approximation of the airplane's total weight. A second approximation for the aircraft weight is necessary to carry the design further. The methods simplest to apply at this stage are empirical in nature, ones which exploit the relative richness of experience in manufacturing commercial airplanes. Such approaches for making a refined weight estimate of the design aircraft are presented by Torenbeek (1982), Raymer (2006), and Roskam (1999), among others. More analytically based estimation methods have been presented by Ardema et al. (1996), among others. Torenbeek's approach is mainly followed here in pursuing the weight estimation process and generally the initial weight estimate W_g found previously is used as the scale factor in applying this method. Of course, the revised gross weight is generally found to differ from the initially chosen value. This is so because the weights of the individual components are themselves taken to be functions of the gross weight. It is generally found that the difference between the initial W_g and the revised W_g is small enough to permit a second iteration with a reasonable expectation of convergence. Thus, the initial gross weight estimate, denoted by $W_{g,1}$, leads to a revised gross weight, $W_{g,2}$. This value may now be used as the starting point in a new iteration of the component weight estimation process which leads to a third estimate, $W_{g,3}$. When the difference is such that $(W_{g,3} - W_{g,2})/W_{go,2}$ is less than, say, 0.005 (i.e., 0.5%) the iteration process may be halted and the last value will be considered the final revised gross weight. Of course, all the component weights will be those corresponding to the final choice of the gross weight.

8.2 Limit load factor

The load factor is defined as $n = L/W$ with $n > 0$ denoting wings pulled up and $n < 0$ denoting wings pulled down. A load factor $n = 1$ denotes steady level flight with $L = W$. Load factors different from $n = 1$ are caused by maneuvers such as turns, dives, climbs, etc. as shown in Figures 8.1 and 8.2.

The structural strength required of the airplane components is determined by the design maximum load factor specified for the airplane and will vary with the function of the airplane, with fighters having load limits set not by structural strength achievable, but rather by the ability of pilots to withstand the accelerations causing the load factor (generally $n < 9$). In any maneuver the maximum lift that can be generated is $L_{max} = n_{max}W$ which leads to the result that

$$n_{max} = \frac{C_{L,max}q}{(W/S)} = \frac{C_{L,max}\rho_{sl}}{2(W/S)}V_E^2 \tag{8.1}$$

We may use Equation (8.1) to illustrate the safe operating regime for an aircraft on a so-called V-n diagram as shown in Figure 8.3. Note that V_E is the equivalent

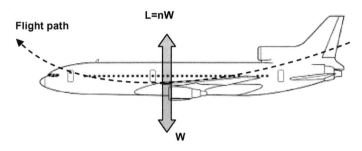

FIGURE 8.1

Load factor during pull-up maneuver.

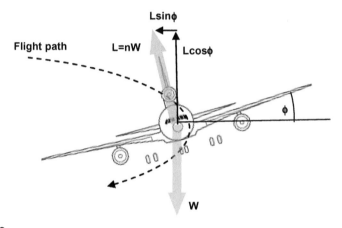

FIGURE 8.2

Load factor during a level coordinated turn.

airspeed and all velocities in this chapter will be in knots equivalent airspeed unless otherwise indicated. The limit normal load factor shown in the figure depends upon the gross weight as given in FAR Section 25.337 according to the following equation:

$$n_{limit} = 2.1 + \frac{24{,}000}{(W_{to} + 10{,}000)} \tag{8.2}$$

This is for $2.5 \le n_{limit} \le 3.8$. For aircraft weighing more than 50,000 lbs or less than 4118 lb, n_{limit} is constant and equal to 2.5 or 3.8, respectively.

8.3 The design dive speed

FAR Section 25.335 specifies that the design dive equivalent airspeed V_D be equal to or greater than 125% of the design cruise equivalent airspeed V_C. For

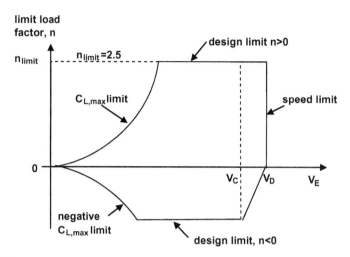

FIGURE 8.3

V-n diagram for an airplane showing limits of flaps-up operation.

turboprops and other low- to moderate-speed aircraft a value of 130–140% may be used at this stage in the design process. However, for high-subsonic-speed aircraft the pertinent factor is the Mach number when determining the design dive speed and the former requirement is superseded by other requirements. FAR Part 25 Section 25.335 requires that the design dive equivalent airspeed be selected such that the design cruise Mach number $M_C \leq 0.8 M_D$ or else that certain minimum speed margins be maintained. Two methods for demonstrating that such margins are met are prescribed. One specifies a dive maneuver for calculating V_D and the other provides for a probabilistic determination of the speed margins achieved. Both methods are beyond the scope of the design process considered here. However, it is also specified that for altitudes where the dive speed is limited by compressibility effects it is sufficient to assume that $M_D = M_c + 0.10$ where M_c is the cruise Mach number. This factor of 0.10 is reasonable, but arbitrary, and may be reduced to 0.07 if weight problems begin to accumulate for the design. However, FAR 25.335 notes that this factor may not be reduced to a value less than 0.05. It should be remembered that the speeds discussed above are taken as equivalent air speeds (EAS) rather than true airspeed, where $V_E = \sigma^{1/2} V$. The equivalent airspeed is a measure of the dynamic pressure experienced by the aircraft and V_D is therefore a measure of the maximum dynamic pressure experienced in flight.

The cruise speed V_C (knots equivalent airspeed) must also be sufficiently greater than V_B (knots equivalent airspeed), the design speed for maximum gust intensity, to provide for inadvertent airspeed increases likely to occur as a result of severe atmospheric turbulence. In any event, $V_C > V_B + 1.32 U_{ref}$ where U_{ref} is an altitude-dependent reference speed given, in ft/s equivalent airspeed, by

$$0 \leq z \leq 15{,}000 \text{ ft} : U_{ref} = 56 - \frac{z}{15{,}000}$$

$$15{,}000 \text{ ft} \leq z \leq 50{,}000 \text{ ft} : U_{ref} = 44 - 7.72 \left(\frac{z}{15{,}000} - 1 \right)$$

The design speed for maximum gust intensity is given by

$$V_B \geq V_s \left[1 + \frac{a K_g V_C U_{ref}}{498 \left(\frac{W}{S} \right)} \right]$$

The factors appearing in the equation for V_B are as follows:

$$K_g = \frac{0.88 \mu}{5.3 + \mu}$$

$$\mu = \frac{2}{\rho_{sl} \sigma \, a g c_{MAC}} \left(\frac{W}{S} \right)$$

$$a = \frac{\partial C_L}{\partial \alpha}$$

Note that K_g, μ, and a are all dimensionless and the constants in these equations account for the different dimensions quoted for U_{ref} and V_C.

8.4 Wing group weight

The wing group incorporates the full structure of the wing including spars, stringers, airfoil sections, skin, and the wing box carry-through structure which passes through the fuselage. The wing weight fraction, W_{wg}/W_{zf}, depends upon the design limit normal maneuvering load factor through $n_{ult} = 1.5 n_{limit}$. Since the wing weight is approximately 8% of the aircraft's weight it is suggested that for aircraft weights in the range where n_{limit} is variable (see Figure 8.3) the wing weight fraction be varied with the limit normal load factor within the overall iteration process described previously. Torenbeek (1982) offers the following equation for initially estimating the weight of the wing group

$$W_{wg} = k_{wg} W_{zf}^{0.7} S^{0.3} n_{ult}^{0.55} t_{r,\max}^{-0.3} \left(\frac{b}{\cos \Lambda_{c/2}} \right)^{1.05} \left(1 + \sqrt{\frac{6.25 \cos \Lambda_{c/2}}{b}} \right) \quad (8.3)$$

This equation is written for lengths in feet and weights in pounds; the quantities W_{zf} and $t_{r,\max}$ denote aircraft zero-fuel weight and wing root maximum thickness,

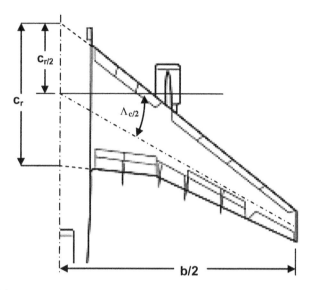

FIGURE 8.4

Schematic diagram of the wing group and its notation. Engines, pylons, and nacelles are not included in the wing group.

respectively. Torenbeek (1982) suggests $k_{wg}=0.0017$, but data comparisons to be described subsequently suggest that $k_{wg}=0.0021$ yields better results. The fuel fraction $M_f=W_f/W_g$ required for the mission was found in Chapter 2, so that $W_{zf}=(1-M_f)W_g$. The other factors in Equation (8.3) have likewise been determined for the design aircraft in previous chapters. A schematic diagram of the wing group and the associated notation is shown in Figure 8.4.

This wing weight expression includes high lift devices, ailerons, and spoilers but not wing-mounted engines. Torenbeek (1982) suggests that for 2 or 4 wing-mounted engines the wing weight given by Equation (8.3) should be reduced by 5% or 10%, respectively. The actual weight of the propulsion group, that is, the weight of the engines and associated equipment is calculated separately. The wing-mounted engines reduce the root bending moment produced by the wing lift and therefore permit some reduction in the wing carry-through structure. As might be expected, the wing weight given by Equation (8.3) is directly proportional to the actual length, rather than the span, of the wing. Reducing the root thickness, on the other hand, increases wing weight because additional strengthening would be required to resist wing bending moments at the root of the thinner wing. All the terms in Equation (8.3) are known for the design from the detailed configuration drawing and data that were developed thus far.

A different correlation for the weight of the wing group is presented by Oman (1977):

$$W_{wg} = k'_{wg} S^{0.7} A^{0.47} \left(\frac{W_g n_{ult}}{1000}\right)^{0.52} \left(0.3 + \frac{0.7}{\cos \Lambda_{c/2}}\right) \left[\frac{1+\lambda}{(t/c)_{r,max}}\right]^{0.4} \quad (8.4)$$

Table 8.1 Configuration and Weight Data for Eight Jet Transports Taken from Ardema et al (1996) and Manufacturers' Data

	B720	B727	B737	B747	DC-8	MD-11	MD-83	L-1011
W_g (lb)	202,000	160,000	111,000	713,000	335,000	602,500	140,000	409,000
W_{wg} (lb)	23,528	17,860	10,687	88,202	35,330	62,985	15,839	46,233
W_{fus} (lb)	19,383	17,586	11,831	72,659	24,886	54,936	16,432	52,329
S (ft^2)	2460	1587	1005	5469	2927	3648	1270	3590
$(t/c)_t$	0.0902	0.0900	0.1120	0.0780	0.1050	0.0930	0.1200	0.0900
$(t/c)_r$	0.1551	0.1540	0.1260	0.1794	0.1256	0.1670	0.1380	0.1300
A	6.96	7.67	8.21	6.96	7.52	7.50	9.62	6.98
λ	0.33	0.26	0.22	0.26	0.20	0.26	0.24	0.30
$\Lambda_{c/4}$	35.00	32.00	25.00	37.17	30.60	35.00	24.16	35.00
v_f (gal)	16060	7680	4720	48445	23400	38815	5840	31642
l_{fus} (ft)	130.5	116.7	96.9	225.2	153.0	192.4	135.5	177.7
d_{fus} (ft)	14.2	14.2	12.7	20.2	13.5	19.8	11.4	19.6
F_{NC}	1.81	2.00	1.92	2.13	2.00	1.67	1.61	1.76
F_{TC}	2.86	2.83	2.36	3.29	2.94	2.27	2.73	2.96

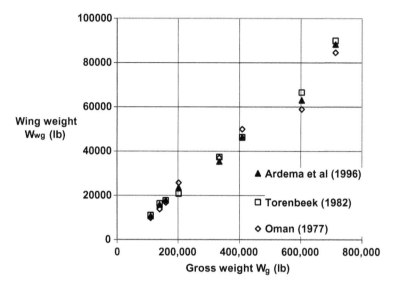

FIGURE 8.5

Wing weights for a range of jet transport aircraft compared to the correlations given by Torenbeek (1982) in Equation (8.3) with $k_{wg}=0.0021$ and by Oman (1977) in Equation (8.4) with $k'_{wg}=0.475$.

Oman (1977) suggests $k'_{wg}=0.416$, but data comparisons to be described subsequently suggest that $k'_{wg}=0.475$ yields better results. Though the appearance is different this equation actually depends upon n_{ult}, $t_{r,max}$, b, and aircraft weight in a manner very similar to Equation (8.3). Thus the wing group weight for the design aircraft may be computed at this point using either Equations (8.3) or (8.4) and it is useful to examine the predictions each makes.

Ardema et al. (1996) carried out detailed studies of wing and fuselage weight for the eight jet transports listed in Table 8.1. They reported actual component weights which are compared in Figure 8.5 with the correlation Equations (8.3) and (8.4). As mentioned previously, slightly different coefficients from those originally proposed are used here and we see that both correlations are reasonably accurate, but neither shows consistent superiority. Although the aircraft considered included both wing-mounted and fuselage-mounted engines, no adjustments were made for the wing group weight data presented in Figure 8.5. Indeed, examination of the results shown in Figure 8.5 suggests that an average between Equations (8.3) and (8.4) provides better results than either alone, with a deviation of only about $\pm 5\%$.

8.5 Fuselage group weight

The fuselage group includes the cabin, the nose cone, the tail cone, the internal structure, and all the covering skin. The fuselage weight is difficult to estimate because it is a complex structure with many openings, support attachments, floors, etc., but it is strongly dependent on the gross shell area, S_g. This is the surface area of the complete fuselage treated as an ideal surface, that is, with no cutouts for windows or wing and tail attachments. CAD packages typically provide auxiliary information like gross shell area but for cylindrical cabin sections of fuselages with high fineness ratio, $F = l_{fus}/d_{fus} > 5$, the gross wetted area may be estimated with the use of Equations (3.14) and (3.15):

$$S_g = \pi d_{fus} l_{fus} \left(1 - \frac{F_{NC}}{3F} - \frac{F_{TC}}{2F} \right) \tag{8.5}$$

The gross wetted area of the fuselage is proportional to the volume υ enclosed by that area. Oman (1977) presents data supporting this idea and S_g and υ may be quite accurately related by the expression

$$S_g = 6.054\upsilon^{0.764}$$

A correlation equation for the fuselage weight is given by Torenbeek (1982) as

$$W_{fus} = k_{fus} S_g^{1.2} \sqrt{V_D \frac{l_h}{b_{fus} + h_{fus}}} \tag{8.6}$$

In this equation $k_{fus} = 0.0227$, the lengths are in feet, the weight is in pounds, and the design dive speed, V_D, is in knots equivalent airspeed. The length l_h is the distance between the root quarter-chord points of the horizontal tail and the wing with b_{fus} and h_{fus} denoting the maximum width and height of the fuselage, respectively. Torenbeek suggests that to this basic weight, for a conventional pressurized cabin, 7% should be added and if the engines are mounted on the aft fuselage another 4% should be added. It is clear that the fuselage weight should be at least directly proportional to the surface area of the fuselage because the fuselage is essentially a hollow pressurized shell. Similarly, fuselage bending moments are proportional to V_D and l_t so that increasing these parameters should require additional strengthening of the fuselage, and therefore additional weight. It may be surprising that the design normal load factor does not appear in the fuselage weight equation, and indeed, Torenbeek (1982) does present a more detailed weight estimation method which does include n_{limit}. It is suggested that pressure forces acting on the fuselage shell are more significant than the fore and aft bending moments acting at the wing-fuselage juncture.

Because the location of the wing and the tail surfaces is not yet fixed we may approximate the length l_t as half the fuselage length, $l_{fus}/2$. Likewise, because the fuselage is close to being circular we may take $b_{fus} + h_{fus} = 2d_{fus}$. Then Equation (8.6) becomes

$$W_{fus} = k_{fus} S_g^{1.2} \sqrt{V_D \frac{F_{fus}}{4}} \tag{8.7}$$

Oman (1977) offers a different correlation as follows:

$$W_{fus} = k'_{fus} n_{ult}^{0.52} W_g^{0.33} l_{fus}^{0.76} (b_{fus} + d_{fus})^{1.2} \tag{8.8}$$

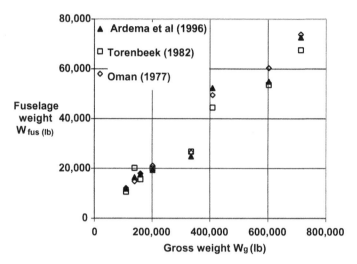

FIGURE 8.6

Fuselage weights for a range of jet transport aircraft compared to the correlations given by Torenbeek (1982) in Equation (8.7) with $k_{fus}=0.032$ and by Oman (1977) in Equation (8.8) with $k'_{fus}=0.0837$.

As in the previous equation, because the fuselage is almost circular we may take $b_{fus}+h_{fus}=2d_{fus}$. In Equation (8.8) the ultimate load factor does appear and again all weights are in pounds and lengths in feet.

Fuselage weight estimates using Equations (8.7) and (8.8) for the aircraft in Table 8.1 are shown in Figure 8.6. Equation (8.7) with the coefficient $k_{fus}=0.032$ appears to yield the best results, although Torenbeek (1982) in Equation (8.6) uses 0.0227. Similarly the coefficient $k'_{fus}=0.0837$ appears to give the best results, though Oman (1977) reports 0.0796. There is more scatter in the fuselage weight estimates than in the wing weight estimates. Once again, examination of the data in Figure 8.6 indicates that the most consistent agreement occurs when using the average of the two methods, with errors within the range of $\pm10\%$. It is suggested that the fuselage weights, like the wing weights, be estimated by averaging the results of the two methods to aid in selecting an appropriate estimate.

8.6 Landing gear group weight

The landing gear includes all rolling components, struts, and activating mechanisms. The landing gear weight may be estimated as a function of gross weight alone, but the relationship is not linear. This is not a problem because the weight fraction of the landing gear is nearly constant at about 3.8–4.5% of the gross weight for aircraft whose weight exceeds 10,000 lbs. The approximation given by Torenbeek (1982) is

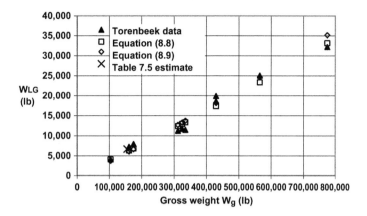

FIGURE 8.7

Landing gear weights for a range of jet transport aircraft presented by Torenbeek (1982) compared to the correlations given by Equations (8.9) and (8.10) and by the estimate presented in Table 7.5.

presented below for airliner-type aircraft; the subscript MG refers to the main gear while the subscript NG refers to the nose gear, and the weights are all in pounds.

$$W_{MG} = 40 + 0.16\,W_g^{3/4} + 0.019\,W_g + 1.5 \times 10^{-5}\,W_g^{3/2}$$
$$W_{NG} = 20 + 0.10\,W_g^{3/4} + 2 \times 10^{-6}\,W_g^{3/2}$$

These relations are used to determine the ratio of landing gear weight in terms of the gross weight

$$W_{LG} = W_{MG} + W_{NG} = 60 + 0.26\,W_g^{0.75} + 0.19\,W_g + 1.7 \times 10^{-5}\,W_g^{1.5} \quad (8.9)$$

Oman (1977) offers a different approximation which is based on landing weight. However, we have shown, in Figure 4.6, that $W_l/W_g = 0.77$ is a reasonable representation of nominal landing weight so that Oman's correlation may be written as follows:

$$W_{LG} = 0.00891\,W_g^{1.12} \quad (8.10)$$

The results for both approximations, Equations (8.9) and (8.10), are shown in Figure 8.7 along with weight data for various aircraft presented by Torenbeek (1982). It appears that both approximations yield similar results. Also shown in Figure 8.7 is the estimate obtained in Section 7.5 for the landing gear weight of a 155,000 lb gross weight aircraft. This estimate appears slightly high in comparison, but note that a 10% reduction in the landing strut weight in Table 7.5 would bring the total landing gear weight estimate down to about 4% of the gross weight. The landing gear strut weight is directly dependent upon the landing gear height so when that height is known with some confidence the weight buildup in Table 7.5 is preferable to the correlations presented in Equations (8.8) and (8.10), though the latter may be used as a convenient check on the results.

8.7 Tail group weight

The tail group includes the horizontal and vertical tail assemblies including the control surfaces and skin. This group also represents a small fraction of the takeoff weight, about 2–3%, but that weight does have an effect on center of gravity location because of the long moment arms. For airliner-type aircraft the weight of the tail surfaces is mainly dependent on the design dive speed and the tail planform areas.

Torenbeek (1982) presents curves illustrating the functional relationships for the weight of the horizontal and vertical tail surfaces. Fitting those curves and summing the horizontal and vertical tail weights yields the following approximate equation for the tail group weight:

$$
W_t = k_h S_h \left[2 + 4.15 erf \left(\frac{S_h^{0.2} V_D}{10^3 \sqrt{\cos \Lambda_h}} - 0.65 \right) \right]
$$
$$
+ k_v S_v \left[2 + 4.15 erf \left(\frac{S_v^{0.2} V_D}{10^3 \sqrt{\cos \Lambda_v}} - 0.65 \right) \right]
$$

(8.11)

The design dive equivalent speed V_D is in knots and the surface areas of the horizontal and vertical tails are in square feet. The coefficients k_h and k_v account for different tail configurations. For example, current practice for airliners is to have variable incidence tails, and $k_h = 1.1$, while a fixed horizontal stabilizer would have $k_h = 1.0$, reflecting the lighter structure typical of fixed equipment. For fuselage-mounted vertical tails $k_v = 1.0$, while for T-tails $k_v = 1 + 0.15 \frac{S_h h_h}{S_v b_v}$. In this last equation the quantities h_h and b_v correspond to the height of the horizontal tail above the fuselage centerline and the height of the tip of the vertical tail above the fuselage centerline, respectively. The quantity $erf(x)$ is the error function and is tabulated in Appendix E as well as in various mathematics reference books. The definition of the error function is

$$
erf(x) = \frac{2}{\sqrt{\pi}} \int_0^x e^{-\eta^2} d\eta
$$

(8.12)

Oman (1977) suggests the following correlations for the horizontal and vertical tail group weights which do not include any dependence on $V_{D,E}$:

$$
W_h = 0.00563 W_g^{0.6} S_h^{0.469} \left(\frac{A_h}{\cos^2 \Lambda_{h,c/2}} \right)^{0.539} \left(\frac{1 + \lambda_h}{(t/c)_{h,max}} \right)^{0.692}
$$

(8.13)

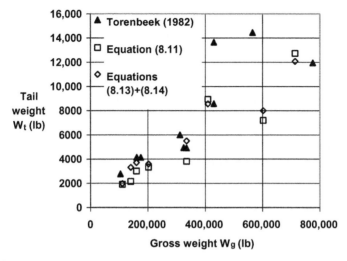

FIGURE 8.8

Tail group weights for a range of jet transport aircraft presented by Torenbeek (1982) compared to the correlations given by Equation (8.11) and by the sum of Equations (8.13) and (8.14).

$$W_v = 0.0909 W_g^{0.333} S_v^{0.7} \left(\frac{A_v}{\cos^2 \Lambda_{v,c/2}} \right)^{0.35} \left(\frac{1 + \lambda_v}{(t/c)_{v,\max}} \right)^{0.5} \left(1 + \frac{h_h}{b_v} \right)^{0.43} \quad (8.14)$$

The tail group weights for the aircraft in Table 8.1, as given by Equation (8.11) and by the sum of Equations (8.13) and (8.14), are shown in Figure 8.8. Also shown in Figure 8.8 are the tail group weights for various jet transports reported by Torenbeek (1982).

Though the two methods described are generally in reasonable agreement with each other, there still is substantial scatter in comparison to the weights reported by Torenbeek (1982). In particular, the two data points showing tail group weights of about 14,000 lb are 50% higher than one would expect based on the data trend. These data are for two versions of the Douglas DC-10, an aircraft with a third engine mounted between the fuselage and the bottom of the vertical tail. It may be the case that the tail-mounted engine required additional strengthening of the vertical tail structure and this caused the total weight of the tail group to be substantially larger than conventional vertical tail installations. It is worth pointing out that the tail group weights for the other approximately 400,000 lb aircraft are in reasonable agreement with the correlations, even though that one is a Lockheed L-1011 which also carries a third engine. However, the engine is actually buried in the aft fuselage; only the inlet duct is part of the vertical tail assembly. As a result the degree of strengthening of the vertical tail structure is not as great as when the engine is actually part of the vertical tail system.

Another interesting feature of Figure 8.8 is that the two predictions of tail group weight for the approximately 600,000 lb aircraft, a McDonnell-Douglas MD-11, don't agree very well with each other and both lie well below the expected data trend. It is interesting to note that the MD-11 was a later version of the DC-10 and was very similar in size and configuration. The most likely explanation for the low prediction of the MD-11 tail group weight is that the area of its horizontal tail surface is substantially smaller than one would expect. Indeed, the horizontal tail area of the MD-11 is about 33% smaller than that of the DC-10, the aircraft from which it is derived. If the larger horizontal tail area is used, both predictions would rise to be in the expected trend region of 9000–10,000 lb. But, as mentioned previously, the prediction methods presented do not take into account unconventional tail structures, such as those which support engines. Once again, it seems prudent to use an average of both prediction methods for conventional tail assemblies. If a tail-mounted engine is considered it appears necessary to increase the predicted tail weight by about 40% for preliminary design purposes.

8.8 Propulsion group weight

The propulsion group includes the engine and associated equipment like thrust reversers and water injection systems. Engine weights are reasonably well correlated to their takeoff static thrust by Equation (8.15), as illustrated by Figure 8.9.

$$W_e = 2.7F_{to}^{0.75} \qquad (8.15)$$

The data shown are for 26 turbofan engines appearing in the database compiled by Svoboda (2000). As can be seen, the error using this correlation is within ±10% but actual engine weights should be used whenever available.

Both Torenbeek (1982) and Oman (1977) present detailed estimation methods for the propulsion group. In addition to the detailed method for estimating the weight of the propulsion group, Torenbeek offers a simpler approximation for podded jet engines equipped with thrust reversers and water injection systems. The propulsion group weight is assumed to be linearly proportional to the total engine weight. The weight data for various transport aircraft collected in Torenbeek (1982) and shown in Figure 8.10 suggest the following form:

$$W_{pg} = 10(n_e W_e)^{0.8} \qquad (8.16)$$

In this equation n_e is the number of engines and W_e is the weight of one engine in pounds. It is noted that Torenbeek (1982) assumes a linear relation given by $W_{pg} = 1.35 n_e W_e$ which agrees with Equation (8.16) at lower total engine weights but is somewhat conservative at the higher engine weights.

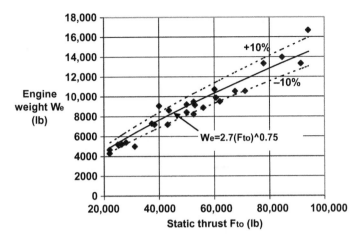

FIGURE 8.9

Variation of engine dry weight with takeoff thrust rating for 26 turbofan engines listed by Svoboda (2000) compared to the correlation of Equation (8.15).

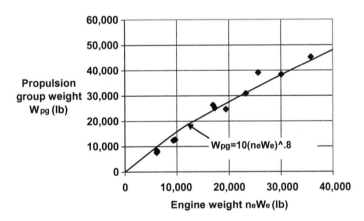

FIGURE 8.10

Propulsion group weights for a range of jet transport aircraft presented by Torenbeek (1982) compared to the correlation of Equation (8.16).

8.9 Nacelle group weight

A typical high bypass turbofan engine and its surrounding nacelle are shown in Figure 8.11 along with some relevant dimensions. The nacelle provides a streamlined housing for the engine and the pylon is the structural member which supports

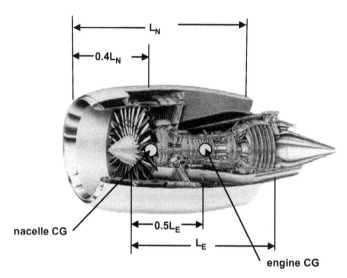

FIGURE 8.11

General Electric GE90 high bypass turbofan engine showing its placement within a nacelle and some relevant dimensions.

the engine and transmits the thrust of the engine to the aircraft. The weight of the nacelle may be written in terms of the nacelle contents. A correlation is offered by Oman (1977) for nacelles housing turbofan engines in the form

$$W_n = 35.45n_n \left[\frac{2.33(1.1W_e)S_{n,wet}}{10,000} \right]^{0.59} \tag{8.17}$$

In Equation (8.17) the weights of the nacelle group W_n and of the engine W_e are in pounds, the wetted area of the nacelle is in square feet, and the quantity n_n denotes the number of nacelles.

The pylon supports the nacelle and engine assembly and is shown in Figure 8.12. The actual length of the nacelle, the pylon, and their placement on the wing influence the drag of the combination and these factors are usually determined in detail in the advanced stages of design. For preliminary design purposes the correlation proposed by Oman (1977) for the total weight of the n_p pylons may be used:

$$W_p = 24.11n_p S_p^{0.381} \left[\frac{1.46n_{ult}(1.1W_e)L_n d_n}{10^6 \cos \Lambda_p} \right]^{0.952} \tag{8.18}$$

Then the weight of the nacelle group is the sum of the nacelle weight and the pylon weight $W_{ng} = W_n + W_p$. However, this sum is not well correlated to the engine weight. On the other hand, if the sum of the actual propulsion group and nacelle group weights is formed it correlates much better with engine weight, as shown in Figure 8.13, where now

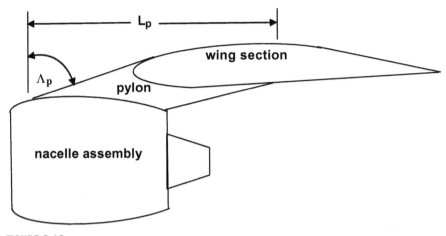

FIGURE 8.12

Sketch of the pylon which attaches the engine and nacelle assembly to the wing.

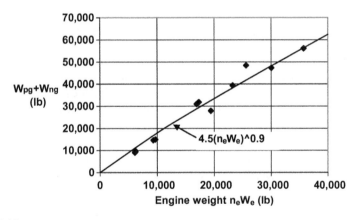

FIGURE 8.13

Propulsion group plus nacelle group weights for a range of jet transport aircraft presented by Torenbeek (1982) compared to the correlation of Equation (8.19).

$$W_{ng} + W_{pg} = 4.5(n_e W_e)^{0.9} \tag{8.19}$$

It is probably reasonable to merely use the correlation of Equation (8.19) to estimate the combined weight of the propulsion and nacelle group for the preliminary weight estimation, although the individual equations may also be used if a reasonable configuration for the engine-nacelle-pylon assembly can be sketched.

8.10 **Flight controls group weight**

The flight controls group is comprised of the actuation systems for the basic maneu-vering controls like ailerons, spoilers, rudder, elevator, and adjustable stabilizer and the actuation systems for the high lift system including Fowler flaps, and extendable leading edge slats. Sample data for flight controls group weights are shown in Figure 8.14 as a function of gross weight and are from Torenbeek (1982) and Kroo (2008). The weight of the flight control group is reasonably well correlated by the following relation:

$$W_{fc} = 1.44\, W_g^{0.625} \tag{8.19}$$

Here the weights are measured in pounds and the variation of the correlation with gross weight is also shown in Figure 8.14. The correlation works reasonably well for 17 of the 23 commercial transports considered, even down into the low gross weight range typical of turboprop-powered aircraft. A large discrepancy exists in the middle of the gross weight range, around 300,000 lb, where several first-generation narrow-body aircraft, B707s and DC-8s, show values for W_{fc} about half that given by Equation (8.19) and one later wide-body aircraft, an A300, shows a value for W_{fc} about half again as great as that given by Equation (8.19). Examination of the variation of the flight controls group weight with wing area shows the same behav-ior: most follow a power-law variation given by $W_{fc} = 7S^{0.8}$, while the value of W_{fc} for the same aircraft mentioned, whose wing areas fall in the middle range of about 3000 ft^2, again fall well below or above the correlation. No clear indication of the reason for this scatter in the middle of the weight and wing area range is available. It is recommended that the more conservative result of Equation (8.19) should be used in the preliminary design project.

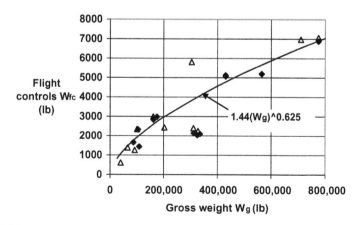

FIGURE 8.14

Variation of the flight controls group weight with takeoff weight. Open symbols are data from Torenbeek (1982) and closed symbols are data from Kroo (2008). The correlation of Equation (8.19) is also shown.

Table 8.2 Airliner Flight Controls and Associated Constants Suggested by Oman (1977)

Surface control	k_{fc}
Ailerons	0.065
Spoilers	0.035
Rudder	0.035
Elevator	0.045
Adjustable stabilizer	0.010
Fowler flaps	0.085
Translating slats	0.075
Krueger flaps	0.055
Total	0.405

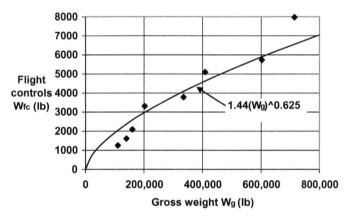

FIGURE 8.15

Comparison of data calculated using Equation (8.20) with $k_{fc}=0.405$ from Table 8.2 for the aircraft in Table 8.1 with the correlation Equation (8.19).

A correlation for the flight controls group is offered by Oman (1977). It is based on aircraft characteristics and is given by

$$W_{fc} = k_{fc} \frac{q_D^{0.16} \sqrt{S} \left(l_{fus} + \frac{b}{\cos \Lambda_{c/2}} \right)}{A^{0.85}} \qquad (8.20)$$

Here S is the wing planform area (in ft^2), q_D is the dynamic pressure (in lb/ft^2) corresponding to the design dive speed, l_{fus} is the fuselage length in feet, b is the wingspan in feet, and $\Lambda_{c/2}$ is the wing sweepback of the 50% chord. Oman (1977) points out that the proportionality constant k_{fc} takes on different values for each of the different aerodynamic control systems and that the individual coefficients may be summed to provide a single value. Typical flight control surfaces for a commercial airliner and the suggested value of k_{fc} suggested by Oman are listed in Table 8.2.

Carrying out this calculation of Equation (8.20) using the characteristics of the aircraft described in Table 8.1 and $k_{fc} = 0.405$ as suggested by Table 8.2 leads to the data points shown in Figure 8.15. The simple correlation curve of Equation (8.19) is also shown on Figure 8.15 and it follows the calculated data quite closely. It is preferable to use Equation (8.20) with $k_{fc} = 0.405$ for preliminary design purposes because it incorporates more specific detail. However, it is advisable to also compute the flight controls group weight using the simple correlation of Equation (8.19) as a check.

8.11 Auxiliary power unit group weight

The auxiliary power unit (APU) is a small gas turbine engine mounted in the tail cone of an aircraft to provide autonomous electrical and mechanical power for the following:

- Starting power for the main engines.
- Pneumatic power for cabin air conditioning systems.
- Shaft power for other pneumatic and hydraulic systems.
- Backup electrical and pneumatic power for in-flight operations and emergencies.
- Electric and pneumatic power for ground operations with the engines shut down.

The APU enhances capability by permitting the aircraft to carry out various functions requiring electric power on the ground without need for a ground-based generating unit to be available at each airport. A schematic diagram of a conventional APU is presented in Figure 8.16 which shows that inlet air is drawn into the APU and divided into two streams. One stream flows through a centrifugal compressor feeding high-pressure air to a combustor which burns fuel drawn from the main fuel tanks. The APU may use up to 2% of the fuel consumed during a typical flight. The hot combustion gases drive a two-stage axial flow turbine, producing shaft power and then exhausting through an exhaust nozzle typically located at the aft end of the fuselage tail cone. The other air stream passes through a second centrifugal compressor driven off the main power shaft. There the air pressure is raised to about 50 psia and then exits to power pneumatic and cabin environmental control systems. The main power shaft also drives the electric generator which feeds electric systems when the electric power from the main engine-driven generators is unavailable.

General characteristics of some modern APUs for typical airliners are described in Table 8.3. Equations for estimating the weight of an installed APU system have been proposed by several investigators.

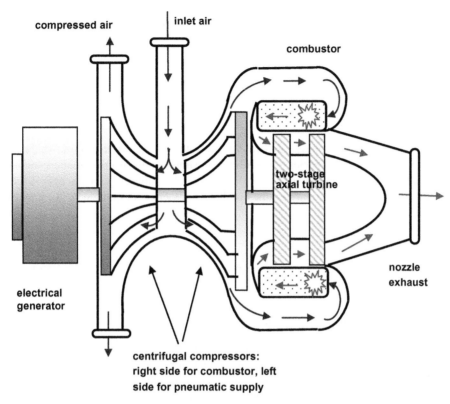

compressed air

inlet air

combustor

two-stage axial turbine

nozzle exhaust

electrical generator

centrifugal compressors: right side for combustor, left side for pneumatic supply

FIGURE 8.16

Schematic diagram of a typical APU shown driving an electric generator while supplying pressurized air for pneumatic systems and cabin environmental control.

Table 8.3 APU Weight and General Characteristics of Some Airliners

Aircraft	Gross Weight W_g (lb)	Passengers N_p	Electric Power (kW)	APU Weight (lb)	Weight of Engines $n_e W_e$ (lb)
Dash8-100	36,300	38	40	115	1872
ERJ 145	46,275	50	37	120	3162
B737-600	145,500	110	60	282	10,250
A320-200	169,800	120–180	90	308	10,500
B787	503,000	250	450	525 (est.)	25,644
B777-200ER	656,000	300	120	730	33,288
B747-8I	975,000	467	180	835	49,600
A380-800	1,234,600	525	240	950 (est.)	59,200

8.11.1 Torenbeek's correlation

Torenbeek (1982) suggests that the installed APU group weight is directly proportional to the weight of the APU itself, that is, $W_{apug} = k_{apug} W_{apu}$ where the range of the proportionality coefficient is given as $2 < k_{apug} < 2.5$. The following correlation for the APU weight correlation is offered:

$$W_{apu} = 16 \dot{w}_{ba}^{0.6} \tag{8.21}$$

This result is based on $\dot{w}_{ba}$, the weight flow rate of bleed air, in lb/min, which the APU compresses and supplies to the aircraft to support onboard pneumatic systems, particularly cabin environmental control. A value of 1.1 lb/min per passenger is recommended and with N_p denoting the number of passengers Equation (8.20) may be written as

$$W_{apu} = 16(1.1N_p)^{0.6} \tag{8.22}$$

Then the estimate for the APU group weight due to Torenbeek (1982) becomes

$$W_{apug} = 16 k_{apug} (1.1N_p)^{0.6} \tag{8.23}$$

8.11.2 Modified correlation

Comparing the predictions for APU weight given by Equation (8.22) with the actual APU weights in Table 8.3 shows that the original Torenbeek suggestion increasingly underestimates the APU weight as the number of passengers increases. A modified correlation of the same form which gives better results is given by

$$W_{apu} = 8(1.1N_p)^{0.75} \tag{8.24}$$

To form the group weight of the APU we again multiply the APU weight by the proportionality constant k_{apug}, that is, the likely APU group weight is approximately some multiple of the actual unit weight. Then the APU group weight is

$$W_{apug} = k_{apug} W_{apu} = k_{apug} 8(1.1N_p)^{0.75} \tag{8.25}$$

8.11.3 Kroo's correlation

Kroo (2008) also offers an approximation for the APU group weight based solely on the number of passengers as follows:

$$W_{apug} = 7N_p \tag{8.26}$$

This approximation is simple yet may tend to overestimate the APU group weight as the number of passengers increases.

8.11.4 Relating APU group weight to aircraft gross weight

The fundamental variable for the design effort is the gross weight of the aircraft. However, Equations (8.23), (8.25), and (8.26) offer estimates of the APU group

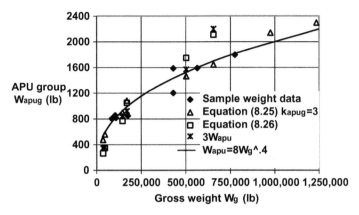

FIGURE 8.17

APU group weight as a function of aircraft gross weight according to various correlations. Weight data with open symbols are from Torenbeek and those with closed symbols are from Kroo (2008).

weight based on the number of passengers, which is related to the cabin volume and therefore to the size of the aircraft. A major job of the APU is providing air to the cabin for pressurization and environmental control making the size of the cabin a major consideration in sizing the APU group. Furthermore, because weight minimization is of paramount importance in aircraft design it may be reasonably expected that the overall density of the aircraft, weight per unit enclosed volume, is kept very close to a minimum. Therefore the gross weight of the aircraft should be a good indicator of the cabin volume, and accurate information on gross weight is usually the easiest to find. Similarly, the typical passenger capacity N_p of an aircraft is usually readily available, as are the gross weight and the engine weight.

The APU dry weight for the aircraft listed in Table 8.3 may be found, or reasonably estimated, from information in the trade literature. The nominal value of N_p for each aircraft, as taken from Table 8.3, may be used in the correlation Equations (8.25) and (8.26) given above to calculate estimates for the weight of the APU group. These results are plotted as open symbols on Figure 8.17 along with the results of an equation for the APU group weight based on gross weight given by

$$W_{apug} = 8(W_g)^{0.4} \tag{8.27}$$

The value for k_{apug} in Equation (8.25) as shown in Figure 8.17 is taken as 3. Also shown in the figure is the value for the APU group weight for the aircraft in Table 8.3 if we assume it to be three times the APU weight itself.

Sample weight statements presented in Kroo (2008) include APU group weights and these are also plotted as a function of aircraft gross weight W_g on Figure 8.17. The correlation equations show somewhat lower values for the higher gross weight aircraft than the sample APU group weights, which are for first- and second-generation airliners, not for more modern aircraft. It is expected that the correlations, which are based upon current APU technology, represent a conservative estimate for the APU group weight and are reasonable for use in the preliminary design study. Of course, if additional details for current APU installations become available the above results should be adjusted accordingly.

8.12 Instrument group weight

The group weights for instrumentation to monitor and control the aircraft are shown in Figure 8.18 and may be estimated by the relation $W_{ig} = 0.55 W_g^{0.6}$ as shown in Figure 8.18.

$$W_{ig} = 0.55 W_g^{0.6} \tag{8.28}$$

This approximation appears to yield a conservative estimate for the instrumentation group. It is noted once again that information reported for the first-generation jet transports is substantially lower than the trend shown by all the other aircraft. However, the weights involved are relatively small in magnitude so that the data scatter involves hundreds, not thousands, of pounds.

For the instrument group weight Oman (1977) suggests a linear relationship with the gross weight. Considering the available data shown in Figure 8.18 a reasonable approximation of this form might be as follows:

$$W_{ig} = 200 + 0.003 W_g \tag{8.29}$$

Examination of Figure 8.18 suggests that the correlation provided by Equation (8.28) is preferable to that of Equation (8.29) because it better captures the data trend at the higher gross weights.

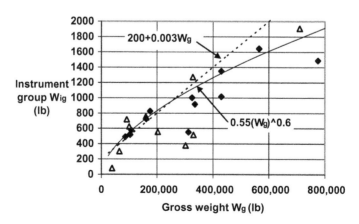

FIGURE 8.18

Variation of the instrument group weight with takeoff weight. Sample weight data with open symbols are from Torenbeek (1982) and those with closed symbols are from Kroo (2008).

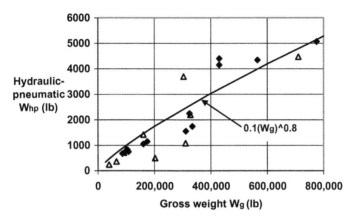

FIGURE 8.19

Variation of the hydraulic/pneumatic group weight with gross weight. Sample weight data with open symbols are from Torenbeek (1982) and those with closed symbols are from Kroo (2008).

8.13 Hydraulic and pneumatic group weight

Systems for applying forces are generally hydraulic or pneumatic in nature. Hydraulic and pneumatic group weight may be estimated by fairing a curve through available sample weight data shown in Figure 8.19, resulting in the following correlation:

$$W_{hp} = 0.1\, W_g^{0.8} \tag{8.30}$$

It should be noted that electrical actuation is increasing in importance, as described subsequently in Section 8.14. It is worthwhile to point out that once again there is wide scatter in the group weight around the 300,000 lb gross weight range. There also appears to be a different trend in the data above and below this gross weight level.

Oman (1977) provides a correlation equation for hydraulic system weight that is based on several important airplane characteristics as follows:

$$W_{hp} = k_{hp} \left[\left(\frac{S_s q_D}{1000} \right)^{1.3125} + \left(l_{fus} + \frac{b}{\cos \Lambda_{c/2}} \right)^{1.06125} \right]^{0.849} + k'_{hp} \tag{8.31}$$

In Equation (8.31) $S_s = S + S_h + S_v$ is the sum of the wing and tail planform areas (in ft²), q_D is the design dive speed dynamic pressure (in lb/ft²), l_{fus} is the fuselage length in feet, b is the wingspan in feet, and $\Lambda_{c/2}$ is the wing sweepback of the 50% chord. Carrying out this calculation using the characteristics of the aircraft described in Table 8.1 and $k_{hp} = 1$ and $k'_{hp} = 0$ leads to the data points shown in Figure 8.20. The simple correlation curve of Equation (8.30) is also shown on Figure 8.20 and it follows the calculated data quite closely. It is preferable to use Equation (8.31) with $k_{hp} = 1$ and $k'_{hp} = 0$ for preliminary design purposes because it incorporates more specific detail. However, it is advisable to also compute the hydraulic-pneumatic group weight using the simple correlation of Equation (8.28) as a check.

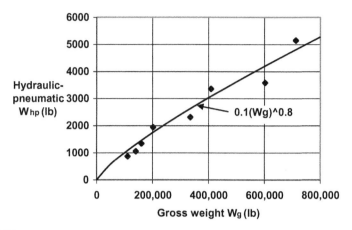

FIGURE 8.20

Comparison of data calculated using Equation (8.31) for the aircraft in Table 8.1 with the correlation provided by Equation (8.30).

8.14 Electrical group weight

The electrical system of a modern airliner is exemplified by that of the B777. It is a "traditional" hybrid providing both 115 Vac and 28 Vdc electric power. The power sources include two 120 kVA/115 Vac/400 Hz generators driven by the main engines with four 950 W permanent magnet generators (PMG) integrated into the two backup generators. In addition, there is one 120 kVA, 115 Vac, 400 Hz Auxiliary Power Unit (APU)-driven generator located in the tail cone and one 7.5 kVA ram air turbine (RAT) which can be extended out into the airstream for emergency wind-driven power. There are also batteries for the main engines, the APU, and the flight controls. To provide conversion and distribution functions there are four 120 Amp DC transformer rectifier units (115 Vac–28 Vdc), battery chargers and inverters, centralized distribution panels, thermal circuit breakers and electro-mechanical relays, and contactors with built-in current sensing and control electronics. The electrical systems of modern aircraft are becoming increasingly important in aircraft and to curb weight higher voltage systems are coming into use.

In developing an expression for the electrical group weight, Oman (1977) assumes that the major influence is the fuel system (less tanks) group weight W_{fs} and the avionics group weight W_{av}. The suggested correlation is given in the following form:

$$W_{el} = k_{el}(W_{fs} + W_{av})^{0.473} + k'_{el}$$

It is reasonable to expect that both W_{fs} and W_{av} are functions of the aircraft gross weight, and we shall see in Section 8.15, that this assumption is supported by weight

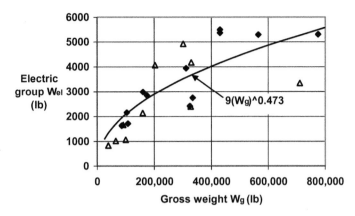

FIGURE 8.21

Variation of the electrical group weight with gross weight. Sample weight data with open symbols are from Torenbeek (1982) and those with closed symbols are from Kroo (2008).

data, at least for the avionics group. From this equation and the data on electrical group weight shown in Figure 8.21 a simple correlation may be given by

$$W_{el} = 9W_g^{0.473} \qquad (8.32)$$

Equation (8.32) is recommended as an approximation to be used in the preliminary design weight refinement process.

8.14.1 The all-electric airplane

The Boeing 787 transformed almost all aircraft systems which formerly depended on bleeding high-pressure air from the compressor sections of the main engines into one which depends completely on electric power generated onboard. The systems affected include pneumatic engine starters, wing ice protection, cabin pressurization, and hydraulic pumps. Only the anti-ice system for the engine inlets still uses engine bleed air. By replacing pneumatic with electric systems in the airplane, a substantial reduction in mechanical complexity is achieved. One of the primary functions of a conventional APU was driving a large pneumatic load compressor to generate the high-pressure air used to start the main engines. Now, however, engines may be started by driving their generators as starter motors. In this fashion the B787 has been able to eliminate both the pneumatic starter from the engine and the load compressor from the APU resulting in improved start reliability and power availability. Though the B787 APU can provide nearly half a megawatt of electrical power on the ground it does so at a low operating noise level.

One innovative application of the more-electric systems architecture on the 787 is the move from hydraulic to electrical actuation of the brakes. Electric brakes significantly reduce the mechanical complexity of the braking system and eliminate problems of leaking brake hydraulic fluid, leaking valves, and other hydraulic failures. In general, electric systems are much easier to monitor for health and system status than hydraulic or pneumatic systems; the brakes take full advantage of this by continuous onboard monitoring of the brakes for fault detection and brake wear.

8.14.2 "Greener" APUs

"More electric" airplanes need large onboard power-generating capacity yet must meet the requirements for "greener aircraft," that is, reduced fuel burn, pollutant emissions, and noise. There is general interest in moving away from fossil fuel gas turbine-powered APUs and toward new energy-generating systems. Fuel cells promise to fulfill this need because of their high efficiency and reliability, low environmental impact, and adaptability to various fuels. They are electrochemical systems that generate electric power as well as usable heat. They do use fuel but instead of burning it they facilitate electrochemical reactions that produce electricity. In many respects a fuel cell is like a battery, in that there is an electrolyte separating two electrodes. The introduction of a fuel, e.g., hydrogen, at one electrode and oxygen at the other, permits chemical reactions to occur in the electrolyte which create a flow of charged particles across the cell in turn promoting a flow of electricity in the exterior circuit connected to the cell as well as producing usable heat. A big advantage of fuel cells is that they are quiet, reliable, have little adverse environmental effect compared to the burning of fossil fuels, and their electrochemical conversion process is highly efficient, converting more than 85% of the total fuel energy. The use of hydrogen as the fuel results in much cleaner energy production because the combination of oxygen from the air and the hydrogen fuel produces only water. Not only is energy made available for the electrical systems with little environmental impact but the water produced may be used to meet onboard requirements. Aspects of the development of this technology for aircraft applications may be found, for example, in DOD-DOE (2011).

8.15 Avionics group weight

Avionics systems include the equipment for communication, control, monitoring, navigation, weather, and collision-avoidance, including the autopilot. The avionics group weight appears to be relatively linear and may be estimated by

$$W_{av} = 600 + 0.005 \, W_g \tag{8.33}$$

Torenbeek (1982) suggests that the avionics group weight for jet transports is a function of aircraft empty weight and the maximum range according to the following equation:

$$W_{av} = 0.575 \, W_e^{0.55} R^{0.25} \tag{8.34}$$

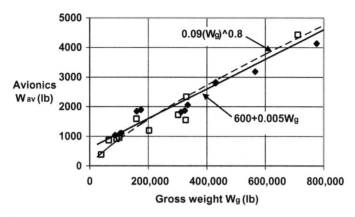

FIGURE 8.22

Variation of the avionics group weight with takeoff weight. Equations (8.33) and (8.34) are shown along with sample weight data. Open symbols are from Torenbeek (1982) and closed symbols are from Kroo (2008).

We have shown in Chapter 2 that the empty weight is proportional to the gross weight and we may assume that the maximum range (in nm) is also roughly proportional to the gross weight. Then Equation (8.34) might be written as

$$W_{av} \sim W_g^{0.8}$$

The available weight data and the results of the approximation offered by Equation (8.33) are shown in Figure 8.22. Also shown in Figure 8.22 is the correlation

$$W_{av} = 0.09 \, W_g^{0.8} \tag{8.35}$$

Both of the correlations are essentially linear in the range of interest so that either may be used for preliminary design purposes. However, the empty weight and the maximum range are known for the proposed aircraft so Equation (8.34) should be used to determine the avionics group weight with the other equations used as a check.

8.16 Equipment and furnishing group weight

Accommodations for passengers include seating, galley equipment, lavatory equipment, in-flight entertainment equipment, emergency equipment, etc. The equipment group weight therefore depends upon the size of the aircraft. Oman (1977) assumes that the equipment and furnishings group depends upon the payload weight W_{pl} and the number of passengers N_p according to the following form of correlation equation:

$$W_{ef} = k_{ef} W_{pl} + k'_{ef} N_p^{1.165} + k''_{ef} \tag{8.36}$$

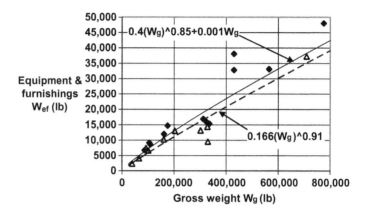

FIGURE 8.23

Variation of the equipment and furnishings group weight with takeoff weight. Sample weight data with open symbols are from Torenbeek (1982) and those with closed symbols are from Kroo (2008).

The payload, or cargo, weight is shown in Figure 3.18 to be proportional to the gross weight with a constant of proportionality that lies in the range 0.067–0.113. Similarly in Figure 1.9 it was shown that the takeoff weight is approximately proportional to the number of passengers according to the relation $W_g = 221.5N_p^{1.361}$. Thus we may write $N_p = 0.0189W_g^{0.735}$. Then, using this information with $k_{ef}'' = 0$, we may rewrite Equation (8.36) as

$$W_{ef} = 0.001W_g + 0.4W_g^{0.85} \tag{8.37}$$

Here we have chosen the constants to yield a reasonable representation of the sample weight data shown in Figure 8.23.

Torenbeek (1982) presents a simple rule of thumb for commercial jet transports as follows:

$$W_{ef} = 0.211W_{zf}^{0.91} \tag{8.38}$$

Here the W_{zf} is the maximum zero-fuel weight, which we may approximate as $W_{zf} \sim 0.77W_g$ so as to yield

$$W_{ef} \approx 0.166 W_g^{0.91} \tag{8.39}$$

This result is also shown in Figure 8.23 and both Equations (8.37) and (8.39) provide reasonable results. It is suggested that both methods be used in the preliminary design as a precaution for avoiding errors. Torenbeek (1982) also presents a brief description of a more detailed method for estimating the equipment and furnishings group weight which involves accounting for each of the individual components which comprise the equipment and furnishings group.

8.17 Air conditioning and anti-icing group weight

The air conditioning group is mainly involved with cooling the avionics and the cabin. Oman presents an estimate based on these two considerations in the following form:

$$W_{ac} = k_{ac}W_{av} + k'_{ac}(d_{fus}N_r)^{0.72} + k''_{ac}$$

The diameter of the fuselage is proportional to the number of seats abreast N_a while the number of rows N_r is proportional to the ratio of the number of passengers N_p to the number of seats abreast N_a. Thus $d_{fus}N_r \sim N_p$ and this is basically proportional to the volume of the cabin that must be air-conditioned. In the previous section we indicated that the number of passengers is related to the gross weight of the airplane by the relation $N_p = 0.0189W_g^{0.735}$, and in the section before that we showed that $W_{av} \sim W_g$. Combining these results gives an estimate for the air-conditioning group weight as follows:

$$W_{ac} = k_{ac}(600 + 0.005W_g) + k'_{ac}\left[\left(\frac{W_g}{221.5}\right)^{0.735}\right]^{0.72} + k''_{ac}$$

This equation is of the form

$$W_{ac} = k_1 + k_2W_g + k_3W_g^{0.53} \tag{8.40}$$

The anti-icing group weight is mainly a function of the length of leading edge that must be protected and Oman (1977) suggests the following relationship:

$$W_{ai} = k_{ai}b_s^{0.95} + k'_{ai}$$

Here $b_s = \frac{b}{\cos \Lambda_{c/2}}$ is the structural length of the wing measured along the mid-chord. The wingspan of a modern commercial aircraft is quite well correlated with the gross weight according to the relation

$$b_s = 0.83W_g^{0.42}$$

Therefore the anti-icing group weight may be written in terms of gross weight as follows:

$$W_{ai} = k_{ai}\left(0.83W_g^{0.42}\right)^{0.95} + k'_{ai} = k_4W_g^{0.4} + k'_{ai} \tag{8.41}$$

The available data for anti-icing group weight have a larger degree of scatter than all the other groups studied and it appears that the air-conditioning group and anti-icing group are best taken together because the combination yields a much better correlation. The sum of Equations (8.40) and (8.41) is of the form

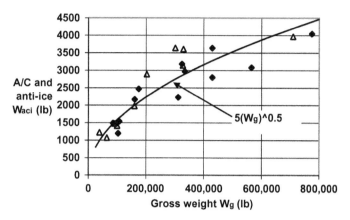

FIGURE 8.24

Variation of the air conditioning and anti-icing group weight with takeoff weight. Sample weight data with open symbols are from Torenbeek (1982) and those with closed symbols are from Kroo (2008).

$$W_{aci} = W_{ac} + W_{ai} = k_2 W_g + k_3 W_g^{0.43} + k_4 W_g^{0.4} + k_5$$

This fairly busy equation is supposed to describe the available weight data shown in Figure 8.24. As can be seen, the data scatter is still fairly substantial so it is recommended that absent any other detailed weight information a simple correlation curve faired through the existing data is to be favored, and the following equation is recommended:

$$W_{aci} = 5\sqrt{W_g} \qquad (8.42)$$

8.18 Wing group center of gravity

A first approximation to the location of the CG of the wing may be obtained by considering the wing to be a flat plate of uniform thickness, as shown in Figure 8.25. The center of gravity (CG) lies along the line midway between the leading and trailing edges with a location from the centerline given by

$$y_{CG} = \frac{b}{6}\left(\frac{c_r + 2c_t}{c_r - c_t}\right) = \frac{b}{6}\left(\frac{1 + 2\lambda}{1 - \lambda}\right) \qquad (8.43)$$

Here c_r, c_t, and $\lambda = c_t/c_r$ are the root chord, tip chord, and taper ratio, respectively. For a typical taper ratio of 1/3, the spanwise CG location would be at approximately $0.42(b/2)$ and would lie at the mid-chord at that spanwise location.

Torenbeek (1982) provides a means of estimating the CG of the wing group which, of course, is not a plate of uniform thickness. Using that estimation method yields a CG location as presented in the schematic diagram of the wing shown in Figure 8.26.

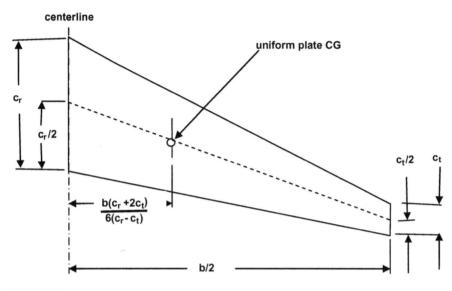

FIGURE 8.25

Center of gravity location for a uniform thickness plate.

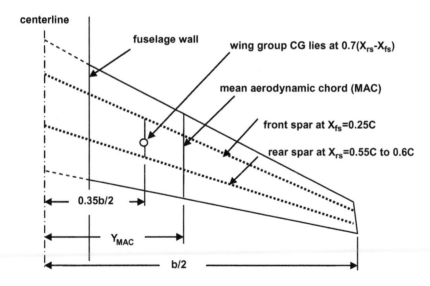

FIGURE 8.26

Schematic diagram of wing layout for estimating the location of the wing CG.

In this diagram it is clear that the wing CG falls between the front and rear spars, the main load-carrying members. Note that the spanwise location of the wing group CG is about 20% further inboard than the simple uniform plate solution, but is located at about the mid-point of the local chord, as in the uniform plate solution. Obviously, because the complete wing group is symmetrical about the longitudinal axis the CG lies on that axis, that is, $y=0$. However, it is the longitudinal (x-axis) location of the CG which is important for stability and control considerations and, again because the wing group is symmetrical about the x-axis, the longitudinal location of the CG shown for the half-span in Figure 8.26 is the same for the entire wing group.

Now, the CAD drawing prepared previously for the projected design airplane has a nominal longitudinal location of the wing group and the CG of the airplane will be determined once the wing is finally located with respect to the fuselage. This location will be determined subsequently by requiring the CG of the entire aircraft to be properly located with respect to the aerodynamic center of the airplane so as to ensure adequate longitudinal static stability.

8.19 Fuselage group center of gravity

We may derive an appreciation for the fuselage center of gravity by making some simple observations about the nature of the fuselage. First, we know that the nose cone and the tail cone are mainly shells and only carry a minor portion of the aircraft operating weight. Second, the cabin, though also a shell, carries all the passengers and the payload. Assume that the fuselage has a conventional layout like that in Figure 8.27.

We may estimate the CG location of the fuselage by taking moments about the nose to arrive at the following equilibrium relation:

$$W_{fus}X_{fus} = W_{NC}x_{NC} + W_C x_C + W_{TC}x_{TC} \tag{8.44}$$

The quantity X_{fus} is the location of the fuselage CG as measured from the nose of the aircraft. The quantities W_{NC}, W_C, and W_{TC} denote the weights of the nose cone, cabin, and tail cone, respectively. The quantities x_{NC}, x_C, and x_{TC} denote the axial distance from the nose to the CG of the nose cone, cabin, and tail cone, respectively. We make the assumption that all the weight of the fuselage, W_f, is concentrated in the cabin so that Equation (8.44) may be rewritten as

$$W_{fus}X_{fus} \approx W_{fus}x_C = W_{fus}\left(F_{NC}d_{fus} + \frac{1}{2}F_C\right) \tag{8.45}$$

Then the non-dimensional distance from the nose to the fuselage CG is

$$\frac{X_{fus}}{l_{fus}} \approx \frac{F_{NC}}{F} + \frac{1}{2}\frac{F_C}{F} \tag{8.46}$$

In Figure 3.13 the following correlation between cabin fineness ratio and fuselage fineness ratio was given: $F=0.9F_C + 5$. Using this relation in Equation (8.46) yields

$$\frac{X_{fus}}{l_{fus}} \approx \frac{F_{NC}}{F} + \frac{F-5}{1.8F} = \frac{1}{F}\left(F_{NC} + \frac{F-5}{1.8}\right) \tag{8.47}$$

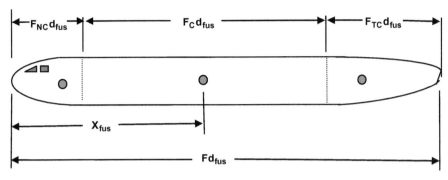

FIGURE 8.27

Schematic diagram of generic fuselage layout in terms of fineness ratio and fuselage diameter. Circular symbols denote CG locations of the three fuselage elements.

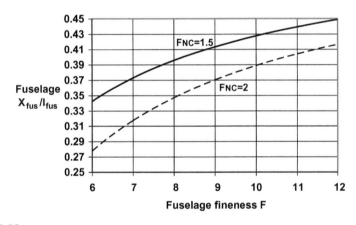

FIGURE 8.28

Variation of fuselage longitudinal CG location for two different nose fineness ratios.

The fuselage CG given by the simple relation of Equation (8.47) is plotted in Figure 8.28.

The fuselage center of gravity (CG), as estimated by Torenbeek (1982), is illustrated in Figures 8.29 and 8.30 for wing-mounted and aft fuselage-mounted engines, respectively. It is clear that the simple approach yields a fair estimate of the CG location. One may carry the simple analysis a step further using the CAD results that should have been developed during the design. The weights of the nose cone, cabin, and tail cone elements can be estimated more accurately using the shell areas of each to more accurately apportion the weights for the moment equilibrium calculation of Equation (8.44).

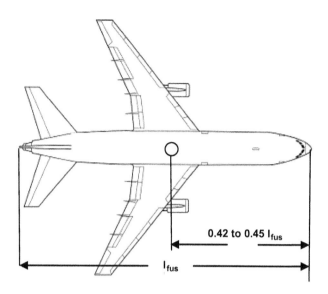

FIGURE 8.29

Approximate location of CG of fuselage group alone for wing-mounted engines.

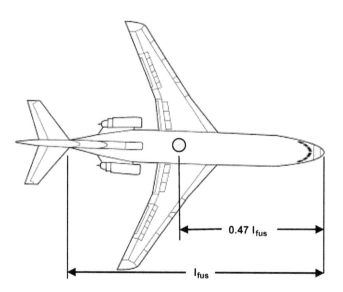

FIGURE 8.30

Approximate location of CG of fuselage group alone for fuselage-mounted engines.

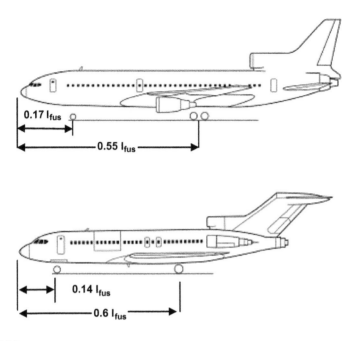

FIGURE 8.31

Approximate locations of landing gear components as functions of fuselage length for different engine mounting configurations.

8.20 Landing gear group center of gravity

The nose gear is placed near the nose of the aircraft and the main landing gear must be placed aft of the overall CG of the complete aircraft. A first approximation would place the nose and main landing gear at the approximate locations shown in Figure 8.31, depending upon the engine mounting configuration. Using the estimated weights of the nose and main landing gear determined previously in this chapter, the results for the general layout of the landing gear developed in Chapter 7, and the fuselage dimensions found using the methods of Chapter 3 one may approximate the location of the CG of the complete landing gear system.

8.21 Tail group center of gravity

The CG of the tail group is dependent on the nature of the tail configuration and Torenbeek (1982) provides some estimates of the CG location for conventional and T-tail arrangements, as shown in Figures 8.32 and 8.33. These approximate locations may be used along with the estimated weights of the tail surfaces to develop the location of the CG of the entire aircraft.

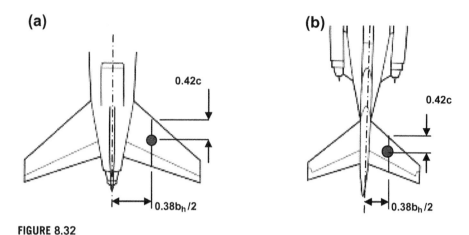

FIGURE 8.32

Approximate location of the CG location of the horizontal tail for (a) wing-mounted engines and (b) fuselage-mounted engines. The term b_h denotes the span of the horizontal tail.

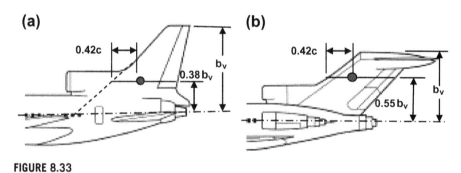

FIGURE 8.33

Approximate CG location of the vertical tail for (a) wing-mounted engines and (b) fuselage-mounted engines. The term b_v denotes the span of the vertical tail as defined in the figure.

8.22 Propulsion group center of gravity

The engine CG should be obtained from the engine manufacturer, or from an estimate based upon the general configuration of the engine using actual dimensions. The nacelle housing the engines may be assumed to have a CG located 40% of the length of the nacelle, as measured from the lip of the nacelle. Typical locations for the engine CG and the nacelle CG are shown in Figure 8.11.

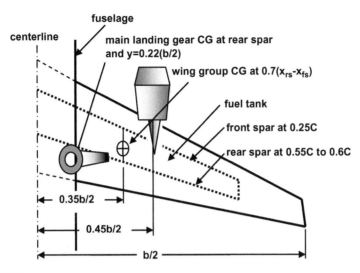

fuselage

centerline

main landing gear CG at rear spar and y=0.22(b/2)

wing group CG at 0.7(x$_{rs}$-x$_{fs}$)

fuel tank

front spar at 0.25C

rear spar at 0.55C to 0.6C

0.35b/2

0.45b/2

b/2

FIGURE 8.34

Composite sketch of wing group, fuel tank, wing-mounted engine, and landing gear from which a collective CG may be determined.

8.23 Aircraft center of gravity

The center of gravity (CG) of the aircraft is of paramount importance with respect to stability and control. This aspect of the design process actually is carried out in concert with the weight estimation process. Center of gravity data for each component whose weight is estimated should be collected concurrently.

One method for proceeding with the determination of the center of gravity of the complete airplane involves dividing the airplane into two groups: the combined fuselage group that includes the fuselage and the tail surfaces, and the combined wing group that includes the wing, fuel tanks, engines, nacelles, and landing gear. A detailed view of the wing group is given in Figure 8.34.

The fuel tanks may be considered to be contained in the space between the front and rear spars. The total weight of fuel required is known and can be converted to a volume. Then fuel tank volume v_f for the prismatic shape required is

$$v_f = \frac{1}{3}l_{ft}\left[S_1 + S_2 + \frac{1}{2}(S_1 + S_2)\right] \qquad (8.48)$$

The quantities S_1 and S_2 are the areas of the inboard and outboard bases of the tank and they are known from the geometry of the wing. Thus the required length for the tank may be calculated. The various CG locations of the components of the combined wing group may be used to produce a weight of the combined wing group and its CG location as measured from the leading edge of the mean aerodynamic chord.

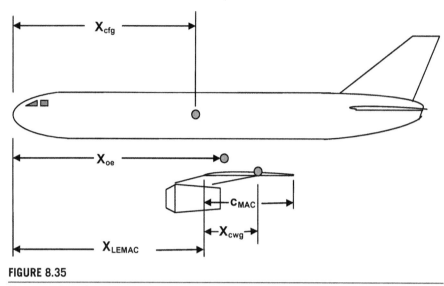

FIGURE 8.35

Elevation view of the two mass groups used in determining the center of gravity of the complete airplane.

Side and plan views of these two groups with appropriate dimensions are shown in Figures 8.35 and 8.36.

The quantity x_{cfg} denotes the location of the center of gravity of the combined fuselage group as measured from the nose of the airplane. The quantity x_{cwg} denotes the location of the center of gravity of the combined wing group as measured from the leading edge of the mean aerodynamic chord (c_{MAC}) of the wing. Finally, the quantity x_{oe} denotes the location of the center of gravity of the operating empty weight (W_{oe}) of the entire airplane, as measured from the nose of the airplane. The center of gravity of each combined group lies along the centerline of the airplane, as shown in Figure 8.36. For each axial position of the combined wing group there will be a different center of gravity location for the airplane. It is desired to determine the location of the center of gravity of the airplane such that the basic requirement of longitudinal static stability is met.

To find x_{oe} for any location of the combined wing group we begin by taking moments about the nose of the aircraft, which yields

$$W_{oe}x_{oe} = W_{cfg}x_{cfg} + W_{cwg}(x_{LEMAC} + x_{cwg}) \tag{8.49}$$

Setting $x^* = x_{oe} - x_{LEMAC}$ and solving for x_{LEMAC} leads to the following result:

$$x_{LEMAC} = x_{cfg} + \left(\frac{W_{cwg}}{W_{cfg}}\right)x_{cwg} - \left(1 + \frac{W_{cwg}}{W_{cfg}}\right)x^* \tag{8.50}$$

Equation (8.50) permits placement of the wing on the fuselage for a given center of gravity location. As described in Chapter 6, the displacement of the center of

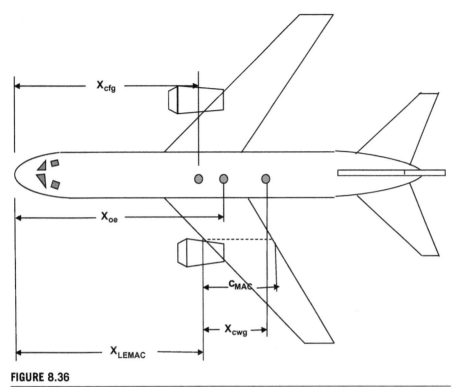

FIGURE 8.36

Plan view of the two mass groups for determining the center of gravity of the complete airplane.

gravity of the airplane forward of its neutral point determines the degree of the airplane's longitudinal static stability. If the two points coincide the stability is neutral, while if the center of gravity falls aft of the neutral point the airplane will be unstable. It is desirable in a commercial passenger transport to have sufficient static stability for comfort and robustness of safety margins while maintaining a level of maneuvering agility suitable to its mission. Considering the location of the center of gravity of the airplane to be at its farthest aft position when at the operating empty weight W_{oe} is placed at a value x^*/c_{MAC} of, say, 0.3 fixes the position of the wing on the design aircraft. Then, as the aircraft is loaded with fuel, passengers, and cargo the center of gravity should move forward. Of course, as the fuel is consumed during flight the center of gravity should move aft. It is possible to construct a loading diagram which illustrates the position of the center of gravity of the aircraft as it is loaded, as shown in Figure 8.37. The starting point is the farthest aft center of gravity location, shown here as $0.35c_{MAC}$, and is the location for the operational empty weight. As passengers board the cabin and cargo is loaded in the holds, the weight increases and the center of gravity moves forward. The fuel tanks in the swept back wings must not permit the center of gravity to shift rearward.

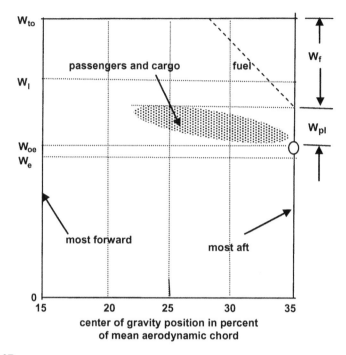

FIGURE 8.37

Notional loading diagram showing shift in center of gravity location as aircraft is loaded.

As can be inferred from Figure 8.37, a reasonable location of the center of gravity to meet the needs of typical passenger aircraft is to have the center of gravity location at about 25–30% of c_{MAC}, the mean aerodynamic chord of the wing. Recall that in Chapter 6 the static margin should be about $0.05–0.10c_{MAC}$ and that it is desirable to have the center of gravity aft of the aerodynamic center of the wing, which is typically near $0.25c_{MAC}$. Therefore, center of gravity positions more forward than about $0.3c_{MAC}$ in Figure 8.35 would require downloads on the horizontal tail.

With the successful conclusion of the CG evaluation the position of the wing with respect to the fuselage is determined and the travel of the CG for different operating conditions can be determined. The addition of the maximum payload, where the passenger load may be considered to be distributed throughout the aircraft and the cargo containers loaded appropriately and located in the positions is determined in Chapter 2. The fuel load may be changed to the landing condition in order to determine the extent of CG travel during the flight. Then the entire aircraft may be drawn and presented as a complete unit in a three-view representation.

A more detailed consideration of the center of gravity location of the aircraft may be carried out by locating the various equipment group weights calculated previously

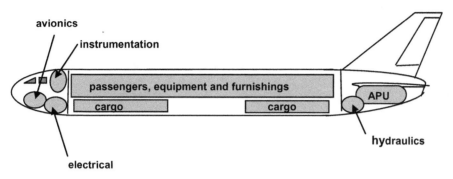

FIGURE 8.38

Typical locations of the different aircraft systems within the fuselage.

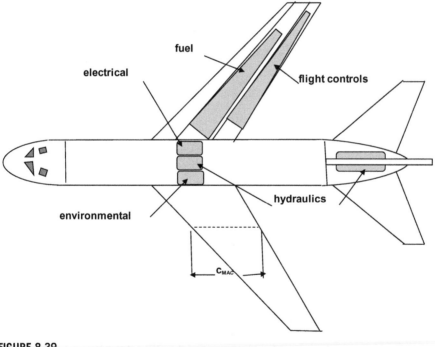

FIGURE 8.39

Typical locations of the different aircraft systems within the wings.

at typical points throughout the aircraft. The distribution of these items is suggested by the illustrations in Figures 8.38 and 8.39.

Logging in the location of these items and their weights as suggested by Table 8.4 can provide a means for a more accurate CG determination.

Table 8.4 Table of Aircraft Weight Breakdown by Groups

Aircraft Designation: _____			
Group	Weight (lbs)	X_{CG} (in.)	Comments
Wing group			
Tail group			
Body group			
Landing gear group			
Surface controls group			
Nacelle group			
Propulsion group			
Flight controls group			
Auxiliary power unit			
Instrument group			
Hydraulics & pneumatics			
Electrical system			
Avionics system			
Equip. & furnishings			
A/C & anti-icing systems			
Empty weight (W_e)			Summation
Operational items			
Operational empty weight (W_{oe})			
Payload weight (W_{pl})			From Chapter 2
Fuel weight (W_F)			From Chapter 2
Gross weight (W_g)			Summation

8.24 Design summary

The weight of the various components and their center of gravity locations determined by the methods described in this chapter should be described in the narrative. Sufficient information should be included so as to permit another engineer to be able to understand and carry out the same calculations. The results obtained should be presented as a table of group weights like that shown in Table 8.4, along with any relevant comments. A diagram illustrating CG locations and travel for different operating conditions should be provided. A three-view of the design aircraft showing pertinent dimensions should be made available as well as a three-view CAD diagram to illustrate the overall design, in the manner illustrated in Figure 8.40.

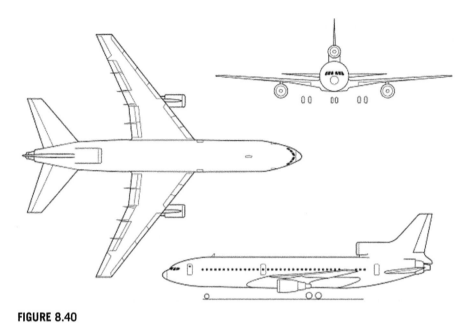

FIGURE 8.40

Typical three-view presentation of an airplane design.

8.25 **Nomenclature**

A	aspect ratio
b	span
c	chord length
C_L	lift coefficient
$C_{L\alpha}$	wing lift curve slope
d	diameter
F_{to}	takeoff thrust
g	acceleration of gravity
h	height
k	constant
L	lift or airplane length
l	length
M	Mach number
M_f	fuel weight fraction
N_p	number of passengers
n	load factor or number
p	pressure

q	dynamic pressure
R	range
S	wing planform area
S_h	horizontal tail planform area
S_v	vertical tail planform area
S_g	gross structural shell area
t	thickness
V	velocity
V_D	design dive equivalent velocity
V_C	cruise equivalent velocity
V_h	horizontal tail volume coefficient
V_v	vertical tail volume coefficient
W	weight
W_{zf}	zero-fuel weight
z	altitude
α	angle of attack
Λ	sweepback angle
λ	taper ratio
ρ	density
σ	atmospheric density ratio
U	volume

8.25.1 Subscripts

ac	air conditioning
aci	air conditioning plus anti-icing
ai	anti-icing
$apug$	auxillary power
av	avionics
c	cruise or cabin
cfg	combined fuselage group
cwg	combined wing group
e	engine
ef	equipment and furnishing
el	electrical
fc	flight control
fus	fuselage
g	gross
h	horizontal tail
hp	hydraulic and pneumatic
ig	instrumentation group
LE	leading edge
LG	landing gear

limit	limit
MAC	mean aerodynamic chord
max	maximum
MG	main landing gear
n	nacelle
NC	nose cone
ng	nacelle group
NG	nose landing gear
p	pressure or pylon
pg	propulsion group
r	root
sl	sea level
t	tail or tip
to	takeoff
ult	ultimate
v	vertical tail
wg	wing group

References

Ardema, M.D., Chambers, M.C., Patron, A.P., Hahn, A.S., Miuras, H., MNoore, M.D., 1996. Analytical Fuselage and Wing Weight Estimation of Transport Aircraft. NASA Technical Memorandum, 110392.

DOD-DOE, 2011. Report of the DOD-DOE Workshop on Fuel Cells in Aviation. Fuel Cell Technologies Program, U.S. Department Of Energy. <ww1.eere.energy.gov/hydrogenandfuelcells/pdfs/aircraft_report.pdf>.

Kroo, I., 2008. Aircraft Design: Synthesis and Analysis. <www.adg.stanford.edu/aa241/AircraftDesign.html>.

Oman, B.H., 1977. Vehicle Design Evaluation Program. NASA CR 145070.

Raymer, D., 2006. Aircraft Design: A Conceptual Approach, fourth ed. American Institute of Aeronautics and Astronautics, Reston, VA.

Roskam, J., 1999. Airplane Design, Part V: Component Weight Estimation. Design & Analysis Corp., Lawrence, KS.

Svoboda, C., 2000. Turbofan engine database as a preliminary design tool. Aircraft Design 3, 17–31.

Torenbeek, E., 1982. Synthesis of Subsonic Airplane Design. Kluwer Academic Publishers, Dordrecht, The Netherlands.

CHAPTER OUTLINE

Commercial Airplane Design Principles. http://dx.doi.org/10.1016/B978-0-12-419953-8.00009-7
© 2014 Elsevier Inc. All rights reserved.

349

9.1 Introduction

The total drag of an airplane is considered to consist of four different types in the subsonic regime of flight as illustrated in Figure 9.1. These are:

- friction drag due to shear stresses,
- form drag due to normal stresses,
- induced drag, or drag-due-to-lift,
- compressibility, or wave drag.

Friction drag is due to the shear stresses brought about by viscosity at the surface of the body. Form drag arises from the pressure field around the body caused by the body shape and the displacement effect of the boundary layer as well as separation of the flow from that body shape because of viscous effects. The induced drag arises as a consequence of the production of lift and represents the "cost" of producing lift by pushing a body through a fluid. The wave drag is a pressure drag that arises because the compressibility of the air permits formation of compression and expansion waves due to the body shape. For consistency, the DATCOM methods presented by Hoak et al. (1978) will be most often followed although other approaches and results will be evaluated.

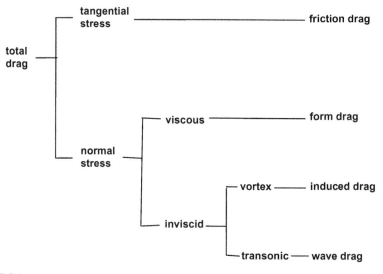

FIGURE 9.1

Components of airplane drag.

9.2 **Skin friction drag**

Skin friction drag can be produced by either laminar or turbulent boundary layer
flow. At the flight speeds and altitudes at which aircraft fly it is usually conservative
to assume that the boundary layer flow is fully turbulent over the entire airplane. On
some aircraft, particularly those made largely of composite material with few joints
or fasteners, a considerable extent of laminar boundary layer flow may exist prior to
the transition point. For such aircraft the above assumption may be overly conservative. Since most aircraft, small or large, are still made using riveted aluminum skin/
stringer/former construction for more than 50% of the total airframe weight, it is
considered prudent in a preliminary design to assume the fully turbulent skin friction
value over the entire airplane. A somewhat more accurate, though simple, approach
is to assume laminar flow over the region from a forward stagnation point or line to
the first surface joint and then assume fully turbulent flow downstream. Of course, if
proven boundary layer codes are available they may be used, but to get the full benefit of the increased effort and detail the geometry of the surfaces considered should
be known to a corresponding level of detail.

It is worth noting that next-generation airliners like the Boeing 787 and the Airbus
380 use much more composite materials in their design than previous designs. Latest
estimates suggest that the B787 uses up to 50% composites in the airframe and the
A380 uses up to 25%. Newer aircraft like the Airbus 350 and the Boeing 777X will
also employ increasing amounts of composite materials. The major motivation for
using advanced composite materials like Carbon Fiber Reinforced Plastics (CFRP),
however, is to save weight, not to reduce skin friction drag.

The aeronautical literature usually provides the engineer with two forms of the
skin friction coefficients and these can easily be confused by the novice. They are the
local skin friction coefficient and the average skin friction coefficient, as calculated
for one side of a flat plate. The local value is the skin friction coefficient at a particular point on the plate and the average, as the name suggests, is integrated value
of the skin friction coefficient from the leading edge of the plate to any point $x=l$
along the plate.

9.2.1 **Laminar flow**

The local and integrated skin friction coefficients are defined first for laminar flow.
White (2006), shows the local skin friction coefficient to be

$$c_{f,lam} = \frac{\tau(x,0)}{\frac{1}{2}\rho V^2} = \frac{0.664}{\sqrt{\frac{\rho Vx}{\mu}}} = \frac{0.664}{\sqrt{Re_x}} \tag{9.1}$$

The average, or integrated, skin friction coefficient is

$$C_{F,lam} = \frac{\int_0^l c_{f,lam}dx}{\int_0^l dx} = \frac{1.328}{\sqrt{Re_l}} \tag{9.2}$$

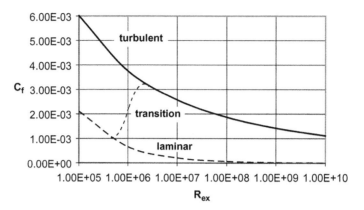

FIGURE 9.2

The variation of the local skin friction coefficient with Reynolds number based on x is shown for laminar flow using Equation (9.1), and turbulent flow using Equation (9.3). A notional transition range between the two is shown.

Note that for the local skin friction coefficient the Reynolds number is based on the local distance, x, from the leading edge of the plate, while for the average skin friction coefficient the Reynolds number is based on the length, l, of the plate. Recall that both are calculated for a flat surface, that is, one side of an actual thin flat plate. The skin friction coefficient in laminar flow is a function also of Mach number, but in the range of flight Mach numbers typical of commercial jet transports, $M < 1$, the differences due to compressibility are relatively small, as described subsequently, and are often neglected for preliminary design purposes.

9.2.2 Turbulent flow

The laminar boundary layer over a smooth flat plate is known to become unstable to small disturbances at a sufficiently high Reynolds number, typically in the range of $1.5 \times 10^5 < Re_x < 1.5 \times 10^6$. In this range the laminar flow develops periodic and then fully random turbulent fluctuations in the flow. This region is called the transitional range. The local skin friction coefficient then changes from the laminar form of Equation (9.1) to the turbulent form, like that given by White (2006):

$$c_{f,turb} = \frac{0.455}{\left[\ln\left(0.06\,Re_x\right)\right]^2} \tag{9.3}$$

Curves of the local skin friction coefficient as given by Equations (9.1) and (9.3) are shown in Figure 9.2. Also shown is a purely notional representation of the typical behavior of the skin friction coefficient in the transitional range. The onset and duration of the transitional range is highly dependent on local conditions and is still the subject of substantial research. Using the definition in Equation (9.2), von

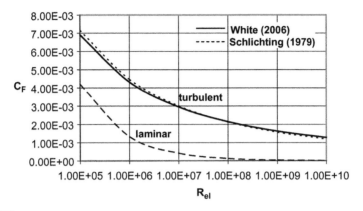

FIGURE 9.3

Average turbulent skin friction as given by Equations (9.5) and (9.6) is shown along with the average skin friction coefficient for incompressible laminar flow as given by Equation (9.2).

Karman determined an expression for the average skin friction coefficient for incompressible flow over a flat plate as follows:

$$\frac{0.243}{\sqrt{C_{F,turb}}} = \log_{10}\left(\text{Re}_l\, C_{F,turb}\right) \tag{9.4}$$

Schlichting (1979) fitted the following more manageable equation:

$$C_{F,turb} = \frac{0.455}{\left(\log_{10} \text{Re}_l\right)^{2.58}}; \; 10^6 < \text{Re}_l < 10^9 \tag{9.5}$$

White (2006) presents a somewhat different form for the same Reynolds number range as follows:

$$C_{F,turb} = \frac{0.523}{\left[\ln\left(0.06\,\text{Re}_l\right)\right]^2} \tag{9.6}$$

White (2006) also points out that Prandtl's use of the 1/7th power law for the boundary layer velocity profile, suggested by results from pipe flow experiments, yields a remarkably simple and accurate formula for the local turbulent skin friction coefficient:

$$c_{f,turb} = \frac{0.027}{\text{Re}_x^{1/7}}$$

The associated average skin friction coefficient is

$$C_{F,turb} = \frac{0.0315}{\text{Re}_x^{1/7}} \tag{9.7}$$

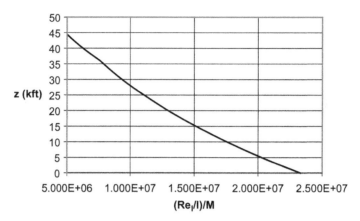

FIGURE 9.4

The ratio of unit Reynolds number to Mach number $(Re_l/l)/M$ is shown as a function of altitude.

The results given by Equations (9.5) and (9.6) are within a percent or two of each other and are illustrated in Figure 9.3. The simpler form of Equation (9.7) yields results around 5% higher than Equation (9.5). Also shown in Figure 9.3 is the average skin friction coefficient for incompressible laminar flow as given by Equation (9.2). The Reynolds number Re_l is based on the length l of the plate and the flow properties outside the viscous boundary layer and may be written as follows:

$$Re_l = \frac{\rho V l}{\mu} = \left(\frac{\rho a}{\mu}\right) M l$$

The Reynolds number is directly proportional to the length of the plate and the flight Mach number and indirectly proportional to the flight altitude. Atmospheric properties for the 1976 US Standard Atmosphere are presented in Appendix B and should be used in all calculations for consistency. The quantity $\rho a/\mu$ has the units of inverse length and is determined solely by atmospheric properties. For the 1976 Standard Atmosphere the quantity Re_l/Ml is shown in Figure 9.4. For typical takeoff conditions of $M = 0.225$ and $z = 0$ ft the unit Reynolds number $Re_l/l = 1.57 \times 10^6$/ft, while for typical cruise conditions of $M = 0.8$ and $z = 36{,}000$ ft the unit Reynolds number $Re_l/l = 1.82 \times 10^6$/ft. The small difference between the coefficients of the unit Reynolds under typical commercial airplane operations suggests that the length-based Reynolds number will be similarly close throughout the flight corridor.

9.2.3 Compressibility effects on skin friction

The Reynolds number depends upon the Mach number as well as on the shear stress on the wall τ. Therefore the skin friction coefficient may also be affected by compressibility. There are over 50 theories for turbulent flow over a flat plate at arbitrary

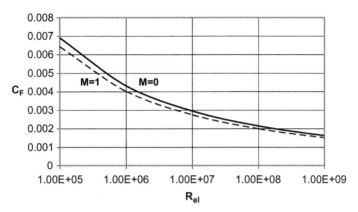

FIGURE 9.5

Average turbulent skin friction coefficient for compressible flow.

Mach number and surface temperature, according to White (2006). He points out that few approaches show a high degree of success for a wide range of flow conditions and he expresses most satisfaction with the "van Driest II" theory. This theory is somewhat cumbersome to employ, and for a good exposition of how to apply this theory see Hopkins (1972). Curves of the average turbulent skin friction coefficient for a flat plate at $M=0$ and $M=1$ are presented in Figure 9.5. The case for $M=0$ is the usual incompressible result as given by Equation (9.6) while that for $M=1$ is adapted from Hopkins (1972) for an adiabatic flat plate in a stream with static temperature outside the boundary layer $T_e=400$R (222 K), which is approximately the temperature in the stratosphere where commercial jet transports cruise. The stipulation of an adiabatic wall recognizes that the time spent in cruising flight is sufficiently long to assure that the surface reaches an equilibrium temperature at which no further heat transfer takes place. In Figure 9.5 it can be seen that the skin friction coefficient depends both upon the value of the Reynolds number and the Mach number, but obviously the Mach number dependence is small in the subsonic flight regime, which is the region of interest for commercial jet transports. For the present preliminary design purposes it is adequate to use the correction proposed by Morrison (1976) to account for compressibility. Using that data, a curve fit for the ratio of the value of the actual skin friction coefficient to the incompressible value is given as follows:

$$\frac{C_F}{C_{F,inc}} = 1 - 0.072M^{1.5} \tag{9.8}$$

9.2.4 Surface roughness effects on skin friction

The results described above were developed for aerodynamically smooth surfaces. The roughness of a surface is characterized by the equivalent sand roughness height h and the ratio of the characteristic length l of the surface to the equivalent sand

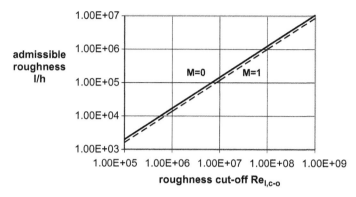

FIGURE 9.6

The cut-off value of the Reynolds number $Re_{\ell,c-o}$ is shown as a function of the admissible roughness parameter ℓ/h.

Table 9.1 Equivalent Sand Roughness of Several Surfaces

Surface	Equivalent Sand Roughness h (in)
Aerodynamically smooth	0
Polished metal	0.02×10^{-3} to 0.08×10^{-3}
Natural sheet metal	0.16×10^{-3}
Smooth matte paint	0.25×10^{-3}
Camouflage paint	0.40×10^{-3}
Dip-galvanized metal	6×10^{-3}

roughness height h is called the admissible roughness parameter, l/h. Curves of the admissible roughness parameter l/h as a function of the so-called cut-off Reynolds number for a flat plate at Mach numbers between 0 and 1 are quite close and may be approximated as follows:

$$M = 0 : Re_{l,c-o} = 29.46 \left(\frac{l}{h}\right)^{1.072} \tag{9.9}$$

$$M = 1 : Re_{l,c-o} = 36.82 \left(\frac{l}{h}\right)^{1.072} \tag{9.10}$$

The general nature of the behavior of the cut-off Reynolds number is shown in Figure 9.6. A table of representative values of equivalent sand roughness appears in Table 9.1.

Table 9.2 Representative Reynolds Numbers Based on the Mean Aerodynamic Chord (c_{MAC}) of Two Representative Airliners

Aircraft	b (ft)	h (in)	c_{MAC} (ft)	l/h	$Re_{l,c-o}$ M=0	$Re_{l,c-o}$ M=1	Re_l M=0	Re_l M=1
B777-200	200	1.60 E-04	26	1.95 E06	1.63E08	2.04E08	3.64E07	4.73E07
B737-700	112.6	1.60 E-04	13.5	1.01 E06	8.07E07	1.01E08	1.89E07	2.46E07

The procedure is to first determine the equivalent sand roughness h from Table 9.1 based on the nature of the surface in question and then to determine the roughness parameter l/h using the length l of the component being analyzed. Then, using Equations (9.9) and (9.10), depending upon the speed range under consideration, determine the cut-off Reynolds number. Alternatively, one may enter the admissible roughness parameter in Figure 9.6 and estimate the corresponding cut-off Reynolds number. Inserting the Reynolds number Re_l of the flat plate into Equation (9.6) and using Equation (9.8), depending on the speed range, permits determination of the average skin friction coefficient C_F for one side of the plate, as long as $Re_l < Re_{l,c-o}$. If $Re_l > Re_{l,c-o}$, then the skin friction coefficient is fixed at the value corresponding to $Re_{l,c-o}$.

Consider the case of two representative commercial airliners as described in Table 9.2, assuming the surface skin is made of natural aluminum sheet metal. Here $M=0$ and 1 refer to the nominal takeoff and cruise conditions, respectively. The average skin friction as a function of Reynolds number based on the mean aerodynamic chord length ($l=c_{MAC}$) is depicted in Figure 9.7 for $M=0$; the results for $M=1$ are similar.

We see from Table 9.2 that both in takeoff and cruise the Reynolds number based on the mean aerodynamic chord length is less than the cut-off Reynolds number. Therefore, the appropriate average skin friction coefficients are those corresponding to the actual value of Reynolds number based on the mean aerodynamic chord length. Calculations based on fuselage length for the two aircraft in Table 9.2 show similar trends. For aircraft with high-quality surface finishes the length-based Reynolds number will typically be less than the roughness-based cut-off Reynolds number.

9.2.5 The equivalent flat plate for calculating component drag

The concept of the boundary layer is based on the idea that it is thin, that is, the extent of the boundary layer normal to the body surface is much smaller than the characteristic dimension of the surface, where the characteristic dimensions are its length and its axial and transverse radii of curvature. To a thin boundary layer then, the local surface appears to be flat. Therefore, for preliminary design purposes we assume that the boundary layer develops over an airplane component, i.e., wing or fuselage, at exactly the same rate as if that component were a so-called equivalent flat plate. To appreciate how thin the boundary layer is, let us consider a flat plate

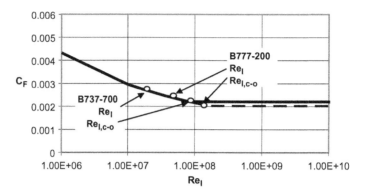

FIGURE 9.7

General nature of the average turbulent skin friction coefficient variation with Reynolds number and Mach number for the aircraft in Table 9.2 at $M=0$.

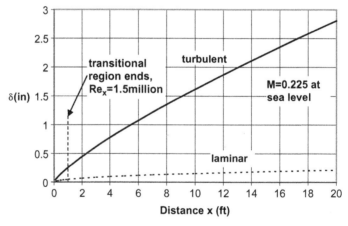

FIGURE 9.8

The boundary layer thickness δ is shown as a function of the distance x from the leading edge of the plate for laminar flow and turbulent flow at low Mach number corresponding to takeoff at sea level.

with a length equal to a value between that of the mean aerodynamic chord of the two aircraft described in Table 9.2, that is, $l=20$ ft. For the low-speed case, $M \ll 1$, the boundary layer thickness δ is shown in Figure 9.8 as a function of the distance x from the leading edge of the plate for laminar flow and turbulent flow.

In Figure 9.8 the end of the transition region is noted at a distance equal to 5% of the length of the plate. Therefore, for applications typical of commercial jet transport operation, the laminar and transitional regions are generally ignored in preliminary design, and the boundary layers over airplane components are assumed to be fully turbulent over their entire length. This is a conservative assumption and is consistent with the accuracy of the other assumptions used throughout the analysis. For completeness, the boundary layer thickness relations for incompressible laminar and turbulent flow over a flat plate, as given by White (2006), are as follows:

$$\frac{\delta}{x} \approx \frac{5}{\left(\text{Re}_x\right)^{1/2}} : \text{laminar flow} \tag{9.11}$$

$$\frac{\delta}{x} \approx \frac{0.37}{\left(\text{Re}_x\right)^{1/5}} : \text{turbulent flow} \tag{9.12}$$

We may then calculate the skin friction drag of a streamlined aircraft component as the product of the average turbulent flat plate skin friction coefficient, the total surface area of the component exposed to the flow (the "wetted" area), and the free stream dynamic pressure, q, as given below:

$$D_f = C_F S_{wet} q \tag{9.13}$$

In detailed studies of airplane performance more elaborate calculations may be made. However, for preliminary design studies the above is satisfactory. For our purposes we will consider the average turbulent skin friction coefficient to depend on the Reynolds number based on length, Re_l, where l is the characteristic length of the component under consideration. For example, in the case of a wing or a tail surface the length is taken as $l = c_{MAC}$, the mean aerodynamic chord for the wing or tail and in the case of an airfoil $l = c$, the chord of the airfoil. Similarly, for a fuselage $l = l_b$, the fuselage length and for a nacelle $l = l_n$, the length of the nacelle.

9.3 Form drag

In inviscid flow there is no drag due to normal stresses, that is, the pressure field. However, in viscous flow, as illustrated in Figure 9.1, the normal stresses contribute to form, or profile, drag. The fluid viscosity retards the flow near a surface through the action of frictional, or tangential, stresses. The depletion of momentum in the boundary layer serves to make it less capable of proceeding against an adverse pressure gradient and ultimately leads to separation of the flow. The departure of the boundary layer from following the surface alters the pressure field from its inviscid distribution resulting in an additional drag component, the form drag. In the preliminary design stage the form drag is generally not calculated directly, but is instead obtained from wind tunnel measurements and empirical equations based upon correlations of those measurements. The most thorough, though dated, collection of results of this type is presented by Hoerner (1958).

Computational fluid dynamics (CFD) codes are capable of calculating the flow field around aircraft components and even whole aircraft, as described in Appendix

C. Obviously, the more accurate the results desired, the more detailed the geometry and flow conditions need to be. A good deal of time and expense is therefore required to estimate the drag of the complete airplane, primarily because of the complexity of the geometry. Prior to the development of effective and general CFD codes for calculating the flow of an aircraft or its components, airframe manufacturers relied on experience, semi-empirical analyses, wind tunnel experiments, and flight tests, in increasing order of expense. The most widely used semi-empirical approach involved estimating the form drag of the various airplane components as a multiple of the skin friction drag. Experience has shown that the total zero-lift drag coefficient calculated in this fashion provides a reasonable estimate of the airplane drag for preliminary design purposes. It should be pointed out that such methods work best for well-constructed airplanes of conventional design, such as commercial airliners. With the increase in fuel prices and the growing concern regarding air pollution and greenhouse gas production, all airplane manufacturers are paying considerably greater attention to drag reduction techniques.

9.4 Drag build-up by components

The zero-lift drag of a typical airplane component, say the wing, in low-speed flight is estimated as the sum of the friction and profile drag contributions as follows:

$$D_{0,w} = D_{f,w} + D_{p,w}$$

For flows that are not near separation it is assumed that the form or profile drag D_p may be expressed as a multiple, k, of the friction drag D_f so that the low-speed zero-lift drag of the wing may be expressed as follows:

$$D_{0,w} = C_{F,w}S_{wet,w}q + k_wC_{F,w}S_{wet,w}q$$

Then the zero-lift drag may be written as

$$D_{0,w} = K_wC_{F,w}S_{wet,w}q$$

Here, $K_w = (1+k_w)$ is a form factor for the wing which includes profile and skin friction drag, and $C_{F,w}$ is the average turbulent skin friction coefficient of a flat plate of length equal to, in this case, the mean aerodynamic chord of the wing, c_{MAC}. The quantity $S_{wet,w}$ is the wetted area of the wing and q is the free stream dynamic pressure, $q = \frac{1}{2}\rho V^2 = \frac{1}{2}\gamma p M^2$. The zero-lift drag coefficient of the wing is based on the wing planform area S and is defined as

$$C_{D,0,w} = \frac{D_{0,w}}{Sq}$$

Then the zero-lift drag coefficient of the wing is related to the skin friction coefficient as follows:

$$C_{D,0,w} = K_wC_{F,w}\frac{S_{wet,w}}{S} \tag{9.14}$$

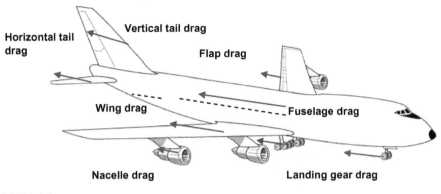

FIGURE 9.9

Drag components making up the total commercial airplane drag.

This equation may now be generalized to apply to the entire aircraft as follows:

$$C_{D,0} = \sum_{i=1}^{n} K_i C_{F,turb_i} \frac{S_{wet_i}}{S} \qquad (9.15)$$

The individual aircraft components considered are identified by the subscript i. In the cruise configuration the components usually number five ($n = 5$) corresponding to (1) wing, (2) fuselage, (3) horizontal tail, (4) vertical tail, and (5) the engine nacelles, including pylons and their interference effects. In the landing and takeoff configuration the landing gear (nose and main) and wing flap drag components would be added as well. The list increases for externally carried antenna, externally carried fuel or stores, etc. The major drag components for commercial aircraft are summarized in Figure 9.9.

It should be realized that the actual zero-lift drag coefficient may be substituted for any one of the components used above if known. For example, if standard NACA airfoil sections are used in the wing design the airfoil drag coefficient may be obtained from Appendix A, Abbott et al. (1945), or from Abbott and Von Doenhoff (1959). Note that the data shown in those references are typically for low speed at Reynolds numbers $Re_l = 3$, 6, and 9 million for smooth surfaces and 6 million for standard surface roughness. Flight values for Re_l are often much larger, as described in Section 5.4.5, and this distinction must be kept in mind when determining drag.

The local zero-lift drag coefficient of the airfoil, $c_{d,0}$, varies in correspondence with the free stream Mach number and the local Reynolds number, which is directly proportional to the local chord c and varies with distance along the span for a tapered wing. The local zero-lift drag coefficient of the airfoil, $c_{d,0}$, can also change in the spanwise direction if different airfoil sections are used from station to station. It is obvious that if a rectangular wing planform is chosen, and only a single airfoil used along the span, the wing and airfoil zero-lift drag coefficients are approximately equal provided the Reynolds numbers are equal and the fuselage is not considered.

As an example of using available airfoil data, we may estimate the wing drag by integrating the local airfoil drag coefficient over that portion of the span actually exposed to the flow, that is, from $y = y_b$ to $y = b/2$:

$$C_{D,0} = \frac{\int_{y_b}^{b/2} c_{d,0}(y)c(y)dy}{\int_{y_b}^{b/2} c(y)dy}$$

9.5 Wing and tail drag

A planar body, like the wing, the horizontal tail, and the vertical tail may be approximated by an equivalent thin flat plate with the same planform shape as the actual body. Then the friction drag coefficient is defined by

$$C_{D,f} = \frac{D_f}{qS} = \frac{2}{S}\int_0^{b/2}\int_0^{c(y)} 2\left(\frac{\tau}{q}\right)dx\,dy = \frac{4}{S}\int_0^{b/2}\int_0^{c(y)} c_f dx\,dy$$
$$= \frac{4}{S}\int_0^{b/2} C_F(y)\,c(y)dy \tag{9.16}$$

In Equation (9.16) the first factor of 2 accounts for the symmetry of the wing about the centerline and the second factor of 2 accounts for both sides of the equivalent flat plate being subjected to the wall shear stress τ. Note that Equation (9.2) shows that the integral of the local skin friction coefficient c_f over the chord in Equation (9.16) is equal to the product of c and C_F. Formulations of the average skin friction coefficient C_F for laminar and turbulent flows were discussed in Section 9.2. Based on the discussion of the transition of the boundary layer in Section 9.2.5 a conservative approach for practical airliner applications is to assume that the boundary layer is fully turbulent over the entire wing or tail surface. Then without loss in generality we may use the simple power-law form for C_F given by Equation (9.7) in Equation (9.16) to obtain

$$C_{D,f} = \frac{4}{S}\int_0^{b/2} \frac{0.0315}{\left[\frac{\rho V c(y)}{\mu}\right]^{1/7}} c(y)\,dy \tag{9.17}$$

Moving all constants outside the integral sign in Equation (9.17) and introducing c_{MAC} and the normalized span coordinate $\eta = y/(b/2)$ yields

$$C_{D,f} = 2\frac{c_{MAC}b}{S}\frac{0.0315}{\left[\frac{\rho V c_{MAC}}{\mu}\right]^{1/7}}\int_0^1\left(\frac{c}{c_{MAC}}\right)^{6/7} d\eta \tag{9.18}$$

This may be rewritten as follows:

$$C_{D,f} = 2\frac{c_{MAC}b}{S}\left(\frac{0.0315}{Re_{MAC}^{1/7}}\right)\int_0^1\left(\frac{c}{c_{MAC}}\right)^{6/7} d\eta \tag{9.19}$$

Note that the first term in parentheses in Equation (9.19) is the average skin friction coefficient based on the mean aerodynamic chord length and may be defined as follows:

$$C_{F,MAC} = \frac{0.0315}{Re_{MAC}^{1/7}}$$

Then Equation (9.19) may be written in the following form:

$$C_{D,f} = 2C_{F,MAC} \left[\frac{c_{MAC}b}{S}\right] \int_0^1 \left(\frac{c}{c_{MAC}}\right)^{6/7} d\eta \tag{9.20}$$

Equation (9.20) shows that the drag coefficient due to friction is proportional to a single value for the skin friction coefficient, $C_{F,MAC}$, and the proportionality factor depends only upon the planform geometry of the wing. We may examine the effect of the remaining terms in Equation (9.20) by assuming a conventional wing with straight leading and trailing edges for which we have the following relations:

$$\frac{c_{MAC}b}{S} = \frac{4}{3}\frac{1+\lambda+\lambda^2}{(1+\lambda)^2} \tag{9.21}$$

$$\frac{c}{c_{MAC}} = \frac{3}{2}\frac{c}{c_r}\frac{1+\lambda}{1+\lambda+\lambda^2} \tag{9.22}$$

$$\frac{c}{c_r} = 1 - (1-\lambda)\eta \tag{9.23}$$

Substituting Equations (9.21), (9.22), and (9.23) into Equation (9.20) yields

$$C_{D,f} = 2C_{F,MAC}\left\{\frac{4}{3}\left(\frac{3}{2}\right)^{6/7}\left(\frac{7}{13}\right)\left(\frac{1+\lambda+\lambda^2}{1+\lambda}\right)^{1/7}\left(\frac{1-\lambda^{13/7}}{1-\lambda^2}\right)\right\} \tag{9.24}$$

The term in braces is equal to unity within a maximum of 1.6% for $0 < \lambda < 1$ and therefore, to a very good approximation, we may consider the frictional drag coefficient of a wing or tail surface to be given simply by

$$C_{D,f} = 2C_{F,MAC} = C_{F,MAC}\frac{S_{wet}}{S}$$

We note that for the equivalent flat plate $S_{wet}/S = 2$ and the frictional drag coefficient given above is almost exactly that suggested by Equation (9.13). The integrated or average skin friction coefficient $C_{F,MAC}$ is a function of the flight Mach number and the Reynolds number based on the mean aerodynamic chord of the wing or tail surface in question and using the compressibility correction of Equation (9.8) we may write:

$$C_{F,MAC} = C_F(Re_{MAC}, M) = \left(1 - 0.072M^{1.5}\right)C_F(Re_{MAC}, 0)$$

Table 9.3 Coefficients for Equation (9.25)

Source	ϕ_1	ϕ_2
DATCOM (1978)	R_{LS}	$1.2\left(\frac{t}{c}\right)_{max} + 100\left(\frac{t}{c}\right)_{max}^4$
Hoerner (1958)	1.0	$2\left(\frac{t}{c}\right)_{max} + 60\left(\frac{t}{c}\right)_{max}^4$
NAA (1952)	1.0	$2.5\left(\frac{t}{c}\right)_{max} + 7 \times 10^{-5}\left(\frac{t}{c}\right)_{max}^{-2}$
Torenbeek (1982)	1.0	$2.7\left(\frac{t}{c}\right)_{max} + 100\left(\frac{t}{c}\right)_{max}^4$
Jenkinson et al. (1999)	1.0	$3.3\left(\frac{t}{c}\right)_{max} - 0.008\left(\frac{t}{c}\right)_{max}^2 + 27\left(\frac{t}{c}\right)^3$
Morrison (1976)[a] SC airfoil	1.0	$4.2\left(\frac{t}{c}\right)_{max}$
Morrison (1976)[b] NACA 65	1.0	$3.53\left(\frac{t}{c}\right)_{max}$

[a]supercritical airfoil.

[b]NACA 65-series airfoil.

Any of the appropriate integrated skin friction formulations given in Section 9.2 may be used to carry out the calculations; the power-law form was used here for clarity in showing how the drag based on the average skin friction coefficient of the mean aerodynamic chord provides an accurate result with no further integration required.

As discussed in Section 9.4 the total drag coefficient of the wing and the tail surfaces, friction plus form drag, may be developed on the basis of a form factor correction applied to the skin friction drag coefficient. In this case the drag coefficient may be written as

$$C_{D,w} = KC_F \frac{S_{wet}}{S} = \phi_1 \left(1 + \phi_2\right) C_{F,w} \frac{S_{wet,w}}{S} \tag{9.25}$$

The choice of two form factor constants to represent the K, a general coefficient ϕ_1 and a shape factor ϕ_2 involving the maximum thickness ratio of the wing or tail section, to account for the pressure drag, is dictated by the nature of different approximations proposed by various investigators, as will become apparent subsequently. Several different forms for the coefficients ϕ_1 and ϕ_2 have been proposed as shown in Table 9.3. A comparison of the shape coefficient ϕ_2 suggested by the various authors is shown in Figure 9.10. The coefficient ϕ_1 was included because the DATCOM method involves R_{LS}, an additional lifting surface correction which accounts for the spanwise flow that is a secondary effect of sweepback. This correction factor is shown in Figure 9.11 for several Mach numbers.

A comparison of the drag coefficient as estimated by Equation (9.25) and the coefficients in Table 9.3 is shown in Figure 9.12. For the purposes of the comparison the flow over the surface was assumed to be turbulent from the leading edge on, a value of $C_{F,w} = 0.003$ was chosen, and it was assumed that the wetted area of the wing $S_{wet,w}$ is equal to twice the planform area S. The thickness to chord ratio

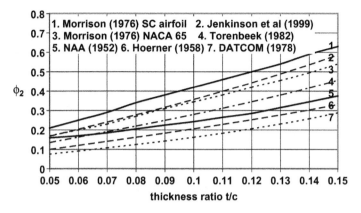

FIGURE 9.10

Pressure drag modifying factor ϕ_2 for wing and tail sections as a function of section maximum thickness ratio $(t/c)_{max}$ according to several sources.

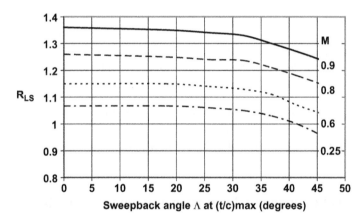

FIGURE 9.11

R_{LS}, an additional lifting surface correction which accounts for the spanwise flow on a wing, is shown as a function of sweepback angle of the maximum t/c location with Mach number as a parameter.

is evaluated in the free stream direction, which accounts for the sweepback of the wing. For the purposes of the comparisons in Figure 9.12, ϕ_1 for the DATCOM method was estimated to be $R_{LS}=1.25$, which corresponds to typical jet transport flight conditions of $M=0.8$ and $\Lambda_{(t/c),max}=25°$. It is apparent that the results are reasonably close, with the DATCOM results being the most optimistic. The shapes of the drag curves are similar but the actual values are substantially different. The results of Torenbeek (1982) fall in the middle of the range with the remaining results within about $\pm12\%$.

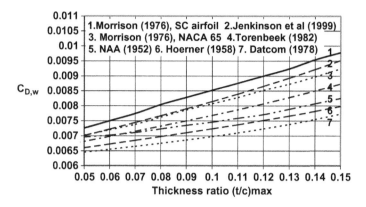

FIGURE 9.12

Drag coefficients for airfoils as predicted by several methods. The comparison is based on fully turbulent flow with $S_{wet}=2S$ and $C_F=0.003$.

It must be recalled that the drag estimation method employed assumes that the flow over the planar wing is turbulent right from the leading edge. Abbott and Von Doenhoff (1959) present minimum drag coefficient data for the NACA-63, -64, -65, and -66 series airfoils as a function of maximum thickness ratio. For airfoils in the range $8\% < ((t/c)_{max} < 15\%$, the minimum drag coefficient range is $0.033 < C_D < 0.049$ for those with smooth surfaces and $0.087 < C_D < 0.010$ for those with leading edge roughness. The effect of transition is therefore of some importance in estimating the wing or tail drag with a reasonable degree of accuracy and this is discussed in the next section.

9.5.1 Effect of boundary layer transition on wing and tail drag

Boundary layers in stagnation regions and in rapidly accelerating flows, such as found in the forward regions of airfoils, are relatively stable to disturbances. Assuming that the boundary layer on an airfoil is laminar up to some chordwise distance x_t, at which transition to turbulence begins and the flow immediately becomes fully turbulent (an idealization) the drag coefficient of a planar wing-like body as given by Equation (9.16) becomes

$$C_{D,f} = \frac{4}{S} \int_0^{b/2} \left[\int_0^{x_t} c_{f,lam}\, dx + \int_{x_t}^{c} c_{f,turb}\, dx \right] c\,dy$$

A reasonable assumption is that the flow over the leading edge of the wing is laminar and makes the transition to fully turbulent flow at a fixed distance x_t from the leading edge. This distance corresponds to a transitional Reynolds number $Re_t = \rho V x_t / \mu$

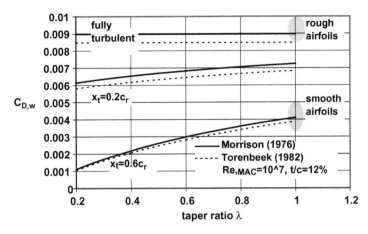

FIGURE 9.13

The variation of the zero-lift drag coefficient for wing-like bodies as a function of planform taper ratio is shown according to the equivalent flat plate method. The shaded regions indicate the range of values typical of NACA airfoils.

such that x_t is constant with spanwise distance. Under these conditions and following the approach used at the start of Section 9.5 the frictional drag coefficient may be written as follows:

$$C_{D,f} = \left\{ C_{F,lam} \left(\frac{c_{MAC}b}{S} \right) \left(\frac{Re_t}{Re_{MAC}} \right)^{1/2} + C_{F,turb} \left[1 - \left(\frac{c_{MAC}b}{S} \right) \left(\frac{Re_t}{Re_{MAC}} \right)^{6/7} \right] \right\} \frac{S_{wet}}{S}$$

(9.26)

Using Equations (9.21)–(9.23) in Equation (9.26) we may examine the case of a wing with a 12% thick airfoil at $Re_{MAC} = 10^7$ under conditions of fully turbulent flow, transition at $x_t = 0.2c_r$, and transition at $x_t = 0.6c_r$. The results according to the approximations for ϕ presented by Morrison (1976) for a NACA 65-series airfoil and by Torenbeek (1982) are shown in Figure 9.13 as a function of taper ratio λ, with $\lambda = 1$ representing either a straight wing or an airfoil. Also shown in the figure as shaded regions at $\lambda = 1$ are the range of results for smooth and rough NACA 6-series airfoils presented by Abbott and Von Doenhoff (1959) as discussed in the previous section. The fully turbulent cases shown agree quite well with the rough airfoil data while the results for transition at $x_t/c_r = 0.6$ agree well with the smooth airfoil data. It is suggested that these methods are appropriate to use in the preliminary design analysis. Determining the transition Reynolds number, and thus the transition location, is quite difficult. Considering completely turbulent flow is a simple choice but it may be more conservative than necessary. If a more optimistic scenario is acceptable a transition at $x_t/c_r = 20\%$ appears reasonable.

9.5.2 Wing and tail wetted areas

The wetted areas of wings and tails may be determined by using the airfoil coordinates to calculate the local perimeter $p(y)$. Forming the local area increment $\Delta S_{wet} = p(y) \Delta y$ and then summing over the exposed span yields S_{wet}. An approximation for the perimeter of an airfoil is given by Torenbeek (1982) as

$$p = 2c \left[1 + 0.25 \left(\frac{t}{c} \right)_{max} \right]$$

If $p(y)dy$ is integrated over the exposed span of a wing with straight edges using Equations (9.21) and (9.22), we obtain

$$\frac{S_{wet}}{S} = 2 \left[1 + \left(\frac{t}{c} \right)_{max} \right] \left[1 - \eta_b \frac{2 - (1 - \lambda) \eta_b}{1 + \lambda} \right]$$

9.6 Fuselage drag

The elements comprising the fuselage drag have been discussed in some detail in Chapter 3 and Equation (3.12) was developed to estimate the drag contribution of the fuselage and is repeated again below in terms of the drag coefficient based on frontal area:

$$C_{D,b,o} = 4 \frac{S_{wet}}{A_o} FC_F \, (Re, M) \left(1 + \frac{1.5}{F^{3/2}} + \frac{7}{F^3} \right) \tag{9.27}$$

The ratio of wetted area to frontal area for conventional fuselages with circular cabin cross-section may be approximated, using the results of Chapter 3, by the following expression:

$$\frac{S_{wet}}{A_o} = 4 \left(F - \frac{F_{NC}}{3} - \frac{F_{TC}}{2} \right) \tag{9.28}$$

In Equation (9.28) $F \geq 4.5$, but the fineness ratios for typical airliners are always greater than this value and the equation is applicable. There are other estimates available but they all have the same basic structure. Torenbeek (1982), Hoerner (1958), and Jenkinson et al. (1999) offer the following form:

$$C_{D,b,o} = 4 \frac{S_{wet}}{A_o} FC_F \, (Re, M) \left(1 + \frac{a}{F^{3/2}} + \frac{b}{F^3} \right) \tag{9.29}$$

The coefficients a and b are given in Table 9.4 and, though different, yield close results for fineness ratios in the range of commercial jet transports. In the DATCOM approach, Hoak et al. (1978), the drag coefficient equation has the form

$$C_{D,b,o} = 4 \frac{S_{wet}}{A_o} FC_F \left(1 + \frac{60}{F^3} + 0.0025F \right) \tag{9.30}$$

Table 9.4 Coefficients for Equation (9.29)

Source	a	b
Torenbeek (1982)	2.2	3.8
Hoerner (1958)	1.5	7
Jenkinson et al. (1999)	2.2	−0.9

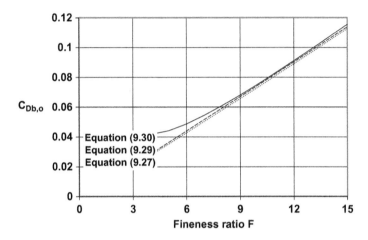

FIGURE 9.14

Fuselage drag coefficient based on frontal area as a function of fineness ratio for conventional fuselages. The case shown is for a skin friction coefficient $C_F=0.002$ and Equation (9.29) uses a and b from Torenbeek (1982).

The results for Equations (9.28)–(9.30) are shown in Figure 9.14 for the case of equal skin friction coefficient values of $C_F=0.002$. It is clear that they all yield essentially the same result for fineness ratios typical of airliner fuselages.

It must be remembered that the fuselage drag coefficients shown are based on frontal area. The drag coefficient for the complete aircraft is always based on the wing planform area S, so the fuselage drag coefficients based on frontal area must be corrected accordingly. The drag of the fuselage may be written as

$$D_b = C_{D,b,o}qA_o = C_{D,b}qS \qquad (9.31)$$

Here q is the dynamic pressure $\rho V^2/2$, $C_{D,b,o}$ is the fuselage drag coefficient based on frontal area $A_o=\pi(d_b)^2/4$ while C_{Db} is the fuselage drag coefficient based on wing planform area S. Therefore, the drag coefficient required for the drag build-up by components is

$$C_{D,b} = C_{D,b,o}\frac{A_o}{S} \qquad (9.32)$$

We wish to keep the form

$$C_{D,0,b} = \frac{K_b C_{F,b} S_{wet,b}}{S}$$

Therefore the form factor for the fuselage, using the DATCOM formulation of Equation (9.30), is given by

$$K_b = 1 + \frac{60}{F^3} + 0.0025F \tag{9.33}$$

9.7 Nacelle and pylon drag

A typical nacelle and pylon installation for a wing-mounted turbofan engine is schematically illustrated in Figure 9.15a. These nacelles have relatively shallow curvature and therefore their drag is due primarily to skin friction on the wetted (outer) surface. The nacelle drag for turbofan engines of moderate to high bypass ratio is considered to be the drag of the fan shroud or cowl alone because the drag of the center body, which encloses the hot section of the engine, is already included in the thrust rating. Because the quoted static thrust of the engine includes the effects of the fan flow over the inside of the cowl, the wetted area for the nacelle should be just outside of the fan cowl because that portion is actually wetted by the free stream flow.

The nacelle is essentially an unswept wing bent into a cylindrical shape with a chord equal to the cowl length l_n. The wetted area may be approximated by $S_w = \pi(d_{n,max}) l_n$ and the wing form factor Equation (9.26) may be used with t/c replaced by $(d_{n,max} - d_{in})/2l_n$. In the case of wing-mounted turbofan engines the

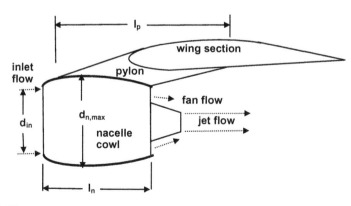

FIGURE 9.15a

Nacelle-pylon assembly for a wing-mounted turbofan showing relevant flow field and dimensions.

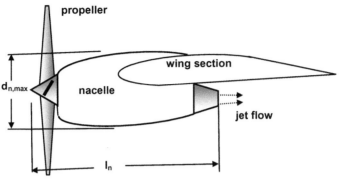

FIGURE 9.15b

Nacelle for a wing-mounted turboprop showing relevant flow dimensions.

drag introduced by considering the pylon alone is not sufficient. Instead it is dependent upon the location of the nacelle with respect to the wing leading edge because the interference drag produced can be substantial if a poor location is selected. Indeed, this is one of the detailed situations of complex flow interference that have proven the value of CFD analyses. Using CFD computations of complete wing-nacelle-pylon assemblies may be carried out for many various geometric configurations in order to select that with the best performance. Such detailed studies are well beyond the scope of this preliminary design phase. Here it is sufficient to follow the trends exhibited by the market survey aircraft in laying out the nacelle and pylon. Then to account for the pylon drag one may use Torenbeek's (1982) suggestion and multiply the wing-like form factor in Table 9.3 for the nacelle by 1.25 and use the result as the form factor for the nacelle-pylon combination. Then the nacelle-pylon assembly form factor is

$$K_n = 1.25 \left[1 + 1.2 \left(\frac{d_{n,\max} - d_{in}}{2l_n} \right) + 100 \left(\frac{d_{n,\max} - d_{in}}{2l_n} \right)^4 \right] \qquad (9.34)$$

In the case of fuselage-mounted turbofan engines the interference penalty is not as great because the pylon mounting doesn't interfere with the wing. Of course one of the advantages claimed for fuselage-mounted engines is that the wing is completely clean and doesn't suffer from interference drag problems. However this sort of mounting places the relatively short-span pylon in a higher-velocity field and one may replace the interference factor of 1.25 in Equation (9.34) and replace it with the factor 1.1.

Turboprop engine nacelles, which are typically mounted on the wing as shown in Figure 9.15b, may be considered to act in the same manner as a fuselage with a length equal to l_n and an equivalent diameter equal to $d_{\max}$ and may be treated with the same equations.

9.8 Landing gear drag

The wing flap and landing gear drag may be obtained (for example) from curves in Perkins and Hage (1949). They use the equivalent parasite area:

$$f = C_D S = C_{D,c} A_c \tag{9.35}$$

In Equation (9.35) C_D is the zero lift drag coefficient of the component in question (here the landing gear) based on wing planform area, S, while $C_{D,c}$ is its drag coefficient based on some component reference area A_c. Then

$$C_{D,LG} = \frac{f_{LG}}{S}$$

Perkins and Hage (1949) show a curve for estimating the parasite drag area for tricycle landing gear based on takeoff weight, which can be fitted by the following equation:

$$C_{D,LG} = 4.05 \times 10^{-3} \frac{W_{to}^{0.785}}{S} \tag{9.36}$$

The same equation is also presented by Torenbeek (1982), while Mair and Birdsall (1987) quote data from the Engineering Sciences Data Unit (1987) which gives the same form of the equation, but with different constants, as follows:

$$C_{D,LG} = 1.79 \times 10^{-3} \frac{W_{to}^{0.785}}{S} \tag{9.37}$$

They also point out that when the flaps are not extended the landing gear drag is higher:

$$C_{D,LG} = 3.30 \times 10^{-3} \frac{W_{to}^{0.785}}{S} \tag{9.38}$$

The reduced drag coefficient at full flaps is supported by some flight tests on commercial airliners and is thought to be a result of reduced airspeeds over the landing gear.

If a preliminary layout of the landing gear has been developed it is possible to use the data presented in Hoerner (1958), for example, to build up the drag contribution of wheels, struts, doors, etc. to arrive at an approximate value for the drag coefficient contribution of the landing gear assembly.

9.9 Flap and slat drag

There is also additional drag when the flaps are extended because of the larger wake produced in this condition. This is a boundary layer separation effect and occurs in addition to the induced drag accompanying the production of lift. The flap drag may also be treated by the parasite drag area approach of Perkins and Hage (1949) or by a similar method proposed by Hoerner (1958). A preferable method for estimating the drag coefficient increment due to flap extension is that presented by McCormick (1995). The equation for the drag increment due to slotted flaps, based on wing planform area, is as follows:

$$C_{D,flap} = 0.9 \left(\frac{c_{flap}}{c} \right)^{1.38} \left(\frac{S_{wflap}}{S} \right) \sin^2 \delta_{flap} \tag{9.39}$$

For plain or split flaps the coefficient in Equation (9.39) is increased to 1.7 from 0.9. Here c_{flap}/c represents the flap chord to wing chord ratio and the quantity S_{wflap}/S is the ratio of wing area affected by the trailing edge flap deflection (including both port and starboard wings) to the total wing area, as is shown in Figure 5.2.

On the other hand, slats provide an increase in lift, but with essentially no drag penalty. This attribute, along with their relatively simple deployment mechanism, makes them a common adjunct to high lift systems for commercial aircraft.

9.10 Other drag sources

The major drag-producing components of the typical commercial jet transport have been addressed but there are other smaller drag sources that deserve attention. Among these are

- Protuberances such as antennas, instrument probes, etc.,
- Leakage sites such as seals between moving components, exhaust vents for cabin climate control, gaps, holes, cracks, etc.,
- Intake scoops,
- Interference drag between components.

Other sources of information on drag estimation may be found in the books by Roskam (1971), Nicolai (1975), and Raymer (1989). A recent drag estimation method for complete aircraft has been presented by Gur et al. (2010).

9.11 Calculation of the zero-lift drag coefficient neglecting wave drag

The zero-lift drag coefficient neglecting wave drag due to compressibility effects for $M > 0.6$ may be found in a systematic fashion by following the procedures discussed in the previous sections. A general outline of how one might proceed to develop the drag of the entire airplane is illustrated in Table 9.5 and described in the following steps. Note that three such tables must be prepared; one each for flight configuration: cruise, landing, and takeoff operations. The inviscid wave drag $C_{D,w}$ is treated separately in a subsequent section.

Table 9.5 Airplane Zero-Lift Drag Coefficient Buildup Neglecting Wave Drag

Date:					Project Team:			

Configuration: takeoff, landing, or cruise
Weight $W =$
Aspect ratio $A =$
Altitude $z =$
Temperature $T =$
Kinematic viscosity $v = \mu\rho =$
Transition at $x_t =$

Aircraft designation:
Wing area $S =$
Flap angle $\delta_{flap} =$
Speed $V =$
Sound speed $a =$
Mach number $M =$
Transition at $Re_t =$

1	2	3	4	5	6	7	8	9
Component	Reference Length, l	Wetted Area, S_{wet}	Re_l	Fineness Ratio F or $(t/c)_{max}$	Form Factor K	C_F	KC_F S_{wet}	$\Delta C_{D,0}$ $=KC_F$ S_{wet}/S
(Units)								
Wing	$c_{MAC} =$							
Flaps								
Horizontal Tail	$c_{MAC,h} =$							
Vertical Tail	$c_{MAC,v} =$							
Fuselage	$l_b =$							
Nacelles and pylons	$l_n =$							
Landing Gear								
Sum of Components								
Miscellaneous (5% of sum)								
Total								

1. A three-view CAD drawing of the design aircraft should be provided to illustrate the geometry of the proposed configuration.
2. The wetted area S_{wet} is the contour surface area in contact with the fluid. The wing, horizontal tail and vertical tail wetted area includes only the *exposed* wing and tail area and is approximately equal to twice this exposed area (the perimeter is slightly greater than twice the length of the chord line). The fuselage wetted area excludes those areas covered by the wing and empennage airfoil cross-sections. The nacelle wetted area is the area of the exposed outer surface of the nacelle and the exposed area of the pylon. It does not include

portions exposed to the fan flow or the nozzle exhaust. Most CAD packages provide surface areas as an output option.

3. The relevant characteristic length of each component should be inserted into column 2 of the table. These include the mean aerodynamic chords of the wing and tail surfaces and the lengths of the nacelles and the fuselage.

4. The Reynolds numbers of the components should be calculated and presented in column 4 of this table. The Reynolds numbers should be based upon the characteristic length of the component, the flight velocity, and the properties of the atmosphere at the altitude being considered.

5. The skin friction coefficients should be calculated according to the discussion in previous sections. For each component account must be taken of the following: (a) the boundary layer transition point, (b) the cut-off Reynolds number, which depends on surface roughness, and (c) compressibility effects in the boundary layer.

6. The form factors K and wetted areas S_{wet} for the wing and tail surfaces are presented and discussed in Section 9.5.

7. The form factor K and wetted area S_{wet} for the fuselage are presented in Section 9.6. The fineness ratio of the body of revolution fuselage is defined in terms of the overall length l_b as $F_b = l_b/d_{max}$. If the cross-section is other than circular, an effective diameter is derived from the maximum cross-sectional area as follows: $d_{eff} = \sqrt{4A_o/\pi}$.

8. The form factor K and wetted area S_{wet} for the nacelles are presented in Section 9.7. Also discussed in that section is the drag contribution due to the nacelle pylons.

9. The landing gear contribution to the drag is discussed in Section 9.8. For example, Equation (9.36) may be used to obtain C_D for the landing gear. Again, columns 2 through 8 of the drag summary table may be left blank and there will be entries for the landing gear contributions only in the tables constructed for landing and takeoff, while those for the cruise condition will be blank.

10. The drag contribution of the flaps is discussed in Section 9.9. If Equation (9.39) for C_D of the flaps is used, there is no need to fill in columns 2 through 8 of the drag summary table. Obviously, there will be entries for the flap contributions in the tables constructed for landing and takeoff, while the flap contributions in cruise will be blank. No entry for drag due to slats is provided because their contribution is typically much smaller than the flap contribution.

11. For interference effects and miscellaneous drag contributions due to other sources, a value of 6% of the sum of the previous entries should be used; of course there is no need to fill in columns 2 through 8.

The drag coefficient calculation table must be repeated twice more, once for the takeoff configuration at the takeoff speed, $V = V_{to}$, and once for the landing configuration at $V = V_l$. There appears to be an apparent contradiction in the presentation of Table 9.5, namely, the flight condition for cruise is shown as $M = 0.8$ but the C_D value calculated in the table is essentially for $M \leq 0.6$. The low-speed drag coefficient values determined for the cruise condition will subsequently be corrected for the effects of compressibility.

9.12 Compressibility drag at high subsonic and low transonic speeds

As the Mach number increases toward unity it becomes increasingly important to account for the compressibility of the gas. In Chapter 5 the effects of compressibility on the lifting characteristics of wings were discussed. The critical Mach number of an airfoil was defined as the free stream value for which the flow first reaches the local sonic speed at some point on the airfoil. Further increase in the free stream Mach number enlarges the region of supersonic flow over the airfoil and leads to the development of shock waves. The abrupt pressure rise due to the shock wave tends to separate the boundary layer. This in turn leads to the so-called drag divergence, where the drag coefficient has a steep rise starting from the drag divergence Mach number M_{dd}, which is arbitrarily defined as the point on a plot of drag coefficient as a function of Mach number where $dC_{Dw}/dM = 0.1$. Recall that the idea behind the development of supercritical airfoil sections described in Chapter 5 was to increase the drag divergence Mach number.

This drag rise effect is shown schematically for flow over a thin airfoil in Figure 9.16. Another, less common, definition of the drag divergence Mach number is sometimes stated as the Mach number at which the wave drag coefficient is 20 drag counts, that is 20 thousandths, or 0.020. This difference in definition may have an appreciable effect on the value of M_{dd} as it does here where the drag divergence Mach number would be about 0.81 rather than 0.74. Because the idea of drag rise

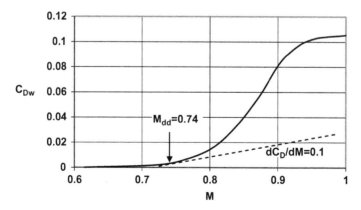

FIGURE 9.16

The typical variation of zero-lift drag coefficient as a function of Mach number for a wing. The drag divergence Mach number is shown as the point where $dC_D/dM = 0.1$.

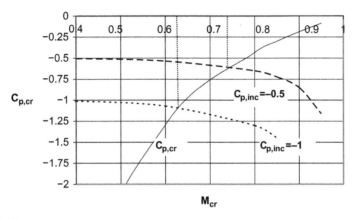

FIGURE 9.17

The critical pressure coefficient is shown as a function of critical Mach number. The determination of the critical Mach number corresponding to two different values of incompressible pressure coefficient is illustrated.

immediately conjures up the idea of a slope, we will adhere to the idea of the drag divergence Mach number being given by the Mach number at which $dC_{Dw}/dM = 0.1$.

The critical Mach number is the indicator of incipient drag rise and is always less than the drag divergence Mach number $M_{cr} < M_{dd}$. The maximum speed of the flow over an airfoil occurs at the point of minimum pressure and the pressure distribution over an airfoil is easily measured, as described, for example, by Goldstein (1996). The pressure coefficient at the critical Mach number is given by Liepmann and Roshko (1957) as follows:

$$C_{p,cr} = \frac{p - p_\infty}{q_\infty} = \frac{2}{\gamma M_{\infty,cr}^2} \left[\left(\frac{2}{\gamma + 1} + \frac{\gamma - 1}{\gamma + 1} M_{\infty,cr}^2 \right)^{\frac{\gamma}{\gamma - 1}} - 1 \right]$$

For the case of air where $\gamma = 1.4$ this becomes

$$C_{p,cr} = \frac{10}{7 M_{\infty,cr}^2} \left[\left(\frac{5}{6} + \frac{1}{6} M_{\infty,cr}^2 \right)^{\frac{7}{2}} - 1 \right] \tag{9.40}$$

The variation of critical Mach number with critical pressure coefficient according to Equation (9.40) is shown in Figure 9.17. Then to find the critical Mach number for an airfoil which has a known minimum value for the pressure coefficient in incompressible flow, $C_{p,inc}$, one may plot the corrected pressure coefficient $C_p = C_{p,inc}/\beta$ as a function of Mach number. The point where that curve crosses the curve of $C_{p,cr}$ defines the critical Mach number for that airfoil. Two examples of such a procedure are also shown on Figure 9.17. The $C_{p,inc} = -1$ case reaches a lower pressure than the $C_{p,inc} = -0.5$ case and therefore reaches the sonic speed at a lower Mach number ($M_{cr} = 0.63$) than that case ($M_{cr} = 0.74$).

Small perturbation theory shows that $C_p \sim -\varepsilon$ where ε is the (small) body thickness or slope. This means that the thinner the airfoil or the smaller its slope (angle of attack, or, equivalently, its lift coefficient), the smaller the magnitude of its C_p, and therefore the higher its critical Mach number will be. Thus, we can see that an airfoil's drag divergence characteristics are presaged by its critical Mach number characteristics and therefore on $(t/c)_{\max}$ and C_L.

The critical Mach number for a slender planar body like a wing is always lower than that for a slender body of revolution, like a fuselage, when both have the same fineness ratio. In the same fashion, wing-like bodies exhibit higher critical Mach numbers as their aspect ratio decreases. In other words, as the wing becomes more three-dimensional, the critical Mach number increases, thereby delaying the drag divergence phenomenon. These effects are described by Sears (1954). Because the wing displays the drag increase at cruise speeds of interest, $0.75 < M < 0.85$, while the high fineness ratio fuselage ($l/d = 10$) and the low aspect ratio and relatively thinner tail surfaces ($3 < A_{tail} < 4$) do not exhibit substantial drag rise until higher Mach numbers are reached, we will consider the wing to be the sole contributor to the wave drag. It is tacitly assumed here that there is negligible interference drag produced between these components as the cruise speed is attained. However, this depends upon the area rule of high-speed flight which is described next.

9.13 The area rule

The area rule was first demonstrated experimentally by Whitcomb (1952) and had an important effect on airplane design in subsequent years. The basic premise of the area rule is that the wave drag of an aircraft configuration depends mainly on the longitudinal variation of its cross-sectional area. The consequence of this premise is that a body of revolution having the same cross-sectional area distribution as a given airplane generates the same wave drag as the airplane. This equivalent body concept is illustrated in Figure 9.18 and was first promulgated for Mach numbers close to unity, but has been extended to supersonic flows by Jones (1953).

The important consideration here is the variation of the cross-sectional area of the complete airplane as a function of axial distance from the nose. As a first approximation, at high subsonic speeds the aircraft will affect the flow field in much the same manner as would an axially symmetric body with the same area distribution. If that area distribution is relatively smooth, without abrupt changes in slope, then the assumption that the wing is the primary element in producing compressibility drag is reasonable. On the other hand, rapid changes in area distribution will lead to important interference effects and the drag increase will be large. This "area rule" was an important lesson learned in the early efforts at transonic flight and led to the development of the "coke-bottle" fuselage shape of early century series fighter aircraft. The reduction of the "waist size" of the fuselage was to smooth out the cross-sectional area distribution of the aircraft. A general discussion of the area rule and its effects is given by McCormick (1995). Therefore it is important to estimate the cross-sectional area distribution of the

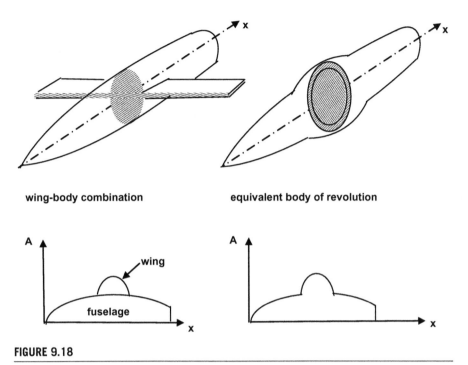

wing-body combination   **equivalent body of revolution**

FIGURE 9.18

The equivalent body concept in terms of longitudinal area distribution.

design aircraft to ensure that a reasonable level of wave drag is achieved and that the assumption of negligible interference drag between aircraft components is allowable. If the wing and fuselage combination has a longitudinal cross-sectional area distribution marked by substantial gradients the interference drag may become large. In a conventional jet transport aircraft, a so-called tube and wing configuration, the leeway for redistributing area is limited, although sweepback and taper ratio may be modified to help smooth out the area distribution as suggested by Figure 9.19.

9.14 Calculation of the wave drag coefficient
9.14.1 Perkins and Hage method

One of the earliest corrections for determining the wave drag coefficient to add to the base value of the zero-lift drag coefficient $C_{D,0,inc}$ was used in conjunction with NACA airfoils, as described, for example, by Perkins and Hage (1949). Abbott and VonDoenhoff (1959) present typical experimental results for the drag divergence Mach number M_{dd} of a NACA airfoil, say a NACA 66-210, and the general behavior is shown as a function of the low-speed airfoil lift coefficient c_l in Figure 9.20. Also shown is the theoretical variation of the critical Mach number with c_l for the same airfoil. The

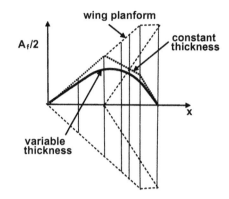

FIGURE 9.19

Smoothing of cross-sectional area of wing through the use of sweepback, taper, and thickness changes. The wing planform is superimposed to aid in illustrating cross-sectional area variation.

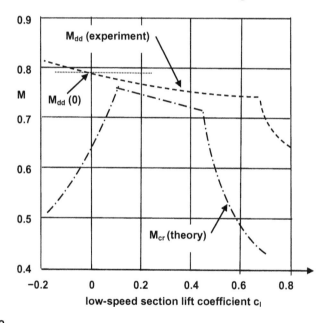

FIGURE 9.20

Experimental drag divergence Mach number and theoretical critical Mach number as a function of the low-speed lift coefficient. The dotted line represents M_{dd} for $c_l=0$.

dotted line in the figure represents the value of M_{dd} at $c_l=0$ and the curve of M_{dd} for $c_l>0$ seems to lie about halfway between the $M_{dd}(0)$ line and the theoretical curve for the critical Mach number M_{cr}. This result forms the basis for the method described by Perkins and Hage. The method is based on experimental data for unswept wings so that corrections for sweepback must be applied, as discussed in a subsequent section.

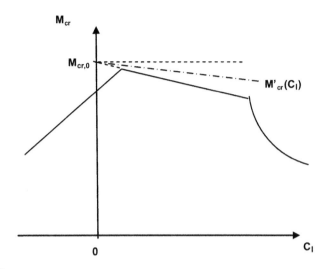

FIGURE 9.21

Typical M_{cr} versus C_l curve for NACA 6-series airfoils from Abbott et al. (1946).

First, the appropriate curve of the critical M_{cr} for the selected NACA airfoil must be obtained from Abbott et al. (1945) or Appendix D where data for several NACA airfoils are reproduced. The general nature of the variation of the critical Mach number with lift coefficient is shown in Figure 9.21. Note that the critical Mach number drops off linearly with lift coefficient over the design lift coefficient range so that an approximate description may be written as

$$M_{cr} = M_{cr,0} - kC_l \qquad (9.41)$$

Here we have assumed that the linear portion of the critical Mach number curve may be extrapolated back to $C_l = 0$, where $M_{cr} = M_{cr,0}(t/c)$. It may be noted that the coefficient k is approximately constant for all the thickness ratios of a given airfoil. The method described by Perkins and Hage (1949) calls for bisecting the angle the "roof top" makes with the horizontal through the point $M_{cr,0}$ as shown in Figure 9.21. This approach is consistent with the trend of experimental results, as described previously and illustrated in Figure 9.20. Then the dash-dot line defined as M'_{cr} is formed so that

$$M'_{cr} = M_{cr,0} - k'C_l \qquad (9.42)$$

All the NACA 6-series airfoils show similar behavior, as can be seen from the representative examples in Appendix D. The zero-lift critical Mach number is found to be a function of the maximum thickness ratio, or $M_{cr,0} = f(t/c)$, as shown in Figure 9.22. The relationship is linear within ±2% when given by

$$M_{cr,0} = 0.89 - 1.3 (t/c) \qquad (9.43)$$

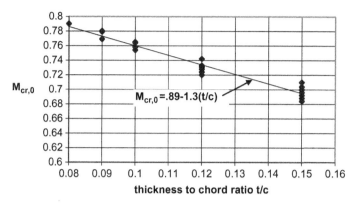

FIGURE 9.22

Variation of $M'_{cr,0}$ with t/c for all the NACA 6-series airfoils in Table 9.6.

Table 9.6 Average Values of k' for the NACA Airfoils
NACA 63-209 to 215: $k=0.095$
NACA 64-208 to 215: $k=0.080$
NACA 65-209 to 215: $k=0.071$
NACA 66-209 to 215: $k=0.050$
NACA 63-412 to 415: $k=0.080$
NACA 64-412 to 415: $k=0.0680$
NACA 65-410 to 415: $k=0.066$

The slope k' of the M'_{cr} relation given in Equation (9.43) has greater variation than does $M'_{cr,0}$, but it seems to be correlated with the airfoil pressure distribution. The average values of k' have the values shown in Table 9.6. The actual values of k' for different t/c in each series are generally well within $\pm 10\%$ of the average values, with the 64-2xx series showing about double that spread. Note, for reference, that k', the slope of the fictitious M'_{cr} curve, is half the value of the slope of the actual M_{cr} curve.

The correlation curve shown in Figure 9.23 relates the difference between the free stream Mach number and the fictitious critical Mach number M'_{cr} to the increment of wave drag C_d. Choosing a value for C_l, use the actual critical Mach number plots or Equations (9.41) and (9.42) to calculate corresponding values for C_{dw} based on Figure 9.23 or the approximation to the Perkins and Hage (1949) curve given by the following equation:

$$C_{dw} = 9.5(M - M'_{cr})^{2.8} + 0.00193 \qquad (9.44)$$

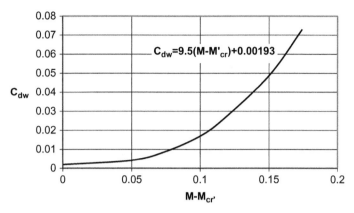

FIGURE 9.23

Compressibility correction curve for airfoils; adapted from Perkins and Hage (1949). The curve may be approximated by Equation (9.44), as given in the figure.

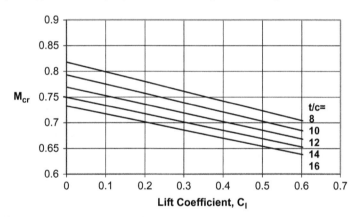

FIGURE 9.24

Critical Mach number curves for "peaky" airfoils. Adapted from Shevell (1989).

9.14.2 Shevell Method

Shevell (1989) presents curves of critical Mach numbers for "peaky" airfoils of the type used in commercial airliners in the 1980s. They are shown in Figure 9.24 and the M_{cr} curves can be approximated by the same form of linear relation as in Equation (9.41):

$$M_{cr} = M_{cr,0} - kC_l \tag{9.45}$$

For Shevell's data, the $C_l=0$ intercept may be written as

$$M_{cr,0} = 0.9 - (t/c) \tag{9.46}$$

In addition, the factor $k = 0.17 \pm 0.016$. One may use the curves discussed in connection with Figure 9.24 to represent the critical Mach numbers of more modern supercritical airfoils, according to Shevell, by adding 0.06 to the critical Mach number so found. He also notes that the thickness to chord ratio, t/c, shown in the figure is a weighted one and is given by the ratio of $\int t\,dy$ to that of $\int c\,dy$, both taken over the half-span.

$$(t/c) = \frac{\int_0^{b/2} t\,dy}{\int_0^{b/2} c\,dy} \tag{9.47}$$

For a linear thickness distribution, $t(y) \sim y$, the result is

$$(t/c) = \frac{t_r + t_t}{c_r + c_t} \tag{9.48}$$

It should be pointed out that care should be taken in using either the NACA 6 series airfoil data in Abbott and Von Doenhoff (1959), Shevell's (1989) "peaky" airfoil results or any other airfoil data because if sweepback is considered, the effective thickness to chord ratio is larger in the direction normal to the quarter-chord line and this should be taken into account. Thus far we have not discussed correcting for sweepback. The wave drag calculation for an airfoil may then be made on the basis of Lock's fourth power which is as follows:

$$C_{dw} = 20(M - M_{cr})^4 \tag{9.49}$$

This empirical relation, developed some time ago by Lock and reported in Hilton (1951), resulted from fitting experimental data on transonic drag rise. Inger (1993) developed a sound theoretical basis for Equation (9.49) by considering the relationship between wave drag and entropy increase. Note that Equation (9.49) is of the same form as Equation (9.44) which was used in the Perkins and Hage method, except that here the actual critical Mach number M_{cr} appears instead of M'_{cr}. For a given section thickness, as defined previously for this method, and lift coefficient, one may select the corresponding value of M_{cr} and use it in Equation (9.49) to compute the wave drag coefficient.

9.14.3 DATCOM method

Calculation of the zero-lift wave drag coefficient by the DATCOM method is described by Hoak et al. (1978). It is based on von Karman's transonic similarity rule and experimental correlations due to McDevitt (1955). A discussion of the development of the transonic similarity rules is presented, for example, by Liepmann and Roshko (1957). As opposed to the previous methods which are strictly for airfoils, this method incorporates the effect of the characteristics of thickness, body slope, angle of attack, and aspect ratio discussed previously, but does not consider sweepback. Figure 9.25 shows a plot of $\beta(t/c)^{-1/3}$ as a function of $C_{Dw}(t/c)^{5/3}$ for two values of $A(t/c)^{1/3}$. We see that the thickness to chord ratio and the aspect ratio are involved, as is the Prandtl-Glauert factor β. The DATCOM method is presented for zero lift so the angle of attack effect is omitted.

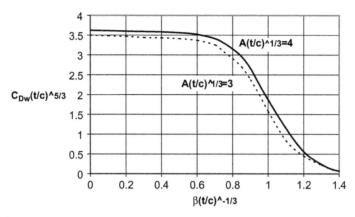

FIGURE 9.25

Basic similarity graph relating wave drag coefficient, thickness to chord ratio, and aspect ratio for use in the DATCOM method.

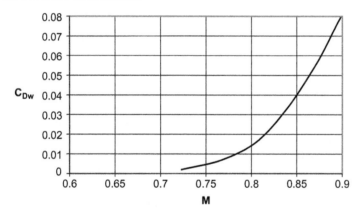

FIGURE 9.26

The zero-lift wave drag coefficient as a function of Mach number for an unswept wing with $A=8.11$ and $t/c=0.12$ according to the DATCOM method.

The values of $A(t/c)^{1/3}=4$ and 3 were selected for illustration because they are appropriate to the range of aspect ratios and thickness to chord ratios commonly encountered in jet transports. For these cases of $A(t/c)^{1/3}=4$ and 3 we may think of wings with $(t/c)=12\%$ and $A=8.11$ and 6.08 or wings with $(t/c)=10\%$ and $A=8.61$ and 6.46.

We may consider the case of $A(t/c)^{1/3}=4$ and use Figure 9.25 to extract the axis information. Then choosing a 12% thick airfoil in a wing with an aspect ratio of 8.11 we may construct Figure 9.26 which shows the wave drag coefficient as a function of free stream Mach number for the case of an unswept wing. The DATCOM method uses transonic similarity theory so there is no reference to the critical Mach number as there was in the two previous methods.

9.14.4 Korn equation method

Another empirical method based on experience has been successfully adapted to aircraft configuration design by Mason (1990). The Korn equation, named after David Korn, who had pioneered studies of computational transonic flow at the Courant Institute of NYU, expresses the idea that the drag divergence Mach number of an airfoil can diminish from the ideal of unity as a result of the effects of thickness to chord ratio and lift coefficient. The Korn equation may be put in the form

$$M_{dd} = \kappa - (t/c) - 0.1C_l \tag{9.50}$$

If $t/c = C_l = 0$ the drag divergence Mach number would ideally be unity, that is, $\kappa = 1$. One measure of an airfoil's merit would then be how closely κ approaches unity. In practice, $M_{dd} > M_{cr}$ so that one may rewrite Equation (9.50) as follows:

$$M_{dd} = M_{cr} + \Delta M = \kappa - (t/c) - 0.1C_l \tag{9.51}$$

Using the definition for drag divergence Mach number $dC_{dw}/dM = 0.1$ in Equation (9.49) we find that

$$M_{dd} = M_{cr} + \left(\frac{0.1}{80}\right)^{\frac{1}{3}} = M_{cr} + 0.108 \tag{9.52}$$

Thus, Equation (9.51) may now be written as

$$M_{cr} = (\kappa - 0.108) - (t/c) - 0.1C_l \tag{9.53}$$

We see that this is exactly the same form as the previous expressions developed for the critical Mach number with some relatively small differences in the magnitude of the various coefficients. Gur et al. (2010) suggest that $\kappa = 0.95$ for supercritical airfoils and $\kappa = 0.87$ for conventional airfoils. To calculate the wave drag coefficient using the Korn equation method one would follow the same approach as in the Shevell method, obtaining M_{cr} from Equation (9.43) and C_{dw} from Equation (9.49).

9.14.5 Evaluation of the methods

We may take a typical case and compare the results obtained by the methods previously presented. For the case of a conventional airfoil we may consider the methods of Perkins and Hage, DATCOM, and the Korn equation for $\kappa = 0.87$. Considering a case of zero lift and $t/c = 0.12$ we find the results shown in Figure 9.27. It is seen that the results for the Korn equation method and the DATCOM method agree quite well up to $M = 0.9$, while the Perkins and Hage method appears to underestimate the drag coefficient by a substantial amount. However, it should be noted that the DATCOM result shown in Figure 9.27 is actually for a wing with A=8.11 while the other results are for airfoils.

In the case of advanced airfoils we may compare the results of the Korn equation method with those of the Shevell method for peaky airfoils and these are shown in Figure 9.28. The Shevell method results for so-called peaky airfoils appear to be more optimistic than those for the Korn equation method. This

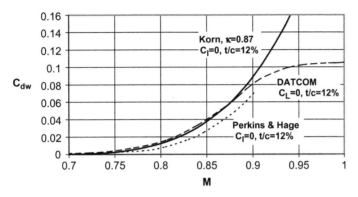

FIGURE 9.27

Comparison of two methods for C_{dw} as a function of M for a conventional NACA 6-series airfoil with $t/c=12\%$ at $C_l=0$. The DATCOM result is for a wing of $A=8.11$ and is shown for comparison.

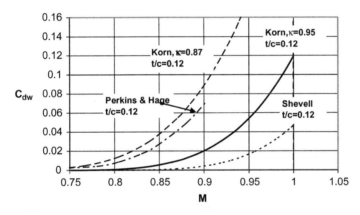

FIGURE 9.28

Comparison of wave drag coefficients at zero lift for advanced airfoils with $t/c=0.12$. The Korn equation result for conventional airfoils, $\kappa=0.87$, is shown for reference.

appears to be due to the relatively high values for $M_{cr,0}$ for the Shevell method, as shown in Figure 9.29.

For reference purposes, the Korn equation result for conventional airfoils, $\kappa=0.87$, is also shown in Figure 9.28. The Perkins and Hage method results are also shown in Figure 9.28 and because they have high values of $M_{cr,0}$ the drag results are lower than the corresponding DATCOM results. In light of these results it seems reasonable to conclude that the use of the Korn equation method is preferable to the other methods because it is simple to apply, has some theoretical backing, agrees with the DATCOM method for conventional airfoils, and provides conservative results for wings with no sweepback. On the other hand, serious consideration should be given to the use of the DATCOM method because it includes relevant wing characteristics, though it is somewhat more complicated.

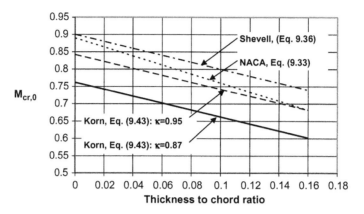

FIGURE 9.29

Comparison of the zero-lift critical Mach number as a function of thickness to chord ratio for all the methods considered.

9.15 **Effects of sweepback**

As shown in Chapter 6, the aspect ratios and thickness ratios of the horizontal and vertical tails are typically chosen to be lower than the wing values. This has the effect of raising their critical Mach numbers and delaying the compressibility drag rise compared to the wing. The fuselage, being a very slender body of high fineness ratio, is analogous to a very small aspect ratio wing and therefore has a high critical Mach number. Thus the wing is primarily responsible for the compressibility drag rise of the aircraft. The use of sweptback wings, like the use of thin airfoil sections, aids in delaying the drag rise associated with compressibility. Consider a segment of length b_y of an infinite wing whose leading edge is swept back at an angle Λ_{LE} as shown in Figure 9.30. The flow over the wing is two-dimensional and frictionless. The effects of the three-dimensionality are treated subsequently.

The pressure occurring at any point on the wing may be put in non-dimensional form by introducing the usual pressure coefficient

$$C_p = \frac{p - p_\infty}{q_\infty} = \frac{2}{\gamma M_\infty^2}\left(\frac{p}{p_\infty} - 1\right)$$

However, the pressure coefficient may also be written in terms of the normal component of the free stream Mach number as follows:

$$C_{p,n} = \frac{2}{\gamma M_n^2}\left(\frac{p}{p_\infty} - 1\right) = \frac{2}{\gamma M_\infty^2 \cos^2\Lambda}\left(\frac{p}{p_\infty} - 1\right) = \frac{C_p}{\cos^2\Lambda}$$

The non-dimensional pressure force associated with the pressure coefficient in the direction normal to the wing surface is equal to $C_p dS = C_{p,n}\cos^2\Lambda dS$. The

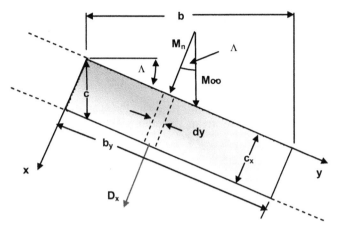

FIGURE 9.30

A segment of an infinite swept wing in a free stream flow at Mach number M.

elemental lift coefficient of the wing is equal to the integral of the difference between the lower and upper surface pressure coefficients taken over the planform area S. That is,

$$dC_L = \left(C_{p,lower} - C_{p,upper}\right) dS$$

Therefore, the lift coefficient based on the free stream Mach number $C_L = C_{L,n}\cos^2\Lambda$, where $C_{L,n}$ is the lift coefficient based on the normal component of the Mach number M_n. The elemental drag coefficient, on the other hand, depends upon the local body slope τ and is given by

$$dC_D = \left(\tau_{upper}C_{p,lower} + \tau_{lower}C_{p,upper}\right) dS$$

The body slope in the direction of the normal component of the Mach number $\tau_n = (dt/dx)_n$ so that the body slope in the free stream direction $\tau = \tau_n\cos\Lambda$. Therefore, the drag coefficient based on the free stream Mach number $C_D = C_{D,n}\cos^3\Lambda$, where $C_{D,n}$ is the drag coefficient based on the normal component of the Mach number M_n.

We may look at this development from another point of view. The flow over an infinite swept wing is equivalent to one in which the wing is moving parallel to itself at $M\sin\Lambda$ so that the drag force is dependent only on the normal component of the Mach number $M_n = M\cos\Lambda$. This force is denoted by the subscript x and is given by

$$D_x = C_{D,x}q_xc_xb_y = C_{D,x}qcb\left(\frac{q_x}{q}\right)\left(\frac{c_x}{c}\right)\left(\frac{b_y}{b}\right)$$

In practice, it is the drag force pertinent to the free stream Mach number M which is of interest and it is given by

$$D = D_x \cos \Lambda = C_{D,x} \cos \Lambda \, (qcb) \left(\frac{q_x}{q}\right) \left(\frac{c_x}{c}\right) \left(\frac{b_y}{b}\right)$$

Substituting the relations between the chord and the span in the direction normal to the leading edge and in the free stream direction yields

$$D = C_{D,x} \cos \Lambda \, (qcb) \left(\frac{q_x}{q}\right) \cos \Lambda \left(\frac{1}{\cos \Lambda_{LE}}\right)$$

Now the dynamic pressure ratio $q_x/q = (V_x/V)^2 = \cos^2 \Lambda$ and therefore the drag is

$$D = C_{D,x} qcb \left(\cos^3 \Lambda\right) \tag{9.54}$$

Since the definition of the drag in the free stream direction is $D = C_D qS$, the drag coefficient of the swept wing is $C_D = C_{D,x} \cos^3 \Lambda$. However, the swept wing has the same thickness independent of the sweep while the thickness-to-chord ratio depends on the angle of sweep according to the relation $(t/c_x) = (t/c) \cos \Lambda$. For example, a 12% thick airfoil in a wing that is swept back $30°$ has an effective thickness ratio $t/c_x = 13.86\%$ in the direction normal to the leading edge.

Thus it seems reasonable to compute the drag coefficient $C_{D,x}$ for the wing at a series of assumed Mach numbers normal to the leading edge. Then the effective drag coefficient increment in the free stream direction may be obtained by multiplying the drag coefficient by the cube of the cosine of the leading edge sweepback angle

$$C_D = C_{D,x} \cos^3 \Lambda \tag{9.55}$$

The appropriate Mach number to use would be, according to the analysis presented previously, the normal Mach number divided by the $\cos \Lambda$:

$$M = \frac{M_n}{\cos \Lambda} \tag{9.56}$$

However, Torenbeek (1982) suggests that dividing by $\cos^{1/2} \Lambda$ instead of the geometrically correct value yields more realistic answers in practical cases. Thus one would use

$$M = \frac{M_n}{\sqrt{\cos \Lambda}} \tag{9.57}$$

We will see this correction for the Mach number in the DATCOM method in Equation (9.58). We will see a similar correction to the drag coefficient in Equation (9.55) where the exponent of the cosine term is reduced from 3 to 2.5 in Equation (9.59). This correction to theory arises from the three-dimensional effects caused by the presence of the fuselage at the inboard station and the wing tips at the outboard station. Some of these effects will be discussed more fully subsequently.

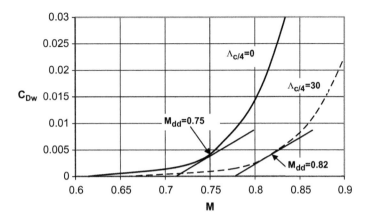

FIGURE 9.31

Comparison of the zero-lift wave drag coefficients for an unswept and a swept wing according to the DATCOM method.

9.15.1 Practical sweepback corrections for the estimation methods

The zero-lift wave drag coefficient as a function of Mach number for an unswept wing with $A=8.11$ and $t/c=0.12$ according to the DATCOM method is shown in Figure 9.26. If the quarter-chord line is now swept back $30°$ the following corrections are introduced in the DATCOM method:

$$M_{\Lambda_{c/4}} = \frac{M_{\Lambda_{c/4}=0}}{\sqrt{\cos \Lambda_{c/4}}} \tag{9.58}$$

$$C_{Dw,\Lambda_{c/4}} = C_{Dw,\Lambda_{c/4}=0} \left(\cos \Lambda_{c/4}\right)^{2.5} \tag{9.59}$$

The results for the zero-lift wave drag coefficient for $\Lambda_{c/4}=30°$ and $\Lambda_{c/4}=0$ are compared in Figure 9.31. Note that the drag divergence Mach number has increased to approximately $M_{dd}=0.82$ from 0.75 for the unswept wing. Recall that this method is based on transonic similarity methods and the corrections shown are outcomes of that approach.

Mason (1990) adapted the Korn equation method to swept wings by inserting sweepback corrections similar to those developed at the beginning of this section into Equation (9.53) to arrive at the following:

$$M_{cr} \cos \Lambda_{c/2} = \kappa - 0.108 \cos \Lambda_{c/2} - \frac{0.1C_L}{\cos^2 \Lambda_{c/2}} - \frac{(t/c)}{\cos \Lambda_{c/2}} \tag{9.60}$$

Note that Equation (9.60) is merely Equation (9.50) corrected for the effects of sweepback on lift coefficient and Mach number. The sweepback angle varies with the chord location chosen to define it. Mason (1990) uses the sweep of the half-chord line in his formulation. The difference between the leading edge sweepback angle used previously and the sweepback angle at the quarter- and half-chord lines tends to be relatively small for high aspect ratio and low taper ratio wings common to jet transports.

For the selected airfoil (κ), sweepback (Λ), and thickness ratio (t/c) for the swept wing under consideration, one may calculate M_{cr} for various values of lift coefficient (C_L) using Equation (9.60), where the Mach number and thickness to chord ratio are those in the free stream direction. Then the wave drag coefficient in the free stream direction may be obtained by using Lock's fourth-power law, Equation (9.49). We may compare the Korn equation method to the DATCOM method for calculating the zero-lift wave drag when the wing is swept back. Results for the case of a 12% thick conventional NACA 6-series airfoil with an aspect ratio of 8.11 and $\Lambda_{c/4} = 30°$ calculated with the Korn equation method using $\kappa = 0.87$ and $\Lambda_{c/4} = 30°$ (note that for this case $\Lambda_{c/2} = 27.3°$, assuming a nominal taper ratio $\lambda = 0.3$) are compared to the DATCOM results of Figure 9.31 in Figure 9.32. Once again the Korn equation method agrees favorably with the DATCOM method.

9.15.2 Other three-dimensional wing effects

We have presented some of the basic features of compressible flow over idealized swept back wings. In the earliest stages of design it is useful to develop a degree of understanding of the nature of the flow and to establish a reasonable estimate of the drag coefficient for preliminary design and comparison purposes. It should be clear that the drag rise is so sharp in the desirable $0.75 < M < 0.95$ range that small changes can have substantially larger effects so that it is difficult to predict drag with great accuracy with relatively simple methods.

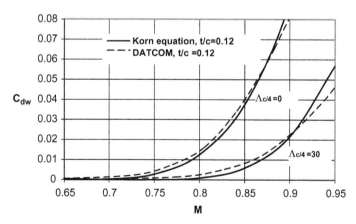

FIGURE 9.32

Comparison of the wave drag coefficient as predicted by DATCOM for a wing with A=8.11 to that for an airfoil using the Korn Equation (9.60) for 0 and 30o sweepback.

Practical wings have three-dimensional aspects in addition to sweepback that complicate the estimation of drag in the transonic flow regime, like taper ratio, thickness variation, wing twist, wing tips, fuselage junction, etc. To deal with these issues in the later design stages there are CFD codes that can be used to predict the flow over complicated wings with quite a high degree of confidence but these codes require substantial skill, experience, and time to apply effectively.

9.16 The drag coefficient of the airplane

We may compute the total drag coefficient for the airplane at various flight Mach numbers by forming the sum of the basic drag zero-lift coefficient $C_{D,0}$ as developed in Table 9.5, the wave drag coefficient C_{Dw} as obtained by one of the methods in Section 9.14.4, and the lift-induced drag coefficient $C_{D,i} = kC_L^2$:

$$C_D = C_{D,0} + C_{Dw} + kC_L^2$$

The factor k in the induced drag term is often written in the form that arises naturally in the lifting line analysis for wings, as discussed in Appendix C, that is,

$$k = \frac{1+\delta}{\pi A}$$

The induced drag coefficient is a minimum when the term $\delta = 0$ and the spanwise lift distribution is elliptic. Most well-designed wings do show a parabolic relation between lift and drag such that $C_D = C_{D,0} + kC_L^2$, particularly for low to moderate angles of attack. To account for the deviation from the ideal elliptic distribution the coefficient k was redefined as

$$k = \frac{1}{\pi e A}$$

Torenbeek (1982) points out that the quantity e was introduced by Oswald in 1932 and it has become generally known as the Oswald span efficiency factor. Since that time there have been many attempts to provide a means for calculating e as a function of wing geometry and flight conditions. Nita and Scholz (2012) present an exhaustive study of the many models for the Oswald efficiency that have been proposed and arrive at a semi-empirical result that is easy to apply and that demonstrates reasonable fidelity with the experimental data currently available. Their model for the Oswald efficiency is written as follows:

$$e = e_t e_b e_D e_M = \frac{1}{1 + Af(\lambda')} \left(e_b e_D e_M \right)$$

The first term, e_t, is a function of aspect ratio, sweepback, and taper ratio and, is called the theoretical efficiency because it has a basis in the various theories for the Oswald efficiency that have been studied over the years. The quantity $\lambda' = 1 - \Delta\lambda$ is a correction to the taper ratio λ which accounts for effects due to sweepback as follows:

$$\Delta\lambda = 0.45 \exp\left(-0.375\Lambda_{c/4}\right) - 0.357$$

The quantity f arises from a curve fit and is given by

$$f\left(\lambda'\right) = 0.0524\left(\lambda'\right)^4 - 0.15\left(\lambda'\right)^3 + 0.166\left(\lambda'\right)^2 + 0.0706\left(\lambda'\right) + 0.0119$$

The correction terms are as follows:

(a) $e_b = 1 - 2\left(\dfrac{d_b}{b}\right)^2$

(b) $e_M = 1 - 1.52 \times 10^{-4}\left(\dfrac{M}{0.3} - 1\right)^{10.82} : M > 0.3$

$e_M = 1 : M \leqslant 0.3$

(c) $e_0 = 0.873$, jet transport

$e_0 = 0.804$, turboprop transport

The correction factor e_b accounts for the effect of the fuselage diameter d_b on the spanwise lift distribution, the correction factor e_0 is a statistically determined factor which accounts for zero-lift drag effects, and the correction factor e_M accounts for the strong effect of compressibility at high subsonic Mach numbers. The results of this method are compared to experimental results compiled by Nita and Scholz (2012) in Figure 9.33.

We now have sufficient information to determine the drag coefficient of the airplane including corrections for compressibility and thereby create an array of results like that shown, purely for illustrative purposes, in Table 9.7. Using this table we may plot C_D versus M for each value of C_L. An example of such a plot is shown in

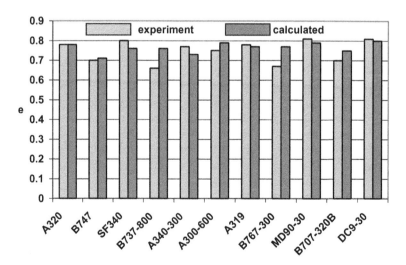

FIGURE 9.33

Comparison between experimental and calculated values of the Oswald efficiency factor e. Mach numbers are in the range $0.76 < M < 0.84$ except for the Saab SF340 turboprop and the MD90-30, both of which are evaluated at $M = 0.3$.

Table 9.7 Corrected Drag Coefficients for Each Lift Coefficient for the Selected Values of κ, (t/c), and $\Lambda_{c/2}$ and the Corresponding C_{D0}

M	κ	t/c	$\Lambda_{c/2}$	C_L	Mcr	M-Mcr	C_{Dw}	$C_{Dw}+C_{Di}+C_{D0}$[a]
0.65	0.87	0.12	27.3	0	0.719	−0.069	4.55E-04	0.0165
0.70	0.87	0.12	27.3	0	0.719	−0.019	2.64E-06	0.0160
0.75	0.87	0.12	27.3	0	0.719	0.031	1.83E-05	0.0160
0.80	0.87	0.12	27.3	0	0.719	0.081	8.58E-04	0.0169
0.85	0.87	0.12	27.3	0	0.719	0.131	5.88E-03	0.0219
0.90	0.87	0.12	27.3	0	0.719	0.181	2.14E-02	0.0374
0.95	0.87	0.12	27.3	0	0.719	0.231	5.69E-02	0.0729
0.65	0.87	0.12	27.3	0.1	0.705	−0.055	1.81E-04	0.0166
0.70	0.87	0.12	27.3	0.1	0.705	−0.005	1.08E-08	0.0165
0.75	0.87	0.12	27.3	0.1	0.705	0.045	8.33E-05	0.0165
0.80	0.87	0.12	27.3	0.1	0.705	0.095	1.64E-03	0.0181
0.85	0.87	0.12	27.3	0.1	0.705	0.145	8.89E-03	0.0253
0.90	0.87	0.12	27.3	0.1	0.705	0.195	2.90E-02	0.0455
0.95	0.87	0.12	27.3	0.1	0.705	0.245	7.23E-02	0.0887
0.65	0.87	0.12	27.3	0.2	0.691	−0.041	5.42E-05	0.0179
0.70	0.87	0.12	27.3	0.2	0.691	0.009	1.58E-07	0.0178
0.75	0.87	0.12	27.3	0.2	0.691	0.059	2.50E-04	0.0181
0.80	0.87	0.12	27.3	0.2	0.691	0.109	2.87E-03	0.0207
0.85	0.87	0.12	27.3	0.2	0.691	0.159	1.29E-02	0.0308
0.90	0.87	0.12	27.3	0.2	0.691	0.209	3.85E-02	0.0563
0.95	0.87	0.12	27.3	0.2	0.691	0.259	9.06E-02	0.1084
0.65	0.87	0.12	27.3	0.3	0.676	−0.026	9.59E-06	0.0202
0.70	0.87	0.12	27.3	0.3	0.676	0.024	6.29E-06	0.0202
0.75	0.87	0.12	27.3	0.3	0.676	0.074	5.90E-04	0.0207
0.80	0.87	0.12	27.3	0.3	0.676	0.124	4.68E-03	0.0248
0.85	0.87	0.12	27.3	0.3	0.676	0.174	1.82E-02	0.0384
0.90	0.87	0.12	27.3	0.3	0.676	0.224	5.01E-02	0.0702
0.95	0.87	0.12	27.3	0.3	0.676	0.274	1.12E-01	0.1324
0.65	0.87	0.12	27.3	0.4	0.662	−0.012	4.24E-07	0.0234
0.70	0.87	0.12	27.3	0.4	0.662	0.038	4.14E-05	0.0234
0.75	0.87	0.12	27.3	0.4	0.662	0.088	1.20E-03	0.0246
0.80	0.87	0.12	27.3	0.4	0.662	0.138	7.24E-03	0.0306
0.85	0.87	0.12	27.3	0.4	0.662	0.188	2.49E-02	0.0483

Table 9.7 Continued

M	κ	t/c	$\Lambda_{c/2}$	C_L	Mcr	M-Mcr	C_{Dw}	$C_{Dw}+C_{Di}+C_{D0}$[a]
0.90	0.87	0.12	27.3	0.4	0.662	0.238	6.41E-02	0.0875
0.95	0.87	0.12	27.3	0.4	0.662	0.288	1.37E-01	0.1609

[a] C_{D0} is to be taken from the results of Table 9.5 (here an arbitrary value of 0.016 has been used merely as an example) and $C_{Di}=C_L{}^2/\pi eA$, $A=8$, and $e=0.85$.

Figure 9.34. The low-speed values of C_D correspond to the incompressible values calculated previously for the cruise configuration ($C_D=C_{D,0}+kC_L{}^2$) and are independent of Mach number until the drag rise starts at higher subsonic speeds. A cross-plot of the data shown in the form of C_L vs. C_D for each Mach number provides the drag polar plots for the aircraft. The curves in Figure 9.34 are used subsequently to determine the thrust required for the performance analysis, i.e., $F_{req}=D=C_D qS$. The thrust required may be simply calculated using spreadsheet tables of the form shown in Table 9.8, which is divided into two Mach number ranges, $0.25<M<0.6$ and $0.65<M<0.85$. The first range is simplified since compressibility effects can be ignored. Note also that the lower limit on the Mach number ($M=0.25$) is a function of altitude, because the lift coefficient will exceed C_{Lmax} at low Mach numbers and high altitudes.

9.17 Thrust available and thrust required

To aid in the determination of maximum and minimum airspeeds at the cruise altitude, graphs of the required thrust, F_{req}, and the available thrust, F_{av}, versus Mach number should be constructed, like that shown in Figure 9.35. The required thrust for cruising flight at a given altitude is obtained from column 5 of Table 9.8.

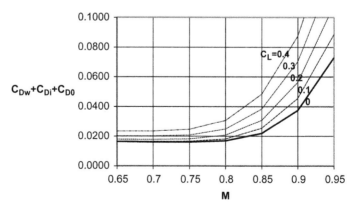

FIGURE 9.34

Sum of wave drag, induced drag, and profile drag coefficients versus Mach number for various values of C_L as taken from Table 9.7.

Table 9.8 Thrust Required

Aircraft designation: Smith 2000-150
Takeoff weight: 150,000 lbs
Aspect ratio $A = 8$
Wing Area S: 1200 ft^2
Altitude: 35,000 ft
$p/p_{sl} = \delta = 0.2351$
$C_{D,0}$ (incompressible): 0.0197
$e = 0.85$

	M	$q = \frac{1}{2}\gamma p_{sl}\delta M^2$	$C_L = (W/S)/q$	C_D (see notes in the first column)	$F_{req} = C_D qS$
$0.25 \leq M \leq 0.6$ (incompressible) $C_D = C_{D,0} + C_L^2/\pi e A$					
	0.25				
	0.30				
	0.35				
	0.40				
	0.45				
	0.50				
	0.55				
	0.60				
$0.6 < M < 0.85$ (compressible) C_D from graph of C_D vs. M for various C_L. Use finer increments in M					
	0.65				
	0.70				
	0.775				
	0.800				
	0.825				
	0.850				

The variation of available cruise thrust with altitude z is often taken as being proportional to the atmospheric density ratio, i.e., $F(z) = \sigma F_{to}$. A review of data presented by Svoboda (2000) suggests that a better approximation for the cruise thrust is $F(z) = \delta F_{to}$, where δ is the atmospheric pressure ratio. Of course if the actual value of thrust available at altitude is known from engine manufacturer data it should be used in the calculations. The thrust of a turbojet is essentially constant with Mach number

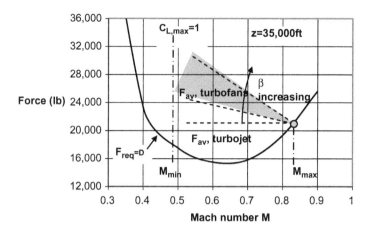

FIGURE 9.35

The variation of thrust available and thrust required with Mach number for a hypothetical jet transport at 35,000 ft altitude. Available thrust from turbofans depends on bypass ratio β.

and turbofans show a drop-off in thrust with Mach number which is appreciable as the bypass ratio increases. Using the cruise thrust available permits one to find the maximum Mach number at a given altitude, as shown in Figure 9.35. The turbofans would show a somewhat higher thrust below the cruise speed, as indicated.

The example shown in Figure 9.35 is a hypothetical, but typical, case of an aircraft with the following characteristics: $W=250{,}000$ lb, $C_{D,0}=0.02$, $S=2500$ ft^2, $A=8$, $e=0.85$, $F_{to}=88{,}000$ lb (two engines each with 44,000 lb of thrust at takeoff). The aircraft is assumed to be flying at an altitude where $\delta=0.24$, about 35,000 ft. Note that the minimum possible Mach number is not determined by the drag (i.e., thrust required) but instead by the maximum lift coefficient achievable. The maximum altitude may also be determined by considering such graphs for higher and higher altitudes. When the maximum and minimum Mach numbers are equal, i.e., when the curve of F_{req} is tangent to that of F_{avail}, one is at the maximum sustainable altitude, and is called the absolute ceiling. The service ceiling, on the other hand, is that altitude at which there is sufficient excess thrust to permit a rate of climb of 100 ft/min. The hypothetical aircraft considered in Figure 9.35 may be analyzed for an altitude of 40,000 ft with the results shown in Figure 9.36. Note that there is very little excess thrust available so that 40,000 ft is close to the maximum ceiling for the aircraft.

It is important to note that most textbook treatments of drag effects on range performance tacitly assume that $C_{D,0}$ is a constant, whereas above $M=0.6$ it must be recognized that $C_{D,0}=C_{D,0}(M)$. A detailed treatment of compressibility effects on cruise performance has been presented by Torenbeek (1997). Therefore one must be careful in using equations developed on the basis of constant zero-lift drag coefficient.

In addition, there is generally a drag penalty associated with trimming the aircraft at any given flight condition. A basic example of this was explained in Section 6.2.1

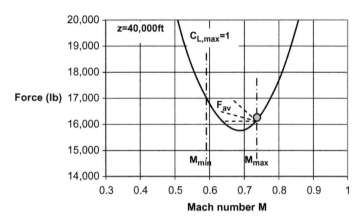

FIGURE 9.36

The variation of thrust available and thrust required with Mach number for a hypothetical jet transport at 40,000 ft altitude.

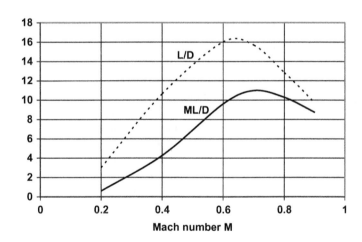

FIGURE 9.37

Variation of the lift to drag ratio L/D and the range factor ML/D with Mach number.

where it was shown that the wing may have to produce more lift than that needed to overcome the weight of the aircraft alone because negative lift on the horizontal tail surfaces may be required to keep the pitching moment equal to zero. We may approximate the trimmed lift coefficient of the wing as follows:

$$C_{L,w} \approx \frac{W}{qS} - \left(C_{m0} + C_{m,f}\right) \frac{c_{MAC}}{l_h}$$

One may also use the results of Table 9.8 to generate a graph of $M(L/D)$ as a function of M. The maximum value of the range factor $M(L/D)$, rather than the maximum value of L/D alone, will determine the Mach number for maximum range. This issue arises in the calculation of the velocity for best range in the performance chapter of the design report. Note that at this point in the design, the drag polar (C_L vs. C_D) for each Mach number and altitude has been calculated so there is no need to resort to the approximate equations often quoted in textbooks on performance. The Mach number variation of the lift to drag ratio L/D and of the range factor ML/D is illustrated in Figure 9.37 for the hypothetical aircraft discussed previously. It is clear that the Mach number for maximum range is different from the Mach number for maximum lift to drag ratio. A detailed review of the current capabilities of CFD analyses for determining the high subsonic flow characteristics of commercial aircraft is presented in Johnson et al. (2005).

9.18 Design summary

The geometric and weight configuration of the aircraft was completed in the previous chapter. There is now sufficient detailed information to permit calculation of the drag of the airplane by the drag buildup method described in this chapter. The component drag analysis should be described in detail and should include the effects of both Reynolds number and Mach number in order to make a practical estimate suitable for assessing the performance potential of the aircraft which will be discussed in the next chapter. The results for this chapter should be summarized in a form like that of Table 9.8 which includes both the low-speed and high-speed lift and drag performance of the design aircraft. A number of tables of this sort should be prepared so as to provide sufficient information on lift, drag, and thrust for use in the performance analyses to be carried out in the next chapter. Such results are needed at least for takeoff, landing, and cruise. Plots of available and required thrust as a function of Mach number at different altitudes should also be presented and the service determined. Similarly, plots of L/D and ML/D as a function of Mach number for different altitudes should also be presented.

9.19 Nomenclature

A	aspect ratio
A_o	frontal area
a	local sound speed or coefficient in Equation (9.29)
b	wingspan or coefficient in Equation (9.29)
c	chord length
C_D	drag coefficient
C_{Dw}	wave drag coefficient
c_d	airfoil section drag coefficient
c_f	local skin friction coefficient

C_F	average, or integrated, skin friction coefficient
C_L	lift coefficient
c_l	airfoil lift coefficient
C_m0	wing pitching moment coefficient
$C_{m,f}$	fuselage pitching moment coefficient
C_p	pressure coefficient
D	drag
d	diameter
e	span efficiency factor
F	fineness ratio
F_{av}	thrust available
F_{req}	thrust required
f	equivalent or parasite drag area
h	roughness height
K_f	fuselage form factor
K_n	nacelle form factor
K_w	wing form factor
k	roughness height or constant in Equation (9.41)
k'	constant in Equation (9.42)
L	lift
l	length
l_h	horizontal tail moment arm
M	Mach number
M_{cr}	critical Mach number
p	pressure or perimeter
q	dynamic pressure
Re	Reynolds number
R_{LS}	lifting surface factor
S	wing planform area
S_{wf}	wing planform area affected by flaps
S_{wet}	wetted area
t	thickness
V	velocity
W	weight
x	chordwise distance
y	spanwise distance from centerline
z	altitude
α	angle of attack
β	Prandtl-Glauert factor $(1-M^2)^{1/2}$ or bypass ratio of a turbofan
δ	atmospheric pressure ratio or boundary layer thickness
δ_f	flap deflection
ϵ	body thickness ratio or slope
η	non-dimensional quantity $y/(b/2)$
ϕ_1	general coefficient for wing drag

ϕ_2	shape factor for wing pressure drag
γ	ratio of specific heats
κ	transonic factor, Equation (9.41)
Λ	wing sweepback angle
λ	wing taper ratio
μ	viscosity
ρ	density
σ	atmospheric density ratio
τ	wall shear stress

9.19.1 Subscripts

b	body or fuselage
c	component
$c/4$	quarter-chord
$c/2$	mid-chord
$c\text{-}o$	cut-off
cr	critical
dd	drag divergence
f	friction or flap
i	induced or index of an element in a series
in	inlet
inc	incompressible
LG	landing gear
l	characteristic length
lam	laminar
MAC	mean aerodynamic chord
max	maximum
min	minimum
n	nacelle or normal component
p	pressure
r	wing root
sl	sea level
t	wing tip or transition point
to	takeoff
$turb$	turbulent
w	wing
x	direction along chord
y	direction normal to chord
0	zero-lift
∞	free stream

References

Abbott, I.H. et al., 1945. Summary of Airfoil Data. NACA Technical Report No. 824, 1945. <http://ntrs.nasa.gov/archive/nasa/casi.ntrs.nasa.gov/19930090976_1993090976.pdf>.

Abbott, I.H., VonDoenhoff, A.E., 1959. Theory of Wing Sections. Dover, NY.

Engineering Sciences Data Unit, 1987. Undercarriage Drag Prediction Methods. ESDU 79015.

Goldstein, R.J. (Ed.), 1996. Fluid Mechanics Measurements, Second Edition. Taylor and Francis, Washington, DC.

Gur, O., Mason, W.H., Schetz, J.A., 2010. Full-configuration drag estimation. Journal of Aircraft 47 (4), 1356–1367.

Hilton, H.W., 1951. High Speed Aerodynamics. Longmans, Green, London.

Hoak, D.E. et al., 1978. USAF Stability and Control DATCOM. Flight Control Division, Air Force Flight Dynamics Laboratory, Wright-Patterson Air Force Base, Ohio 45433.

Hoerner, S.F., 1958. Fluid Dynamic Drag. Published by the author.

Hopkins, E.J., 1972. Charts for Predicting Turbulent Skin Friction from the van Driest II Method. NASA TN D-6945.

Inger, G.R., 1993. Application of Oswatitsch's theorem to supercritical airfoil drag calculation. Journal of Aircraft 30 (3), 415–416.

Jenkinson, L.R., Simpkin, P., Rhodes, D., 1999. Civil Jet Aircraft Design. AIAA, Reston, VA.

Johnson, F.T., Tinoco, E.N., Jong Yu, N., 2005. Thirty years of development and application of CFD at Boeing Commercial Airplanes, Seattle. Computers and Fluids 34, 1115–1151.

Jones, R.T., 1953. Theory of Wing-Body Drag at Supersonic speeds. NACA RM A53H18a.

Liepmann, H., Roshko, A., 1957. Elements of Gas Dynamics. Wiley, New York.

Mair, W.A., Birdsall, D.L., 1987. Aircraft Performance. Cambridge University Press, NY.

Mason, W.H., 1990. Analytic Models for Technology Integration in Aircraft Design. AIAA/AHS/ASEE Aircraft Design, Systems and Operations Conference Paper 90-3263, Dayton, OH.

McCormick, B.H., 1995. Aerodynamics, Aeronautics, and Flight Mechanics. Wiley.

McDevitt, J.B., 1955. A Correlation by Means of Transonic Similarity Rules of Experimentally Determined Characteristics of a Series of Symmetrical and Cambered Wings of Rectangular Planform. NACA TR 1253.

Morrison, W.D., 1976. Advanced Airfoil Design: Empirically Based Transonic Aircraft Drag Buildup Technique. NASA CR-137928.

NAA, 1952. External Drag Evaluation. North American Aviation, Inc. Report ADL-52-2.

Nicolai, L.M., 1975. Fundamentals of Aircraft Design. University of Dayton Press, Dayton, Ohio.

Nita, M., Scholz, D., 2012. Estimating the Oswald Factor from Basic Aircraft Geometrical Parameters. Document ID:281424, Deutscher Luft- und Raumfahrtkongress.

Perkins, C.D., Hage, R.E., 1949. Airplane Performance Stability and Control. Wiley, NY.

Raymer, D.P., 1989. Aircraft Design: A Conceptual Approach. American Institute of Aeronautics and Astronautics, Washington, DC.

Roskam, Jan, 1971. Method for Estimating Drag Polars of Subsonic Airplanes, Roskam Aviation and Engineering Corp. Lawrence, Kansas, 66044.

Schlichting, H., 1979. Boundary Layer Theory. McGraw-Hill, NY.

Sears, W.R., 1954. Small Perturbation Theory. In: General Theory of High Speed Aerodynamics, vol. 6. Princeton University Press, Princeton, New Jersey.

Shevell, R., 1989. Fundamentals of Flight. Prentice-Hall, Englewood Cliffs, NJ.

Svoboda, C., 2000. Turbofan engine database as a preliminary design tool. Aircraft Design 3, 17–31.

Torenbeek, E., 1982. Synthesis of Subsonic Airplane Design. Kluwer Academic Publishers, The Netherlands.

Torenbeek, E., 1997. Cruise performance and range prediction reconsidered. Progress in the Aerospace Sciences 33, 285–321.

Whitcomb, R.J., 1952. A Study of the Zero-Lift Drag-Rise Characteristics of Wing-Body Combinations near the Speed of Sound. NACA RM L52H08.

White, F.M., 2006. Viscous Fluid Flow, third ed. McGraw-Hill, NY.

Aircraft Performance

Commercial Airplane Design Principles. http://dx.doi.org/10.1016/B978-0-12-419953-8.00010-3
© 2014 Elsevier Inc. All rights reserved.

10.1 **The range equation**

In the cruise configuration we can estimate airplane performance characteristics assuming steady and approximately level flight in the x-z plane, where x is the local horizontal direction and z is the local vertical direction as shown in Figure 10.1. Treating the airplane as a point mass the equilibrium equations along and normal to the flight path are given as follows:

$$F_n = D + W \sin \gamma$$

$$L = W \cos \gamma$$

The angle γ is the flight path angle, that is, the angle between the flight trajectory and the local horizontal, or $\tan \gamma = dz/dx$.

The weight of the aircraft is readily described in terms of the fixed weight (the operating empty weight plus the payload weight), $W_{OE} + W_{PL}$, and the time-varying weight (the fuel weight), W_F, as follows:

$$W = W_{oe} + W_{pl} + W_f(t) \tag{10.1}$$

Here we focus on turbofan-powered aircraft; turboprop-powered aircraft are treated separately in a subsequent section. For turbofan engines the rate at which fuel is consumed in quasi-steady operation is assumed to be linearly related to the net thrust produced so that the rate of change of the weight of the aircraft is

$$\frac{dW}{dt} = \frac{dW_f}{dt} = -C_j F_n \tag{10.2}$$

The quantity C_j is the specific fuel consumption in pounds of fuel consumed per hour per pound of thrust produced and F_n is the total net thrust of the engines. This

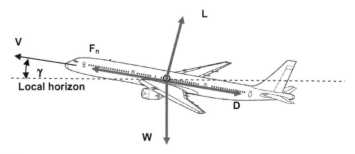

FIGURE 10.1

Forces on the aircraft in quasi-equilibrium flight.

equation may also be written as

$$\frac{dW}{dt} = -\frac{F_n}{I} \tag{10.3}$$

The quantity I is the specific impulse, another measure of fuel efficiency, and it is measured in pounds of thrust produced per pound of fuel consumed per second. Thus, the relationship between specific fuel consumption is

$$I = \frac{3600}{C_j} \tag{10.4}$$

The rate of fuel consumption may now be described as follows:

$$\frac{dW_f}{dt} = -C_j F_n = -\frac{F_n}{I} = -C_j(D + W \sin \gamma)$$

Therefore

$$dt = -\frac{W}{C_j(D + W \sin \gamma)} \frac{dW_f}{W} = -\frac{1}{C_j \left(\frac{D}{W} + \sin \gamma\right)} \frac{dW}{W} \tag{10.5}$$

The ratio D/W may be written as

$$\frac{D}{W} = \frac{D \cos \gamma}{L} = \cos \gamma \left(\frac{L}{D}\right)^{-1}$$

Here L/D is the lift to drag ratio, or aerodynamic efficiency. We now must make some assumption as to the flight path. In this quasi-steady approximation it is taken for granted that the flight path angle $\gamma \ll 1$, so that

$$\sin \gamma = \gamma + \gamma^3/3 + \cdots \simeq \gamma \text{ and } \cos \gamma = 1 - \gamma^2/2 + \cdots \simeq 1$$

Therefore the lift and weight are always in balance, i.e., $L = W$, to $O(\gamma^2)$. Since the airplane's weight is continually decreasing as fuel is burned, the airplane will slowly rise during the course of the flight. This rise will be shown to be consistent with the assumption that $\gamma \ll 1$.

Furthermore, the horizontal speed $V = dx/dt$, and the flight Mach number is $M = V/a$ so that the equation for the time differential may be rearranged as follows:

$$\frac{dW}{W} = -\frac{C_j}{aM} \left(\frac{1}{L/D} + \gamma\right) dx$$

The range achievable for turbofan aircraft is given by

$$R = \int_0^R dx = \int_{W_4}^{W_5} -\frac{aM(L/D)}{C_j(1 + \gamma L/D)} \frac{dW}{W} \tag{10.6}$$

In our notation for the mission phases, the quantities W_4 and W_5 denote the aircraft weight at the start and end of the cruise segment, respectively. The simplest solution to this equation is achieved for flight where the quantity multiplying dW/W in the integrand in Equation (10.6) is constant. Flight at constant Mach number provides constant ML/D. Jet aircraft have made cruising above the weather, that is, above the troposphere, a common occurrence. Flight in this region of the tropopause and the start of the stratosphere (roughly 30,000 ft $< z <$ 40,000 ft) is characterized by nearly constant temperature (412R $> T >$ 393R) so that the speed of sound, a, is also essentially constant (995 fps $> a >$ 968 fps). Furthermore, the specific fuel consumption C_j for jet engines in this flight region is also roughly constant (at about 0.5–0.6 lbs fuel/lb thrust/hr for modern turbofan engines at cruise Mach numbers between 0.75 and 0.85) and γ, the flight path angle, is, as will be shown subsequently, very small (on the order of 10^{-4}) and therefore essentially constant over the cruise range. Using a consistent set of units we find, for L/D set to a reasonable value of approximately 15, that the term $\gamma(L/D)$ is about 10^{-3} and therefore may be neglected with respect to unity in the bracketed term in the denominator of the integrand of the range equation. Additional detailed discussion of range prediction is presented by Torenbeek (1997).

10.1.1 Factors influencing range

Then the range in nautical miles, with a, the sound speed in mph (say 581 kts at typical cruise altitudes), is given by

$$R = \frac{aM(L/D)}{C_j}\left(\ln \frac{W_4}{W_5}\right)$$

This is one form of Brequet's range equation. Because $W_5 = W_4 - W_{f,cr}$, where $W_{f,cr}$ is the weight of fuel consumed during the cruise, the range may also be written in the alternative form

$$R = -\frac{aM(L/D)}{C_j}\ln\left(1 - \frac{W_{f,cr}}{W_4}\right) \tag{10.7}$$

This suggests what may appear obvious: To get long range carry more fuel! However this is not always a good design solution. If we assume that $W_{f,cr}$ is essentially the total fuel weight W_F, then the range Equation (10.7) becomes:

$$R = \frac{aM(L/D)}{C_j}\ln\left(1 + \frac{W_f}{W_{oe} + W_{pl}}\right) \approx \frac{aM(L/D)}{C_j}$$
$$\times \left(\frac{W_f}{W_{oe} + W_{pl}} - \frac{W_f^2}{2(W_{oe} + W_{pl})^2} + \cdots\right)$$

Note that for small values of $\frac{W_f}{W_{oe} + W_{pl}}$, changes in that parameter yield linear changes in range while for larger values this advantage drops off. Certainly it is not desirable in general to increase range by decreasing the payload. And, at some point, increases in W_f or reductions in W_{oe} are less effective in increasing range than are increases in L/D or reductions in C_j. From another perspective, this illustrates that a point is reached where sacrifices in payload and/or structural weight are not very efficient in producing range. Increases in L/D arise from reducing drag, particularly induced, or lift-dependent drag, as discussed by Kroo (2001).

As pointed out previously, the reduction in weight of the airplane as it burns fuel causes it to slowly rise. Solving the range equation for W_5/W_4 yields

$$\frac{W_5}{W_4} = \exp\left(-\frac{RC_j}{aM(L/D)}\right)$$

For our assumption of flight at constant M and L/D, the lift coefficient C_L also remains constant and the form of the range equation shown immediately above becomes

$$\frac{W_5}{W_4} = \frac{C_{L,5}q_5 S}{C_{L,4}q_4 S} = \frac{C_{L,5}p_5 M_5^2}{C_{L,4}p_4 M_4^2} = \frac{p_5}{p_4} = \exp\left(-\frac{RC_j}{aM(L/D)}\right)$$

The atmospheric pressure may be approximated by an exponential function (see Appendix B) so that

$$\frac{p_{sl}\exp(-z_5/H)}{p_{sl}\exp(-z_4/H)} = \exp\left(-\frac{z_5-z_4}{H}\right) = \exp\left(-\frac{RC_j}{aM(L/D)}\right)$$

This gives then for the flight path angle

$$\tan\gamma \approx \gamma = \frac{z_5-z_4}{R} = \frac{C_j H}{aM(L/D)} \tag{10.8}$$

For typical cruise values of the parameters in the previous equation, the number of feet rise in altitude is approximately equal to the number of miles covered in range, that is, Δz in feet $\sim R$ in nautical miles.

10.2 Takeoff performance

The takeoff may be considered in two parts: a ground run and an air run, as shown schematically in Figure 10.2. The simplest description of the takeoff process is that the engine thrust is increased to the takeoff level at $x=0$ and the brakes released to begin acceleration down the runway. At some point the pilot commands rotation of the aircraft, lifting the nose wheel from the ground and achieving the takeoff angle of attack. The aircraft lifts completely from the ground and begins climbing. The point at which the aircraft reaches an altitude of 35 ft (10.7 m) is considered the end of the takeoff run. This is the usual situation for takeoff and its ground and air segments are considered next. Subsequently, the modifications to safely deal with a takeoff emergency, such as an engine failure, will be discussed.

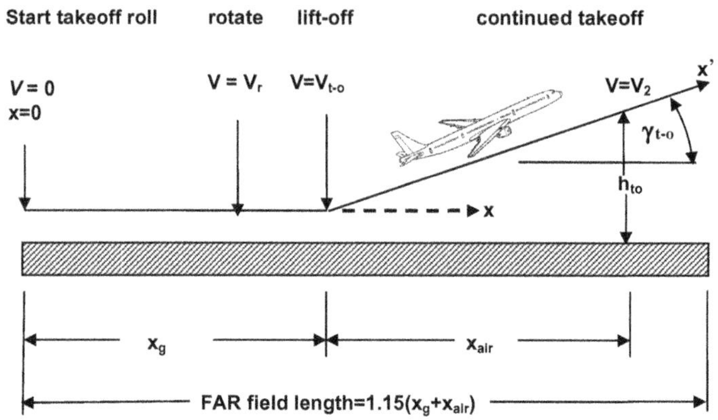

FIGURE 10.2

Takeoff procedure showing ground run, air run, and FAR field length with all engines operating.

10.2.1 Ground run

During the ground run we assume that all forces contributing to the acceleration of the aircraft are parallel to the ground resulting in the following equation of motion

$$m\frac{dV}{dt} = mV\frac{dV}{dx} = (F_n - D - D_{roll}) \tag{10.9}$$

The weight in this equation is the takeoff weight W_3 which can be obtained using Table 2.1 as follows:

$$W = W_3 = \frac{W_1}{W_0}\frac{W_2}{W_1}\frac{W_3}{W_2}W_0$$

The weight ratios in Table 2.1 yield $W = 0.975W_0$, or the takeoff weight is about 2.5% less than the maximum gross weight. The net thrust for an engine is the gross thrust minus the ram drag

$$F_n = F_g - F_r = F_g - \rho A_\infty V^2$$

For the case of takeoff the gross thrust F_g is the static thrust of the engine, the thrust value usually quoted in the manufacturer's specifications. It is often referred to in the literature as the takeoff thrust. However, it is the net thrust which provides the accelerating force for the aircraft and it drops off as speed increases because of the ram drag effect. The ram drag term F_r represents the loss in momentum due to taking air into the engine at the flight speed V through inlets which capture streamtubes of total area A_∞. However, this streamtube area captured by the inlets is a function of flight speed $A_\infty = A_\infty(V)$ and varies considerably at low speeds common to takeoff operations so that using this form for the ram drag becomes cumbersome. The net thrust may instead be written as

$$F_n = F_g - F_r = F_g - \dot{m}V$$

Under optimum conditions for thrust production, that is, where the fan and jet nozzle exit pressures are matched to the local ambient pressure, the gross thrust may be written as

$$F_g = \dot{m}_{core}V_{e,\,jet} + \dot{m}_{fan}V_{e,\,fan} = \dot{m}_{core}V_{e,\,jet}\left(1 + \beta\frac{V_{e,\,fan}}{V_{e,\,jet}}\right) \qquad (10.10)$$

The bypass ratio β measures the mass flow $\dot{m}_{fan}$ which bypasses the central "hot" core of the engine is accelerated mechanically by the fan to the fan exit velocity $V_{e,fan}$ as given below

$$\beta = \frac{\dot{m}_{fan}}{\dot{m}_{core}}$$

The core mass flow $\dot{m}_{core}$ passes through the compressor and is heated by fuel burned in the combustor and then passes through a nozzle which accelerates the core flow to the jet exit velocity $V_{e,jet}$. Noting that $\dot{m} = \dot{m}_{core}(1 + \beta)$ we may rewrite Equation (10.10) as follows:

$$\frac{F_n}{F_g} = 1 - \frac{V}{V_{e,\,jet}}\frac{(1+\beta)}{\left(1+\beta\frac{V_{e,\,fan}}{V_{e,\,jet}}\right)} \qquad (10.11)$$

We may replace the velocities by the corresponding Mach numbers, noting that $M = V\sqrt{\gamma RT}$. Assuming that $\gamma = \gamma_{fan} = 1.4$ and $\gamma_{jet} = 1.33$ we may rewrite Equation (10.11) as

$$\frac{F_n}{F_g} = 1 - \frac{M}{M_{e,\,jet}}\sqrt{\frac{\gamma_0 T_0}{\gamma_{jet}T_{e,\,jet}}}(1+\beta)\left(1 + \beta\frac{M_{e,\,fan}}{M_{e,\,jet}}\sqrt{\frac{\gamma_{fan}T_{e,\,fan}}{\gamma_{jet}T_{e,\,jet}}}\right)^{-1} \qquad (10.12)$$

Here T_0 and γ_0 represent the atmospheric ambient temperature and ratio of specific heats, respectively. It is common to operate commercial turbofan engines with the fan and nozzle exhaust operating nearly choked, and it is reasonable to set $M_{e,jet} = M_{e,fan}$. In addition, using the methods in Sforza (2012) we see that while the fan adds a good deal of mechanical energy to the flow it adds little thermal energy so that we may assume that $T_0 = T_{e,fan}$. Similarly, we may find that a representative stagnation temperature in the exhaust flow of a commercial fan jet engine is around $T_{t,jet} = 475\,C$ (890 F) so that, for a choked nozzle $T_{e,jet} = [2/(\gamma_{jet} + 1)]T_{t,jet} \sim 370\,C$ (890 F) the following approximations may be made:

$$\sqrt{\frac{1.4T_0}{1.33T_{e,\,jet}}} \approx \sqrt{\frac{1.4T_{e,\,fan}}{1.33T_{e,\,jet}}} \approx \frac{2}{3}$$

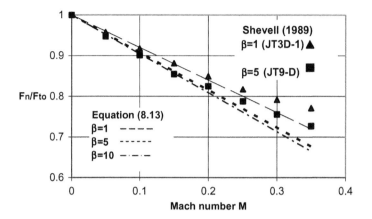

FIGURE 10.3

Variation of the ratio of net thrust to static thrust as a function of Mach number for turbofans with bypass ratio β. Data from Shevell (1989).

Then Equation (10.12) becomes

$$\frac{F_n}{F_g} = 1 - 2M\frac{1+\beta}{3+2\beta} \tag{10.13}$$

The variation of the net to gross thrust ratio of turbofan engines with flight Mach number during the takeoff run is illustrated in Figure 10.3. Here the gross thrust is denoted as the static, or takeoff, thrust F_{to}. It should be noted that the greater the bypass ratio the greater the decrease in the net thrust as a function of Mach number. The approach represented by Equation (10.13) gives good results up to about $M=0.25$, which is at the upper end of the range of takeoff speeds for commercial jet transports. The exhaust jet velocity actually slowly increases as the speed rises so one may enhance the accuracy of Equation (10.13) by adding a term of $O(M^2)$, but the improvement in the takeoff speed range is less than 2%. If one has engine data for the actual thrust variation with Mach number it should be used to calculate the average thrust, unless a numerical solution to the full equation is to be sought. In that case the appropriately fitted engine data should be used.

When $\beta=0$ we have the special case of the turbojet engine. Though pure turbojets are no longer used for commercial jet transports because of the superior fuel economy of turbofans, they do have a place in the history of commercial airliners and for special applications, such as small engines for unmanned aerial vehicles. The thrust behavior for a turbojet is illustrated in Figure 10.4. Note that the net thrust drops off more slowly for turbojets compared to that for turbofans. Once again, the net thrust will level out and begin to increase slightly as the Mach number increases beyond $M=0.3$. The military power case in Figure 10.3 illustrates the increased

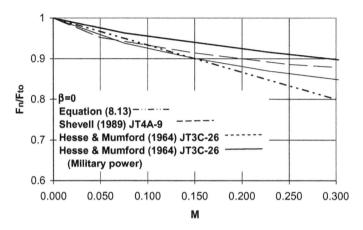

FIGURE 10.4

Variation of ratio of ratio of net thrust to static thrust as a function of Mach number for turbojets. Data from Shevell (1989) and Hesse and Mumford (1964).

performance arising from increasing the jet temperature. However, this increase comes at the expense of fuel economy and engine lifetime.

The aerodynamic drag of the aircraft as it accelerates down the runway is given by

$$D = C_D \, qS \tag{10.14}$$

The drag due to rolling resistance is given by

$$D_{roll} = \mu(W - L) = \mu(W - C_L qS)$$

The aerodynamic lift is given by

$$L = C_L \, qS$$

The difference between the lift and the weight is the normal force on the runway and μ represents the coefficient of rolling friction. Note that the drag coefficient C_D and the lift coefficient C_L are evaluated at the takeoff condition, that is, with high-lift devices, like flaps and slats, deployed, landing gear down, and ground effects taken into account.

The equation of motion, Equation (10.9), for the takeoff run may now be written as follows:

$$\frac{d}{dx}\left(\frac{1}{2}V^2\right) = \frac{g}{W_{to}}\left[F_g\left(1 - 2\frac{V}{a}\frac{1+\beta}{3+2\beta}\right) - \frac{1}{2}\rho V^2(C_D S - \mu C_L S) - \mu W\right]_{to} \tag{10.15}$$

This is a nonlinear first order ordinary differential equation for the specific kinetic energy of the aircraft during the ground run. An analytic solution to this equation may be found although it is quite complex in form. Alternatively, the equation may be solved numerically. However, instead of using the instantaneous net thrust, which is a linear function of velocity, let us consider an average value for the net thrust, the first term in brackets in Equation (10.15), during the ground run, as determined by

$$F_{avg,to} = \frac{F_{to}}{M_{to}} \int_0^{M_{to}} \left(1 - M\frac{1+\beta}{2+\beta}\right) dM = F_{to}\left[1 - \frac{1}{2}M_{to}\frac{1+\beta}{2+\beta}\right] \quad (10.16)$$

The takeoff Mach number $M_{to} = V_{to}/a_0$ where $a_0 = (\gamma RT_0)^{1/2}$ is the local speed of sound at the takeoff site and T_0 is the local atmospheric temperature. Note that the gross thrust in takeoff is the so-called takeoff, or static, thrust F_{to}. Inserting this definition for average net thrust during takeoff into Equation (10.15) transforms it into a linear first order ordinary differential equation in V^2 which may be rearranged as follows:

$$\frac{d}{dx}\left(\frac{1}{2}V^2\right) + c_1\left(\frac{1}{2}V^2\right) - c_2 = 0$$

We are assuming here that the weight of the aircraft and the aerodynamic and friction coefficients change a negligible amount as the velocity increases during takeoff roll and climb-out. The initial condition is that $V=0$ at $x=0$ so that the solution for the takeoff velocity V_{to} at the end of the ground run $x=x_g$ is

$$\frac{1}{2}V_{to}^2 = \frac{c_2}{c_1}[1 - \exp(-c_1 x_g)] \quad (10.17)$$

Here the coefficients of the solution are

$$c_1 = \frac{\rho g S}{W_{to}}(C_D - \mu C_L)_{to}$$

$$c_2 = g\left[\left(\frac{F_{avg}}{W}\right)_{to} - \mu\right]$$

Solving Equation (10.17) for the ground run yields

$$x_g = -\frac{1}{c_1}\ln\left[1 - \frac{c_1}{c_2}\left(\frac{1}{2}V_{to}^2\right)\right] \quad (10.18)$$

The coefficient in the bracketed term in Equation (10.18) is

$$\frac{c_1}{c_2}\left(\frac{1}{2}V_{to}^2\right) = \frac{\frac{1}{2}\rho V_{to}^2}{\left(\frac{W}{S}\right)_{to}}\frac{(C_D - \mu C_L)_{to}}{\left(\frac{F_{avg}}{W}\right)_{to} - \mu} = \frac{1}{C_{L,to}}\frac{(C_D - \mu C_L)_{to}}{\left(\frac{F_{avg}}{W}\right)_{to} - \mu} = \frac{\left(\frac{D}{L}\right)_{to} - \mu}{\left(\frac{F_{avg}}{W}\right)_{to} - \mu}$$

Then Equation (10.18) may be solved for the ground run distance and written as follows:

$$x_g = -\frac{V_{to}^2}{2g} \frac{1}{[(\frac{D}{L})_{to} - \mu]} \ln \left[\frac{\left(\frac{F_{avg}}{W}\right)_{to} - \left(\frac{D}{L}\right)_{to}}{\left(\frac{F_{avg}}{W}\right)_{to} - \mu} \right] \qquad (10.19)$$

We may readily measure x_g and V_{to} so we rearrange Equation (10.19) into the following dimensionless form:

$$\frac{2g x_g}{V_{to}^2} = \frac{1}{[(\frac{D}{L})_{to} - \mu]} \ln \left[\frac{\left(\frac{F_{avg}}{W}\right)_{to} - \left(\frac{D}{L}\right)_{to}}{\left(\frac{F_{avg}}{W}\right)_{to} - \mu} \right] \qquad (10.20)$$

It is common to find the rolling coefficient of friction to be taken as $\mu = 0.02$, see, for example, Torenbeek (1982). However, according to Daugherty (2003), the coefficient of rolling friction can conservatively be assumed to be $\mu = 0.015$ for typical dry runways. This result is found to be independent of speed or tire inflation. The difference in the choice of the coefficient of rolling friction makes little effect on the takeoff roll, typically less than 1.5%. We will not address the case of adverse conditions like wet or icy runways; in such cases the friction coefficient should be altered accordingly.

As can be seen from Equation (10.16) and the results shown in Figures 10.3 and 10.4, the average thrust to weight ratio during the takeoff roll will be somewhat less than the static thrust to weight ratio, which lies in the range $0.22 < (F/W)_{to} < 0.36$ for currently operational airliners, as can be seen from Figure 10.5. With this information we may carry out a parametric study of Equation (10.20) in order to assess its sensitivity to variations in thrust to weight and lift to drag ratios. For the test cases

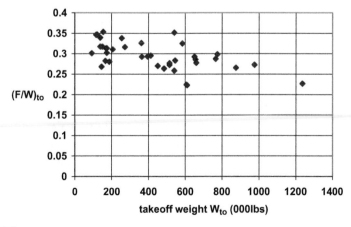

FIGURE 10.5

Variation of static thrust to weight ratio at takeoff for 27 operational commercial jet airliners.

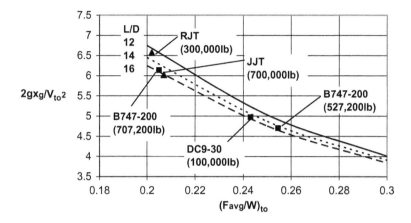

FIGURE 10.6

Variation of non-dimensional ground run distance with average thrust to weight ratio for different values of the takeoff lift to drag ratio. The B747 data are from flight test and the other data are from test-based simulations.

shown in Figure 10.6 the average thrust to weight ratio was deduced from Equation (10.16) using the static thrust information provided or that advertised for the engines employed. The results of such a study are shown in Figure 10.6, where it is seen that the sensitivity of ground distance to (L/D) is not as great as that due to the average thrust to weight ratio. The apparently large ratios of lift to drag, $12 < L/D < 16$, are typical of those generated while the aircraft is operating in ground effect. Both factors are saddled with some degree of uncertainty although the bounds on each are fairly well understood.

In addition to the lack of precise data on lift to drag ratio and thrust to weight ratio is the uncertainty introduced by the style of piloting. Also shown in Figure 10.6 are some flight test results along with some simulation results based on flight test data. The Boeing 747 flight test data are from Hanke and Nordwall (1970) while the Douglas DC-9 data are based on simulation studies by McCullers (1993). In addition, some results of pilot simulation studies are also shown. Snyder et al. (1973) presented a simulator study concerned with that portion of the takeoff from brakes off up to attainment of the 35 ft (10.7 m) obstacle height for FAA certification purposes. A simulator operated by NASA pilots generated results for a reference jet transport (RJT) and a jumbo jet transport (JJT). The characteristics of the JJT were generally similar to those of the RJT; observed differences resulted primarily from the larger moments of inertia and reference length which made the yawing motion following engine failure less abrupt and more easily controlled. Some of the pertinent data taken, or estimated, from the flight test and simulator studies cited above are presented in Table 10.1.

Table 10.1 Characteristics of Flight Test and Simulator Aircraft

Aircraft	β	W_{to} (lbs)	M_{to}	F_{to} (lbs)	n_e	F_{avg}/W_{to}	$(F/W)_{to}$	$(F_n/W)_{to}$
B747-200	5	707,200	0.254	41,076	4	0.2051	0.2323	0.1779
B747-200	5	527,200	0.211	37,164	4	0.2545	0.2820	0.2271
DC-9	1.74	100,000	0.223	14,500	2	0.2627	0.2900	0.2353
Reference Jet	0	300,000	0.240	16,500	4	0.2024	0.2200	0.1848
Jumbo Jet	0	700,000	0.240	39,500	4	0.2077	0.2257	0.1896

Insofar as the effect of piloting technique on the takeoff distance is concerned, the following comments from Snyder et al. (1973) should be noted. In the simulator one pilot flew takeoffs representative of standard commercial operations, while a second pilot flew takeoffs more typical of those conducted in certification testing, where minimizing takeoff distance to the 35 ft (10.7 m) height is a primary consideration. The first pilot used low rotation rates which required 10–15 s for achieving the climb attitude. The second pilot carried out the rotation more rapidly, reaching the climb attitude in less than 5 s. Thus, substantial differences in takeoff distance may arise purely from piloting technique.

10.2.2 Air run

After the aircraft rotates near the end of the ground run x_g, the climb may be assumed to proceed along an arc transitioning to a constant climb angle, γ_{to}, at the location of the 35-ft obstacle defining the end of the air run. In Figure 10.2 we approximated the curved trajectory by a straight path described by the coordinate x', which starts from the lift-off point, where $V = V_{to}$. After lift-off there is no further effect of drag due to rolling friction D_{roll}, but now there is a component of aircraft weight $W_{to}\sin\gamma_{to}$ which acts like a drag force so that Equation (10.9) may be rewritten as

$$\frac{d}{dx'}\left(\frac{1}{2}V^2\right) = \frac{g}{W_{to}}(F_n - D - W_{to}\sin\gamma_{to}) \tag{10.21}$$

Substituting for the thrust and drag forces as in the case of the ground run yields

$$\frac{d}{dx'}\left(\frac{1}{2}V^2\right) = \frac{g}{W_{to}}\left[F_n - \frac{1}{2}V^2\rho C_D S - W_{to}\sin\gamma\right]_{to} \tag{10.22}$$

Here, however, we take note of the fact that the velocity changes very little between V_{to}, the takeoff value, and V_2, the value at the 35-ft obstacle, typically between 3% and 6%. We therefore assume that the right-hand side of Equation (10.22) is approximately constant, as evaluated at the takeoff velocity V_{to}, and integration immediately yields

$$\frac{V^2}{2g} - \frac{V_{to}^2}{2g} = x' \frac{F_{n,to}}{W_{to}} - x' \left(\frac{C_D q}{W/S}\right)_{to} - x' \sin \gamma_{to} \tag{10.23}$$

The initial condition is that $V = V_{to}$ at $x' = 0$ while at the end of the air run the climb-out velocity over the 35-ft obstacle is denoted as V_2. At the end of the air run we set $x' = x_{air} \cos \gamma_{to} \approx x_{air}$, assuming that the climb-out angle is small and that $\cos \gamma_{to}$ is approximately unity. Note that $x' \sin \gamma_{to} = h_o$, the height of the theoretical obstacle that must be cleared by the aircraft as it climbs out. Then Equation (10.23) may then be written as

$$\frac{1}{2g} V_2^2 - \frac{1}{2g} V_{to}^2 = \left[\frac{F_{n,to}}{W_{to}} - \left(\frac{C_D q}{W/S}\right)_{to}\right] x_{air} + h_{to} \tag{10.24}$$

Our interest here is in the air distance x_{air} and solving for it yields

$$x_{air} = \frac{\frac{1}{2g} V_{to}^2 \left[\left(\frac{V_2}{V_{to}}\right)^2 - 1\right] + h_{to}}{\frac{F_{n,to}}{W_{to}} - \left(\frac{C_D q}{W/S}\right)_{to}} \tag{10.25}$$

The equation for quasi-equilibrium flight during climb out requires that the flight path angle be given by

$$\sin \gamma = \frac{F_n - D}{W} = \frac{F_n}{W} - \frac{1}{L/D} \tag{10.26}$$

Note that the denominator on the right-hand side of Equation (10.25) is just the difference between the thrust to weight ratio at takeoff and the inverse of the lift to drag ratio:

$$\frac{F_{n,to}}{W_{to}} - \left(\frac{C_D q}{W/S}\right)_{to} = \left(\frac{F_n}{W}\right)_{to} - \left(\frac{1}{L/D}\right)_{to} \approx \sin \gamma_{to} \tag{10.27}$$

Substituting Equation (10.27) into Equation (10.25) yields

$$x_{air} = \frac{\frac{1}{2g} V_{to}^2 \left[\left(\frac{V_2}{V_{to}}\right)^2 - 1\right] + h_{to}}{\sin \gamma_{to}} \tag{10.28}$$

Using the approximation that $1.03 < V_2/V_{to} < 1.06$ results in values between 0.06 and 0.12 for the quantity $[(V_2/V_{to})^2 - 1]$. Using a representative value of 0.1 in Equation (10.28) yields

$$x_{air} = \frac{\frac{1}{20g} V_{to}^2 + h_{to}}{\sin \gamma_{to}} \tag{10.29}$$

These results for the air and ground run are combined to yield the total takeoff distance

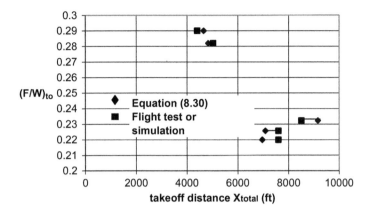

FIGURE 10.7

Predictions of total takeoff distance as a function of takeoff thrust to weight ratio compared to the results for several flight tests and simulations.

$$x_{to} = x_g + x_{air} = \frac{-V_{to}^2}{2g\left[\left(\frac{D}{L}\right)_{to} - \mu\right]} \ln\left[\frac{\left(\frac{F_{avg}}{W}\right)_{to} - \left(\frac{D}{L}\right)_{to}}{\left(\frac{F_{avg}}{W}\right)_{to} - \mu}\right] + \frac{V_{to}^2 + 20gh_{to}}{20g\left[\left(\frac{F_n}{W}\right)_{to} - \left(\frac{D}{L}\right)_{to}\right]}$$

(10.30)

The value for x_g is obtained from Equation (10.19) and that for x_{air} from Equation (10.30). Predictions for total takeoff distance are compared to the results of flight test and piloted simulations in Figure 10.7. Keep in mind that takeoff distances are dependent upon piloting technique and certification trials may result in shorter takeoff distances than those achieved in commercial operations.

10.2.3 Approximate solution for takeoff

For preliminary design purposes, where details of the aircraft being designed are not yet accurately known, it is reasonable to apply some further simplifications. For representative transport aircraft in the takeoff configuration the drag to lift ratio is likely to be in the range $1/12 < (D/L)_{to} < 1/16$ and we may assume a representative value of 0.075. This low value of $(D/L)_{to}$ is due mainly to the beneficial effects of the wing being close to the ground, as discussed by Suh and Ostowari (1988). Then, with the effects of rolling friction described by $\mu = 0.015$, Equation (10.30) becomes

$$x_{to} \approx \frac{-V_{to}^2}{0.12g} \ln\left[\frac{\left(\frac{F_{avg}}{W}\right)_{to} - 0.075}{\left(\frac{F_{avg}}{W}\right)_{to} - 0.015}\right] + \frac{V_{to}^2 + 20gh_{to}}{20g\left[\left(\frac{F_n}{W}\right)_{to} - 0.075\right]}$$

(10.31)

The result shown in Equation (10.31) is the approximate takeoff distance used in Chapter 4, Equation (4.8). Once the dimensions and characteristics of the design aircraft have been determined, as carried out in Chapters 2 through 9, one must carry out a more accurate solution, such as that given by Equation (10.31) or by numerically solving Equations (10.15) and (10.22).

A simple rule of thumb for estimating the total takeoff distance arises from considering the basic equation of motion, Equation (10.9), with the net thrust minus the aerodynamic drag taken as a constant multiple of the average thrust given by Equation (10.16), that is, $D_{avg} = k(F_{avg})_{to}$. Neglecting rolling friction on the ground and the work done to raise the aircraft to the 35 ft altitude at the end of the field one may integrate the equation of motion and solve for the total takeoff distance as follows:

$$x_{to} \approx \frac{V_{to}^2}{2g} \frac{1}{k \left(\frac{F_{avg}}{W} \right)_{to}} \tag{10.32}$$

Shevell (1989) offers a simple correlation that works reasonably well in predicting the takeoff distance requirements for air transports in Federal Air Regulation (FAR) Part 25 and it corresponds to the result in Equation (10.32) if we take $k = 1/2$.

10.2.4 Speeds during the takeoff run

There is a hierarchy of speeds in the takeoff process that are carefully defined in the FAR for the purposes of certification. For our purposes we note the salient characteristics of each speed, in ascending order, as follows:

V_{sr} = reference stall speed for 1 g load factor which is the velocity determined by $C_{L,max}$.

V_{mcg} = minimum speed for maintaining control on the ground by rudder alone.

V_{mc} = minimum speed for maintaining control in the air.

V_{ef} = speed at which engine failure occurs; must be greater than V_{mc}.

V_1 = critical speed; must be greater than V_{ef} plus the speed gained during the interval between engine failure and first response of pilot.

V_r = rotation speed, speed at which nose wheel is lifted off the ground and aircraft is rotated to the liftoff attitude.

V_{mu} = minimum unstick speed, speed at which aircraft can safely lift from the ground and continue the takeoff.

V_{lof} = velocity at liftoff; this is usually called the takeoff velocity V_{to}.

$V_{2,min}$ = minimum velocity at the 35-ft obstacle and greater than $1.08 V_{sr}$ for modern jet transports as well as greater than $1.1 V_{mc}$.

V_2 = velocity over the 35-ft obstacle; must be greater than V_r plus the speed increment attained before reaching a height of 35-ft above the takeoff surface.

V_{fto} = final takeoff speed, transition to the en route configuration complete, typically at 1500 ft altitude; must be greater than $1.18 V_{sr}$.

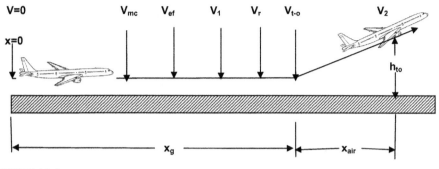

FIGURE 10.8

Hierarchy of speeds during takeoff run. Velocities shown are defined at the beginning of Section 10.2.4.

The hierarchy of the major speeds for our purposes is illustrated in Figure 10.8. In Chapter 4 we were particularly concerned with the takeoff distance x_{to} and the takeoff velocity

$$V_{to} = \sqrt{\frac{2(W/S)_{to}}{\rho C_{L,to}}} \qquad (10.33)$$

These two parameters and the thrust to weight ratio were used to help define the design space for the aircraft being designed. The concept of the critical velocity V_1 was used there to estimate the balanced field length for the design aircraft. Some takeoff reference speeds typical of jet transports taken from Snyder et al. (1973) are presented in Table 10.2.

Table 10.2 Typical Takeoff Reference Speeds

Aircraft Type	Takeoff Weight (lb)	Mass (kg)	Flap Position (degrees)	V_1 (knots)	V_r (knots)	V_2 (knots)
Reference jet transport ($S = 2750\,\text{ft}^2$)	300,000	136,000	15	138	153	164
	300,000	136,000	25	134	147	158
	200,000	90,600	15	114	131	149
	200,000	90,600	25	114	129	146
Jumbo jet transport ($S = 6422\,\text{ft}^2$)	700,000	317,100	15	138	153	164
	700,000	317,100	25	134	147	158
	460,000	208,400	15	114	131	149
	460,000	208,400	25	114	129	146

10.2.5 Continued takeoff with single engine failure

In the event of an engine failure during the takeoff roll the pilot must decide whether to continue the takeoff or instead abort the takeoff and decelerate to a stop on the runway. Obviously, if the engine failure occurs when the aircraft is traveling very slowly, the aircraft should be kept on the ground and brought to a stop at some safe location off the runway. Conversely, if the engine failure occurs when the aircraft is close to the takeoff speed the takeoff should be continued. The designer must provide a means for deciding whether it is safer to abort the takeoff or continue it. The situation when the choice is to continue the takeoff with one engine out is depicted schematically in Figure 10.9.

The critical velocity, usually denoted as V_1, is the velocity at which action is taken, not that at which the decision to act is taken. The time between the recognition of an engine failure, which occurs at V_{ef}, and the critical velocity V_1, when action is taken is required to be more than one second. Generally this time period, which is set by the reaction time of the pilot, is taken to be about 3 s. If the pilot's decision is to continue the takeoff with one engine inoperative, the distance to the lift-off speed V_{to} and to the subsequent climb-out to h_{to}, a 35 ft height above the runway, will obviously be longer than with all engines operating.

The average net thrust for the one-engine-out case is shown in Figure 10.10 for a twin-engine aircraft. The behavior of the net thrust for three- and four-engine aircraft is the same except for the scale of the drop in thrust. Between the start of the takeoff roll and the achievement of the takeoff velocity the average thrust power expended is kept essentially equal to the actual thrust power. Therefore, though the actual thrust is cut in half (for a twin-engine configuration) at V_{ef}, the average thrust drops by more

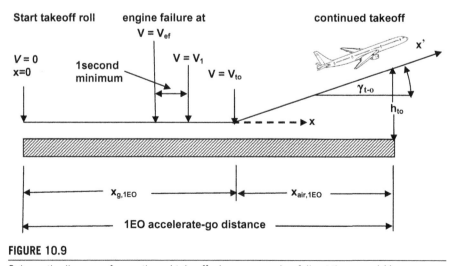

FIGURE 10.9

Schematic diagram of a continued takeoff when one engine fails at the speed V_{ef}.

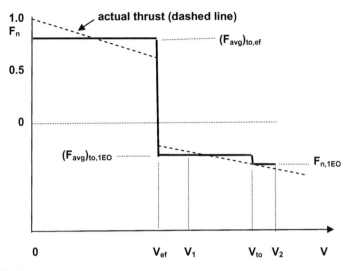

FIGURE 10.10

Average net thrust as a function of velocity for the one-engine-out scenario for a twin-engine aircraft.

than half because the net thrust is a decaying function of speed in the takeoff regime. The net thrust during climb-out is kept equal to the value at the takeoff velocity V_{to} because the change in speed is small and the linear thrust variation assumed here tends to underpredict the thrust as the takeoff velocity is approached. The average thrust during the first segments prior to the engine failure event is

$$(F_{avg})_{to,ef} = F_g \left[1 - \frac{1}{2} M_{ef} \left(2\frac{1+\beta}{3+2\beta} \right) \right] \tag{10.34}$$

If n_e denotes the number of engines, the average thrust after one of them fails is

$$(F_{avg})_{to,1EO} = \frac{n_e - 1}{n_e} F_g \left[1 - \frac{1}{2}(M_{ef} + M_{to}) \left(2\frac{1+\beta}{3+2\beta} \right) \right] \tag{10.35}$$

The net thrust in the one-engine-out condition is

$$(F_n)_{to,1EO} = \frac{n_e - 1}{n_e} F_g \left[1 - 2M_{to}\frac{1+\beta}{3+2\beta} \right] \tag{10.36}$$

Obviously, having more than two engines enhances safety of operation because more thrust is available. On the other hand, having more engines increases the likelihood of a failure. In addition, there are additional costs in maintenance and overhaul with additional engines. For this reason, emphasis has been placed on increasing the thrust level and reliability of turbofan engines so that safe operation can be realized with only two engines. This increased reliability of turbofan engines also made it possible to fly long distances over water where no emergency landing fields may be nearby.

For the first segment of the one-engine-out takeoff one chooses the velocity at which the engine failure event occurs, V_{ef}, and then calculates the associated location of the event, x_{ef}, from the ground roll, Equation (10.20), using the average thrust given in Equation (10.34):

$$\frac{2g x_{ef}}{V_{ef}^2} = \frac{1}{\left[\left(\frac{D}{L}\right)_{to} - \mu\right]} \ln\left[\frac{\left(\frac{F_{avg}}{W}\right)_{to,ef} - \left(\frac{D}{L}\right)_{to}}{\left(\frac{F_{avg}}{W}\right)_{to,ef} - \mu}\right] \tag{10.37}$$

After the failure event the aircraft will continue to gain speed, but at a lower rate because of the loss of a significant fraction of the net thrust. Assuming a value of the critical velocity V_1 permits the calculation of the distance traveled during the short reaction time increment Δt (typically about 3 s) according to an average velocity during the reaction period as follows:

$$x_1 = x_{ef} + \frac{1}{2}(V_1 + V_{ef})\Delta t \tag{10.38}$$

Because this is the case of a continued takeoff with one engine out, we proceed to calculate the distance required in going from x_1 at speed V_1 to the lift-off point $x_{g,1EO}$ at the known speed for liftoff, V_{to}, according to Equation (10.20). During this phase the average thrust for the one-engine-out case is taken from Equation (10.35) which leads to

$$x_{g,1EO} = x_1 - \frac{V_{to}^2}{2g}\frac{1}{\left(\frac{D}{L}\right)_{to} - \mu} \ln\frac{\left(\frac{F_{avg}}{W}\right)_{to,1EO} - \left(\frac{D}{L}\right)_{to}}{\left[\left(\frac{F_{avg}}{W}\right)_{to,1EO} - \mu\right] - \left[\left(\frac{D}{L}\right)_{to} - \mu\right]\left(\frac{V_1}{V_{to}}\right)^2} \tag{10.39}$$

Now the air distance may be calculated from Equation (10.21), again with the thrust appropriate to one engine being inoperative, that is, using Equation (10.36), as follows:

$$x_{air,1EO} = \frac{\frac{1}{20g}V_{to}^2 + h_{to}}{\sin\gamma_{to}} = \frac{\frac{1}{20g}V_{to}^2 + h_{to}}{\left(\frac{F_n}{W}\right)_{to,1EO} - \left(\frac{D}{L}\right)} \tag{10.40}$$

The total continued takeoff distance with one engine inoperative is the sum of the ground and air runs, $x_{to,1EO} = x_{g,1EO} + x_{air,1EO}$. As the value chosen for V_{ef} increases and V_1 approaches V_{to}, the ground run with one engine inoperative will approach that for all engines operating. Thus the increased takeoff distance will be due to the reduced capability during the air run with one engine out.

10.2.6 Aborted takeoff with single engine failure

As mentioned previously, in the event of an engine failure during the takeoff roll, the pilot may decide to abort the takeoff, decelerate, and bring the aircraft to a safe stop on the runway. The situation when the choice is to abort the takeoff after one engine fails is depicted schematically in Figure 10.11. In the case of the aborted takeoff the pilot will apply the necessary braking procedures in order to get the maximum permissible deceleration while maintaining adequate control of the airplane's motion.

The portion of the aborted takeoff run up to the critical velocity V_1 is calculated in the same fashion as that for the continued takeoff, so that the distance to x_1 is the same in both cases. However, now the pilot decides to abort the takeoff so braking procedures must be applied to decelerate the aircraft. The equation of motion, Equation (10.9), may be written for the case of braking on the ground as follows:

$$\frac{dV}{dt} = \frac{d}{dx}\left(\frac{V^2}{2}\right) = -g\left(\frac{F_{rev}}{W} + \frac{D}{W} + \frac{F_{brake}}{W}\right) \tag{10.41}$$

The contributions to the total braking force are the reverse thrust F_{rev} available from the engines still operating, the aerodynamic drag force D, and the braking force F_{brake}. The combination of these forces generally provides a level of deceleration of about 0.3–0.5 g. The simplest representation of these forces is as follows:

$$F_{rev} = k_{rev}F_g \tag{10.42}$$

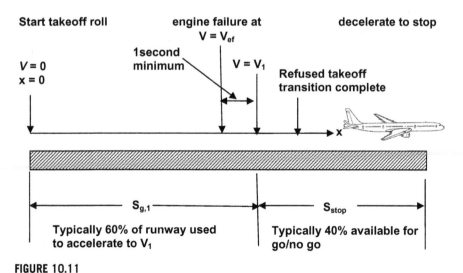

FIGURE 10.11

Schematic diagram of an aborted takeoff when one engine fails at the speed V_{ef}.

$$D = C_D \frac{1}{2}\rho S V^2 \tag{10.43}$$

$$F_{brake} = \mu_{brake}(W - L) = \mu_{brake}\left(W - \frac{1}{2}C_L \rho S V^2\right) \tag{10.44}$$

Here k_{rev} represents the thrust reverser effectiveness, that is, the fraction of static thrust that can be applied by engaging the thrust reversers, C_D is the total drag coefficient including the effect of lift dumpers or spoilers deployed to reduce lift generation, and μ_{brake} is the braking coefficient which represents the effectiveness of the wheel braking system.

The practical considerations by airlines on the use and benefits of thrust reversers are discussed by Yetter (1995). Airlines use thrust reversers to provide an additional stopping force on wet and slippery runways and to provide additional safety and control margins during aborted takeoffs, landings, and ground operations. It is common practice to deploy thrust reversers on every landing and for maximum effectiveness they are deployed as soon as possible after touchdown. The amount of reverse thrust used is left to the pilot's discretion but airlines suggest the use of "normal" reverse thrust levels in order to reduce brake loads. Because of the added safety margins, the airlines consider thrust reversers to be cost effective and essential to achieving the maximum level of aircraft operating safety.

The overall thrust reverser effectiveness is defined as the ratio of measured reverse thrust to the total nozzle forward thrust that would be produced at corresponding fan and core nozzle pressure ratios. Yetter et al. (1996) reported that conventional thrust reverser effectiveness is approximately $k_{rev}=0.32$. Then Equation (10.41) becomes

$$\frac{d}{dx}\left(\frac{V^2}{2}\right) = -g\left[k_{rev}\left(\frac{F_g}{W}\right) + \mu_{brake} + \frac{1}{2}\frac{\rho(C_D - \mu_{brake}C_L)}{\left(\frac{W}{S}\right)}V^2\right] \tag{10.45}$$

Assuming that all the coefficients are independent of the velocity this is a first order linear differential equation for V^2 that has the solution

$$V^2 = \left(V_1^2 + \frac{c_2}{c_1}\right)\exp[-c_1(x - x_1)] - \frac{c_2}{c_1} \tag{10.46}$$

Here

$$c_1 = \left[1 - \mu_{brake}\left(\frac{L}{D}\right)\right]\frac{C_D \rho g}{\left(\frac{W}{S}\right)} \tag{10.47}$$

$$c_2 = 2g\left[k_{rev}\left(\frac{F_g}{W}\right) + \mu_{brake}\right] \tag{10.48}$$

Then the stopping point of the aircraft x_{stop} is given by the location where $V=0$ or

$$x_{stop} = x_1 + \frac{1}{c_1} \ln \left(1 + \frac{c_1}{c_2} V_1^2 \right)$$ (10.49)

It is clear that the higher the critical velocity V_1, the longer the runway needed to stop the aircraft safely.

10.2.7 The balanced field length

The velocity-time history of the different takeoff scenarios is illustrated in Figure 10.12. The standard takeoff proceeds smoothly through increasing speeds, lifting off at V_{to} and climbing out to the 35 ft obstacle height at V_2. The takeoff distance for any of the three cases is the area under the corresponding velocity-time curve in Figure 10.12 between times $t=0$ and $t=t_2$, and this may be written as follows:

$$S_{to} = \int_0^{t_2} V dt$$

It is obvious from Figure 10.12 that in the case of an engine failure at velocity V_{ef} the distance traveled from the start of the takeoff roll to the critical velocity V_1 is the same for both the continued takeoff with one-engine-out case and the aborted takeoff case. However, beyond the critical velocity V_1, the distance traveled depends on the magnitude of V_1. For example, when the critical velocity V_1 is much smaller than the

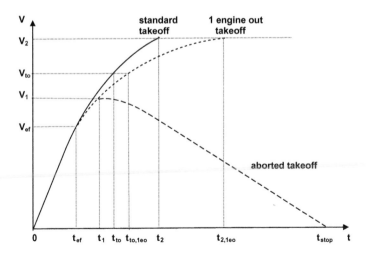

FIGURE 10.12

Schematic illustration of the velocity-time history of a standard takeoff, a continued takeoff with one engine out, and an aborted takeoff.

climb-out velocity V_2, the distance required to abort the takeoff and come to a safe stop on the runway will be shorter than that required to continue and complete the takeoff. This is true because the airplane will be rolling slowly and can be readily decelerated by braking, whereas the reduced thrust due to the failure of one engine will reduce the acceleration capability of the airplane and require a much longer distance to get up to takeoff velocity. Conversely, when the critical velocity V_1 is close to the climb-out velocity V_2, the distance required to continue and complete the takeoff with one engine inoperative will be shorter than that required to abort the takeoff and bring the airplane to a safe stop on the runway. This follows from the understanding that braking at high speeds is limited by brake heating and tire integrity, while continuing the acceleration to V_2 is more readily attainable.

Instead of considering the limiting cases of aborting takeoffs at low V_1 and continuing takeoffs at high V_1 it is useful to determine the critical velocity for which the distance required to continue the takeoff is equal to the distance required to safely abort it. This distance is called the balanced field length. In the case of a balanced field length the areas under the one-engine-out takeoff and aborted takeoff curves are equal. Another way to look at the balanced field length is to plot the distance required to continue the takeoff with one engine inoperative, the sum of Equations (10.39) and (10.40), as a function of the critical velocity V_1, and compare it to the required stopping distance, Equation (10.49). A generic plot of this type is shown in Figure 10.13. The critical velocity defines the balanced field length as being that distance where the stopping distance is equal to the continued takeoff distance.

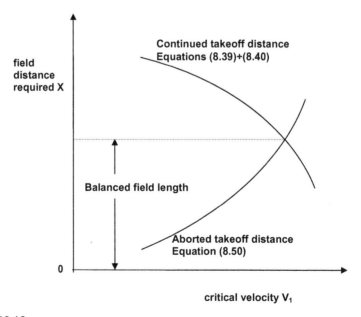

FIGURE 10.13

Illustrative plot of the field distances required to continue a takeoff or to abort it when one engine fails as a function of critical velocity.

10.3 Climb

It is typical to climb as rapidly as possible to the cruising altitude for a number of reasons, including limiting the noise annoyance caused in the vicinity of the airport. The climb is carried out at constant indicated, or calibrated, airspeed in order to afford the pilot the simplest means of monitoring the climb. This airspeed is within 10% of the equivalent airspeed for the usual flight envelope of commercial airliners, the difference arising from a correction to account for compressibility effects. The onboard measurement of airspeed is discussed in greater detail in the last section of this chapter. If the aircraft climbs at constant calibrated airspeed, and it is assumed that this is approximately the equivalent airspeed, then the true airspeed is not constant, but increasing with altitude, that is, there is some acceleration. Since $V = V_E/\sqrt{\sigma}$ and V_E is approximately constant the non-dimensional magnitude of the acceleration is

$$\frac{1}{g}\frac{dV}{dt} = \frac{1}{g}\left(\frac{dV}{dz}\right)\left(\frac{dz}{dt}\right) = \frac{1}{g}\left[\frac{d}{dz}\left(\frac{V_E}{\sqrt{\sigma}}\right)\right](V_{ROC}) = \frac{V_E V_{ROC}}{2gH\sqrt{\sigma}} \quad (10.50)$$

In Equation (10.50) V_{ROC} is the rate of climb and $H = 29,000$ ft is the scale height of the assumed exponential atmosphere in which $\sigma = \exp(-z/H)$. Then Equation (10.50) may be written as

$$\frac{1}{g}\frac{dV}{dt} = 1.503 \times 10^{-8}\frac{V_E V_{ROC}}{\sqrt{\sigma}} \quad (10.51)$$

In Equation (10.51) the numerical coefficient on the right-hand side takes care of the units if the equivalent velocity of flight V_E is expressed in knots while V_{ROC} is expressed in ft/min, the conventional units for the rate of climb. Taking relatively large values for the velocities V_E and V_{ROC}, say 400 kts and 6000 ft/min respectively, yields a non-dimensional acceleration $(dV/dt)/g = 0.036\exp(z/2H)$ and, at 20,000 ft altitude $(dV/dt)/g = 0.05$. Thus, even for the extreme case this value of acceleration can still be considered small enough compared to unity to be neglected in a preliminary calculation of the rate of climb. An illustration of the climb variables is shown in Figure 10.14.

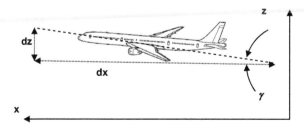

FIGURE 10.14

Schematic diagram of an aircraft in climb at constant angle γ.

Under this assumption of negligible vertical acceleration the balance of forces permits us to put the vertical speed of the aircraft in terms of the time rate of change of altitude, z. Thus the rate of climb V_{ROC} is given by

$$V_{ROC} = V_c \sin \gamma = V_c \frac{F_n - D}{W} = V_c \left[\delta(F/W)_{to} - \frac{1}{(L/D)_c} \right] \qquad (10.52)$$

The rate of climb is measured in feet per minute (fpm), so the climb velocity V_c must be corrected appropriately. Note that in order to climb there must be excess thrust. As pointed out in Chapter 9, the altitude at which the excess thrust limits the rate of climb to 100 fpm is called the service ceiling for the aircraft. The climb is generally performed with the throttles at the maximum continuous power setting.

A major factor in the climb formulation is the variation of thrust with speed and altitude. For turbojets and turbofans the former is less important than the latter. Typically the thrust is assumed to be constant with Mach number, though this is less accurate as the bypass ratio of the engine increases. The relationship between sea level takeoff thrust and the net thrust produced at the cruise speed and altitude appears to be reasonably approximated by $F_n/F_{g,to} = \delta$, the atmospheric pressure ratio. A fairly good analytic expression for the pressure ratio is given by $\delta = \exp(-z/24{,}000)$, with z in feet. This expression can be used to facilitate analytic integration of the rate of climb equation, if desired.

10.3.1 Climb profile

A typical flight profile in terms of the climb velocity V_c for the ascent of a commercial airliner may be expressed as follows:

Sea level $< z < 10{,}000$ ft	$V_{E,c} = 250$ kts.
$10{,}000 < z < 25{,}000$ ft	$V_{E,c} = 320$ kts.
$25{,}000 < z <$ cruise altitude	$V_c = a_c M_c = a_{cr} M_{cr}$.

An illustration of the different climb segments, along with the pressure, density, and temperature profiles in the atmosphere is given in Figure 10.15. The speed variation with altitude during climb is shown in Figure 10.16.

The climb is usually carried out at constant equivalent airspeed (EAS) up to about 25,000 ft so that the pilot can more easily monitor and control the ascent. This means that the true airspeed of the aircraft is continually rising during the climb to 25,000 ft. Thereafter the climb is carried out at constant Mach number, which is close to a constant speed climb since the sound speed only drops around 5% by the time 35,000 ft is reached.

Note that the lift coefficient needed, assuming that the climb angle is shallow and $L \sim W$, is $C_L = 2(W/S)/(\rho_{sl} V_{E,c}^2)$. The true airspeed therefore will be increasing during the climb so C_L will be lower than the design value for cruise, at least for segments 1 and 2, and therefore L/D in those two segments will be less than the maximum and this will lengthen the time to climb.

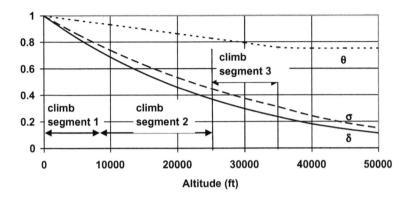

FIGURE 10.15

Typical climb segments shown along with the non-dimensional thermodynamic properties of the standard atmosphere.

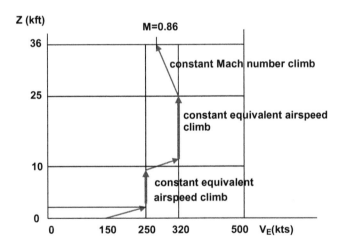

FIGURE 10.16

Equivalent speed variation during climb as a function of altitude.

10.3.2 Time to climb

The rate of climb equation for time elapsed in climb from altitude z_1 to altitude z_2 may be expressed as follows:

$$
t = \int_{z_1}^{z_2} \frac{dz}{101.36 \frac{V_{E,c}}{\sqrt{\sigma}} \left[\delta \left(\frac{F}{W} \right)_{to} - \frac{1}{L/D} \right]}
\tag{10.53}
$$

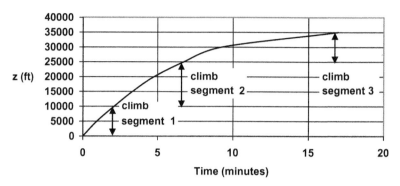

FIGURE 10.17

Time to climb of a typical case for a turbofan airliner.

Here, with z in feet and $V_{E,c}$ in knots, the time to climb, in minutes, is determined. The appropriate (constant) value of $V_{E,c}$ (in kts) must be used in the three altitude segments to obtain the time interval in minutes.

For a hypothetical airplane with $(F/W)_{to}=0.3$, $M_{cr}=0.8$, and L/D as given below, the climb performance is as follows:

Sea level $<z<10{,}000\,\mathrm{ft}$:	$V_{E,c}=250\,\mathrm{kts}$, $L/D=14.3$, $\Delta t=2.07\,\mathrm{min}$
$10{,}000<z<25{,}000\,\mathrm{ft}$:	$V_{E,c}=320\,\mathrm{kts}$, $L/D=14.3$, $\Delta t=4.65\,\mathrm{min}$
$25{,}000<z<\mathrm{cruise\ altitude}$:	$V_{E,c}=474\,\mathrm{kts}$, $L/D=15.2$, $\Delta t=10.23\,\mathrm{min}$
Total time to climb	$\Sigma(\Delta t)=16.95\,\mathrm{min}$

As mentioned previously, values of L/D during segments 1 and 2 are chosen to be lower than the assumed cruise value of 15.2. The climb history for this example case is shown in Figure 10.17. Here it is seen that the last segment of the climb takes more time than the first two combined. This is due to the drop in engine thrust with altitude, as mentioned previously. Note that the denominator in Equation (10.53) can become very small as δ drops with altitude. This makes the integrand large and in turn increases the time required to climb higher. The integration of Equation (10.53) may be carried out analytically using the exponential atmosphere approximation but it is rather involved. Here it is simpler to carry out the integration numerically using the actual standard atmospheric properties.

10.3.3 Distance to climb

The differential of the distance to climb is given by

$$x_c = \int_{z_1}^{z_2} \frac{dx/dt}{dz/dt}dz = \int_{z_1}^{z_2} dz \frac{V_{E,c}}{V_{E,c}[\delta(F/W)_{to} - 1/(L/D)_c]} \tag{10.54}$$

This equation may be readily integrated numerically since the information for the integrand is already available from Equation (10.52). The aircraft trajectory for the example in Figure 10.15 is illustrated in Figure 10.18. This assumes that the aircraft

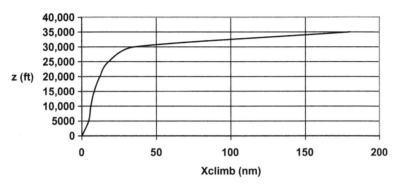

FIGURE 10.18

Trajectory of the aircraft climb profile illustrated in Figure 10.15.

travels in a plane and that airspeed in climb is equal to the true ground speed. Such an assumption does not account for winds.

10.3.4 Fuel to climb

A reasonable approximation to the fuel used in climbing to the cruise altitude, i.e., segment 4 of the mission profile described in Chapter 2, is given by $W_{f,used,4} = C_j F_n \Delta t$, using consistent units. However, because the thrust decreases with altitude it is preferable to calculate the fuel usage directly by integration, as shown in Equation (10.55), where C_j is divided by 60 to keep the time units consistent as minutes.

$$\frac{W_{f,used4}}{W_{to}} = \int_{z_1}^{z_2} \frac{\frac{C_j}{60}\delta \left(\frac{F}{W}\right)_{to} dz}{101.36\frac{V_{E,c}}{\sqrt{\sigma}}\left[\delta \left(\frac{F}{W}\right)_{to} - \frac{1}{L/D}\right]} \tag{10.55}$$

Thus, rather than use the approximate values in Table 2.3 for the fuel weight fraction used in climb one may calculate it directly using the actual characteristics of the design aircraft.

10.4 Descent

In a fashion similar to that followed in treating climb, one may determine the rate of descent V_{ROD} from the rate of change of altitude as follows:

$$V_{ROD} = \frac{dz}{dt} = V_d \frac{-D}{W} = -V_d \frac{1}{(L/D)_d} \tag{10.56}$$

The rate of descent is also measured in feet per minute (fpm), so V_d, the flight speed at which descent is carried out, must be corrected appropriately. In descent the idle thrust power setting is typically used so that the thrust may be neglected in the calculation.

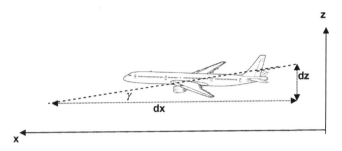

FIGURE 10.19

The aircraft in descent at constant angle γ.

10.4.1 Descent profile and performance

Similarly, a typical flight profile for the descent of a commercial airliner may be described as follows:

Cruise altitude $> z > 10{,}000$ ft $\qquad V_{E,d} = 250 + (V_{E,cr} - 250)(z/z_{cr})$ with V in kts

$10{,}000$ ft $> z >$ sea level $\qquad V_{E,d} = 250$ kts

An illustration of the aircraft in descent is shown in Figure 10.19. For the final portion of the descent of the representative aircraft under consideration the rate of descent is as follows:

$$V_{ROD} = -250\,\text{kts}\,\frac{-1}{15.2} = -1667\,\text{ft/min} \tag{10.57}$$

The rate of descent is also measured in feet per minute (fpm), so $V_{descent}$ must be corrected appropriately. In descent the idle thrust power setting is used so that the thrust may be neglected in the calculation. The deployment of lift dumpers, or spoilers, on the upper surface of the wing can modulate L/D and thereby control the rate of descent as required by air traffic control.

10.4.2 Time to descend

We assume that the airspeed is reduced linearly with altitude so that $V = 250 + fz$, where $f = [(V_{cr} - 250)/z_{cr}]$ and the cruise speed is V_{cr} and is given in knots. The rate of descent Equation (10.56) for time elapsed in descending from altitude z_1 to altitude z_2 may be integrated to yield:

$$\Delta t = -9.87 \times 10^{-3}\frac{(L/D)}{f}\left\{\log_e\left[\frac{1 + kz_2}{1 + kz_1}\right]\right\} \tag{10.58}$$

Here $k = [(V_{cr}/250) - 1]/z_{cr}$ and, with V_{cr} in knots and z in feet, the time interval is in minutes. The lift to drag ratio is assumed constant in descent, but since flight speed is decreasing and density is increasing the actual value during descent will be less than the cruise value. The units must be consistent so a value of time elapsed in minutes (or hours) can be correctly calculated. For the hypothetical airplane considered pre-viously we find $f=0.0064$ and $k=2.56\times10^{-5}$ so that descent from $z_1=35{,}000\,\text{ft}$ to $z_2=10{,}000\,\text{ft}$, assuming $(L/D)_d=14$ requires $\Delta t=0.148\,\text{h}=8.9\,\text{min}$. From $z_1=10{,}000\,\text{ft}$ to $z_2=0$ ft the equation for Δt is just

$$\Delta t = \frac{(z_2 - z_1)}{101.36V_d}\left(\frac{L}{D}\right)_d \tag{10.59}$$

For this portion of the descent $\Delta t=5.52\,\text{min}$ and the total time for descent is $8.9+5.52=14.42\,\text{min}$.

10.4.3 Distance to descend

The distance to descend may be found in the same fashion as the distance to climb was found. The rate of descent is reformulated as follows:

$$dx_d = \frac{dx/dt}{dz/dt}dz = V_d\frac{dz}{RD} = \frac{z_{cr} - z_{airport}}{6080}\left(\frac{L}{D}\right)_d \tag{10.60}$$

Thus, the distance to descend is determined approximately by the glide ratio in descent, $(L/D)_d^{-1}$. For the example aircraft considered previously we have $x_d=14(35{,}000-0)/6080=80.6$ nautical miles.

10.4.4 Fuel to descend

Since the engines are at lower power settings in the descent phase it is difficult to accurately determine the fuel used in the preliminary design stage. The fuel used may be estimated by working from the weight fraction estimate for stage 10 in Table 2.3. Another approach, suggested by Torenbeek (1982), is to calculate the fuel consumed in cruising a distance equal to x_d. It is suggested to calculate both and use the larger of the two as a conservative estimate.

10.5 Landing

The landing process begins about 8–10 miles away from the airport and the sequence of events is described schematically in Figure 10.20. The final stage of the land-ing occurs at the runway as previously shown in Figure 4.4. There the landing field length is denoted by $x_l=x_a+x_g$, that is, the sum of an air run and a ground run. These two portions of the landing process are treated sequentially below.

10.5.1 Air run

The air run may be considered to start at the $h_l=50$-ft obstacle, so that in an x-y coor-dinate system with $x=0$ at $y=h_l$, the approach speed equals V_a, which, according

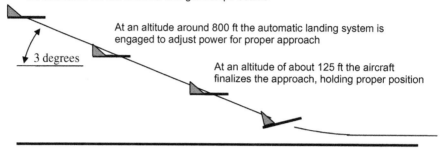

At about 10 miles out, flying at around 160 kts, the pilot lines up the aircraft on the ILS localizer and glide slope beams

At an altitude around 800 ft the automatic landing system is engaged to adjust power for proper approach

3 degrees

At an altitude of about 125 ft the aircraft finalizes the approach, holding proper position

Over the symbolic 50-ft obstacle at the end of the runway the throttles close automatically and the nose is raised, initiating the flare

FIGURE 10.20

Initial stages of the landing process leading up to the flare and actual landing. The angle of the flight path and all distances are exaggerated for clarity.

to FAR Part 25, must be at least $1.3V_s$. The aircraft executes a gradual flare of large radius R, turning the flight path angle from γ around $2°$ or $3°$ down to zero at which time the aircraft has slowed to V_l, that is, to a value around $1.2–1.25V_s$ and settled onto the runway, ending the air run at $x=x_a$.

A simplified analysis of the air run considers the flight path angle to change so slowly that the angular acceleration may be ignored in the force balance, leading to the following approximate equations:

$$F_n + W \sin \gamma - D = m\frac{dV}{dt} \tag{10.61}$$

$$L - W \cos \gamma = m\frac{d}{dt}\left(R\frac{d\gamma}{dt}\right) \approx 0 \tag{10.62}$$

During the flare, the deceleration along the trajectory may be written as $\frac{dV}{dt} = V\frac{dV}{dx}$. The distance traveled during the air run from $V=V_a$ at $x=0$, to $V=V_l$ at $x=x_{air}$, may be found by integrating the equations

$$F_n - D + W\gamma = mV\frac{dV}{dx} = \frac{1}{2}\frac{W}{g}\frac{dV^2}{dx} \tag{10.63}$$

$$L - W = 0 \tag{10.64}$$

In these equations it was assumed that the flight path angle is small enough to properly replace the sine and cosine functions with the leading terms of their expansions. This leads to the following equation

$$\frac{dV^2}{dx} = 2g \left(\frac{F_n - D}{W} + \gamma \right) \tag{10.65}$$

The weight is approximately equal to the lift and the engines are kept near idle thrust during the flare so that the first term in parentheses is an effective drag to lift ratio, that is, $\frac{F_n - D}{L} = -(\frac{D}{L})_{eff}$ and may be considered constant during the flare. If it is assumed that the actual, slowly varying, small flight path angle may be replaced by an average value, say γ_a, then the equation may be integrated to yield

$$V_l^2 - V_a^2 = 2g \left[-\left(\frac{D}{L}\right)_{eff} + \gamma \right] x_{air} \tag{10.66}$$

Equation (10.66) may be rearranged to yield the air run as follows:

$$x_{air} = \frac{1}{2g} \left[\frac{V_l^2 - V_a^2}{-\left(\frac{D}{L}\right)_{eff} + \gamma_a} \right]$$

In terms of the stall characteristics of the aircraft, this equation becomes

$$x_{air} = \frac{\left(\frac{W}{S}\right)_l}{g\rho_{sl}\sigma C_{L,l,max}} \left[\left(\frac{D}{L}\right)_{eff} - \gamma_a \right]^{-1} \tag{10.67}$$

Note that $(D/L)_{eff}$ may be approximated by

$$\left(\frac{D}{L}\right)_{eff} \approx \frac{c_{D,l}}{c_{L,l}} = \frac{\left(c_{D,0} + \frac{c_L^2}{\pi eA} \right)_l}{c_{L,l}} = 1.44 \frac{c_{D,0,l}}{c_{L,max,l}} + \frac{c_{L,max,l}}{1.44\pi eA}$$

In the absence of lift and drag characteristics, as in the first stages of preliminary design, one may approximate the effective drag to lift ratio to be around $(D/L)_{eff} = 1/6$ for airliners in the landing configuration, while the average flight path angle γ_a may be taken to be around $1.5°$ or 0.026 radians and the value of x_a should be positive. Then the total landing air run would be given by

$$x_{air} = 7.11 \frac{\left(\frac{W}{S}\right)_l}{g\rho_{sl}\sigma C_{L,max,l}} \tag{10.68}$$

10.5.2 Ground run

On the simplest level, one may assume that the average deceleration during the ground run is

$$\frac{dV}{dt} = \frac{d}{dx}\left(\frac{1}{2}V^2\right) = -a_{avg} = const.$$

Integrating from the $V=V_l$ (the landing speed) at $x=0$ to $V=0$ at the end of the landing where $x=x_g$ yields $x_g = \frac{V_l^2}{2a_{avg}}$. For a case of $V_l=240\,\text{km/h}=149\,\text{mph}=218.77\,\text{ft/s}$ and $a_{avg} = 0.31g$ we find the ground run to be $x_g=2395\,\text{ft}$. The distance covered during the ground run in terms of the stall characteristics of the aircraft becomes

$$x_g = \frac{1.44 \left(\frac{W}{S}\right)_l}{a_{avg}\rho_{sl}\sigma C_{L,\max,l}} \qquad (10.69)$$

At the next level of approximation, the dynamics of the situation require that

$$\frac{1}{g}\frac{dV}{dt} = \frac{1}{2g}\frac{dV^2}{dx} = \left(\frac{F}{W}\right)_l - \left(\frac{D}{W}\right)_l - \left(\frac{F_{brake}}{W}\right)_l$$

In this equation we do know that $\left(\frac{D}{W}\right)_l = \frac{C_{D,l}\rho_{sl}\sigma V_l^2}{2\left(\frac{W}{S}\right)_l}$, where the drag coefficient will depend on the specific landing configuration (flap and slat deflection and possibly aileron droop) and spoiler deflection. The braking force may be written in terms of the normal force the wheels exert on the runway according to $F_{brake}=\mu_{brake}(W-L)$, where the braking coefficient of friction lies in the range of $0.4<\mu_{brake}<0.6$ for concrete runways. Then we may expand L so that we have

$$\left(\frac{F_{brake}}{W}\right)_l = \mu_{brake}\left(1-\frac{L}{W}\right) = \mu_{brake}\left[1 - \frac{C_{L,l}\rho_{sl}\sigma V_l^2}{2\left(\frac{W}{S}\right)_l}\right] \qquad (10.70)$$

The thrust in landing is the idle thrust so we may take $(F/W)_l=$constant and our deceleration equation is

$$a = \frac{1}{2}\frac{dV^2}{dx} = g\left[\left(\frac{T}{W}\right)_l - \mu_{brake}\right] - g\left[1 - \mu_{brake}\left(\frac{C_L}{C_D}\right)_l\right]\frac{C_{D,l}\rho_{sl}\sigma V_l^2}{2\left(\frac{W}{S}\right)_l} \qquad (10.71)$$

Physically, this equation is of the form $a=a_1+a_2$, where a_1 is the deceleration due to thrust reversal (for which $F/W<0$) and braking, and a_2 is the deceleration due to aerodynamic drag less the effect of aerodynamic lift on reducing the normal force between the wheels and the runway. Equation (10.71) is of the form

$$Y' + AY + B = 0 \qquad (10.72)$$

Here $Y=V^2$ and A and B are given by

$$A = \left[1 - \mu_{brake}\left(\frac{L}{D}\right)_l\right]\frac{C_{D,l}\rho_{sl}\sigma g}{\left(\frac{W}{S}\right)_l}$$

$$B = -2g\left[\left(\frac{T}{W}\right)_l - \mu_{brake}\right]$$

The solution in terms of the ground distance is

$$V^2 = \left[V_l^2 + \frac{B}{A}\right]e^{-Ax} - \frac{B}{A} \tag{10.73}$$

The total ground distance is then determined by setting $V=0$ and is given by

$$x_g = \frac{1}{A}\ln\left(1 + \frac{A}{B}V_l^2\right) \tag{10.74}$$

The lift to drag ratio in landing depends upon the combination of high lift devices acting in concert with spoilers. However, it must be recognized that once the aircraft touches down the lift must eventually go to zero, if for no other reasons than the facts that the airplane levels out, reducing its angle of attack, and that it is slowing down and eventually comes to rest. Indeed, touchdown of the aircraft generally automatically initiates deployment of the spoilers to dump the lift and let the aircraft settle down on the ground, thereby also increasing the rolling resistance which helps slow the aircraft down. Thus the $(L/D)_l$, as shown in the parameter A, and which comes about from Equation (10.70), is perhaps 6 or 7 at touchdown and will likely be dropping to zero quite soon after touchdown. Therefore it is reasonable to set it to zero or some small average value, if desired.

Furthermore, the parameter $(F/W)_l$ could be set to some reasonable negative value to represent the application of thrust reversers during some stage of the landing process. The actual value of the thrust to weight ratio to apply depends upon the capability of thrust reversers, if employed. If thrust reversers are not used it is reasonable to set $(T/W)_l=0$. Finally, there should be some free roll time to account for the reaction time of the pilot and the time until application of the thrust reversers is permissible, and this could be several seconds of travel at the touchdown speed. Thus the landing might be considered to take place in three stages: (1) touchdown and free roll slowdown for several seconds, (2) Continued deceleration under the action of the thrust reversers at a modest level of F/W, say 0.1 or so, and (3) Final deceleration with thrust reversers off and full braking action. During all of this L/D could be considered zero.

Another, simpler, estimate of landing distance x_l, the sum of the air distance x_a from the 50-foot obstacle h_l and the ground distance x_g, is given by Shevell (1989) as follows:

$$\begin{aligned}
x_l &= x_{air} + x_g \\
x_{air} &= \frac{1}{\frac{1}{(L/D)_l} - \left(\frac{T}{W}\right)_l}\left[50 + \tfrac{1}{2g}(V_a^2 - V_l^2)\right] \\
x_g &= \frac{V_l^2}{2a_{avg}} \\
a_{avg} &= \mu g\left(1 - \tfrac{L}{W}\right)_{eff} + g\left(\tfrac{D}{W}\right)_{eff}
\end{aligned} \tag{10.75}$$

Here a_{avg} is again an average deceleration caused by the effects rolling friction drag and the effective aerodynamic drag during the rollout over the ground run. As the aircraft touches down, spoilers or lift dumpers are automatically deployed and the lift decreases to zero so that the normal force is equal to the landing weight of the aircraft alone. The values for the coefficient of friction, μ, range from $0.4 < \mu < 0.6$ for dry concrete, $0.2 < \mu < 0.3$ for wet concrete, and as little as $\mu = 0.1$ for ice. The average deceleration generally lies in the range $0.3 < a_{avg} < 0.4$, as shown previously in Figure 4.5. The quantity V_{50} is the airspeed over the 50-foot obstacle, and is required by FAR Part 25 to be at least $1.3V_l$.

A further requirement in FAR Part 25 is that the landing field length be a factor of 8/5 of the demonstrated landing distance x_l. Shevell (1989) offers an approximation for the FAR landing distance in the form FAR $x_l = A + BV_s^2$. The values for A and B are as follows: for main gear struts with two wheels each, $A = 4000$, $B = 0.333$, and for main gear struts with four-wheel bogies, $A = 1400$, $B = 0.4$.

10.6 Turboprop-powered aircraft
10.6.1 Turboprop range

We may follow the development of Section 10.1 where now the rate at which fuel is used in quasi-steady flight is assumed to be linearly related to the power produced by the gas turbine engine so that the rate of change of the weight of the aircraft is

$$\frac{dW}{dt} = \frac{dW_f}{dt} = -C_{tp}P \tag{10.76}$$

The quantity C_{tp} is the turboprop-specific fuel consumption in lbs fuel per hour per horsepower and P is the total power of the engines in horsepower. Thus the rate of fuel consumption for V measured in knots may be described as follows:

$$\frac{dW_f}{dt} = -C_{tp}P = -C_{tp}\frac{F_n V}{326\eta_p} = -C_{tp}(D + W\sin\gamma)\frac{V}{326\eta_p}$$

Therefore

$$dt = -\frac{(326\eta_p)W}{C_{tp}(D + W\sin\gamma)}\frac{dW_F}{W} = -\frac{326\eta_p}{C_{tp}\left(\frac{D}{W} + \sin\gamma\right)}\frac{dW}{W}$$

Using the approximations discussed in Section 10.1 we put

$$\frac{D}{W} \approx \frac{D\cos\gamma}{L} \approx \left(\frac{L}{D}\right)^{-1}$$

Similarly, because the flight path angle is small, we also put

$$\sin\gamma \approx \gamma \ll 1$$

Furthermore, the horizontal speed $V = dx/dt$, so that the equation for the time differential may be rearranged as follows:

$$\frac{dW}{W} = -\frac{C_{tp}}{375\eta_p}\left(\frac{L}{D}\right)^{-1} dx$$

The range achievable is then given by

$$R = \int_0^R dx = \int_{W_4}^{W_5} -\frac{\eta_p(L/D)}{C_{tp}}\frac{dW}{W} \tag{10.77}$$

The quantities W_4 and W_5 denote the aircraft weight at the start and end of the cruise segment, respectively. The simplest solution to this equation is achieved for flight where the quantity multiplying dW/W in the integrand in Equation (10.77) is constant.

The specific fuel consumption C_{tp} for turboprop engines is roughly constant (at about 0.5–0.6 lbs fuel/hp/hr for modern turboprop engines at cruise Mach numbers between 0.45 and 0.6) and the propeller efficiency η_p is generally around 0.80–0.85 and essentially constant over the cruise range. Then the range in nautical miles for a turboprop aircraft may be estimated as follows:

$$R = \frac{326\eta_p}{C_{tp}}\frac{L}{D}\ln\left(\frac{W_4}{W_5}\right) \tag{10.78}$$

10.6.2 Turboprop takeoff ground run

The general ground run equation for an aircraft was given in Section 10.2.1 as Equation (10.9). We may carry through the same approach using the average net thrust during takeoff, Equation (10.16), which for a propeller-powered aircraft requires that the bypass ratio $\beta \to \infty$. In that case the average net thrust is

$$F_{n,avg,to} = F_{to}\left(1 - \frac{M_o}{2}\right) \tag{10.79}$$

This result may be substituted into Equation (10.19) for the ground run distance and written as follows:

$$x_g = -\frac{V_{to}^2}{2g}\frac{1}{\left[\left(\frac{D}{L}\right)_{to} - \mu\right]}\ln\left[\frac{\left(\frac{F_{n,avg}}{W}\right)_{to}\left(1 - \frac{M_{to}}{2}\right) - \left(\frac{D}{L}\right)_{to}}{\left(\frac{F_{n,avg}}{W}\right)_{to}\left(1 - \frac{M_{to}}{2}\right) - \mu}\right] \tag{10.80}$$

The determination of the static thrust F_{to} for a propeller-driven aircraft will be carried out subsequently.

10.6.3 Turboprop takeoff air run

When the aircraft lifts off the ground the primary objective is to climb rapidly to clear the 35-ft obstacle that marks the end of the complete takeoff run. The aircraft speed doesn't increase very much so the net thrust remains almost constant and we may simply apply the approach used for the turbofan engines in Section 10.2.2. The result for the air run is Equation (10.29) in which the net thrust in takeoff for a propeller aircraft is given by Equation (10.13) with $\beta \to \infty$ or

$$F_{n,to} = F_{to}(1 - M_{to}) \tag{10.81}$$

Then Equation (10.29), with the current notation for turboprop aircraft, becomes

$$x_{air} = \frac{V_{to}^2 + 20gh_{to}}{20g\left[\left(\frac{F}{W}\right)_{to}(1 - M_{to}) - \left(\frac{D}{L}\right)_{to}\right]} \tag{10.82}$$

The total takeoff distance is the sum of the ground and air runs $x_{to} = x_g + x_{air}$, that is, the sum of Equations (10.80) and (10.82).

10.6.4 Approximate solution for turboprop takeoff

For preliminary design purposes, where details of the aircraft being designed are not yet accurately known, it is reasonable to apply some further simplifications. For representative transport aircraft in the takeoff configuration the drag to lift ratio is likely to be in the range $1/12 < (D/L)_{to} < 1/16$ and we may assume a representative value of 0.075. This low value of $(D/L)_{to}$ is due mainly to the beneficial effects of the wing being close to the ground. The effects of rolling friction may be described by $\mu = 0.015$, and the takeoff Mach number is $M_{to} = V/a_{sl}\theta^{1/2}$, where a_{sl} is the speed of sound at sea level. Typical values of takeoff Mach number for turboprop aircraft are around $M_{to} = 0.13$. The total takeoff distance for turboprop aircraft may be approximated for preliminary design evaluation by

$$x_{to} = \frac{V_{to}^2}{2g}\left\{-16.67\ln\left[\frac{\left(\frac{F}{W}\right)_{to} - 0.081}{\left(\frac{F}{W}\right)_{to} - 0.016}\right] + \frac{0.115}{\left(\frac{F}{W}\right)_{to} - 0.0862}\right\} + \frac{1.15h_{to}}{\left(\frac{F}{W}\right)_{to} - 0.0862} \tag{10.83}$$

Note that Equation (10.83) may be used with any consistent set of units.

10.6.5 Turboprop propeller characteristics

The static thrust of a propeller-driven aircraft is given by

$$F_{to} = 33,000\frac{C_T}{C_P}\frac{P}{Nd_p} \tag{10.84}$$

The net thrust for a propeller-driven aircraft in flight is given by

$$F_n = 326\eta_p\frac{P}{V} \tag{10.85}$$

Here P, N, d_p, and V denote the shaft power, propeller speed, propeller diameter, and flight velocity; they are given in hp, rpm, ft, and kts, respectively. The coefficients C_T,

C_P, and η_p denote the propeller thrust coefficient, power coefficient, and efficiency respectively, and will be discussed subsequently. Note that the propeller net thrust drops off rapidly with speed and therefore propeller-driven aircraft are speed-limited. Turboprop airliners are generally limited to speeds of around 350 mph. Though it may appear that for a given power the thrust will grow rapidly as $V \rightarrow 0$, the efficiency drops off equally rapidly so that a finite static thrust is indeed achieved.

For good performance the propeller tip speed should be as high as possible, but is generally kept to a Mach number of 0.8, low enough to limit compressibility losses and undue noise generation. Then $V_{tip} = 0.8a = 0.8(\gamma RT)^{1/2}$, and because the highest sound speed will occur at the lowest altitude, that is, where the local temperature is highest, we choose the sea level value of 1116 ft/s. Then $V_{tip} = 893$ ft/s and the propeller tip speed

$$V_{tip} = \frac{1}{2}\omega d_p = \frac{1}{2}\left(2\pi\frac{N}{60}\right)d_p$$

Then the product of propeller rpm and propeller diameter (in ft) is

$$Nd_p = 17,100 \tag{10.86}$$

Then Equations (10.84) and (10.85) may be put in terms of the power to weight ratio as follows:

$$\frac{F_{to}}{W} = 1.93\frac{C_T}{C_P}\left(\frac{P}{W}\right)_{to} \tag{10.87}$$

$$\frac{F_n}{W} = 326\frac{\eta_p}{V}\left(\frac{P}{W}\right) \tag{10.88}$$

The thrust coefficient C_T, power coefficient C_P, and the propeller efficiency η_p are functions of the propeller design and are included in the performance maps provided by the propeller manufacturer. Using the same units, the power coefficient under static ($V=0$) conditions is defined as

$$C_P = 5 \times 10^{10}\frac{P}{\sigma N^3 d_p^2} \tag{10.89}$$

The power coefficient of Equation (10.89) may then be written as

$$C_P = \frac{0.008}{\sigma}\left(\frac{P}{A_p}\right) \tag{10.90}$$

Thus the power coefficient depends upon the local atmospheric density ratio and the power loading of the propeller, that is, the ratio of the shaft power P to the propeller swept, or disk, area A_p. The static thrust efficiency may be expressed as a ratio of the ideal power required to produce a given static thrust to the ideal power required; see, for example, Sforza (2012). This ratio is often called the figure of merit and is given by

$$FOM = \frac{P_i}{P} = 0.798\frac{C_T^{\frac{3}{2}}}{C_P} \tag{10.91}$$

When the takeoff power available is known, the figure of merit may be used for rapid estimation of propeller static thrust using the following expression

$$F_{to} = 10.42[(FOM)\sqrt{\sigma}d_p P_{to}]^{\frac{2}{3}} \tag{10.92}$$

The figure of merit for a well-designed propeller will be around $0.68 < FOM < 0.73$ and a value of 0.7 may be used for preliminary estimates.

10.6.6 Selecting a propeller

The propeller selection analysis follows the development presented by Sforza (2012). One may make a reasonable estimate of the propeller required by first considering the quantity

$$\frac{J}{C_P^{\frac{1}{3}}} = V\left(\rho_{sl}\sigma\frac{d_p^2}{P}\right)^{\frac{1}{3}} \tag{10.93}$$

Here J is an important propeller characteristic called the advance ratio

$$J = 101.4\frac{V}{Nd_p} \tag{10.94}$$

The propeller rotational speed $n = N/60$ in revolutions per second, so J measures the number of diameters the propeller advances during one revolution. It also provides an indication of the relative magnitude of forward flight speed compared to propeller rotational speed. Because the propeller rpm N is essentially constant for typical flight operations, takeoff (low speed) occurs at low J and cruising flight (high speed) at high J. To illustrate these relative magnitudes of J we may use the suggestion for Nd_p given by Equation (10.86) and assume a representative takeoff velocity $V_{to} = 100$ kts, which yields $J = 0.59$. On the other hand, for a cruise velocity $V = 300$ kts the advance ratio $J = 1.78$.

Because most of the fuel is consumed in cruise the propeller should be optimized for best efficiency at the cruise conditions. Analysis indicates that the best efficiency (η_p around 86%) occurs for the range $2.4 < \frac{J}{C_P^{1/3}} < 3.2$ and this occurs when the modified power coefficient parameter $C_{P,X} = \frac{C_P}{X}$ $C_{P,X} = C_P/X$ lies in the range $0.2 < C_{P,X} < 0.4$. It can be shown that $C_P \sim (AF)_{tot}$ where the total activity factor $(AF)_{tot} = B(AF)$, B is the number of blades, and the activity factor for a single blade is given by

$$AF = 10^5 \int_{0.15}^{1} \left(\frac{c}{d}\right) \left(\frac{r}{R}\right)^3 d\left(\frac{r}{R}\right) \tag{10.95}$$

Obviously a blade with a wider chord c, particularly at the outboard portions of the blade, will have a higher activity factor and therefore a higher power coefficient. The typical range for activity factor is about $100 < AF < 150$. The quantity X "adjusts" the C_P for the activity factor. This adjustment factor can be estimated by the following expression:

$$X = 1.49 \left(\frac{AF_{tot}}{1000}\right)^{\frac{5}{4}} \tag{10.96}$$

First, choosing a value for $J/C_P^{1/3}$ from the high-efficiency range mentioned previously permits one to determine the associated value of the propeller diameter d_p from Equation (10.93) because the cruise speed, altitude, and available power are known. It is obvious that the choice of $J/C_P^{1/3}$ should be somewhat on the lower side of the range given, say around 2.6, since that will lead to a smaller-diameter propeller.

With d_p known for the trial propeller one may calculate the power coefficient C_P from Equation (10.93). Then the parameter $X = C_P/C_{P,X}$ may be found by selecting a value of $C_{P,X}$ from the high-efficiency band mentioned previously and the total activity factor may be estimated from Equation (10.96). Typically, for values of $AF_{tot} < 400$ a three-bladed propeller may be preferable while for higher values of AF_{tot} a four-bladed propeller would be the better choice. Even higher values of AF_{tot} may require the use of five or six blades.

The only aspect of the trial propeller that is unknown is the integrated design lift coefficient

$$C_{L,d} = 4 \int_{0.15}^{1} c_{L,d} \left(\frac{r}{r_p}\right)^3 d\left(\frac{r}{r_p}\right) \tag{10.97}$$

This is the local blade airfoil lift coefficient integrated over the radius of the propeller r_p from $r/r_p = 0.15$ to $r/r_p = 1$ weighted by the cube of the local radius. The region $r/r_p < 0.15$ is assumed to be taken up by the propeller hub. It is at this point, with the trial propeller characteristics settled, that one may go to the propeller performance procedures described previously and refine the propeller design. Propeller charts, like that shown in Figure 10.21, are presented as plots of C_P vs. J for specific values of B, AF, and $C_{L,d}$; see, for example, Hamilton Standard (1974). With the cruise power coefficient $C_{p,c}$ and advance ratio J known for a selected activity factor AF, one may enter the propeller chart for that AF and determine the efficiency for a specific design lift coefficient, $C_{L,d}$, as indicated by the dashed arrows in Figure 10.21. Because all but the last parameter are defined, the designer can carry this process out for various values of the design lift coefficient and then plot $C_{L,d}$ vs. η_p to select the

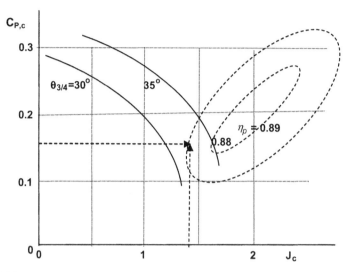

FIGURE 10.21

Typical propeller efficiency plot for a three-bladed propeller with AF = 140 and a design lift coefficient of 0.5. Arrows delineate calculated data entered to find propeller efficiency η_p and blade pitch angle $\theta_{3/4}$.

optimum propeller. This selection process also provides the propeller pitch setting at the three-quarter radius point, $\theta_{3/4}$, for that particular propeller. The pitch angle is the angle between the chord of the propeller and the plane of rotation of the propeller. For further details and examples, see Sforza (2012).

10.7 The air data system

The altitude, airspeed, climb speed, Mach number, temperature, and related flight data are provided through the processing of measured raw data by a computer-controlled instrumentation system called the air data system and described by Collinson (2003). The pilot monitors the airspeed by means of an instrument called the airspeed indicator and its display is called the indicated airspeed. The true airspeed, which is the speed of the aircraft relative to the undisturbed atmosphere through which it is passing, may be found by applying the principles of compressible flow. For isentropic flow along a streamline the relationship between the stagnation pressure p_t and the static pressure p depends upon the Mach number M and the ratio of specific heats of the gas. The pressure coefficient at the stagnation point is given below:

$$C_{p,t} = \frac{p_t - p}{q} = \frac{2}{\gamma M^2}\left[\left(1 + \frac{\gamma - 1}{2}M^2\right)^{\frac{\gamma}{\gamma - 1}} - 1\right]$$

(10.98)

The quantity q is the dynamic pressure

$$q = \frac{1}{2}\rho V^2 = \frac{1}{2}\gamma p M^2 \tag{10.99}$$

Since the atmosphere of interest is air we may set $\gamma = 1.4$ so that the pressure coefficient equation becomes

$$C_{p,t} = \frac{p_t - p}{q} = \frac{10}{7M^2}\left[\left(1 + \frac{M^2}{5}\right)^{\frac{7}{2}} - 1\right] = \left(1 + \frac{M^2}{4} + \frac{M^4}{40} + \frac{M^6}{1600} + \cdots\right) \tag{10.100}$$

This equation shows the effect of compressibility on the measurement of velocity by considering only the difference in stagnation and static pressures. Of course, when the Mach number is small the difference between the stagnation and static pressures is approximately equal to the dynamic pressure and one may compute the true airspeed directly. Flying at a speed of 200 mph at sea level is equivalent to a Mach number of about 0.3 which yields an error of 2.3% over the incompressible value for the pressure coefficient and this is considered negligible. However, for commercial jetliners cruising at $M = 0.8$ the error jumps to about 17% and cannot be ignored.

If one measures p_t with a Pitot tube and pressure sensor and then p with an appropriate static pressure port and pressure sensor one can determine the Mach number from Equation (10.98). Since a static pressure port must be located somewhere on the aircraft it is unlikely to measure the exact static pressure in the undisturbed air and therefore the output of the static pressure sensor would have to be calibrated over the expected speed range in order to correct for this problem. The instrument using just these pressure measurements is called a Machmeter. Measuring the static temperature T then permits the true airspeed (TAS) to be determined using the definition of Mach number: $M = V/a$. The relationship between static and stagnation temperature in isentropic flow is

$$\frac{T}{T_t} = 1 + \frac{\gamma - 1}{2}M^2 = 1 + \frac{M^2}{5} \tag{10.101}$$

The stagnation temperature T_t is relatively easily measured, for example with a thermocouple. Then the true airspeed is $V = M\sqrt{\gamma RT} = 29M\sqrt{T}$ in kts when T is measured in degrees Rankine. The calibrated airspeed (CAS) arises from calculating the velocity from the isentropic relation

$$V^2 = \left(\frac{2\gamma}{\gamma - 1}\right)\frac{p}{\rho}\left[\left(1 + \frac{p_t - p}{p}\right)^{\frac{\gamma-1}{\gamma}} - 1\right] \tag{10.102}$$

If one uses the sea-level pressure and density rather than the actual values in all but the pressure difference term in Equation (10.102) produces the calibrated airspeed V_{CAL} which is described by the following equation

$$V_{CAL}^2 = \left(\frac{2\gamma}{\gamma - 1}\right)\frac{p_{sl}}{\rho_{sl}}\left[\left(1 + \frac{p_t - p}{p_{sl}}\right)^{\frac{\gamma-1}{\gamma}} - 1\right] \tag{10.103}$$

The equivalent airspeed (EAS), which is a measure of the dynamic pressure of the airstream, is defined in terms of the true airspeed as follows:

$$V_E^2 = \sigma V^2 = \frac{\rho}{\rho_{sl}}\left(\frac{2\gamma}{\gamma - 1}\right)\frac{p}{\rho}\left[\left(1 + \frac{p_t - p}{p}\right)^{\frac{\gamma-1}{\gamma}} - 1\right] \tag{10.104}$$

Then the EAS is directly related to the CAS by $V_E = FV_{CAL}$ and the correction factor F is given by

$$F^2 = \frac{\frac{p}{p_{sl}}\left(1 + \frac{p_t - p}{p}\right)^{\frac{\gamma-1}{\gamma}}}{\frac{p_{sl}}{\rho_{sl}}\left(1 + \frac{p_t - p}{p_{sl}}\right)^{\frac{\gamma-1}{\gamma}}} \tag{10.105}$$

Using $\gamma = 1.4$ and introducing the atmospheric pressure ratio $\delta = p/p_{sl}$ we can simplify the correction function as follows:

$$F^2 = \delta\frac{\left(1 - \frac{p_t - p}{p}\right)^{\frac{2}{7}} - 1}{\left(1 - \delta\frac{p_t - p}{p}\right)^{\frac{2}{7}} - 1} \tag{10.106}$$

Thus the compressibility correction is also corrected for altitude and one can compute V_{CAL} and find the correction factors for incorporation into the data-processing algorithms of the air data system. This approach doesn't require measurement of temperature and uses the pressure altitude, as given by δ, which is used to provide the pilot with an altitude reading.

A sample calculation was carried out for the case of flight at an altitude of 35,000 ft and the results are shown in Figure 10.22. The calibrated airspeed, and therefore the indicated airspeed read by the pilot, are approximately equal to the equivalent airspeed and thus provide a dynamically appropriate indication of the aircraft performance since aircraft loads are proportional to the square of the EAS. Notice that in Figure 10.22 the TAS is 70% greater than the CAS. The value of the TAS is in the navigation of the aircraft since it, with information about the prevailing winds, provides the ground speed of the aircraft. The advent of improved sensors and navigational aids like the global positioning system (GPS) has served to alter the avionics suite common to airliners; for more information see Collinson (2003).

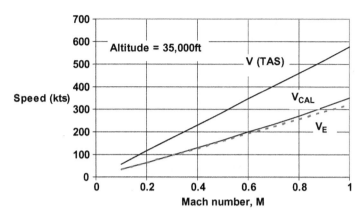

FIGURE 10.22

Illustration of the relationship between true airspeed (TAS), calibrated airspeed (CAS), and equivalent airspeed (EAS) as a function of Mach number at an altitude of 35,000 ft (10.67 km).

10.8 Design summary

A preliminary design of the aircraft has been completed by the methods developed in the preceding chapters. This aircraft should be subjected to a complete analysis of its performance potential using the techniques described in this chapter. The results of such an analysis will point out the strengths and weaknesses of the design and provide indications of what aspects of the design may require further refinement. Often it is necessary to go through a number of iterations through the whole design process to make the necessary improvements. If the design approach has been systematized in one or more computer programs or spreadsheets this iteration process can be readily handled. A compilation of configuration and performance data for a variety of aircraft which can supply useful comparative information is presented by Filippone (2000).

10.9 Nomenclature

A	aspect ratio, area, or constant in Equation (10.73)
AF	propeller activity factor, Equation (10.95)
a	sound speed or acceleration
B	number of propeller blades or constant in Equation (10.73)
c	chord length
C_D	drag coefficient
C_L	lift coefficient
$C_{L,d}$	design lift coefficient, Equation (10.97)

C_j	thrust specific fuel consumption
C_P	propeller power coefficient
C_p	pressure coefficient
C_T	propeller thrust coefficient
C_{tp}	turboprop power specific fuel consumption
D	drag
d	diameter
e	span efficiency factor
f	descent profile slope, Equation (10.58)
F	thrust or correction factor in Equation (10.105)
FOM	figure of merit
F_g	gross thrust
F_n	net thrust
F_r	ram drag
g	acceleration of gravity
H	scale height of the atmosphere
h	obstacle height
I	specific impulse
J	propeller advance ratio
k	constant
L	lift
M	Mach number
m	mass
$\dot{m}$	mass flow
N	rotational speed in rpm
n_e	number of engines
P	shaft power
p	pressure
q	dynamic pressure
R	range or gas constant
r	radius
S	wing planform area
T	temperature
t	time
V	velocity
V_{CAL}	calibrated airspeed
V_E	equivalent velocity
V_{ef}	velocity at engine failure
V_{ROC}	rate of climb
V_{ROD}	rate of descent
V_1	critical velocity
V_2	velocity over the takeoff obstacle
W	weight
X	adjustment factor, Equation (10.96)
x	distance

x'	distance along the (straight) climb-out path
z	altitude
α	angle of attack
β	turbofan bypass ratio
δ	atmospheric pressure ratio
γ	flight path angle or ratio of specific heats
Λ	wing sweepback angle
η	efficiency
μ	friction factor
μ'	effective friction factor
ρ	density
σ	atmospheric density ratio
ω	rotational speed

10.9.1 Subscripts

a	landing approach conditions
air	air run
avg	average value
$brake$	conditions under braking
c	climb conditions
cr	cruise conditions
$crit$	critical
d	descent conditions
e	exhaust conditions
ef	engine failure
eff	effective
f	fuel
g	ground run
i	ideal
l	landing
max	maximum
oe	operating empty
pl	payload
p	propeller
rev	reverse thrust conditions
$roll$	rolling conditions
s	stall
sl	sea level
to	takeoff
t	stagnation
0	zero-lift or free stream conditions
1EO	one engine out condition
4	start of cruise
5	end of cruise

References

Collinson, R.P.G., 2003. Introduction to Avionics Systems, second ed. Kluwer Academic Publishers, Dordrecht, The Netherlands.

Daugherty, R.H., 2003. A Study of the Mechanical Properties of Modern Radial Aircraft Tires. NASA/TM-2003-212415.

Hamilton Standard, 1974. Generalized Method of Propeller Performance Estimation, Hamilton Standard Division of United Technologies, PDB 6101. Revision A, June 1963, interim reissue September 1974.

Hanke, C.R., Nordwall, D.R., 1970. The Simulation of a Jumbo Jet Transport Aircraft, vol. 2. Modeling Data. NASA CR 114494.

Hesse, W.J., Mumford, N.V.S., 1964. Jet Propulsion for Aerospace Applications, second ed. Pitman, NY.

Kroo, I., 2001. Drag due to lift—concepts for prediction and reduction. Annual Reviews in Fluid Mechanics 33, 587–617.

McCullers, L.A., 1993. Takeoff: Detailed Takeoff and Landing Analysis Program Users Guide. NASA Langley Research Center.

Filippone, A., 2000. Drag and performances of selected aircraft and rotorcraft. Progress in Aerospace Sciences 36, 629–654.

Sforza, P.M., 2012. Theory of Aerospace Propulsion. Elsevier, Boston, MA.

Shevell, R.S., 1989. Fundamentals of Flight. Prentice-Hall, Englewood Cliffs, NJ.

Snyder, C.T., Drinkwater, F.J., Fry, E.B., Forrest, R.D., 1973. Takeoff Certification Considerations for Large Subsonic and Supersonic Transport Airplanes Using the Ames Flight Simulator for Advanced Aircraft. NASA TN D-7106.

Suh, Y.B., Ostowari, C., 1988. Drag reduction factor due to ground effect. Journal of Aircraft 25 (11), 1071–1072.

Torenbeek, E., 1982. Synthesis of Subsonic Airplane Design. Kluwer Academic Publishers, Dordrecht, The Netherlands.

Torenbeek, E., 1997. Cruise performance and range prediction reconsidered. Progress in the Aerospace Sciences 33 (5–6), 285–321.

Yetter, J.A., 1995. Why Do Airlines Want and Use Thrust Reversers? NASA Technical Memorandum 109158.

Yetter, J.A., Asbury, S.C., Larkin, M.J., Chilukuri, K., 1996. Static Performance of Several Novel Thrust Reverser Concepts for Subsonic Transport Applications. AIAA 96–2649, 32nd AIAAlASME/SAE/ASEE Joint Propulsion Conference.

Aircraft Pricing and Economic Analysis

11.1 Introduction

Cost evaluation is an important element of the commercial airplane business. An airplane company makes money by selling airplanes to operators who, in turn, make money by selling the use of the airplanes to customers who wish to travel from place to place. The operator, say an airline, determines the value of purchasing a particular airplane by considering five basic elements:

- Capital cost—the cost of buying the airplane.
- Direct operating cost—the cost of using the airplane, including fuel and maintenance.

Commercial Airplane Design Principles. http://dx.doi.org/10.1016/B978-0-12-419953-8.00011-5

- Indirect operating cost—the annualized cost of utilizing the airplane in delivering service.
- Total operating cost—the sum of all costs involved in providing service.
- Cost per seat-mile—the cost of moving a seat, full or not, per mile of route served.

The various items may be treated sequentially in a manner that will permit a cost evaluation for the aircraft under design.

11.2 Capital cost

The airplane being designed must be priced for sale and this has often been a contentious issue, both for commercial and military aircraft programs. This is due, in great measure, to the need to estimate costs based on very little actual data, typically just the mission statement which provides limited performance and physical characteristics of the final airplane being designed, as described by the US Department of Defense (1999). There are two major elements in pricing a program, one is the cost of developing the airplane design and the second is the cost of producing the airplane. These may be broken down further into the following categories directly related to making the airplane: engineering, tooling, manufacturing labor, manufacturing material, flight test, and quality control. There are other indirect costs associated with the operation of the business, including sales and customer service. The detailed development of these costs can be quite complex and although there are open sources for such procedures like reports issued by the US Department of Defense, DOD (1999), and the National Aeronautics and Space Administration, NASA (2012), cost estimation typically involves closely held proprietary procedures particular to individual aircraft companies.

An early comprehensive review of the estimation of airframe costs for military aircraft of all types is found in Large et al. (1976). This report presents generalized equations for estimating development and production costs on the basis of primary performance specifications, like weight and speed. Separate equations are provided for the following cost elements: engineering, tooling, nonrecurring manufacturing labor, recurring manufacturing labor, nonrecurring manufacturing materials, recurring manufacturing materials, flight test operations, and quality control, as well as equations for estimating total program cost and prototype development cost. The equations were derived from cost data on 25 military airplanes that first flew within the time period of 1953–1970 and covered empty weights from 5000 to 279,000 pounds and speeds from 300 to 1300 knots. The resulting equations showed three classes of aircraft: group 1 had weights less than 50,000 pounds and a speed less than 550 knots, group 2 had weights of less than 50,000 pounds and a speed of 1150 knots, and group 3 had weights greater than 50,000 pounds and a speed less than 550 knots. Equations are presented for the total cost for 100 aircraft in thousands of $1975 as follows:

- Group 1: $TC_{100} = 2967W_e^{0.58}$ based on 9 aircraft (A-3, A-4, A-6, RB-66, F-3, F4D, F-100, F-102, T-38).
- Group 2: $TC_{100} = 13.35W_e^{1.16}$ based on 8 aircraft (A-5, B-58, F-4, F-14, F-104, F-105, F-106, F-111).
- Group 3: $TC_{100} = 30.92W_e^{0.96}$ based on 6 aircraft (B-52, C-5, C-130, C-133, KC-135, C-141).

Restricting the data presented for group 3 to those aircraft most like the commercial airliners of interest here, i.e., removing the B-52 bomber and the turboprop C-130, and adding data for the then-new Boeing 747 and Douglas DC-10 suggests the following slight modification to the total cost for 100 units:

- Group 3 revised: $TC_{100} = 26W_e^{0.96}$ based on 6 aircraft (C-5, C-133, KC-135, C-141, B747, DC-10).

The total cost equation above appears in terms of $1975 and may be inflated to $2012 by applying the ratio of the consumer price index (CPI) for 2012 (117.3) to that for 1975 (27.0) to get a multiplier of 4.34. The federal government produces various price deflators illustrating the effect of inflation on various sectors of the economy. Some of the deflators that are of use in the aerospace industry are reproduced here in Table 11.1. More extensive information on the consumer price index (CPI) and other economic factors may be found in BLS (2012).

The inflation of the 1975 cost of the revised Group 3 aircraft by using the CPI growth results in the following equation:

$$TC_{100} = 112.84W_e^{0.96} \text{ in thousands of \$2012}$$

The cost of one aircraft would then be

$$TC = 0.00113W_e^{0.96} \text{ in millions of \$2012} \tag{11.1}$$

Boeing (2013) and Airbus (2013) present prices for aircraft and these are shown as a function of empty weight, along with the historical correlation of Equation (11.1), in Figures 11.1 and 11.2. The prices are essentially nominal values, since the exact price of a given aircraft depends upon special equipment, such as engines and interior configuration, particular to different buyers.

A correlation for the cost of one aircraft, in millions of $2012, based on the data summarized in Figures 11.1 and 11.2 is given by

$$TC = 425erf\left[\frac{W_e - 10^4}{4.5 \times 10^5}\right] \tag{11.2}$$

Once again, the weight is measured in pounds and $erf(x)$ is the error function which is tabulated in Appendix E and can be found in standard mathematics textbooks. The correlations of Equations (11.1) and (11.2) are shown in Figure 11.3 along with the available price figures in millions of $2012 for Boeing and Airbus as well as several regional turbofan and turboprop aircraft.

Table 11.1 Federal Price Deflators for GDP, PPI, and CPI 1978–2012

Year	CY GDP CY2005 = 100	PPI CY2005 = 100	CPI CY2005 = 100
1978	40.5	49.3	33.4
1979	43.8	53.6	37.2
1980	47.8	59.3	42.2
1981	52.3	65.4	46.5
1982	55.5	69.2	49.4
1983	57.7	71.1	51.0
1984	59.9	72.8	53.2
1985	61.7	74.3	55.1
1986	63.1	75.9	56.1
1987	64.8	77.2	58.2
1988	67.0	79.0	60.6
1989	69.6	82.2	63.5
1990	72.3	85.0	66.9
1991	74.8	87.6	69.7
1992	76.6	89.3	71.8
1993	78.3	90.9	74.0
1994	79.9	92.7	75.9
1995	81.6	94.5	78.0
1996	83.2	95.6	80.3
1997	84.6	95.6	82.2
1998	85.6	95.2	83.5
1999	86.8	95.2	85.3
2000	88.7	96.0	88.2
2001	90.7	96.6	90.7
2002	92.2	96.2	92.1
2003	94.1	96.5	94.2
2004	96.8	97.8	96.7
2005	100.0	100.0	100.0
2006	103.2	101.5	103.2
2007	106.2	103.5	106.2
2008	108.6	106.0	110.2
2009	109.5	109.6	109.8
2010	110.0	111.0	111.7
2011	113.4	113.7	113.2
2012	115.4	116.6	117.6

Source: Aerospace Industry Association, AIA (2012) and Bureau of Labor Statistics, BLS (2012).
Notation: CPI = Consumer Price Index, CY = Calendar Year, GDP = Gross Domestic Product Index,
PPI = Producer Price Index for finished goods excluding food and energy.

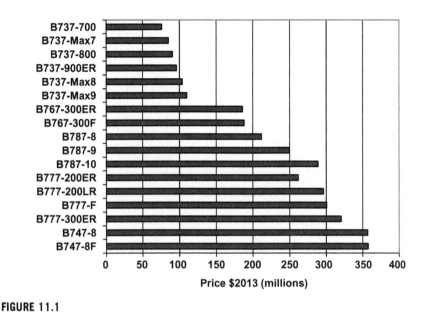

FIGURE 11.1

Prices of Boeing aircraft in millions of $2013 as reported in Boeing (2013).

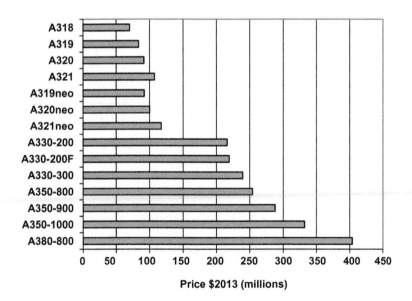

FIGURE 11.2

Prices of Airbus aircraft in millions of $2013 as reported in Airbus (2013).

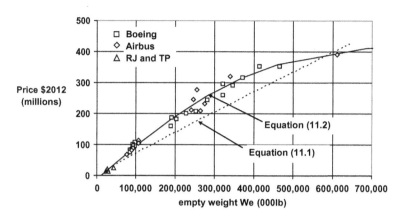

FIGURE 11.3

Boeing and Airbus aircraft prices in millions of $2012 are shown as a function of empty weight. Also shown is the historical correlation curve given in Equation (11.1), as well as the correlation Equation (11.2). The lowest empty weights are for various regional jets and turboprops, and are denoted by RJ and TP, respectively.

It is clear from Figure 11.3 that the historical correlation curve inflated by the CPI, Equation (11.1), predicts prices considerably below the data for the 32 aircraft. However, 29 fall within ±10% of the correlation curve of Equation (11.2), while the prices of three are overpredicted by between 10% and 15%. The larger aircraft, like the A380 and the B747, suggest a true efficiency of scale because these aircraft are operational and the weights and costs are reasonably accurate. The specific cost, in $2012 per pound, for all the aircraft in Figure 11.3 is shown in Figure 11.4. The correlation curve Equation (11.2) is divided through by W_e to obtain the specific cost

$$SC = \frac{TC \times 10^6}{W_e} = \frac{4.25 \times 10^8}{W_e} erf \left[\frac{W_e - 10^4}{4.5 \times 10^5} \right] \tag{11.3}$$

The resulting curve is compared with the price data in Figure 11.4. The trend is once again in reasonably good agreement with the data available. However, in the case of the newest aircraft the data must be considered preliminary. The three Airbus data points above the correlation are for the A350-800, -900, and -1000 for which the empty weights are merely estimates and not well-established figures. Similarly, the results for the B737-800Max and -900Max fall below the correlation but they are also new aircraft for which empty weights are not well known at the moment but merely estimates. The fact that the B737-700Max is closer to the correlation is probably fortuitous. The Airbus 319neo, 320neo, and 321neo, for which empty weights also had to be estimated, fall somewhat below the correlation. The A380's initially projected price of over $300 million kept creeping up with time and now that it is operational the quoted average price is $389.2 million in $2012. The effect of the introduction of a new aircraft on the economics of the company introducing the

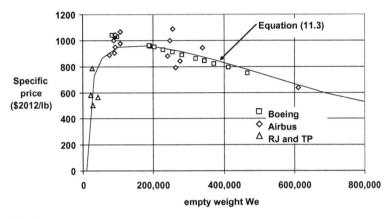

FIGURE 11.4

Specific prices of Boeing and Airbus airliners as a function of empty weight. Equation (11.3) is shown as a solid line. The lowest empty weights are for various regional jets and turboprops, and are denoted by RJ and TP, respectively.

aircraft, as well as on its competitors, is described in Irwin and Pavcnik (2004) for the particular case of the A380.

It is suggested that one use the correlation given above in Equation (11.2), which is based on current Boeing and Airbus aircraft and on limited data for some representative regional jets and turboprop aircraft, to determine an initial cost estimate for the design aircraft. Of course, in future years the cost data should be inflated suitably, or recourse may be made to current pricing prevalent in the industry, as was done here with the Boeing and Airbus data. It is interesting to examine the specific cost of the aircraft in terms of $2012/lb and this is shown in Figure 11.4. The airliners are seen to rapidly increase in specific cost from about $600 up to $1100 per pound with a slow decline back toward $600 per pound as the empty weight increases further.

Note that the specific price of the aircraft appears to have a maximum of around $1000/lb in the empty weight range of about 100,000 lb–250,000 lb. The efficiency of scale seems to be operating at both ends of the practical empty weight spectrum. The current push by regional aircraft manufacturers to enlarge their aircraft into the 100-passenger category is likely to result in higher specific costs which in turn will be reflected in considerably higher initial pricing of such aircraft.

11.3 Direct operating cost

In order to operate the aircraft as a revenue producer there are recurring costs that must be paid. The major elements here are the cost of consumables, like fuel and oil, and the cost of maintenance labor and parts. The professional society of the airliner business in the US, the Air Transportation Association of America, now called

Aircraft for America (A4A), used industry-wide statistical data to develop a standard method for estimating comparative direct operating costs of jet airplanes; see, for example, ATA (1967). Their approach, which is adopted here, determines the direct operating cost per air mile, denoted in the report by C_{am}, which can be readily converted to other units such as cost per flight hour, and involves the calculation of a number of primary variables. Note that these costs used in ATA (1967) are in $1967 and must be adjusted appropriately for the current application. The different variables used in the cost estimation method are discussed in detail and related to those developed through the design elements in the preceding chapters in the following subsections. Note that the ATA (1967) approach often uses statute miles as well as nautical miles and uses miles per hour for speed rather than knots.

11.3.1 Block speed

The block speed, in statute miles per hour, may first be put in terms of previously defined variables as follows:

$$V_b = \frac{R}{t_b} = \frac{R}{t_{gm} + t_4 + t_{10} + t_{cr} + t_{am}} \tag{11.4}$$

The block time is the sum of the time spent in each of the segments denoted in Equation (11.4), that is,

$$t_b = t_{gm} + t_4 + t_{cr} + t_{10} + t_{am} \tag{11.5}$$

The quantity R is the range covered in statute miles and t_4 and t_{10} denote the time, in hours, spent in the climb and descent segments of the mission profile, as was shown in Figure 2.2. These times are determined in the performance evaluations of Chapter 10. The time at cruise altitude, in hours, in Equation (11.5), which includes air traffic allowance, is defined by

$$t_{cr} = \frac{(R + K_a + 20) - (x_4 + x_{10})}{V_{cr}} \tag{11.6}$$

In this equation the distances covered in climb and descent, x_4 and x_{10}, are those values, in statute miles, determined in Chapter 10 and V_{cr} is the cruise velocity in statute miles per hour. The airway distance increment is given by

$$K_a = 7 + 0.015R \quad R \le 1400 \text{ mi}$$
$$K_a = 0.02R \quad R > 1400 \text{ mi} \tag{11.7}$$

The ground maneuver time t_{gm} is defined as being 0.25 h for all aircraft though in the current environment where substantial delays are not uncommon this appears to be an understatement. However, for uniformity it is recommended that this standard be used in all calculations. In the same fashion, the air maneuver time t_{am} is specified as 0.1 h for all aircraft and again this standard should be applied in all calculations.

11.3.2 Block fuel

The reserve fuel required by the aircraft has already been estimated in Chapter 2 and should have been refined in Chapter 10. The block fuel, in pounds, is defined as follows:

$$W_{f,b} = W_{f,gm} + W_{f,am} + W_{f,4} + W_{f,cr} + W_{f,10} \tag{11.8}$$

This quantity is approximately equal to the fuel calculated to be used in the standard flight profile, that is, segments 1–5 and 10 and 11 of Figure 2.2, as given below:

$$W_{f,b} \approx \sum_{i=1}^{5} W_{f,i} + W_{f,10} + W_{f,11} \tag{11.9}$$

In Table 2.4 $W_{f,5}$ is calculated based on R/V_{cr} as the time spent in cruise while $W_{f,cr}$ is based on t_{cr} as given by Equation (11.6) and the results will be slightly different. The ground maneuver fuel $W_{f,gm,}$ which is based on the 15-min ground maneuver time and is basically given by

$$W_{f,gm} = F_{taxi} C_{j,taxi} \left(\frac{14}{60}\right) + F_{to} C_{j,to} \left(\frac{1}{60}\right) \tag{11.10}$$

Here 14 of the 15 min of the ground maneuver time are spent at taxi-level thrust (say, 5% to 10% of takeoff thrust) and the remaining 1 min at takeoff thrust. The air maneuver fuel $W_{f,am,}$ is based on the 6-min air maneuver time at best cruise procedure, which yields

$$W_{f,am} = W_5 \left[e^{\frac{6}{60} \frac{C_j}{L/D}} - 1 \right] \tag{11.11}$$

It is prudent to compare the calculation of the block fuel according to Equation (11.8) with the results of Equation (11.9) as a check.

11.3.3 Flight crew costs

The flight crew cost equation for aircraft with gas turbine engines given in ATA (1967) is based on economic conditions in 1967 and the results must be updated to $2012. It is convenient to simply inflate the equation result by the ratio of the consumer price index (CPI) in 2012 to that in 1967 which is

$$\frac{CPI_{2012}}{CPI_{1967}} = 6.974 \tag{11.12}$$

This yields the following equation for cost (in $2012) per air mile for turbofan aircraft with a two-man crew:

$$C_{am} = \left[0.349 \frac{W_{to}}{1000} + 697 \right] V_b^{-1} \tag{11.13}$$

The corresponding equation for turboprop aircraft with a two-man crew is

$$C_{am} = \left[0.349\frac{W_{to}}{1000} + 439\right]V_b^{-1} \tag{11.14}$$

For international flights with a three-man crew the cost equation becomes

$$C_{am} = \left[0.349\frac{W_{to}}{1000} + 836.4\right]V_b^{-1} \tag{11.15}$$

These equations depend upon aircraft size because flight crews typically receive higher compensation when flying larger aircraft. In order to assess the reasonableness of this simple correction, the cost data provided by MIT (2012) were examined. There it is reported that for the year 2011 the average compensation package including wages, salaries, pensions, and benefits for flight crew, that is, pilots and copilots, was $179,580 across a spectrum of airlines including main legacy carriers and low-cost carriers. The average monthly flying hours logged across the same spectrum in 2011 was reported as 54 h. This yields an annual average of 648 h, which, for an assumed average block speed of 450 mph, yields an annual average of 291,600 air miles flown. Inflating that cost to $2012 results in a compensation package of $186,084 (in $2012). Assuming that the number of air miles flown remains the same for 2012, the cost per air mile (in $2012) for a two-man flight crew is estimated to be

$$C_{am} \approx \frac{2\,(\$186,084)}{291,600\text{ mi}} = 1.28\$/\text{mi} \tag{11.16}$$

Results obtained from Equations (11.13)–(11.16) are compared in Figure 11.5 and suggest that the direct inflation of the costs is consistent with readily available information on crew costs.

The cost for the flight crews as given by Equation (11.16) is based on recent compensation data that are an average over different carriers and aircraft. As a result, there is no dependence on aircraft gross weight shown by Equation (11.16). However, in practice the compensation for flight crew is dependent on the gross weight of the aircraft. It is important to note that crew cost has been quite volatile over the last decade. For example, if we look at the year 2002 and include, to the extent possible, the same carriers we would find a crew compensation of $201,514 in $2012 which represents an 8% increase in compensation cost compared to the actual costs in 2012. There was also a reduction in block hours per month so that the cost per air mile would have been $1.50 in $2012. Thus productivity per pilot has been increased and it seems reasonable to use an average between Equation (11.16) and one of Equations (11.13)–(11.15) to estimate flight crew costs.

11.3.4 Fuel and oil costs

The cost of Jet A fuel cannot currently be estimated accurately by using simple consumer price index inflation. This technique was reasonably accurate up to about 2002

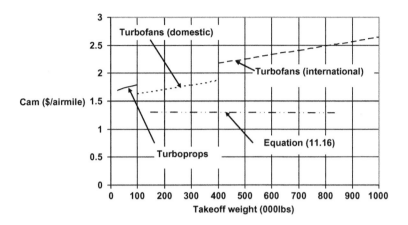

FIGURE 11.5

Comparison of the cost per air mile for the flight crew as given by
Equations (11.13)–(11.16) as a function of aircraft takeoff weight.

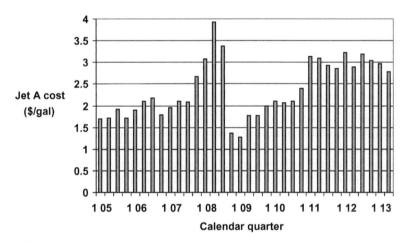

FIGURE 11.6

Cost per gallon of Jet A fuel in constant $/gal as a function of calendar quarter as reported
by IATA (2012).

when Jet A was about $0.54 per gallon but the fuel cost index rose by a factor of
2.67–$1.55 per gallon between then and late 2004. The actual fuel costs are tracked
weekly by various groups, including the US Department of Energy, and are reported
by the International Air Transport Association in IATA (2012). The level of Jet A fuel
cost was reasonably steady until the end of 2007 when it started to rise rapidly, as
shown in Figure 11.6. Thus in calculating the fuel cost it is important to use current
data as can be found in IATA (2012). To convert to weight, the density of Jet A may
be taken as 6.76 lbs per gal at standard conditions.

Turbine lubricating oil prices are more difficult to obtain directly from standard searches and may require more detailed contact with suppliers. An examination of retail prices for turbine oil shows a cost of around $60 per gallon in $2012. To convert to weight, the density of turbine lubricating oil may be taken as 8.1 lbs per gal at standard conditions.

The cost per air mile for jet fuel and oil, where the cost of fuel C_F in $/lb and of lubricating oil C_o is measured in $/lb, is given by

$$C_{am} = \frac{1.02 \left(W_{f,b} C_f + 0.135 n_e C_o t_b \right)}{R} \tag{11.17}$$

11.3.5 Hull insurance

The insured value is assumed to be the full initial cost of the aircraft and the insurance premium rate, r_i, which generally ranges from 1% to 3% per year, is assumed to average 2% over the useful life of the aircraft, typically about 12 years, although many airlines have been operating their fleet for as many as 20 years, as illustrated by the data in Figure 1.1. With the current emphasis on fuel economy and reduced emissions operators are now eager to eliminate older aircraft and buy newer, more advanced aircraft. A side effect of this movement is a drop in the price of used aircraft while disassembling them for resale of their components has become an attractive business. The cost per air mile for hull insurance is given by

$$C_{am} = \frac{r_i \, (TC)}{UV_b} \tag{11.18}$$

In Equation (11.18) r_i is the insurance rate in percent, TC is the total airplane cost in dollars, and U is the annual utilization in block hours per year. A correlation for the utilization factor U is shown in Figure 11.7. For example, an aircraft whose typical block time for a flight is 3 h would have a utilization factor $U = 4000$ h in a year. If the airplane flew one flight per day it would be utilized for 1095 h. This implies that, on the average, the aircraft with $U = 4000$ would be making 3.65 flights per day for a year. If the block time is doubled to 6 h, the utilization factor would be $U = 4400$ h which implies an average of only two flights per day over a year. Equation (11.18) indicates that cost per air mile should decrease with high utilization factors and high block speed. Thus short trips and low speeds, like those for regional aircraft, are more expensive per air mile than long-range flights.

11.3.6 Airframe maintenance labor and materials

The correlation equation for labor cost is directly proportional to the labor rate:

$$C_{am} = \frac{L_{mh,f} t_f + L_{mh,c}}{V_b t_b} r_L \sqrt{M_{cr}} \tag{11.19}$$

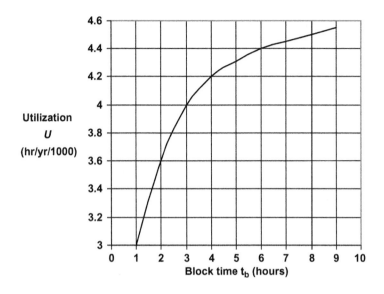

FIGURE 11.7

Annual utilization factor U is shown as a function of the block time.

The number of labor man-hours per flight cycle is given by the following correlation:

$$L_{mh,c} = 6 + 0.05 \frac{W_e}{1000} - \frac{630}{\frac{W_e}{1000} + 120} \tag{11.20}$$

The number of labor man-hours per flight hour is $L_{mh,f} = 0.59 L_{mh,c}$ and r_L is the labor rate in $ per hour. Here W_e is the airplane empty weight in pounds and M_{cr} is the cruise Mach number which is taken as unity for subsonic aircraft.

MIT (2012) reports the average annual salary for airline maintenance personnel to be $77,955 ($2011) while the average number of maintenance workers per plane appears to be around 10 per airplane. Assuming a 2000-h work year and inflating the $2011–$2012 yields a labor cost of about $40 per hour in $2012. This value, or the appropriately inflated value for future years, is to be used for the labor rate in the airplane labor correlation in Equation (11.19). In this equation the flight time t_f should be considered to be the block time less the ground maneuvering time of 0.25 h, that is, $t_f = t_b - 0.25$. In the equation for aircraft material costs it is sufficient to use the simple CPI inflator.

The materials cost for airframe alone is given by

$$C_{am} = \frac{(TC - n_e C_e)(3.08 t_f + 6.24) \times 10^{-6}}{V_b t_b} \tag{11.21}$$

Here the correlating factor is the current capital cost of the airplane less the current cost of the engines $(TC - n_e C_e)$ and because these are current costs no inflator should be applied.

11.3.7 Engine maintenance labor and materials

The engine maintenance costs are proportional to the maximum takeoff thrust and are given by

$$C_{am} = \frac{r_L n_e}{V_b t_b} \left\{ \left[0.6 + 0.027 \left(\frac{F_{to}}{1000} \right) \right] (1.08)^j t_f + \left[0.065 + 0.03 \left(\frac{F_{to}}{1000} \right) \right] \right\} \quad (11.22)$$

In Equation (11.22) $j=0$ and F_{to} is takeoff thrust in pounds for turbofans, while $j=1$ and F_{to} is takeoff power in horsepower for turboprops. It is suggested to use the same labor rate r_L as for the airframe, $40 per hour in $2012. The material cost equation depends upon engine cost C_e, so that information must be sought from the manufacturer. Jenkinson et al. (1999) suggested a correlation for the engine price in the following form (a and b are constants)

$$C_e = a + b \frac{F_{cr}^{0.088}}{C_{j,cr}^{2.58}} \quad (11.23)$$

Updating this correlation to $2012 yields the following approximation:

$$C_e = 1.2 \left(1 + \frac{F_{cr}^{0.088}}{C_j^{2.58}} \right) \quad (11.24)$$

Using the data presented in the database compiled by Svoboda (2000) for cruise thrust and cruise-specific fuel consumption for 26 turbofan engines in Equation (11.24) yields the results illustrated in Figure 11.8. However, because cruise values of thrust and specific fuel consumption are rarely easily available, the price data are shown as a function of takeoff, or static, thrust, which is generally quoted by manufacturers. There is appreciable scatter, but the trend is evident. The open symbols in Figure 11.8 represent prices for various engines quoted from Deagel (2012). The agreement suggests that the correlation of Equation (11.24) is reasonable and should be satisfactory for use in Equation (11.25). However, the difficulty in obtaining accurate information on cruise values of thrust and specific fuel consumption or actual prices forces one to make an estimate based on the required takeoff thrust, which is known for the design aircraft. In this case the data shown in Figure 11.8 may be used in helping to select a preliminary engine price to use Equation (11.25). Of course, actual price data should be used in the analysis whenever possible.

In the case of turboprop engines, the range of required power is typically between 1000 and 500 horsepower. Some data presented by Forecast International (2013) suggest that turboprop engines in that power range are priced at about $330–$450 per horsepower in $2012. The specific power of these engines is in the range of 2–3.3

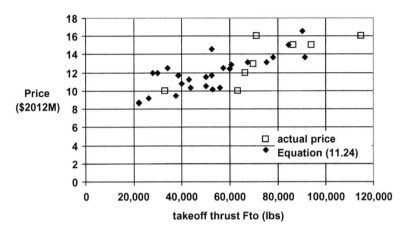

FIGURE 11.8

Estimated prices of turbofan engines in millions of $2012, based on Equation (11.24), are shown as a function of takeoff thrust in pounds. The open symbols denote publically quoted price for various engines.

horsepower per pound. Once again, it is always advisable to attempt to use actual weight and cost information whenever possible.

The engine materials cost per air mile is approximated by

$$C_{am} = \frac{2n_e \left(\frac{C_e}{10^5} \right) (1.25t_f + 1)}{V_b t_b} \qquad (11.25)$$

11.3.8 Maintenance burden

An indirect cost burden associated with the airframe and engine labor costs is levied at a rate of at least 180% of the sum of the direct labor costs in these categories. This reflects other costs not directly attributable to labor on the airframe, such as employee benefits, travel, training, etc.

11.3.9 Depreciation

Recognizing that depreciation of capital value is specific to the particular operator and current economic and competitive conditions, the calculation method uses a simple amortization over a fixed period t_{dep}, typically 12 years, over which the residual value of the airplane is taken to be zero. The depreciation cost is given by

$$C_{am} = \frac{0.9TC - 0.3n_e C_e}{UV_b t_{dep}} \qquad (11.26)$$

11.3.10 Direct operating cost

The values for the contribution to the cost per air mile from the various operational elements evaluated previously in this section must be summed to obtain the total direct operating costs for the airplane. The direct operating cost in $/mile is denoted by C_{am} but is more commonly referred to as DOC. Sometimes it is given in cents per available seat-mile, or DOC_{asm} (cents/seat-mile)$= 100C_{am}/N_p$.

11.4 Indirect operating cost

There are other costs encountered in the commercial airline business which are not connected with the actual flight operations described previously. These are costs associated with airport landing fees, flight attendants, food and beverage service, passenger-related activities like reservations, sales, and baggage handling, general and administrative expenses, etc. In the past the Aircraft Industries Association had developed statistical equations for estimating these indirect operating costs, or IOC. Using those equations, Shevell (1990) generated an equation for IOC based on a study of three aircraft: Boeing 747 (4 engines), Douglas DC-10 (3 engines), and a large twin-engine jet typical of more current airliners. Modifying this approach to eliminate the use of graphs and updating the cost to $2012 leads to the following equation for IOC in $2012/mi:

$$IOC = R^{-0.41}[1.42 \times 10^{-4}W_{to} + (0.13 + 1.4LF)N_p - 4.4] \qquad (11.27)$$

The quantity LF denotes the load factor and is equal to the ratio of paid passengers to N_p, the actual number of seats available. Equation (11.27) should be suitable for preliminary design purposes. Shevell's (1990) approach is based on the observation that the ratio of total operating cost to direct operating cost decreases as range increases. From the definition that the total operating cost is the sum of the direct and indirect costs, the ratio of total to direct operating cost is given by

$$\frac{TOC}{DOC} = 1 + \frac{IOC}{DOC}$$

If this ratio decreases with range, then obviously the ratio IOC/DOC also decreases with range. This seems reasonable since the increase in range is basically an increase in block time and most of the indirect costs, like ticket sales, reservations handling, general and accounting costs, etc., don't depend on block time. Note that Equation (11.27) shows that the IOC grows with aircraft size through both the gross takeoff weight and the number of seats N_p. It may be recalled that in Figure 1.12 it was shown that the takeoff weight may be roughly correlated by the number of passenger seats according to the relation

$$W_{to} \approx 222N_p^{1.361} \qquad (11.28)$$

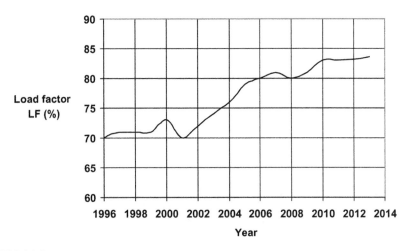

FIGURE 11.9

Annual average load factor for the total US domestic network, including major airlines, low-cost carriers, and others over the period 1996–2013.

The *IOC* also grows with the load factor *LF*, emphasizing that the indirect costs increase as more passengers fill up the available seats. Of course an increase in load factor also increases the revenue collected. Load factors for the total US domestic network, including major airlines, low-cost carriers, and others, were fairly flat at 70–72% over the period from 1996 to 2002, as reported in MIT (2012) and shown in Figure 11.9.

The increasingly competitive nature of the airline business and rising costs placed a number of carriers in jeopardy, precipitating bankruptcies and mergers. The awareness that aircraft had to be more fully loaded with paying passengers prompted airlines to trim the number of flights resulting in a continuous upward slope of the load factor during the period from 2004 to 2007. The economic crisis of 2008–2009 depressed load factors briefly, but then they started upward once more and ended 2010 at about 83% and remained between 83% and 84% up to 2013. Load factors are reported in airlines' monthly traffic reports while revenue per available seat-mile, or unit revenue, is generally reported only in quarterly financial reports. The fares charged may vary significantly in these quarterly periods so the actual revenue will also vary.

Equation (11.27) divided by N_p was used to calculate the indirect operating cost per seat-mile IOC_{asm} for several W_{to} and N_p pairs representing small narrow-body (B737-600), large narrow-body (B757-300), and large wide-body (B747-400) aircraft. The load factors of 60%, 80%, and 100% were considered. The results for the case $LF = 80\%$ are shown in Figure 11.10 along with a simple correlation which agreed well with all of the data given by

$$IOC_{asm} = \frac{IOC}{N_p} = \frac{1.79LF}{R^{0.41}} \tag{11.29}$$

The results in Figure 11.10 are representative of the results for other load factors as given by Equation (11.29). This may be seen by dividing Equation (11.27) by the number of available seats, N_p, and by substituting Equation (11.27) for the takeoff weight to get the indirect operating cost per available seat-mile or IOC_{asm} as given by

$$IOC_{asm} = \frac{IOC}{N_p} = 0.0315\frac{N_p^{0.361}}{R^{0.41}} + \frac{0.13 + 1.4LF}{R^{0.41}} - \frac{4.4}{N_p R^{0.41}} \tag{11.30}$$

The range and number of passenger seats is also related, as is shown in Figure 1.11 where $15 < R/N_p < 40$ with an average around $R/N_p = 25$. The first term on the right-hand side of Equation (11.30) may be written as follows:

$$0.0315\frac{N_p^{0.361}}{R^{0.41}} < 0.0315\left(\frac{N_p}{R}\right)^{0.41} = O\left(10^{-2}\right)$$

Using the same reasoning, the last term on the right-hand side of Equation (11.30) can be shown to be of $O(10^{-4})$ and the dominant term therein is of the form of Equation (11.29), which was deduced from the data. This suggests that the major influence on the indirect operating cost is the load factor and the range.

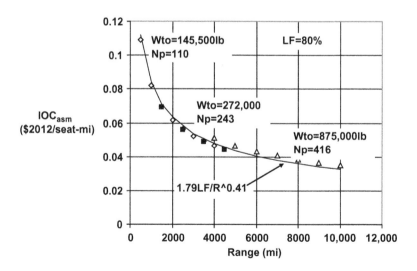

FIGURE 11.10

Variation of the IOC_{asm} for several representative aircraft as a function of range according to Equation (11.29) for a load factor of 80%. The data are well represented by the simple relation $IOC_{asm} = 1.79LF/R^{0.41}$.

11.5 **Breakeven load factor**

The revenue provided by the paying passengers is equal to the average ticket price P paid by the passengers actually carried on the flight so that

$$\text{Income} = (P)(N_p)(LF) \qquad (11.31)$$

The total cost of a flight of range R is proportional to the $TOC = DOC + IOC$, and is given by the following equation:

$$\text{Expense} = R(DOC + IOC) \qquad (11.32)$$

Thus the breakeven load factor is that value for which the income and expense are equivalent, or

$$LF_{be} = (DOC + IOC) \frac{R}{PN_p} \qquad (11.33)$$

Actual load factors for the total US airline network during the critical period of 2000–2008, when fuel prices were rapidly rising, fell below the corresponding breakeven values. It is clear that, on the average, airlines were losing money during the period 2001–2006. Cost cutting and increasing load factor brought a period of prosperity back to the airlines, but at the end of 2008 the economic crisis took hold. Although fuel prices fell at first, so did traffic, and by the beginning of 2009 profitability was effectively terminated by the subsequent surge in fuel prices.

Longer trips and fewer seats will tend to increase the breakeven load factor as will lower average ticket prices. However the DOC and IOC are both dependent on range and number of seats so that the issue must be studied at a finer level of detail to emerge with meaningful assessments of the breakeven load factor. A review of the database of 41 commercial transports with $30 < N_P < 555$ and $800 < R < 9210 \text{mi}$, the average value of the quantity R/N_P is 25, with $12 < R/N_P < 54$. Ticket prices, on the other hand, are sensitive to current economic conditions and the competitive environment.

Fuel costs now outpace labor costs and constitute the major portion of operating cost. The fuel cost as a percentage of passenger revenue is shown in Figure 11.11 over the time period 1995–2011, clearly illustrating the major effect of those fuel costs and why fuel efficiency is such an important driver in aircraft design. However, it is worth noting that the amount of fuel consumed per block has been continually decreasing during the period shown in Figure 11.11 and that fuel efficiency climbed to more than 50 mi per gallon per paying passenger.

The cost and associated revenue in cents per available seat-mile for passenger service, excluding cargo and other transport-related items, is shown in Figure 11.12 for the period 1995–2011. Although both cost and revenue continued to rise over this period, it was the cost of fuel that had the major effect of driving up the cost. For more detail on airline operations see Belobaba et al. (2009).

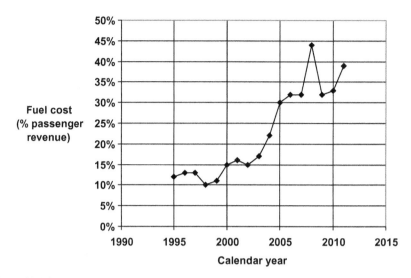

FIGURE 11.11

Fuel cost as a percentage of total passenger revenue as report by MIT (2012).

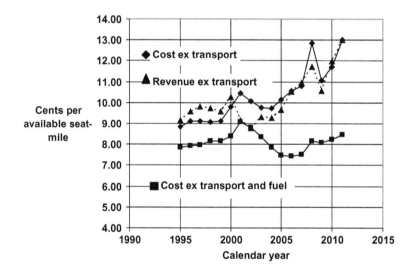

FIGURE 11.12

Unit cost of operation in cents per available seat-mile over the period 1995–2011 as reported in MIT (2012).

The breakeven load factor involves ticket prices and information on average fares, both domestic and international, may be found, for example, in Airlines (2013). They note that the average domestic round-trip fare rose from $186 in 1979 to $378 in

2012 while the US CPI rose by a factor of 3.16 over the same period. The average mileage for the round trip during this period changed little, rising from 1947 to 2356 mi. Thus purchasing a ticket in 1978 with $2012 would mean paying out $588. The average international round-trip fare rose from $782 in 1990 to $1236 in 2012. Thus purchasing a ticket in 1990 with $2012 would mean paying out $1375 for a round-trip international flight of about 7700 mi, about 10% longer than an average 1990 round trip. The specific cost to the passenger is about 16 cents/mi (in $2012) for both the domestic and international average round-trip flights quoted.

11.6 Design project activity

At this point in the design project the preliminary design and the aerodynamic performance of the aircraft are well defined. It is now important to assess the economic performance of the design aircraft using the techniques developed in this chapter. The empty weight of the aircraft, as determined in the refined weight estimate carried out as part of the design project activity in Chapter 8, may be used to estimate the capital cost of the aircraft. The direct operating cost may be estimated using information developed in other parts of the preliminary design process. The range and speed, along with the climb and descent distances, permit calculation of the block time and speed, while the fuel fractions calculated for the different phases of the mission will determine the block fuel. The type of aircraft and the number of crew are used to evaluate the flight crew costs. The expected fuel costs may be determined with the aid of the detail reports on current and projected fuel prices. The hull insurance cost depends upon the capital cost of the aircraft and the assumed useful life of the aircraft. The costs of airframe and engine maintenance and labor and the overhead burden on those costs may be estimated, along with the depreciation. The indirect operating costs will determine the cost per air mile and the breakeven load factor.

11.7 Nomenclature

C_{am}	cost per air mile
C_e	cost of one engine
C_f	cost of fuel
C_j	specific fuel consumption
C_o	cost of lubricating oil
CPI	consumer price index
DOC	direct operating cost
F	thrust
IOC	indirect operating cost
K_a	airway distance increment, Equation (11.7)
$L_{mh,c}$	labor man-hours per flight cycle
$L_{mh,f}$	labor man-hours per flight hour

LF	load factor
M	Mach number
n_e	number of engines
N_p	number of passenger seats
P	ticket price
R	range
r_i	insurance cost rate
r_L	labor cost rate
SC	specific cost, $/lb
TC	total cost of one airplane
t_f	time in flight
TOC	total operating cost
t	time
U	utilization factor
V	velocity
W_e	airplane empty weight
W_f	airplane fuel weight
x	distance

11.7.1 Subscripts

am	air maneuver
asm	per available seat-mile
b	block
be	breakeven
cr	cruise
dep	depreciation
f	fuel
gm	ground maneuver
taxi	taxiing conditions
to	takeoff
4	climb segment
5	cruise segment
10	descent segment
11	landing segment
100	100 aircraft

References

AIA, 2012. Aerospace Industries Association. <http://www.aia-aerospace.org/economics/aerospace_statistics/>.

Airbus, 2013. Airbus Price List. <http://www.airbus.com/presscentre/corporate-information/key-documents/?eID=dam_frontend_puh&docID=14849>.

Airlines, 2013. <http://www.airlines.org/Pages/Annual-Round-Trip-Fares-and-Fees-International--.aspx. http://www.airlines.org/Pages/Annual-Round-Trip-Fares-and-Fees-Domestic.aspx>.

ATA, 1967. Standard Method of Estimating Comparative Direct Operating Costs of Turbine Powered Airplanes. Air Transport Association of America.

Belobaba, P., Odoni, A., Barnhurst, C., 2009. The Global Airline Industry. Wiley, New York.

BLS, 2012. U.S. Bureau of Labor Statistics. <http://stats.bls.gov/cpi>.

Boeing, 2013. Boeing Jet Prices. <http://www.boeing.com/commercial/prices/>.

Deagel, 2012. <http://www.deagel.com>.

DOD, 1999. Joint Industry/Government Parametric Estimating Handbook, second ed. U.S. Department of Defense. Forest International (2013), <http://www.ispa-cost.org/PEIWeb/newbook.htm>, <www.forecastinternational.com/samples/F461_CompleteSample.pdf>.

Forecast International, 2013. <http://www.forecastinternational.com/samples/F461_CompleteSample.pdf>.

IATA, 2012. International Air Transport Association Jet Fuel Price Monitor. <http://www.iata.org/whatwedo/economics/fuel_monitor/Pages/price_analysis.aspx>.

Irwin, D.A., Pavcnik, N., 2004. Airbus versus Boeing revisited: international competition in the aircraft market. Journal of International Economics, citation.

Jenkinson, L.R., Simpkin, P., Rhodes, D., 1999. Civil Jet Aircraft Design. AIAA, Reston, VA.

Large, J.P., Campbell, H.G., Gates, D., 1976. Parametric Equations for Estimating Aircraft Airframe Costs. RAND Corporation, Report R-1693-1-PA&E.

MIT, 2012. Global Airline Industry Program. <http://web.mit.edu/airlinedata/www/default.html>.

NASA, 2012. Cost Estimating Resources. <http://cost.jsc.nasa.gov/>.

Shevell, R. <http://adg.stanford.edu/aa241/cost/cost.html>.

Svoboda, C., 2000. Turbofan engine database as a preliminary design tool. Aircraft Design 3, 17–31.

Airfoil Characteristics

Airfoil development is discussed in Section 5.2 and the associated theoretical fundamentals are developed in Appendix C. For illustrative purposes, data on airfoil lift, drag and moment coefficients are presented for several selected NACA 6-series airfoils taken from the summary of data for over 120 NACA airfoils reported by Abbott et al. (1945) and Abbott and von Doenhoff (1959). The format used for the data shown here is typical of that often available for airfoils. An exhaustive catalog of airfoil data from many different sources is available in UIUC (2013).

The use of articulated flaps to enhance the maximum lift of an airfoil is discussed in Section 5.5. The double-slotted flap was shown to provide a good practical compromise between aerodynamic efficiency and mechanical complexity. In order to illustrate the deployment of a double-slotted flap and the corresponding aerodynamic effects, data from Abbott et al. (1945) are shown here for a NACA 6-series airfoil so equipped.

The Reynolds numbers and Mach numbers for the data shown here are relatively low compared to the flight conditions typically encountered by commercial aircraft, as discussed in Section 5.4. The development of advanced cryogenic wind tunnels has permitted the experimental study of flows at conditions replicating those of practical aircraft, as discussed in Section 5.2. The state of the boundary layer flow over an airfoil is dependent upon the Reynolds and Mach numbers, the condition of the airfoil surface, and the degree of turbulence in the free stream passing over the airfoil. The difference between the aerodynamic characteristics of an airfoil in laminar and turbulent flow is substantial as discussed in Section 9.5. The data shown here provide one illustration of this important difference. It must be kept in mind that laminar flow predominates for those airfoils described as "smooth" while turbulent flow predominates over those airfoil denoted as "rough," that is, with artificially roughened regions near the leading edge. It is apparent from the data that the location of the transition from laminar to turbulent flow over an airfoil largely determines its performance.

NACA 63₁-012

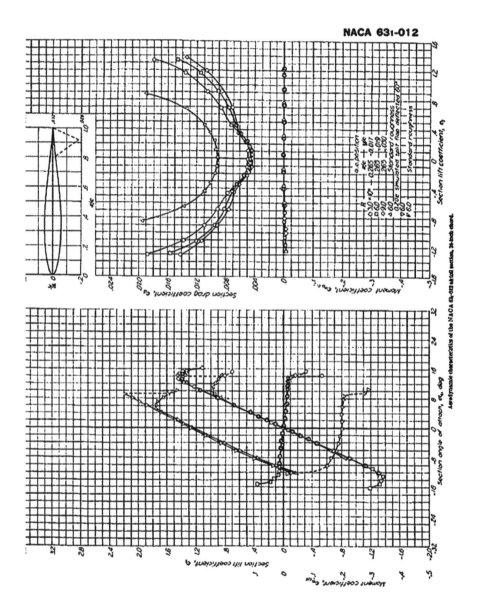

Aerodynamic characteristics of the NACA 63₁-012 airfoil section, 24-inch chord.

NACA 63₁-212

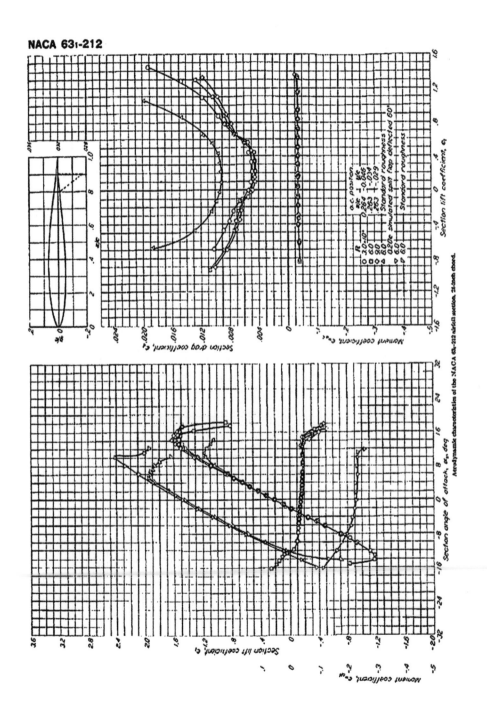

Aerodynamic characteristics of the NACA 63₁-212 airfoil section, 24-inch chord.

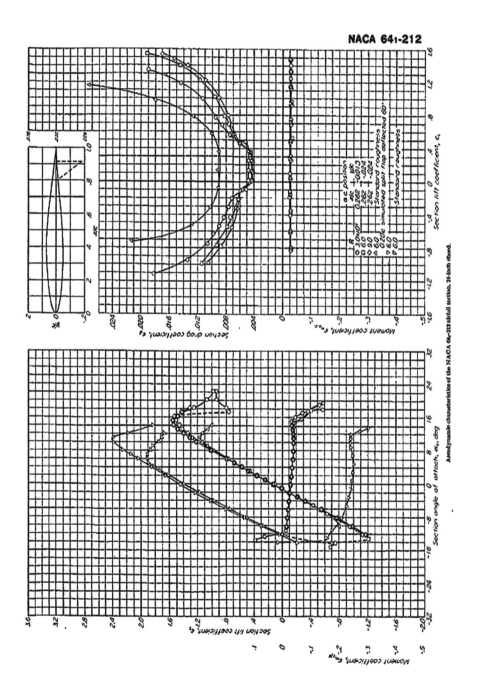

NACA 64₁-212

Aerodynamic characteristics of the NACA 64₁-212 airfoil section, 24-inch chord.

NACA 64₁-412

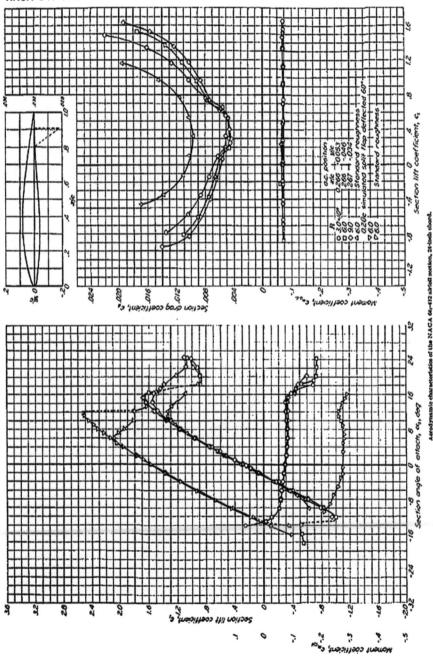

Aerodynamic characteristics of the NACA 64₁-412 airfoil section, 24-inch chord.

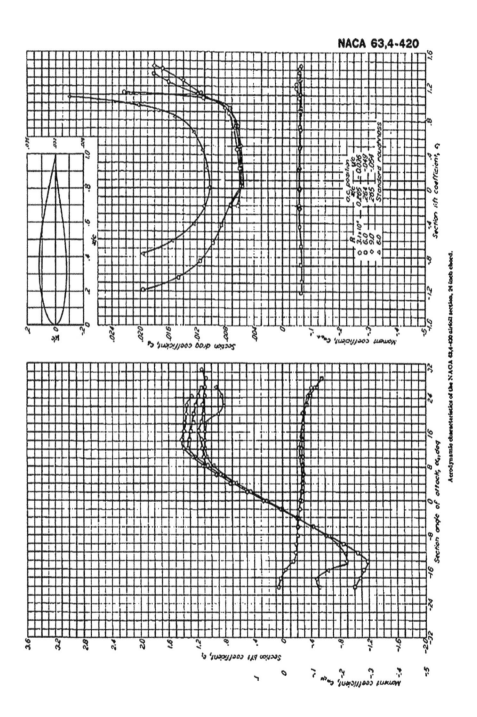

NACA 63,4-420

Aerodynamic characteristics of the NACA 63,4-420 airfoil section, 24 foot chord.

NACA 63,4-420 with flap

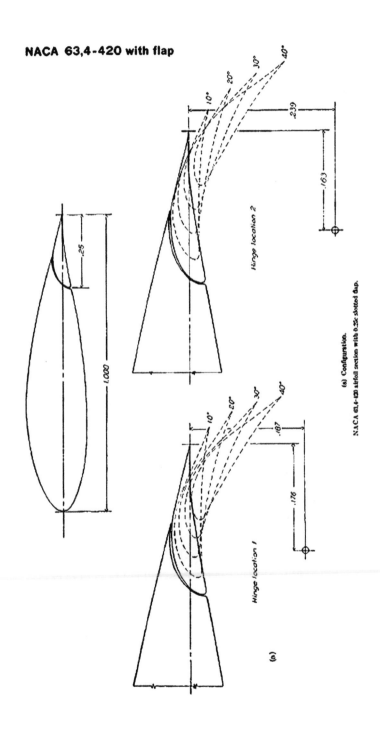

(a) Configuration.

NACA 63,4-420 airfoil section with 0.25c slotted flap.

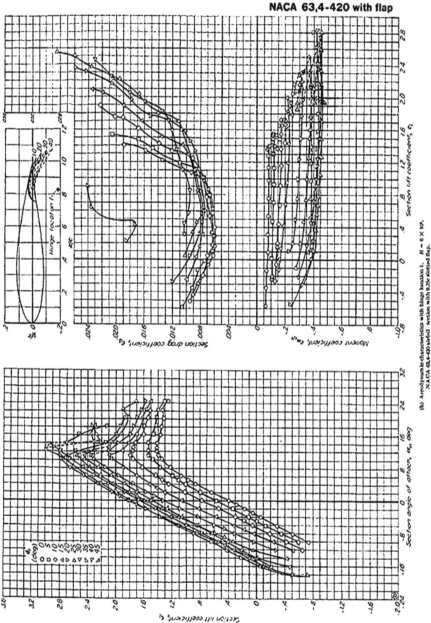

NACA 63,4-420 with flap

(b) Aerodynamic characteristics with hinge location 1. $R = 6 \times 10^6$.
NACA 63,4-420 airfoil section with 0.2c slotted flap.

References

Abbott, I.H., et al., 1945. Summary of Airfoil Data, NACA Technical Report No. 824.

Abbott, I.H., von Doenhoff, A.E., 1959. Theory of Wing Sections. Dover, NY.

UIUC, 2013. Airfoil Data Site, <www.ae.illinois.edu/m-selig/ads.html>.

1976 US Standard Atmosphere Model

B.1 The atmospheric environment

The chemical composition of the sensible atmosphere is essentially constant and by mole fraction it is comprised of 78% nitrogen, 21% oxygen, and 1% other gases. (Argon 0.93%, CO_2 0.03%, and neon, helium, krypton, hydrogen, xenon, and ozone in increasingly smaller amounts.) For most thermo-chemical purposes the atmosphere is considered a binary mixture of 79% nitrogen and 21% oxygen. This fixed composition approximation makes the temperature distribution a reliable means for dividing up the various important regions in the atmosphere. The pressure and density in the regions of specified temperature behavior may be determined from the equation of state and the conditions of hydrostatic equilibrium.

B.2 Equation of state and hydrostatic equilibrium

If one considers the atmosphere to behave as a perfect gas, then

$$p = \rho R T = \rho \frac{R_u}{W_m} T$$

The molecular weight of the mixture of atmospheric gases is essentially constant up to 100 km and is given by $W_m = 28.96$, and the atmospheric gas constant $R = 0.287\,\text{kJ/kg-K}$ (or $1716\,\text{ft}^2/\text{s}^2\text{-R}$). The hydrostatic equation for the atmosphere is

$$dp = -\rho g dz$$

Since the gravitational acceleration depends on altitude a new altitude function may be defined. This is the *geopotential* altitude, h, and it is related to the geometric altitude by the relation

$$g_E dh = g dz$$

Here $g_E = 9.087\,\text{m/s}^2$ (or $32.15\,\text{ft/s}^2$) is the gravitational acceleration at the surface of the Earth, $z=0$. The gravitational acceleration which varies with altitude according to Newton's law of gravitation may be written as

Commercial Airplane Design Principles. http://dx.doi.org/10.1016/B978-0-12-419953-8.00019-X
© 2014 Elsevier Inc. All rights reserved.

$$g = \frac{g_E R_E^2}{(R_E + z)^2} = g_E \frac{1}{\left(1 + \frac{z}{R_E}\right)^2}$$

The radius of the Earth is taken to be $R_E = 6357\,\text{km}$ (3950 miles or 3430 nm) so that for altitudes typical of commercial aircraft ($z < 15\,\text{km}$ or 49,000 ft) the ratio $z/R_E \sim 0.0024$, and therefore the difference between h and z is relatively small. Integrating the relation between h and z yields the geopotential altitude

$$h = z \frac{1}{1 + \frac{z}{R_E}}$$

Thus the difference between the geometric and geopotential altitudes for commercial aircraft is less than 1%. Using the equation of state in the hydrostatic equilibrium equation yields

$$\frac{dp}{p} = -g \frac{dz}{RT} = -g_E \frac{dh}{RT}$$

With an appropriate relation for temperature in terms of geopotential altitude, $T = T(h)$, one may integrate the hydrostatic equation and find $p = p(h)$ and $\rho = \rho(h)$. The 1976 standard atmosphere defines atmospheric layers, each with $T = T_i + \lambda_i(h - h_i)$, where T_i is the temperature of the start of layer i, h_i is the altitude at the start of layer i, and λ_i is the lapse rate, i.e., dT/dh, in that layer. Integration of the hydrostatic equation for non-zero λ yields

$$p = p_i \left[\frac{T_i}{T_i + \lambda_i (h - h_i)} \right]^{\frac{g_0}{R\lambda_i}}$$

In isothermal layers where $\lambda = 0$, the temperature $T = T_i = \text{constant}$, and the pressure is instead given by

$$p = p_i \exp\left[-\frac{g_E}{RT_i}(h - h_i) \right]$$

The temperature at the Earth's surface is taken as $T = 15\,^\circ\text{C} = 288.15\,\text{K}$ (518.7R) and $g_E/R = 34.17\,\text{K/km}$ (0.01874R/ft). For general reference the properties of this atmospheric model in the various layers are given in Table B.1.

The distribution of pressure, in kPa, in the various layers is then given by the following:

Layer 1 (0–11 km):

$$p = 101.3 \left(\frac{288.15}{288.15 - 6.5h} \right)^{\frac{34.17}{-6.5}}$$

Layer 2 (11–20 km):

Table B.1 Definition of the Layers in the 1976 Model Atmosphere[a]

Layer	Geopotential Altitude, h (km)	Geopotential Altitude, h (kft)	Lapse Rate, λ_i (K/km)	Thermal Type
1	0	0	−6.5	Neutral
2	11	36.1	0	Isothermal
3	20	65.6	+1.0	Inversion
4	32	105	+2.8	Inversion
5	47	154	0	Isothermal
6	51	167	−2.8	Neutral
7	71	233	−2.0	Neutral
8	84.85	278	+1.65	Inversion
9	100			

[a] The eighth layer lapse rate is based on the 1962 Standard Atmosphere. The difference between the 1976 and 1962 Standard Atmospheres is small for altitudes $h < 150$ km.

$$p = 22.62 \exp \left(\frac{-34.17\,[h - 11]}{216.65} \right)$$

Layer 3 (20–32 km):

$$p = 5.47 \left(\frac{216.65}{216.65 + [h - 20]} \right)^{\frac{34.17}{1}}$$

Layer 4 (32–47 km):

$$p = 0.8669 \left(\frac{228.65}{228 + 2.8\,[h - 32]} \right)^{\frac{34.17}{2.8}}$$

Layer 5 (47–51 km):

$$p = 0.1107 \exp \left(\frac{-34.17\,[h - 47]}{270.65} \right)$$

Layer 6 (51–71 km):

$$p = 0.06681 \left(\frac{270.65}{270.65 - 2.8\,[h - 51]} \right)^{\frac{34.17}{-2.8}}$$

Layer 7 (71–84.85 km):

$$p = 0.003946 \left(\frac{214.65}{214.65 - 2\,[h - 71]} \right)^{\frac{34.17}{-2}}$$

Layer 8 (84.85–100 km)

$$p = 0.0003724 \left(\frac{186.95}{186.95 + 1.65\,[h - 84.85]} \right)^{\frac{34.17}{1.65}}$$

The density may be found from the equation of state. Note that in Layer 2 the temperature is constant at about $216\,K$ or $-57\,°C$ ($390R$ or $-70\,°F$) in the altitude range of 10–20 km. This region is called the stratosphere and is the domain of high-speed manned flight, from jet airliners to military aircraft up to the Mach 3 SR-71 Blackbird. Because the speed of sound is proportional to the square root of the temperature, and the temperature through the stratosphere has little variation, the speed of sound is relatively constant. It is common to assume a constant value for preliminary design purposes and this can be used with minimal error. A table of useful pressure, density, temperature, sound speed, and kinematic viscosity data for this atmosphere model is given in Tables B.2 and B.3 for English units and in Tables B.4 and B.5 for

Table B.2 Properties of the 1976 US Standard Atmosphere for Altitudes between Sea Level and 50,000 ft in 2000 ft Increments

z (ft)	δ	θ	σ
0	1.0000	1.0000	1.0000
2000	0.9298	0.9862	0.9428
4000	0.8636	0.9725	0.8881
6000	0.8013	0.9587	0.8358
10,000	0.6877	0.9312	0.7384
12,000	0.6359	0.9175	0.6931
14,000	0.5874	0.9037	0.6500
16,000	0.5419	0.8900	0.6089
18,000	0.4993	0.8762	0.5698
20,000	0.4595	0.8625	0.5327
22,000	0.4222	0.8487	0.4975
24,000	0.3875	0.8350	0.4641
26,000	0.3551	0.8212	0.4324
28,000	0.3249	0.8075	0.4024
30,000	0.2969	0.7937	0.3740
32,000	0.2708	0.7800	0.3472
34,000	0.2467	0.7662	0.3219
36,000	0.2243	0.7525	0.2980
38,000	0.2005	0.7519	0.2666
40,000	0.1821	0.7519	0.2422
42,000	0.1654	0.7519	0.2200
44,000	0.1502	0.7519	0.1998
46,000	0.1365	0.7519	0.1815
48,000	0.1239	0.7519	0.1649
50,000	0.1126	0.7519	0.1497

Table B.3 Properties of the 1976 US Standard Atmosphere for Altitudes Between Sea Level and 50,000 ft in 2000 ft Increments (Concluded)

z (ft)	p (lb/ft²)	T (R)	ρ(slug/ft³)	a (ft/s)	a (kts)	v (ft²/s)
0	2116	518.7	2.377E-03	1116	660.9	1.573E-04
2000	1968	511.5	2.241E-03	1109	656.4	1.650E-04
4000	1828	504.4	2.111E-03	1101	651.8	1.733E-04
6000	1696	497.3	1.987E-03	1093	647.2	1.821E-04
10,000	1455	483.0	1.755E-03	1077	637.8	2.014E-04
12,000	1346	475.9	1.648E-03	1069	633.1	2.121E-04
14,000	1243	468.7	1.545E-03	1061	628.3	2.234E-04
16,000	1147	461.6	1.447E-03	1053	623.5	2.356E-04
18,000	1057	454.5	1.354E-03	1045	618.7	2.486E-04
20,000	972.3	447.3	1.266E-03	1037	613.8	2.626E-04
22,000	893.6	440.2	1.183E-03	1028	608.9	2.775E-04
24,000	820.0	433.1	1.103E-03	1020	604.0	2.936E-04
26,000	751.5	426.0	1.028E-03	1012	599.0	3.109E-04
28,000	687.7	418.8	9.565E-04	1003	593.9	3.294E-04
30,000	628.3	411.7	8.891E-04	994.5	588.8	3.495E-04
32,000	573.1	404.6	8.253E-04	985.9	583.7	3.710E-04
34,000	522.0	397.4	7.652E-04	977.1	578.6	3.944E-04
36,000	474.6	390.3	7.084E-04	968.3	573.3	4.196E-04
38,000	424.2	390.0	6.337E-04	967.9	573.1	4.687E-04
40,000	385.3	390.0	5.756E-04	967.9	573.1	5.160E-04
42,000	350.0	390.0	5.229E-04	967.9	573.1	5.680E-04
44,000	317.9	390.0	4.749E-04	967.9	573.1	6.254E-04
46,000	288.8	390.0	4.314E-04	967.9	573.1	6.885E-04
48,000	262.3	390.0	3.919E-04	967.9	573.1	7.580E-04
50,000	238.3	390.0	3.559E-04	967.9	573.1	8.344E-04

SI units. The dynamic viscosity, calculated using Sutherland's law for air, is given below for English and SI units, respectively:

$$\mu = 2.270 \times 10^{-7} \left[\frac{T^{3/2}}{T + 198.6} \right] \frac{\text{lb} - \text{s}}{\text{ft}^2}$$

$$\mu = 1.461 \times 10^{-6} \left[\frac{T^{3/2}}{T + 111} \right] \frac{\text{N} - \text{s}}{\text{m}^2}$$

An important parameter for commercial aircraft is the equivalent velocity $V_E = \sqrt{\sigma} V = \sqrt{\sigma} aM$. The variation of V_E/M with altitude is shown in Figure B.1, along with the variation of the local free stream static pressure with altitude. A

Table B.4 Properties of the 1976 US Standard Atmosphere for Altitudes Between Sea Level and 20 km in 1 km Increments

z (km)	δ	θ	σ
0	1.0000	1.0000	1.0000
1	0.8870	0.9774	0.9075
2	0.7846	0.9549	0.8216
3	0.6919	0.9323	0.7421
4	0.6083	0.9098	0.6687
5	0.5331	0.8872	0.6009
6	0.4656	0.8647	0.5385
7	0.4052	0.8421	0.4812
8	0.3513	0.8195	0.4287
9	0.3034	0.7970	0.3807
10	0.2609	0.7744	0.3369
11	0.2234	0.7519	0.2971
12	0.1908	0.7519	0.2538
13	0.1630	0.7519	0.2168
14	0.1392	0.7519	0.1851
15	0.1189	0.7519	0.1581
16	0.1015	0.7519	0.1351
17	0.08673	0.7519	0.1154
18	0.07408	0.7519	0.0985
20	0.05404	0.7519	0.0719

useful simple approximation for the atmospheric pressure ratio is $p/p_{sl} = \exp(-z/H)$ where $H = 24{,}000$ ft (7.32 km). A similar approximation for the density ratio uses $H = 29{,}000$ ft (8.84 km). Three aircraft and their typical cruise speeds are shown at their typical flight altitudes. The equivalent velocity for each is indicated and we note that the dynamic pressure $q = \frac{1}{2}\rho_{sl}V_E^2$. In Figure B.1 it is seen that a conventional jet transport (B747) flies at an equivalent airspeed about 50% higher than does a conventional turboprop regional airliner (ATR) and therefore experiences more than twice the dynamic pressure. Similarly, a supersonic airliner like the Concorde flies at an equivalent airspeed about 50% higher than does a conventional jet transport and therefore it encounters almost five times the dynamic pressure than does the regional turboprop.

Another important parameter for commercial airliners is the Reynolds number

$$Re = \frac{\rho Vl}{\mu} = \frac{aMl}{\nu}$$

The ratio Re/Ml depends upon altitude alone and its variation is shown in Figure B.2. Three representative aircraft are again shown on that plot at their usual flight altitude and Mach number. The corresponding Reynolds number for each, based on fuselage length l, is given in Figure B.2. Note that the Reynolds numbers are all

Table B.5 Properties of the 1976 US Standard Atmosphere for Altitudes Between Sea Level and 20 km in 1 km Increments (Concluded)

z (km)	p (kPa)	T (K)	ρ (kg/m³)	a (m/s)	a (kts)	ν (m²/s)
0	101.3	288.2	1.225	340.3	660.9	1.461E-05
1	89.85	281.7	1.112	336.4	653.5	1.582E-05
2	79.48	275.2	1.006	332.5	645.9	1.716E-05
3	70.09	268.7	0.9091	328.6	638.2	1.864E-05
4	61.63	262.2	0.8191	324.6	630.5	2.029E-05
5	54.01	255.7	0.7361	320.5	622.6	2.212E-05
6	47.17	249.2	0.6597	316.4	614.6	2.418E-05
7	41.05	242.7	0.5895	312.3	606.6	2.649E-05
8	35.59	236.2	0.5252	308.0	598.4	2.908E-05
9	30.74	229.7	0.4663	303.8	590.1	3.200E-05
10	26.43	223.2	0.4127	299.4	581.7	3.531E-05
11	22.63	216.7	0.3639	295.0	573.2	3.907E-05
12	19.33	216.7	0.3109	295.0	573.2	4.573E-05
13	16.51	216.7	0.2655	295.0	573.2	5.355E-05
14	14.10	216.7	0.2268	295.0	573.2	6.269E-05
15	12.04	216.7	0.1937	295.0	573.2	7.340E-05
16	10.29	216.7	0.1654	295.0	573.2	8.594E-05
17	8.786	216.7	0.1413	295.0	573.2	1.006E-04
18	7.504	216.7	0.1207	295.0	573.2	1.178E-04
20	5.47	216.7	0.0880	295.0	573.2	1.615E-04

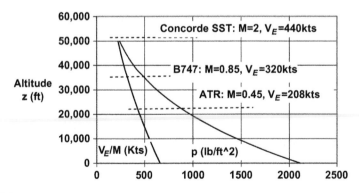

FIGURE B.1

The variation of pressure and V_E/M is shown as a function of altitude for the standard atmosphere. Three representative aircraft are depicted at their flight altitude and Mach number and the corresponding V_E is indicated.

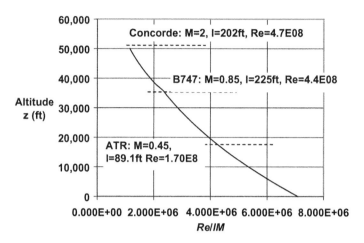

FIGURE B.2

The ratio Re/lM is shown as a function of altitude. Three representative aircraft are depicted at their flight altitude and Mach number. Their Reynolds numbers based on fuselage length l are indicated.

in excess of 100 million. Because the mean aerodynamic chord is on the order of 10–20% of the fuselage length, we see that the Reynolds numbers on the wings of these aircraft are on the order of 10 million. The flow over the wings and fuselage of these aircraft are then going to be turbulent over most of the surface giving rise to frictional drag much higher than would be the case if the flow were laminar. The search for means to maintain larger areas of laminar flow on aircraft components is driven by the potentially large drag reductions that might be realized.

Note that the model atmosphere does not account for day-to-day variations in the characteristics of the atmosphere. More detailed models for narrower geographical and seasonal data are usually given as mid-latitude winter, mid-latitude summer, subarctic winter, subarctic summer, and tropical and will show deviations from the US standard atmosphere. A description of the various models in use is provided by ANSI/AIAA (2004).

B.3 Nomenclature

a	speed of sound
h	geopotential altitude
g	acceleration of gravity
g_E	acceleration of gravity at the Earth's surface $z=0$
l	length
M	Mach number V/a

P	pressure
q	dynamic pressure
R	gas constant
R_E	radius of the Earth
Re	Reynolds number
R_u	universal gas constant
T	temperature
V	velocity
V_E	equivalent velocity $\sqrt{\sigma}V$
W_m	molecular weight
z	geometric altitude
δ	pressure ratio p/p_{sl}
λ	lapse rate dT/dz
μ	dynamic viscosity
ν	kinematic viscosity μ/ρ
p	density
σ	density ratio ρ/ρ_{sl}
θ	temperature ratio T/T_{sl}

B.3.1 Subscripts

i	denotes a specific layer in the atmosphere
sl	denotes sea lecel conditions

Reference

ANSI/AIAA, 2004. Guide to Reference and Standard Atmosphere Models, G-003B-2004. American Institute of Aeronautics and Astronautics, Reston, VA.

Airfoil and Wing Theory and Analysis

C.1 Linearized potential flow

Small disturbance theory assumes that the inviscid velocity field over a thin wing at small angle of attack is only slightly perturbed from the free stream velocity U_∞ so that the velocity components are given by

$$U = U_\infty + u'$$
$$V = v'$$
$$W = w'$$

Here the velocity components u, v, and w are in the free stream direction x, spanwise direction y, and the normal to the x–y plane direction z, respectively. The velocity perturbations are characterized by the following inequalities:

$$\left|u'\right|, \left|v'\right|, \left|w'\right| \ll U_\infty$$

This small disturbance assumption is tantamount to requiring that no shocks appear in the flow. Thus, without the irreversible effects of shock waves and viscosity the flow may be considered isentropic. As shown by Liepmann and Roshko (2001), Ashley (1992), and others, under these conditions the Euler equations which describe the steady flow may be satisfied by a velocity potential

$$\Phi = U_\infty x + \phi(x, y, z)$$
$$u' = \frac{\partial \phi}{\partial x}$$
$$v' = \frac{\partial \phi}{\partial y}$$
$$w' = \frac{\partial \phi}{\partial z}$$

The perturbation potential $\phi = \phi(x, y, z)$ is governed by the following equation:

$$\left(1 - M_\infty^2\right) \frac{\partial^2 \phi}{\partial x^2} + \frac{\partial^2 \phi}{\partial y^2} + \frac{\partial^2 \phi}{\partial z^2} = 0 \qquad (C.1)$$

Commercial Airplane Design Principles. http://dx.doi.org/10.1016/B978-0-12-419953-8.00020-6

This equation is subject to the restriction that M_∞ is neither near unity nor very much larger than unity, that is, the flow is neither transonic nor hypersonic. In such cases the nonlinearities which have been neglected assume sufficient importance to require their retention. Keeping this restriction in mind we may apply the following coordinate transformation to the perturbation potential Equation (C.1):

$$x_0 = \frac{x}{\sqrt{1 - M_\infty^2}}$$ (C.2)

$$y_0 = y$$

$$z_0 = z$$

This coordinate transformation yields a new perturbation potential equation as follows:

$$\frac{\partial^2 \phi}{\partial x_0^2} + \frac{\partial^2 \phi}{\partial y_0^2} + \frac{\partial^2 \phi}{\partial z_0^2} = 0$$

This is Laplace's equation which describes the potential in incompressible flow, that is, for $M_\infty = 0$. Therefore we may use solutions for incompressible flows in the compressible subsonic regime by merely accounting for the stretching of the x-coordinate. Thus, for example, the pressure coefficient in the compressible flow (C_p) is related to the pressure coefficient in the incompressible flow ($C_{p,0}$) by the following:

$$C_p = -\frac{2}{U_\infty} \frac{\partial \phi}{\partial x} = \frac{C_{p,0}}{\sqrt{1 - M_\infty^2}}$$

Because the loading on a wing depends upon integrals of the difference between the pressure coefficients on the upper and lower surfaces of the wing, $(C_{p,u} - C_{p,l})$, the lift coefficient of the wing in the compressible flow (C_L) is related to the lift coefficient of the transformed wing in the incompressible flow ($C_{L,0}$) by the following relation, sometimes called the Prandtl-Glauert rule:

$$C_L = \frac{C_{L,0}}{\sqrt{1 - M_\infty^2}}$$

A thin wing in compressible subsonic flow may be transformed into a corresponding thin wing in incompressible flow by properly stretching the x-coordinate alone, according to the transformations in Equation (C.2). Thus the wingspan b and the profile in the z-direction both remain the same, but the chord of the transformed wing c_0 increases along that span as the Mach number increases according to the following relation:

$$c_0(y_0) = \frac{c(y)}{\sqrt{1 - M_\infty^2}}$$

For a wing with straight leading and trailing edges the aspect ratio of the transformed wing decreases as the Mach number increases as follows:

$$A_0 = A\sqrt{1 - M_\infty^2}$$

The flow over an airfoil is equivalent to that over a wing of constant chord and infinite span for which there are no changes in the y-direction and the lift coefficient for the airfoil is

$$c_l = \int_0^1 (C_{p,l} - C_{p,u}) d\left(\frac{x}{c}\right) = \int_0^1 \frac{(C_{p,l} - C_{p,u})_0}{\sqrt{1 - M_\infty^2}} d\left(\frac{x_0\sqrt{1 - M_\infty^2}}{c_0\sqrt{1 - M_\infty^2}}\right) = \frac{c_{l,0}}{\sqrt{1 - M_\infty^2}}$$

Because the airfoil is the section of a wing of infinite aspect ratio there is no effect of aspect ratio on the lift coefficient or its slope

$$a = \left(\frac{\partial c_l}{\partial \alpha}\right) = \frac{a_0}{\sqrt{1 - M_\infty^2}}$$

Thus we expect that the lift coefficient and the lift curve slope will increase as the subsonic Mach number increases.

C.2 Airfoils in incompressible potential flow

Because the previous section describes the relationship between compressible and incompressible flows over thin wings at small angle of attack, we may just deal directly with incompressible flows. Flows over thin airfoils are described by the two-dimensional form of Equation (C.1) for steady incompressible flow in the x–z plane:

$$\frac{\partial^2 \phi}{\partial x^2} + \frac{\partial^2 \phi}{\partial z^2} = 0$$

Laplace's equation is linear, so solutions to complex problems may be readily formed by superposition of simple solutions such as uniform streams, sources, sinks, and vortices. The generation of lift is due to the presence of circulation in the flow and this may be provided by having a vortex in the flow. The basic representation of a lifting body in such cases is the circular cylinder with circulation immersed in a uniform stream as produced by the appropriate summation of a uniform stream, a doublet (a source-sink pair of equal strength), and a vortex. The integration of the pressure distribution around the circular cylinder gives rise to the following results for lift and drag per unit length:

The circulation is defined by

$$L = \rho V \Gamma$$
$$D = 0$$

(C.3)

$$\Gamma = -\oint_C \vec{V} \cdot dl \qquad\qquad (C.4)$$

Here the integral is taken in the counterclockwise sense around a closed path C. In the case in question the circulation is zero for any path not enclosing the vortex. For paths enclosing the vortex, in particular, the path lying on the circumference of the cylinder, the circulation is non-zero and a lift force is produced on the cylinder. Because Laplace's equation is unchanged by transformation into the complex plane one may use conformal mapping to extend the repertoire of cylinder shapes considered. This is the basis for the generation of the Joukowski airfoil profiles where the circulation is specified by the requirement that the flow in the real plane leave the (sharp) trailing edge of the airfoil smoothly.

Solving Laplace's equation in two-dimensional flow provides the following results:

- Circular cylinders with circulation Γ immersed in a uniform stream of velocity V yield a lift force $L = \rho V\Gamma$ per unit depth.
- Conformal transformation from a circular cylinder with circulation in the complex plane leads to lifting Joukowski airfoils in the real plane.
- Conformal transformation from a prescribed airfoil shape in the real plane leads to a near-circle in the complex plane which can be treated by Theodorsen's method, as described by Abbott and von Doenhoff (1959) to determine the flow about the true circle.

These solutions of Laplace's equations, which can provide the complete flow field over specific geometries, may be replaced by an approximation based on the knowledge that practical airfoils are relatively thin, that is, $t/c \ll 1$ everywhere. The basic assumption of thin airfoil theory is that the thickness may be ignored and the characteristics of the airfoil are determined solely by the mean camber line. In this case, rather than combining a doublet, a vortex, and a uniform stream, each individually solutions to Laplace's equation, to form a circular cylinder, a distribution of infinitesimal vortices of strength γds is assumed to exist along s, the coordinate of the mean line of the airfoil, as shown in Figure C.1. The airfoil itself must be a stream surface so that the mean camber line which replaces the airfoil must be a streamline. The boundary condition to be applied then is that the camber line is impermeable to the flow so that velocities normal to it must be zero.

The "thinness" of the airfoil in this approximation requires that the mean camber line is very close to the chord line such that the distribution of vortices may be considered to be arrayed along the chord line ($\gamma ds \simeq \gamma dx$). Thus, within the thin airfoil approximation the total circulation is given by

$$\Gamma = \int_0^c \gamma \, dx$$

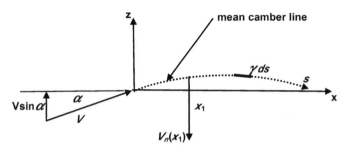

FIGURE C.1

Schematic diagram of the mean camber line and the velocity components normal to the chord line.

The component of velocity normal to the chord at a location x_1 due to the distribution of vortices along the chord is

$$V_n = \int_0^c \frac{\gamma \, dx}{2\pi \, (x - x_1)} \tag{C.5}$$

The requirement that the mean camber line is a streamline may be satisfied by setting

$$V \frac{dz}{dx} = V_n + V\alpha \tag{C.6}$$

It has also been assumed here that the angle of attack is small enough such that $\sin\alpha \sim \alpha$. Details of the approaches mentioned above are discussed by Abbott and von Doenhoff (1959) and they point out that Equations (C.5) and (C.6) may be solved in order to determine the characteristics of thin airfoil sections. They also note that the mapping technique of Theodorsen can provide the same information more accurately and with little more effort.

The interesting thing to note about all these solutions is that they are comprised of sums of simple singular functions that satisfy Laplace's equation. The singularities are placed within the body (conformal mapping techniques) or on a line representing the body (distribution of vortices along the mean camber line). Furthermore, in the thin airfoil approximation the vortex singularities are continuously distributed on a line representing the body and the resulting solution provides an accurate description of the flow field over that line. The approximation resides in the fact that the mean camber line represents a real body with thickness only to a certain degree of accuracy. If one distributes over the actual body shape a finite number of simple functions that satisfy Laplace's equation their sum would also constitute a solution. In satisfying the boundary condition that the surface is impermeable at an equal number of discrete points on the body one can determine the flow over the body to the extent

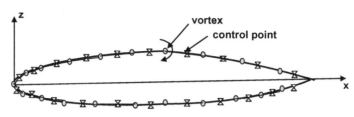

FIGURE C.2

Airfoil shape approximated by a distribution of discrete singularities and associated control points.

that it is well represented by the finite set of collocation points. This distribution of singularities, like doublets or vortices, is shown in Figure C.2. In addition, the Kutta condition, which requires that the flow leaves the trailing edge smoothly, must also be satisfied in order to determine the lift produced.

One expects that the accuracy of the solution will increase as the number of singularities placed along the body surface is increased. This approach is typically called a panel method, because the surface is divided into a series of finite length segments, or two-dimensional panels, on which a particular type of singularity is defined along with a particular point at which the appropriate boundary condition is to be satisfied.

Computational codes for airfoils which incorporate both mapping techniques and panel techniques have become commonly available because of increasingly powerful personal computers. An airfoil code developed by Drela (1989), Xfoil, has become popular among students and is readily available. Another code of interest is due to Epperle (1980) which is also easily available. These codes also include coupled boundary layer analyses to provide drag data. These programs have a number of existing airfoils stored within them and can also treat new airfoil shapes or design airfoils to particular pressure distributions. There are other airfoil codes which are available publicly or commercially, but care should be taken in selecting a code for regular use. Using any code requires time to develop familiarity with its use and the pre- and post-processing capabilities of the code deserve special attention.

C.3 Airfoils in wind tunnels

From Helmholtz's vortex theorems we know that a vortex may not end in the fluid, but only at a boundary. Then mounting a suitably designed length of airfoil-shaped cylinder between the walls of a wind tunnel would approximately satisfy the condition of having the vortex bounded within the airfoil and bounded at each end by the wind tunnel walls. Experiments show that the lift experienced by such a quasi-two-dimensional wing grows proportional to angle of attack α, at least for small angles, say $\alpha < 15°$. Thus the lift coefficient of an airfoil may be written as

$$c_l = \frac{L}{qc} = \frac{\partial c_l}{\partial \alpha} (\alpha - \alpha_0) = c_{l,\alpha} (\alpha - \alpha_0) \tag{C.7}$$

Note that because the airfoil represents the cross-section of an infinite wing the lift coefficient, which is dimensionless, is defined per unit span, thus the reference area for an airfoil is the product of the chord length and a unit length of span. The normal operating range for an airfoil is defined then by the lift curve slope $c_{l,\alpha}$ and the zero-lift angle of attack α_0.

The ideal flow development of the lift does not account for viscosity and therefore predicts zero drag for the airfoil. Of course the drag of an airfoil in a real air flow is not zero, but it is a small fraction of the lift and typically grows in a manner proportional to the square of the lift. This behavior is typically represented by the following equation:

$$c_d = \frac{D}{qc} = c_{d,0} + kc_l^2 \tag{C.8}$$

Here the reference area for the drag coefficient of the airfoil is the same as that used for the lift coefficient and $c_{d,0}$ is called the zero-lift drag coefficient. The coefficient k, along with $c_{d,0}$, defines the aerodynamic efficiency of the airfoil, that is, the lift to drag ratio of the airfoil where

$$\frac{L}{D} = \frac{c_l}{c_d} = \frac{1}{\frac{c_{d,0}}{c_l} + kc_l} \tag{C.9}$$

The maximum value of the lift to drag ratio may be determined by taking the derivative of L/D with respect to c_l and setting the result to zero thereby obtaining

$$c_{l,(L/D)_{\max}} = \sqrt{\frac{c_{d,0}}{k}} \tag{C.10}$$

The drag coefficient at maximum L/D may be found from Equations (C.8) and (C.10) to be

$$c_{d,(L/D)_{\max}} = 2c_{d,0} \tag{C.11}$$

Thus when the drag coefficient due to lift of the airfoil is equal to the zero-lift drag coefficient the maximum L/D is obtained and is given by

$$\left(\frac{L}{D}\right)_{\max} = \frac{1}{2\sqrt{kc_{d,0}}} \tag{C.12}$$

The careful study of airfoil sections by the NACA, as reported, for example, by Abbott et al. (1945) and Abbott and von Doenhoff (1959), provides some generally applicable results for understanding typical trends. For example, experiments indicate that airfoils constructed by usual fabrication techniques generally show zero-lift drag coefficients in the range $0.007 < c_{d,0} < 0.008$ independent of the airfoil section considered. Though providing a smoother surface finish can reduce that value, retaining high levels of surface smoothness under normal operating conditions is unlikely. Of course, more modern fabrication and finishing techniques, along with more refined airfoil section design, may produce individual cases which improve on this nominal value for zero-lift drag coefficient and when detailed information is in hand it is prudent to use those data. If we use a typical value of $k = 0.005$ we may find a useful estimate for the airfoil drag coefficient using Equation (C.12). This rule of thumb results in a value of $(L/D)_{max}$ around 80, which is about midway between the value found in wind tunnel tests for most smooth NACA airfoils ($L/D = 110$) and the value found for the same airfoils with standard (sand) roughness applied in the vicinity of the leading edge surfaces to promote the onset of turbulent flow over the airfoil ($L/D = 60$).

C.4 Airfoils in wings

The simplest wing theory assumes that the span of the wing is divided into a number of airfoil sections of width dy and that the aerodynamic force and moment on each section are functions only of the airfoil shape and the angle of attack. Implicit in this assumption is that the flow is locally two-dimensional, that is, the local flow is unaffected by any other features of the wing. Obviously this fails in the vicinity of wingtips, deflected flaps or ailerons, fuselages, or nacelles. Similarly, wings with sweepback, changes in planform, airfoil section, or twist cause three-dimensional effects which cannot be ignored. More accurate means of predicting the lifting characteristics of wings will be covered in subsequent sections of this appendix.

C.4.1 Selection of root sections

A thicker wing provides space for fuel tanks, landing gear, and associated high lift and control equipment. At the same time the larger web depth of the airfoil can support bending loads at reduced structural weight, permitting larger wingspans to be used. Thus the appropriate design choice would be the thickest root section that is aerodynamically practical for the mission. For the typical range of airfoil thickness ratios it is found that the minimum drag coefficient shows little variation for smooth airfoils so that the choice of thickness ratio is constrained by characteristics other than maximum lift and minimum drag. Indeed, the critical Mach number of the airfoil section provides the major limitation on the selection of thickness ratio for high-subsonic speed commercial aircraft. Root sections must have sufficiently high critical Mach numbers to limit large drag increases due to compressibility effects in high-speed cruise. In this condition due consideration must be made of the increased velocity of flow over the root region arising from flow interference with the fuselage junction.

In the case of turboprop aircraft where the cruise speed is low enough that compressibility effects do not constrain the thickness ratio, attention should turn to the limitation posed by excessive drag coefficients at moderate and high lift coefficients. When thick root sections are under consideration to permit use of higher aspect ratios, the drag coefficients of such sections with the expected surface conditions at moderately high lift coefficients should be evaluated carefully. An optimum aspect ratio should be sought beyond which further increases in aspect ratio and root thickness ratio degrade performance.

Interference effects on the inboard sections of wings are likely to be characterized by surface openings and gaps caused by access doors, landing-gear wells, and the like. As a result, maintaining any substantial streamwise run of laminar flows is less likely on the inboard wing panels than on the outboard panels. Therefore use of airfoil sections permitting extensive laminar flow will provide little drag reduction. The use of airfoils like those in the NACA 63-series will be of greater benefit for thick sections because they are more conservative than those designed for a greater run of laminar flow.

C.4.2 Selection of tip sections

Tip sections should be selected on the basis of high maximum lift coefficient achieved at a high angle of attack as compared with the root section and the tip section should be such as to stall with a gradual loss in lift. High maximum lift coefficient at the tip section is difficult to attain because of the lower Reynolds number at the tip when the wing is tapered. Increasing the tip section maximum lift coefficient may be achieved by increasing the camber, although this is limited by either the critical Mach number requirement or by the need for the section to have low drag coefficient at the high-speed lift coefficient.

For moderately thick root sections, a tip section with a somewhat reduced thickness ratio is desirable. This reduction in thickness ratio gives a margin in critical speed permitting the camber of the tip section to be increased, but this may be limited by the loss in maximum lift coefficient typical of too thin an airfoil section.

C.4.3 Conclusions regarding NACA airfoil sections

The following are among the conclusions drawn from the NACA airfoil data presented by Abbott and von Doenhoff (1959). Most of the data, particularly for the lift, drag, and pitching-moment characteristics, were obtained at Reynolds numbers from 3 to 9×10^6. Airfoil sections permitting extensive laminar flow, such as the NACA 6- and 7-series sections, result in substantial reductions in drag at high speed and cruising lift coefficients as compared with other sections if, and only if, the wing surfaces are fair and smooth. Wings of moderate thickness ratios with surface conditions corresponding to those obtained with typical construction methods may achieve minimum drag coefficients of the order of 0.0080. The values of the minimum drag coefficient for such wings depend primarily on the surface condition rather than on the airfoil section. Substantial reductions in drag coefficient at high Reynolds numbers may be obtained by maintaining smooth wing surfaces, even if extensive laminar flow is not obtained. Leading edge roughness causes large reductions in maximum lift coefficient for both plain airfoils and airfoils equipped with split flaps deflected 60°. The decrement in maximum lift coefficient resulting from standard roughness is essentially the same for the plain airfoils as for the airfoils equipped with the 60° split flaps. The effect of leading edge roughness is to decrease the lift curve slope, particularly for the thicker sections having the position of minimum pressure far back. Therefore it is quite important to keep the wing surfaces clean, particularly the region of the leading edges.

C.5 Flow over finite wings

Since wings are finite in span the flow over them is three-dimensional and the airfoil results described above will not be uniformly applicable across the entire span of the wing. A wing with a very large span should have close to two-dimensional flow near the center of the wing while near the tips the difference in pressure above and below the wing should produce some movement of the air around the wingtips, thereby reducing the lift over those parts of the wing from values expected on the basis of airfoil theory. The lift distribution might look like that shown in Figure C.3. The effect of sweep is to load the outboard sections more heavily than the inboard sections so that the lift distribution appears as shown in Figure C.4. Indeed, it is the increased loading on the outboard panels due to sweep that makes outboard stalling more likely and leads to the need to decrease the local angle of attack of those portions of the wing. This technique is called "washing out" the wing, that is, twisting the wing so as to make those outboard sections have a reduced angle of attack.

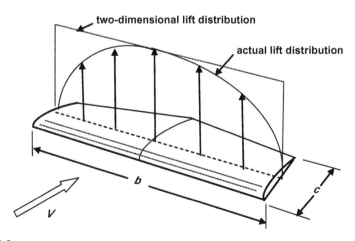

FIGURE C.3

Lift distribution over a finite wing with no sweepback.

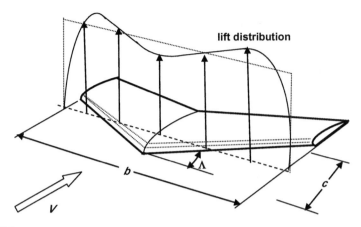

lift distribution

FIGURE C.4

Lift distribution over a sweptback wing.

The movement of air around the wingtips induced by the pressure difference between the upper and lower surfaces is a spinning motion where air moves around the wingtip from the bottom surface to the top surface making the flow three-dimensional. This behavior is schematically illustrated in Figure C.5. The mathematical description of such a flow is quite complex and a simplified model would be of great value. Such a model was provided by Prandtl's lifting line theory and has found success in calculating the flow over finite wings and the ideas upon which it is based have aided in the development of powerful computational tools for modern design purposes.

Recall that the vortex imbedded in the airfoil must either stretch to infinity or stop at a solid surface like a wind tunnel wall. Here neither option presents itself and the vortex is considered to bend back streamwise and proceed downstream to infinity. This model of a so-called "horseshoe" vortex is illustrated in Figure C.6.

The horseshoe vortex is actually a single vortex ring with the bound vortex fixed within the wing, two arms extending far downstream (trailing vortices) where they are joined by the "starting vortex," which is formed at the initiation of the motion of the wing. This horseshoe vortex has a fixed circulation since, according to another theorem attributed to Helmholtz, the strength of a vortex is constant along its length. Then, by the Kutta-Joukowski law, the lift is given by

$$L = \rho V \Gamma b \qquad \text{(C.13)}$$

The horseshoe vortex induces a velocity field of its own according to the Biot-Savart law

$$dV_i = \frac{\Gamma}{4\pi} \frac{R}{r^3} ds \qquad \text{(C.14)}$$

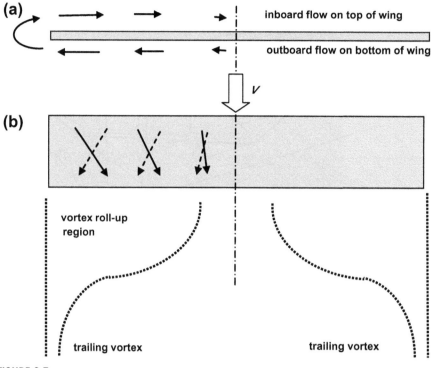

(a)

inboard flow on top of wing

outboard flow on bottom of wing

V

(b)

vortex roll-up
region

trailing vortex   trailing vortex

FIGURE C.5

Schematic illustration of the flow over a finite lifting wing: (a) a front view looking at the leading edge and showing the cross-flow velocity components produced by the higher pressures on the bottom wing surface and lower pressures on the top wing surface; (b) a top view looking down on the wing illustrating the velocity on the top surface (solid arrows) and on the bottom surface (dashed arrows). The swirling flow downstream of the trailing edge of the wing organizes, that is, "rolls up" itself into two concentrated trailing vortices after a distance on the order of the wingspan.

Here dV_i is the velocity induced at a point in the flow field by an elemental length ds of a vortex of strength Γ. The point is located a distance r from the element and at a perpendicular distance R from the element. The induced velocity acts in the same sense as the circulation Γ and is perpendicular to the plane formed by r and ds. To determine the total velocity induced at the point in question Equation (C.14) must be integrated over the complete length of the vortex. It is noteworthy that the trailing vortices influence the velocity field in the vicinity of the wingtips and give rise to inboard flow on the upper surface of the wing and outboard flow on the lower surface, as is shown in Figure C.5.

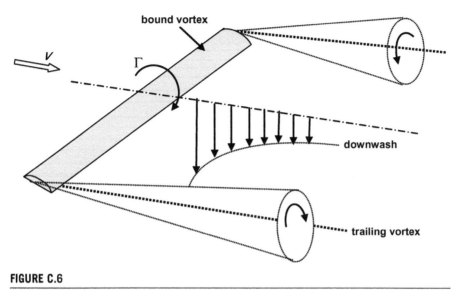

bound vortex

V

Γ

downwash

trailing vortex

FIGURE C.6

A finite wing showing the idealized horseshoe vortex and the associated induced down-wash flow field downstream of the wing.

C.5.1 Prandtl's lifting line theory for wings

The surface of a wing may be considered to be covered in a vortex sheet that leaves the sharp trailing edge of the wing and moves downstream as a flat ribbon of width equal to the span b. The trailing vortex sheet cannot support a force since the pressure is equilibrated across it, but it does induce a velocity field which causes the outermost edges to curl up and the sheet to deform. This complicated flow field was approximated by Prandtl using the following assumptions: (1) the trailing edge is a straight line and the vortex sheet extends downstream to infinity as a flat ribbon parallel to the free stream velocity V, (2) the vortex lines forming the vortex sheet are assumed to be straight lines parallel to V, (3) the vortex lines on the upper and lower surfaces of the wing are assumed to be closely packed straight lines running in the spanwise direction forming, with the straight trailing vortices, a family of horseshoe vortices. This model is illustrated in Figure C.7.

Furthermore, the closely packed spanwise vortex lines are assumed to form a single vortex line of variable strength rising from zero at the wingtip to a maximum at the centerline and (symmetrically) back down to zero at the other wingtip. This single variable-strength vortex line is called the lifting line because it produces the lift of the wing. The circulation around any spanwise station of the wing is given by Stokes theorem

$$\oint_C \vec{V} \cdot d\vec{l} = \int \int_A \vec{\omega} \cdot \hat{n} \, dA \tag{C.15}$$

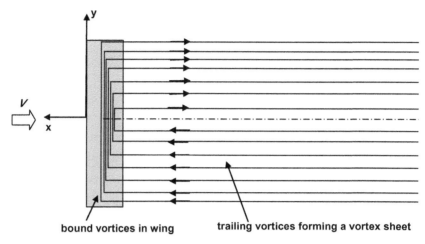

bound vortices in wing   trailing vortices forming a vortex sheet

FIGURE C.7

Prandtl's lifting line model showing some of the infinite number of infinitesimal nested horseshoe vortices lying in the plane of the wing and forming a vortex sheet extending out to infinity.

This equation states that the circulation around a closed path C is equal to the flux of vorticity ω passing through the area A subtended by the path. Then, looking at Figure C.7, if the closed circuit C is taken around the wing in the plane $y=b/2$ the circulation $\Gamma(b/2)=0$, while taking it at other planes $y=$constant <0 we find $\Gamma(y)$ increasing since the circuit C subtends more and more of the nested vortex lines of infinitesimal strength. At $y=0$, owing to symmetry, the circulation $\Gamma(0)$ is a maximum, and as y is further decreased $\Gamma(y)$ falls until the port (left) wingtip is reached, where $\Gamma(-b/2)=0$ again. Thus the circulation varies continually from wingtip to wingtip, rising from zero at one wingtip to a maximum at the centerline and then back down to zero at the other wingtip.

Now to relate the circulation in the three-dimensional case to the lift produced, Prandtl assumed that, for a fairly high aspect ratio wing, the behavior of an infinitesimal portion of the span dy of the wing in the central region of the wing should behave in a fashion similar to that of the airfoil making up that section of the wing. The airfoil characteristics are known in terms of the free stream velocity V and the angle of attack α, the angle between the chord line and the direction of V. The difference here is that the airfoil is part of the wing and is subject to the velocity field of the trailing vortex sheet as well as to the free stream velocity V. The induced velocity produced by the trailing vortex sheet at the assumed location of the lifting line is the downwash w, as shown in Figure C.8. As a result we may write

$$V' = V\sqrt{1 + \left(\frac{w}{V}\right)^2}$$

$$\alpha' = \alpha - \arctan\left(\frac{w}{V}\right) \tag{C.16}$$

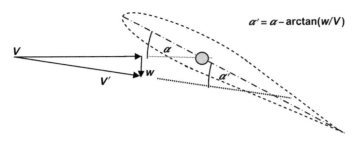

$$\alpha' = \alpha - \arctan(w/V)$$

FIGURE C.8

Airfoil section of the wing showing the induced angle of attack produced by the downwash generated by the trailing vortex sheet. The bundle of elemental vortex lines forming the lifting line is shown as a gray circle and the airfoil is displayed as a dashed contour around it. The induced velocity w is produced by the net effect of the trailing vortex sheet at the location in question.

For small downwash $w/V \ll 1$ we may expand Equation (C.16) as follows:

$$V' = V\left[1 + \frac{1}{2}\left(\frac{w}{V}\right)^2 + \cdots\right] \tag{C.17}$$

$$\alpha' = \alpha - \left(\frac{w}{V}\right) + \cdots$$

Then, neglecting second-order terms we may write

$$V' \approx V$$

$$\alpha' \approx \alpha - \frac{w}{V} \tag{C.18}$$

Thus Prandtl suggested that the behavior of the airfoil in the wing at a given angle of attack is equivalent to the behavior of the airfoil itself evaluated at that angle of attack less the downwash angle w/V. This set of simplifications of a very complicated problem has shown itself to be remarkably accurate in predicting the lifting behavior of wings.

C.5.2 The induced velocity field

Now to find the velocity induced by the vortex sheet we may apply the Biot-Savart law to a segment of any vortex line and then integrate over the length of the segment and then over all vortex lines. First, in keeping with Prandtl's approximate theory we realize that we are really looking for just the downwash induced at the location of the lifting line. It isn't clear exactly where that line lies in the wing but it is sufficient to assume that this is at a well-defined location, such as the aerodynamic center (where the moment coefficient is independent of the angle of attack) or the center of pressure

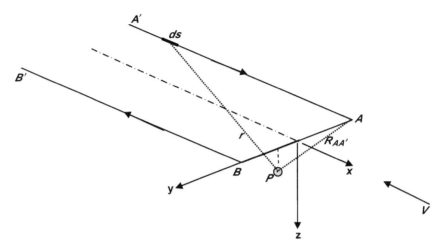

FIGURE C.9

Schematic for determining the downwash dw at a point P in the y,z plane caused by an element ds of a typical horseshoe vortex of strength Γ. The length dr is the distance from the element of the vortex line to the point P while the length $R_{AA'}$ is the perpendicular distance from the vortex line AA' to the point P.

(where the resultant force acts with no moment) of the airfoil. We are interested in the downwash at the lifting line, which corresponds to a velocity in the y,z plane. Consider a general point P in the y,z plane passing through the lifting line, as shown in Figure C.9. The induced downwash velocity caused by the general element is given by von Mises (1959) as

$$dw = \frac{\Gamma}{4\pi r^3} R ds \tag{C.19}$$

Substituting the identity $r^2 = s^2 + R^2$ and integrating Equation (C.19) over the whole length of the leg AA' of the vortex on the negative y-side yields

$$w_{AA'} = \frac{\Gamma R_{AA'}}{4\pi} \int_0^\infty \frac{ds}{r^3} = \frac{\Gamma R_{AA'}}{4\pi} \int_0^\infty \frac{ds}{\left(s^2 + R_{AA'}^2\right)^{\frac{3}{2}}} \tag{C.20}$$

$$w_{AA'} = \frac{\Gamma}{4\pi R_{AA'}} \tag{C.21}$$

This velocity acts normal to $R_{AA'}$ and in the downward direction. The contribution of the leg AB of the horseshoe vortex to the downwash velocity at the point P is zero since the velocity induced is normal to the plane containing the vortex segment

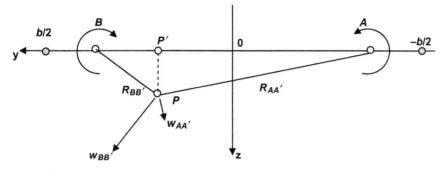

FIGURE C.10

Downwash velocity components due to the one horseshoe vortex illustrated in Fig. C.3.

AB and the point P, that is, $w_{AB}=0$. The contribution of leg BB' is found in the same fashion as for the leg AA' and is given by

$$w_{BB'} = \frac{\Gamma}{4\pi R_{BB'}} \tag{C.22}$$

We may determine the sum of the induced velocities by considering the y,z plane as shown in the diagram in Figure C.10. For the case where we desire the downwash velocity on the lifting line segment AB itself, that is, the point P' in Figure C.10, we see that

$$w_{P'} = \frac{\Gamma}{4\pi}\left(\frac{1}{y - y_A} + \frac{1}{y_B - y}\right) \tag{C.23}$$

In the case of the vortex sheet trailing behind the wing which is made up of a continuous distribution of horseshoe vortices like the one just considered, we may imagine the local circulation on any segment to be given by

$$d\Gamma = \frac{d\Gamma}{dy}dy = \Gamma' \, dy \tag{C.24}$$

Then the downwash velocity induced at the point y_P by an infinitesimal width dy of the vortex sheet may be written as

$$dw(y_P) = \frac{\frac{d\Gamma}{dy}dy}{4\pi\left(y_P - y\right)} = \frac{\Gamma' \, dy}{4\pi\left(y_P - y\right)} \tag{C.25}$$

We may determine the distribution of the downwash along the entire span of the wing by integrating Equation (C.25) to obtain

$$w(y) = \frac{1}{4\pi} \int_{-b/2}^{b/2} \frac{\Gamma'(\eta)}{y - \eta} d\eta \qquad (C.26)$$

This equation may be interpreted with the aid of Figures C.11 and C.12.

The integrand in Equation (C.26) has the following identity

$$\frac{d\Gamma}{y - \eta} = \frac{-d\Gamma}{\eta - y} \qquad (C.27)$$

Thus since the vorticity vector has the upstream direction ($d\Gamma$ positive) on the port side and the downstream direction ($d\Gamma$ negative) on the starboard side Equation (C.27) remains consistent independent of the wing station y_P.

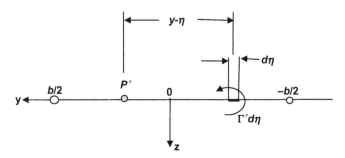

FIGURE C.11

The strip of vortex sheet of width $d\eta$ and located at η on the port wing inducing a downwash at the general point P' located at y.

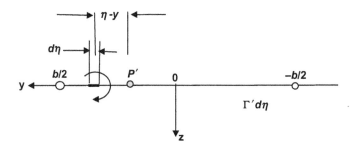

FIGURE C.12

The strip of vortex sheet of width $d\eta$ and located at η on the starboard wing inducing a downwash at the general point P' located at y.

C.5.3 Wing lift and drag

Of course the distribution of the circulation $\Gamma(y)$ along the lifting line is unknown at this point, but we may use the previous assumption concerning the local behavior of an airfoil on the wing to relate $w(y)$ to $\Gamma(y)$. Recall that a segment of the wing dy in width is assumed to act like the airfoil comprising it except that it is responding to the velocity V' rather than the free stream velocity V. Thus the pressure force on the segment according to the Kutta-Joukowski theorem is $dF = \rho V'\Gamma(y)dy$ and it acts in the direction normal to V'. From Equation (C.17) we note that the magnitude of the velocity $V' = V$ to second order in the downwash angle $\varepsilon = \arctan(w/V) \ll 1$ so that it is correct to write $dF = \rho V\Gamma(y)dy$. The lift of the wing is defined as the component of resultant force acting normal to the free stream velocity V. Referring to Figure C.13 we see that the force normal to the free stream velocity, that is, the lift produced by the wing segment is

$$dL = dF\cos\varepsilon \approx dF\left(1 - \frac{\varepsilon^2}{2} + \cdots\right) = \rho V\Gamma(y)dy \qquad \text{(C.28)}$$

However, the same figure shows that there is a component of force in the direction of the free stream velocity, that is, a drag force

$$dD = dF\sin\varepsilon = dF\left(\varepsilon - \frac{\varepsilon^3}{3!} + \cdots\right) = \rho V\Gamma(y)\varepsilon\,dy \qquad \text{(C.29)}$$

Since the downwash angle is small we may express it as follows:

$$\varepsilon = \arctan\left(\frac{w}{V}\right) = \frac{w}{V} - \frac{1}{3!}\left(\frac{w}{V}\right)^3 + \cdots \approx \frac{w}{V} \qquad \text{(C.30)}$$

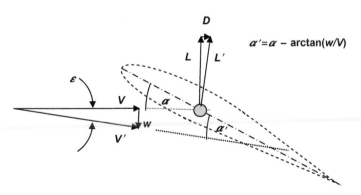

FIGURE C.13

The lift and drag produced on an element of a wing due to the downwash induced by the wing's trailing vortex sheet. The lifting line is shown here as a shaded circle in the wing section.

Then the equation for the drag force induced by the downwash in Equation (C.29) becomes

$$dD = \rho w \Gamma(y) dy \qquad (C.31)$$

Then the lift and induced drag for the entire wing may be found by integration across the entire wing as follows:

$$L = \rho V \int_{-b/2}^{b/2} \Gamma(y) dy \qquad (C.32)$$

$$D = \rho \int_{-b/2}^{b/2} w(y) \Gamma(y) dy \qquad (C.33)$$

In order to calculate both the lift and the drag we must find a relationship between $w(y)$ and $\Gamma(y)$ and that can be done by recalling that any infinitesimal spanwise segment of the wing acts locally like the airfoil except that the lift produced is normal to the direction of the velocity V'. The lift of an airfoil is defined as

$$L = c_l \frac{1}{2} \rho V^2 c(1) \qquad (C.34)$$

Here c_l is the airfoil lift coefficient, c is the airfoil chord length, and the numeral 1 in parentheses denotes a unit span of the (infinite) wing comprised of that airfoil section. However the lift of the airfoil may also be expressed in terms of the circulation by means of the Kutta-Joukowski theorem as follows:

$$L = \rho V \Gamma(1) \qquad (C.35)$$

Once again the numeral 1 in parentheses denotes a unit span of the (infinite) wing comprised of that airfoil section. Equating the two expressions for the lift produced by the airfoil we find

$$\Gamma = \frac{1}{2} c_l c V \qquad (C.36)$$

The lift coefficient for the useful (linear) operating range of the airfoil is given by

$$c_l = a(\alpha - \alpha_0) \qquad (C.37)$$

Here a is the lift curve slope, α is the geometric angle of attack, and α_0 is the zero-lift angle of attack of the airfoil. In the case where the airfoil is part of a finite wing we must account for the downwash and therefore use the actual angle of attack shown in Equation (C.17) so that

$$c_l = a \left(\alpha - \frac{w}{V} - \alpha_0 \right) \qquad (C.38)$$

Note that the zero-lift angle of attack is not influenced by the downwash because there is no downwash induced when no lift is generated. Now the circulation may be written as

$$\Gamma = \frac{1}{2}cVa\left(\alpha - \alpha_0 - \frac{w}{V}\right) = \frac{1}{2}cVa\left(\alpha - \alpha_0\right) - \frac{1}{2}caw \tag{C.39}$$

Equations (C.32) and (C.33) now have the lift and drag related to just one unknown, the spanwise distribution of the induced downwash w. The other parameters are known or specified characteristics of the wing: the free stream velocity V, the spanwise distribution of the chord length $c(y)$, and the lifting characteristics of the airfoil $a(\alpha - \alpha_0)$, which may also be functions of the spanwise location. However, the downwash distribution is not known and we really have two equations, Equation (C.26) and Equation (C.39) in two unknowns, $w(y)$ and $\Gamma(y)$.

C.5.4 Solving for the induced velocity

The problem posed by Equation (C.13) may be recast in terms of new physical and dummy variables as follows:

$$y = -\frac{b}{2}\cos\phi \tag{C.40}$$

$$\eta = -\frac{b}{2}\cos\theta$$

Here the ranges of the new variables are $0 < \phi < \pi$ and $0 < \theta < \pi$ and we must be able to express the circulation distribution in terms of the new variables. We expect the circulation distribution to be symmetrical about $y = 0$ and therefore we may assume a Fourier series expansion of the circulation so that

$$\Gamma(\theta) = 2bV\sum_{n=1}^{\infty}A_n\sin n\theta \tag{C.41}$$

Then the derivative of the circulation distribution is

$$\frac{d\Gamma}{d\theta} = \frac{d\Gamma}{d\eta}\frac{d\eta}{d\theta} = \frac{d\Gamma}{d\eta}\frac{b}{2}\sin\theta = \Gamma'(\theta) \tag{C.42}$$

Using Equation (C.41) in Equation (C.42) yields

$$\Gamma'(\theta) = 4V\sum_{n=1}^{\infty}nA_n\cos n\theta \tag{C.43}$$

Now Equation (C.26) becomes

$$w(\phi) = \frac{V}{\pi}\sum_{n=1}^{\infty}nA_n\int_0^{\pi}\frac{\cos n\theta}{\cos\phi - \cos\theta}d\theta \tag{C.44}$$

The integral in Equation (C.44) may be found in tables of integrals and is given by

$$\int_0^\pi \frac{\cos n\theta}{\cos\phi - \cos\theta} d\theta = \frac{\pi \sin n\phi}{\sin\phi} \tag{C.45}$$

Thus the induced downwash is now

$$w(\phi) = V \frac{\sum_{n=1}^\infty nA_n \sin n\phi}{\sin\phi} \tag{C.46}$$

Using this result in Equation (C.39) yields

$$\Gamma(\phi) = \frac{1}{2}Vc(\phi)a(\phi)\left[\alpha(\phi) - \alpha_0(\phi)\right] - \frac{1}{2}Vc(\phi)a(\phi)\frac{\sum_{n=1}^\infty nA_n \sin n\phi}{\sin\phi} \tag{C.47}$$

Note that all the local airfoil parameters may vary with ϕ, that is, with spanwise location and only V is a constant. On the other hand, one may specify the wing design and thus all the airfoil parameters would be known as a function of spanwise location. Then substituting for the circulation distribution the form given in Equation (C.41) transforms Equation (C.47) into

$$2bV \sum_{n=1}^\infty A_n \sin n\phi = \tfrac{1}{2}Vc(\phi)a(\phi)\left[\alpha(\phi) - \alpha_0(\phi)\right]$$

$$-\tfrac{1}{2}Vc(\phi)a(\phi)\frac{\sum_{n=1}^\infty nA_n \sin n\phi}{\sin\phi} \tag{C.48}$$

This equation may be rearranged into the form

$$\sum_{n=1}^\infty A_n \sin n\phi \left[1 + \frac{nc(\phi)a(\phi)}{4b \sin\phi}\right] = \frac{c(\phi)a(\phi)}{4b}\left[\alpha(\phi) - \alpha_0(\phi)\right] \tag{C.49}$$

If the coefficients of the $\sin(n\phi)$ term were constants, the values of all the A_n terms could be found by using the orthogonality condition of the Fourier series. However, the coefficients are functions of ϕ, that is, of spanwise station, so this technique is not applicable here. If all the coefficients A_n of the Fourier series are not necessary for an answer of acceptable accuracy, then Equation (C.49) could be satisfied at a discrete number N of points thereby producing a system of N linear equations for the N values of A_n. The analysis by De Young and Harper (1948) finds that $N=7$ leads to useful results. Therefore let us choose seven equally spaced values of $\phi=n\pi/8$, $n=1, 2, 3, \dots, 7$. In general the wing will be symmetric about the centerline and furthermore $\sin(n\pi - \phi) = -i^n\sin(n\phi)$. Then writing out Equation (C.49) for $\phi < \pi$ and for the symmetric spanwise station $\pi - \phi$ and subtracting the two equations leads to the result that the even-numbered coefficients $A_2 = A_4 = A_6 = 0$. Thus only four equations in the odd-numbered unknowns A_1, A_3, A_5, and A_7 need to be satisfied.

C.5.5 Sample calculation for an unswept wing

Assume a trapezoidal wing with an unswept quarter-chord line, a taper ratio $\lambda = c_t/c_r$, and an aspect ratio $A = b^2/S$ as shown in Figure C.14. The chord at any station y is given by

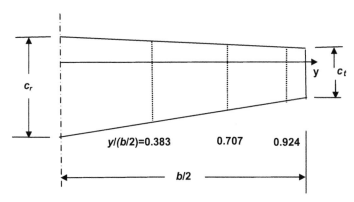

FIGURE C.14

Example case of a trapezoidal wing with $\lambda=0.5$ and $A=7$. The three dotted lines and the wingtip denote the wing stations considered.

$$c(y) = c_r \left(1 - 2\frac{y}{b} + 2\lambda\frac{y}{b}\right) \tag{C.50}$$

The aspect ratio may also be expressed in terms of the root chord c_r and the tip chord c_t or, equivalently, the taper ratio λ as follows

$$A = \frac{b^2}{S} = \frac{b}{\frac{1}{2}(c_r + c_t)} = \frac{2b}{c_r(1+\lambda)} \tag{C.51}$$

Taking the simple case of an untwisted wing ($\alpha=$ constant) with $\lambda=0.5$ and $A=7$, and assuming that the airfoil is a symmetric section ($\alpha_0=0$) with an ideal lift curve slope ($a=2\pi$) we may write the chord distribution as $c(y)=c_r(1-0.5\cos\phi)$ and Equation (C.49) becomes

$$\sum_{n=1,3,5,7} A_n \sin n\phi \left[1 + 0.299n\frac{1 - 0.5\cos\phi}{\sin\phi}\right] = 0.299(1 - 0.5\cos\phi)\alpha \tag{C.52}$$

This equation has four unknowns, A_1, A_3, A_5, and A_7 so we consider four equally spaced values of the spanwise location ϕ, say $\pi/8$, $2\pi/8$, $3\pi/8$, and $4\pi/8$. Substituting these values into Equation (C.52) leads to the following set of equations:

$$\begin{aligned}
0.5437A_1 + 2.0891A_3 + 2.8659A_5 + 1.5088A_7 &= 0.1609\alpha \\
0.9004A_1 + 1.2871A_3 - 1.6737A_5 - 2.0604A_7 &= 0.1933\alpha \\
1.1658A_1 - 0.6831A_3 - 0.8834A_5 + 2.6163A_7 &= 0.2418\alpha \\
1.2992A_1 - 1.8976A_3 + 2.4960A_5 - 3.0944A_7 &= 0.2992\alpha
\end{aligned} \tag{C.53}$$

These equations may be solved, for example, by the Gauss elimination technique to yield

$$
\begin{aligned}
A_1 &= 0.2205\alpha \\
A_3 &= 0.0071\alpha \\
A_5 &= 0.0095\alpha \\
A_7 &= -0.0008\alpha
\end{aligned}
\tag{C.54}
$$

From Equations (C.32) and (C.40) the lift is

$$
L = \rho V \int_{-b/2}^{b/2} \Gamma \, dy = \frac{1}{2} \rho V b \int_0^\pi \Gamma(\phi) \sin\phi \, d\phi
\tag{C.55}
$$

Substituting the circulation distribution of Equation (C.41) into Equation (C.55) yields

$$
L = \rho V^2 b^2 \int_0^\pi \left(\sum_{n=1,3,5,7} A_n \sin n\phi \right) \sin\phi \, d\phi
\tag{C.56}
$$

Note that the integral in Equation (C.56) is non-zero only when $n=1$, for which its value is $\pi/2$ which leads to the result

$$
L = \rho V^2 b^2 A_1 \frac{\pi}{2}
\tag{C.57}
$$

Using the definition of the aspect ratio $A = b^2/S$, where here $A = 7$, and the calculated value $A_1 = 0.2205$, Equation (C.57) may be written as

$$
L = \frac{1}{2} \rho V^2 S \left(2\pi A \frac{A_1}{2} \right) = qS \left(2\pi A \frac{0.2205\alpha}{2} \right) = [(0.7718)\, 2\pi\alpha]\, qS
\tag{C.58}
$$

The result given in Equation (C.58) shows that the lift coefficient for the example wing is about 77% of that which would be calculated on the basis of the airfoil lift coefficient, which was assumed to be the ideal value of $c_l = 2\pi\alpha$.

The induced drag, as given by Equation (C.33), becomes, using Equations (C.41) and (C.46),

$$
D = \rho \int_{-b/2}^{b/2} w\Gamma \, dy = \rho V^2 b^2 \int_0^\pi \sum_{m=1,3,5,7} \frac{mA_m \sin m\phi}{\sin\phi} \left(\sum_{n=1,3,5,7} A_n \sin n\phi \right) \sin\phi \, d\phi
\tag{C.59}
$$

Rearranging Equation (C.59) yields

$$
D = \rho V^2 b^2 \sum_{m=1,3,5,7} \sum_{n=1,3,5,7} mA_m A_n \int_0^\pi \sin m\phi \sin n\phi \, d\phi
\tag{C.60}
$$

The integral in Equation (C.60) is non-zero only when $m = n$ and then it is equal to $\pi/2$ which leads to the following result:

$$D = \frac{1}{2}\pi\rho V^2 b^2 \sum_{n=1,3,5,7} nA_n^2 = qS\pi A \left(A_1^2 + 3A_3^2 + 5A_5^2 + 7A_7^2\right) \quad \text{(C.61)}$$

Factoring out the term A_1 leads to

$$D = \left[\pi A A_1^2 (1 + \delta)\right] qS \qquad \text{(C.62)}$$

The term δ includes the effects of the A_3, A_5, and A_7 terms, which are considerably smaller than the A_1 term as may be seen from Equation (C.54). Now, from Equation (C.58) the lift coefficient of the wing is

$$C_L = \pi A A_1 \qquad \text{(C.63)}$$

From Equation (C.62) the induced drag coefficient of the wing is

$$C_{D,i} = \pi A A_1^2 (1 + \delta) \qquad \text{(C.64)}$$

Note that a subscript i has been affixed to the drag coefficient to express the fact that this drag is only the contribution from the wing-induced downwash. Airfoils have a drag due to pressure and skin friction that varies with angle of attack and the drag due to the wing being finite is in addition to the basic airfoil drag. Thus the drag coefficient induced by the generation of lift may be expressed as follows:

$$C_{D,i} = \frac{C_L^2}{\pi A}(1 + \delta) \qquad \text{(C.65)}$$

For the example case considered the wing lift coefficient and the wing-induced drag coefficient

$$C_L = 1.545\pi\alpha$$
$$C_{D,i} = 0.3403\pi\alpha^2$$

C.5.6 The span loading coefficient

We may assume that each section of the wing is characterized by a local lift coefficient $c_l(y)$. Then, integrating over the span of the wing the total lift coefficient of the wing is given by

$$C_L = \frac{L}{\frac{1}{2}\rho V^2 S} = \frac{1}{S}\int_{-b/2}^{+b/2} c_l(y)c(y)dy = \left[\int_0^1 \frac{c_l c}{c_{av}}d\left(\frac{y}{b/2}\right)\right]$$

Dividing through by C_L yields

$$\int_0^1 \frac{c c_l}{c_{av}C_L}d\left(\frac{y}{b/2}\right) = 1$$

Here we call the integrand the loading coefficient. From Equation (C.56)

$$L = \frac{1}{2}\rho V^2 \left[2b^2 \int_0^\pi \left(\sum_{n=1,3,5,7} A_n \sin n\phi \right) \sin \phi \, d\phi \right]$$

$$= \frac{1}{2}\rho V^2 \left[4b^2 \int_0^1 \left(\sum_{n=1,3,5,7} A_n \sin n\phi \right) d\left(\frac{y}{b/2} \right) \right]$$

Then the lift coefficient is given by

$$1 = \frac{4A}{C_L} \int_0^1 \left(\sum_{n=1,3,5,7} A_n \sin n\phi \right) d\left(\frac{y}{b/2} \right)$$

The loading coefficient for the lifting line result is

$$\frac{cc_l}{c_{av}C_L} = \frac{4A\sum_{n=1,3,5,7} A_n \sin n\phi}{C_L} = \frac{4}{\pi A_1} \sum_{n=1,3,5,7} A_n \sin n\phi \qquad \text{(C.66)}$$

Using the results for the A_i given in Equation (C.54) we may plot the loading coefficient curve as shown in Figure C.15. Also shown in that figure are the results for the same wing using the lifting surface approach of De Young and Harper (1948) which will be described in Section C.6.

Figure C.15 shows the load distribution along the half-span of the wing and the manner in which it falls off as the tip is approached.

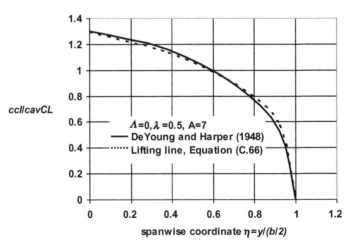

FIGURE C.15

Loading coefficient curve for the example wing as given by the simple lifting line theory and by the lifting surface approach of De Young and Harper (1948).

C.5.7 **Minimum induced drag**

The induced drag in Equation (C.65) is minimized when $\delta=0$, that is, when $A_3=A_5=A_7=0$, in which case the circulation distribution, according to Equations (C.40) and (C.41), would reduce to

$$\Gamma = 2bVA_1 \sin\phi = 2bVA_1\sqrt{1 - 4\left(\frac{y}{b}\right)^2} \tag{C.67}$$

Thus the circulation distribution is elliptic for minimum induced drag. From the description of the downwash velocity given in Equation (C.46) and the requirement that $A_3=A_5=A_7=0$ for minimum induced drag we find that the downwash must be constant, that is,

$$w = VA_1 = \text{const.} \tag{C.68}$$

Then substituting Equation (C.67) into Equation (C.39) we find

$$2bVA_1\sqrt{1 - 4\left(\frac{y}{b}\right)^2} = c(y)\pi V\left[(\alpha - \alpha_0) - \frac{w}{V}\right] \tag{C.69}$$

If the wing is untwisted ($\alpha=$constant with y), then the chord distribution of the wing, that is, the planform shape, must be elliptic. On the other hand, this equation illustrates that a wing may be twisted along the span so as to approximate the minimum induced drag developed by an elliptic circulation distribution. In general, wings with some taper, like the present example case, have small values of $A_3, A_5,$ and A_7 so that the induced drag is not far from the ideal.

C.6 **Lifting surface theory**

The lifting line theory described in the preceding sections has had substantial success in predicting the lift and drag of conventional wings. However, it cannot properly handle swept wings and other theories have been developed to provide this additional flexibility. The idea of moving from a lifting line to a lifting surface model for the wing was a logical extension. The simplest configuration is to imagine a line connecting two points in a vertical plane $y=$constant as defining a section of the wing surface. This line can be part of the chord line of an airfoil section in the plane with the upstream point being the location of a lifting vortex while the downstream point being a control point. The control used is to require that the net velocity normal to the chord line at the control point is zero, that is, there is no flow "through" the chord line at that control point. The velocity vector at that point is formed by the vector sum of the free stream velocity and the velocity induced at that point by the entire vortex system. A schematic of the section and the lifting vortex is shown in Figure C.16.

In the case of an infinite wing only the bound vortex exists, and as shown in Figure C.16, the induced velocity at any point r along the chord line is $w=\Gamma/4\pi r$ and is directed normal to the chord. The component of the free stream velocity normal to the chord at any point along the chord is $V\sin\alpha$ and has the opposite sense to that

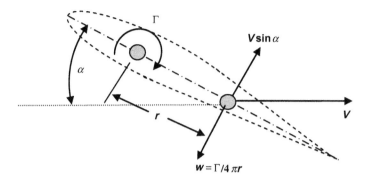

FIGURE C.16

Schematic diagram of bound vortex located at one point on the airfoil chord in the plane y=constant and the velocities normal to the chord as evaluated at the control point, which is located a distance r downstream of the bound vortex.

of the induced velocity, as also shown in Figure C.16. If we wish there to be no net velocity passing through the chord line at this "control" point located a distance r from the bound vortex, then we must have

$$\frac{\Gamma}{4\pi r} = V \sin \alpha \tag{C.70}$$

According to the Kutta-Joukowski law the lift per unit span of the wing L' is given by

$$L' = \rho V \Gamma \tag{C.71}$$

Using the value for Γ from Equation (C.70) in Equation (C.71) and considering only small angles of attack such that $\sin \alpha \simeq \alpha$, yields

$$L' = \rho V^2 4\pi r \sin \alpha = \frac{1}{2}\rho V^2 (4r)(2\pi \sin \alpha) \approx (2\pi\alpha)\frac{1}{2}\rho V^2 (4r)$$

Note that if we consider the distance separating the control point from the bound vortex to be $r=c/2$ the lift per unit span becomes

$$L' = (2\pi\alpha)\frac{1}{2}\rho V^2 c$$

This is the ideal lift per unit span we expect from an infinite wing, so it seems that choosing a control point a distance $c/2$ from the bound vortex will provide the correct inviscid flat plate airfoil lift coefficient $c_l = 2\pi\alpha$.

C.6.1 The Weissinger lifting surface method

Weissinger (1947) proposed applying this simplified flat plate approximation to the case of the swept wing and achieved useful results. De Young and Harper

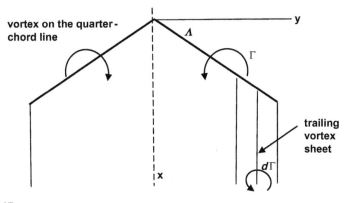

FIGURE C.17

Schematic diagram of the swept lifting line situated on the quarter-chord line of the wing. A vortex sheet trails downstream from the lifting line.

(1948) generalized this concept to a swept wing which is assumed to be represented by a flat plate with the same planform and twist distribution as the actual wing being analyzed. A lifting line is assumed to exist as a straight line coincident with the quarter-chord line of the wing and a vortex sheet trails back behind it as shown in Figure C.17. The boundary condition which will fix the circulation distribution along the span is that there is no net flow passing through the plate at the control points that are situated on the three-quarter-chord line, based on the results for the airfoil described previously. This means that the same lift curve slope of 2π per radian (0.11 per degree) is assumed to apply at each section of the finite wing. It has been pointed out by Abbott and von Doenhoff (1959) that for the NACA 63-, 64-, and 65-series airfoils the lift curve slope is very close to this value in the range of Reynolds number based on chord of 3–9 million and cambers up to 4% of the chord. Values of the lift curve slope for these airfoils were presented in Table 5.1.

Though as many control points as desired may be located along the spanwise direction, De Young and Harper (1948) report that seven points are generally adequate for most wings and that these may be represented by one point on the centerline and three points symmetrically distributed on either side of the centerline. Exactly the same spanwise locations are used by De Young and Harper (1948) as we used in the lifting line example calculation for an unswept wing given in Section C.5.5. In cases where there are steep spanwise gradients in loading, such as occurs with flap deflections, additional vortex and control points would be required.

The method presented by De Young and Harper (1948) involves the determination of the velocities induced at the control points by the lifting line in the wing and by the trailing vortex sheet and then applying the boundary condition of zero net flow through the control points. Only the utilitarian aspects of the approach are given here; details of the method and its derivation may be found in De Young and Harper

(1948) along with a solution method based on a large number of calculations made by the authors. They solved the equations for 200 wings with different planform shapes, that is, straight tapered wings with various values of sweepback, taper ratios, and aspect ratios.

The configuration of the wing is assumed to be as shown in Figure C.18. The root chord is taken as the reference line. When the root chord is at an angle of attack α_r, the angle of attack at any other spanwise control point is the sum of α_r and the local angle of twist ε_v, or $\alpha_v = \alpha_r + \varepsilon_v$. For symmetrically loaded wings, we need only consider the right, or starboard, side of the wing so that, counting from the tip toward the root, $n = 1, 2, 3, 4$ are the vortex points and $v = 1, 2, 3, 4$ are the control points, as indicated in Figure C.18. The basic equation to be solved at each control point is, referring to Figure C.16,

$$\left(\frac{w}{V}\right)_v = \sin \alpha_v \approx \alpha_v \tag{C.72}$$

Equation (C.72), which requires that the local angle of attack is everywhere small enough to justify replacing $\sin\alpha$ by α, may be written as follows:

$$\alpha_v = \sum_{n=1}^{4} a_{vn} G_n \quad v = 1, 2, 3, 4 \tag{C.73}$$

Here the a_{vn} represent the influence coefficients for the symmetric seven-point solution while $G_n = \Gamma_n/bV$ is the normalized circulation at the spanwise station n and is equal to the loading coefficient $(cc_l/2b)_n$. Thus, once the matrix of influence coefficients is known, Equation (C.73) is a set of four simultaneous equations for the four normalized circulations G_n which can be solved once the local angle of attack,

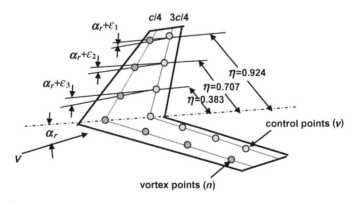

FIGURE C.18

Flat plate wing with twist distribution $\varepsilon = \varepsilon(\eta)$ showing the seven spanwise locations of discrete points (index n) for the bound vortex on the quarter-chord line and the associated control points (index v) on the three-quarter-chord line.

sweep, and chord distribution is specified. The load is then specified at four discrete stations $\eta = y/(b/2) = 0.924, 0.707, 0.383$, and 0 ($n = 1, 2, 3$, and 4, respectively). The influence coefficients are functions of the geometry of the planform alone which permits calculation of the span loading for different angles of attack and spanwise distributions of twist angle ε. Note that the method does not account for airfoil geometry, but instead relies on the theoretical lift curve slope of 2π per radian which, as pointed out previously, is quite adequate.

It is convenient to divide the lift into two components, that due to the distribution of wing twist, called the basic loading, and that due to angle of attack, called the additional loading. The calculation of these two components of the spanwise lift loading is discussed in the two subsequent sections.

C.6.2 Basic loading due to twist

The linearity of the lift curve slope requires that the zero-lift angle of attack of the reference line, that is, the root chord line, be known as α_{r0} so that we may write $C_L = 2\pi(\alpha - \alpha_{r0})$. DeYoung and Harper (1948) show that the lift coefficient may be written in terms of the individual normalized circulations G_n. Setting the wing lift coefficient to zero they develop the following relationship between the zero-lift circulations G_{n0}, the influence coefficients a_{vn}, and the local twist angles ε_v:

$$\varepsilon_1 = \left[a_{11} - a_{41} - 0.765k_1\right] G_{10} + \left[a_{12} - a_{42} - 1.414k_1\right] G_{20} + \left[a_{13} - a_{43} - 1.848k_1\right] G_{30}$$
$$\varepsilon_2 = \left[a_{21} - a_{41} - 0.765k_2\right] G_{10} + \left[a_{22} - a_{42} - 1.414k_2\right] G_{20} + \left[a_{23} - a_{43} - 1.848k_2\right] G_{30}$$
$$\varepsilon_3 = \left[a_{31} - a_{41} - 0.765k_3\right] G_{10} + \left[a_{32} - a_{42} - 1.414k_3\right] G_{20} + \left[a_{33} - a_{43} - 1.848k_3\right] G_{30}$$

$$\text{(C.74)}$$

Here the k_v are given as follows:

$$k_1 = a_{14} - a_{44}$$
$$k_2 = a_{24} - a_{44} \qquad \text{(C.75)}$$
$$k_3 = a_{34} - a_{44}$$

The equation for the root load is

$$G_{40} = -\left(0.765G_{10} + 1.414G_{20} + 1.848G_{30}\right) \qquad \text{(C.76)}$$

The angle of attack for zero lift is given by

$$\alpha_{r0} = \alpha_{40} = \left(a_{41} - 0.765a_{44}\right) G_{10} + \left(a_{42} - 1.414a_{44}\right) G_{20} + \left(a_{43} - 1.848a_{44}\right) G_{30}$$

$$\text{(C.77)}$$

The spanwise loading at the zero-lift angle of attack is called the basic loading and is due solely to the twist distribution of the wing and is independent of angle of attack. Although the net lift of the wing is zero in the case of the basic loading there is positive and negative lift produced at different spanwise stations so that there is an induced drag associated with the local lift production and it is given by

$$C_{D,i0} = \pi A \left[0.668 G_{10}^2 + G_{20}^2 + 2.082 G_{30}^2 + (0.215 G_{10} + 1.442 G_{30}) G_{20} + 1.061 G_{10} G_{30} \right]$$
(C.78)

Because there is no net force, the moment produced by the basic loading is a pure couple. This means that the moment produced is independent of the location of the moment axis and one may pick a y-axis at any x location about which to calculate the pitching moment due to the basic loading. Selecting an arbitrary axis and summing the moment from each wing element De Young and Harper (1948) obtain the zero-lift moment coefficient as

$$C_{m0} = -\frac{Ab}{c_{MAC}} \tan \Lambda_{c/4} \left(0.126 G_{10} + 0.175 G_{20} + 0.106 G_{30} \right)$$
(C.79)

C.6.3 Additional loading due to angle of attack

Because the twist of the wing has been accounted for in the previous section, it is only necessary to consider the effects of changes in angle of attack. In that case we may treat the wing as a flat plate with the same planform and the angle of attack α is the same for all spanwise locations so that Equation (C.73) may be written as follows:

$$1 = \sum_{n=1}^{4} a_{vn} \frac{G_{na}}{\alpha} \quad v = 1, 2, 3, 4$$
(C.80)

Solving this set of four simultaneous equations gives the normalized circulations G_{na}/α per radian at each of the spanwise stations $\eta = y/(b/2) = 0.924, 0.707, 0.383,$ and 0. The lift curve slope of the wing is then given by

$$\frac{dC_L}{d\alpha} = C_{L,\alpha} = \frac{\pi A}{8} \left(\frac{G_{4a}}{\alpha} + 1.848 \frac{G_{3a}}{\alpha} + 1.414 \frac{G_a}{\alpha} + 0.765 \frac{G_{1a}}{\alpha} \right)$$
(C.81)

Note that the dimensionless circulation per radian

$$\frac{G_{na}}{\alpha} = \frac{C_{L,\alpha}}{2A} \left(\frac{c_{la} c}{C_L c_{av}} \right)_n = C_{L,\alpha} \left(\frac{c_{la} c}{2bC_L} \right)_n$$

The induced drag arising from the additional loading due to change in angle of attack is calculated in the same fashion as for the basic loading. The values of the normalized circulations G_{na} are calculated for each value of C_L and used in Equation (C.78) to obtain the induced drag coefficient at that lift coefficient. Another format for the induced drag due to the additional loading is given by

$$\frac{C_{Dia}}{C_L^2} = \frac{\pi A}{2C_{L,\alpha}^2} \left[\begin{array}{l} \left(\frac{G_{1a}}{\alpha} \right)^2 + \left(\frac{G_{2a}}{\alpha} \right)^2 + \left(\frac{G_{3a}}{\alpha} \right)^2 + \frac{1}{2} \left(\frac{G_{4a}}{\beta} \right)^2 \\ - \frac{G_{4a}}{\alpha} \left(0.056 \frac{G_{1a}}{\alpha} + 0.789 \frac{G_{3a}}{\alpha} \right) \\ - \frac{G_{2a}}{\alpha} \left(0.733 \frac{G_{1a}}{\alpha} + 0.845 \frac{G_{3a}}{\alpha} \right) \end{array} \right]$$
(C.82)

The spanwise position of the center of pressure, which is located on the quarter-chord line, due to the additional loading is given by

$$\eta_{c.p.a} = \frac{0.352\frac{G_{1a}}{\alpha} + 0.503\frac{G_{2a}}{\alpha} + 0.344\frac{G_{3a}}{\alpha} + 0.041\frac{G_{4a}}{\alpha}}{0.383\frac{G_{1a}}{\alpha} + 0.707\frac{G_{2a}}{\alpha} + 0.924\frac{G_{3a}}{\alpha} + +0.5\frac{G_{4a}}{\alpha}} \tag{C.83}$$

The additional loading gives rise to a lift force which is assumed to be distributed along the quarter-chord line. Integrating the product of the elemental lift force and its distance from the x-axis yields the bending moment at the wing root. Knowing this moment and the total lift of the half-span permits the determination of the spanwise location at which the lift force acts, that is, at $y_{c.p.}$ the center of pressure. The pitching moment due to the additional loading depends on the choice of the moment axis. Selecting a point along the x-axis, say the center of gravity location $x_{c.g.}$, one may calculate the pitching moment $M_{c.g.}$ about a y-axis through this point $(x_{c.g.}, 0)$ due to the concentrated load which acts at the point $(x_{c.p.}, y_{c.p.})$ as follows:

$$M_{c.g.} = L\left(x_{c.p.} - x_{c.g.}\right)$$

Then the pitching moment coefficient due to the additional loading is simply given by

$$C_{ma} = C_L \frac{x_{c.p.} - x_{c.g.}}{c_{MAC}} \tag{C.84}$$

The axial location of the center of pressure may be found from the wing geometry. Because in this analysis it is assumed that the quarter-chord line is straight, the center of pressure location may be put in terms of the location of the quarter-chord of the wing root and the spanwise location of the concentrated lift force as follows:

$$\frac{x_{c.p.}}{c_{MAC}} = \frac{x_{r,c/4}}{c_{MAC}} - \eta_{c.p.} \frac{b \tan \Lambda_{c/4}}{2c_{MAC}}$$

The spanwise location of the center of pressure $\eta_{c.p.}$ is at the centroid of the span loading distribution. For wings with close to elliptic span loading the center of pressure location is approximately $\eta_{c.p.} = 4/3\pi$. For such wings the aerodynamic center location with respect to the aerodynamic chord is approximately

$$\frac{x_{a.c.}}{c} = \frac{1}{4} + \frac{0.342 - 0.567\lambda - 0.908\lambda^2}{10\left(1 + \lambda + \lambda^2\right)} A \tan \Lambda_{c/4}$$

C.6.4 Gross loading on the wing

The total load on the wing equals the sum of the basic load and the additional load, that is, a constant term and a variable term. Therefore the wing characteristics must be determined for each lift coefficient. Outside of operation near the stall point, the effects of twist are much smaller than those due to sweep, aspect ratio, and taper ratio. It must be kept in mind, however, that aeroelastic effects, where the loading may change the twist distribution of the wing, must be carefully considered so that safe operation is maintained. These effects are outside the scope of this book and for a treatment of this topic reference should be made, for example, to Bisplinghoff et al. (1983).

The span loading is given by the sum of the basic loading and the additional loading, so that the total normalized circulation is $G_n = G_{n0} + G_{na}$. The twist built into the wing determines the basic loading and therefore it is used to fine-tune the spanwise load distribution. The twist may be used to tailor the spanwise loading to closely approximate the optimal elliptic loading or to reduce the possibility of early stalling of the tips. Because we assume no aeroelastic effects the twist of the wing remains constant and the lift of the wing is that due to the additional loading as given by Equation (C.81).

The induced drag may be found from Equation (C.78) using the total normalized circulation $G_n = G_{n0} + G_{na}$. Note that C_{Di} is a function of the loading and changes with the change in lift coefficient so that the induced drag is not the sum of the basic and additional loading-induced drag coefficients, that is, $C_{Di} \neq C_{Di0} + C_{Dia}$. The pitching moment, on the other hand, may be found by adding the pitching moment due to basic loading and that due to the additional loading. The root bending moment may be readily determined by simply considering the total lift being applied at the center of pressure.

C.6.5 Applications of the method

The geometry of the wing is all that is needed to calculate the 16 influence coefficients a_{vn}. Although the matrix of influence coefficients is fixed for a given wing, their actual calculation is rather tedious and is described in some detail by DeYoung and Harper (1948). However, Stevens (1948) tabulated the coefficients for a wide range of aspect ratios, taper ratios, and sweepback angles. For example, the 16 influence coefficients for wings with $\Lambda_{c/4} = 30°$, aspect ratios of 8 and 10, and taper ratios of 0.25 and 0.5 are given in Table C.1 while those for wings with $\Lambda_{c/4} = 0°$, aspect ratios of 8 and 10, and taper ratios of 0.25 and 0.5 are given in Table C.2. In addition, the 16 influence coefficients for wings with $\Lambda_{c/4} = 45°$, aspect ratios of 6 and 8, and taper ratios of 0.25 and 0.5 are given in Table C.3. The increased sweepback angle case is included with lower aspect ratios because, as described subsequently in Section C.6.6, at high subsonic Mach numbers a wing transformed according to the Prandtl-Glauert rule will have increased sweepback and lower aspect ratio, but the taper ratio will be unaffected. If rapid results are desired for values of the wing parameters not presented in the tables it is possible to interpolate or extrapolate the solutions for wings that bracket the actual parameters rather than to interpolate or extrapolate the influence coefficients. Of course, it is best to calculate the influence coefficients for the wing in question wherever possible.

Table C.1 Coefficients $a_{\nu n}$ for the Lifting Surface Method of De Young and Harper (1948) for 30° Sweep

$\Lambda_{c/4}$	30	30	30	30
λ	0.25	0.5	0.25	0.5
A	8	8	10	10
a_{11}	13.09	11.82	14.27	12.49
a_{12}	−3.81	−3.64	−4.02	−3.72
a_{13}	0.45	0.21	0.69	0.34
a_{14}	−0.43	−0.32	−0.54	−0.37
a_{21}	−1.00	−1.17	−0.69	−0.89
a_{22}	7.68	7.32	8.43	7.95
a_{23}	−2.52	−2.45	−2.66	−2.57
a_{24}	0.25	0.20	0.33	0.28
a_{31}	0.03	0.01	−0.02	−0.04
a_{32}	−1.01	−0.95	−0.77	−0.71
a_{33}	5.68	5.79	6.16	6.31
a_{34}	−1.68	−1.68	−1.71	−1.73
a_{41}	−0.12	−0.11	−0.11	−0.09
a_{42}	0.17	0.13	0.12	0.07
a_{43}	−2.17	−1.96	−1.90	−1.64
a_{44}	5.06	5.42	5.50	5.96

The lift curve slope for the wing as determined by the Weissinger method (1947) according to the technique developed by De Young and Harper (1948) is shown in Figure C.19 for two aspect ratios as a function of the compressibility-corrected sweepback angle

$$\Lambda_\beta = \arctan\left(\frac{\tan \Lambda_{c/4}}{\beta}\right) \tag{C.85}$$

Here β is the Prandtl-Glauert factor $\beta = (1 - M^2)^{1/2}$ and clearly for incompressible flow ($M = 0$) the compressibility-corrected sweepback angle $\Lambda_\beta = \Lambda_{c/4}$. The lift curve slope suggested by DATCOM as given in Equation (5.18) is also shown for comparison in Figure C.19. Note that there are consistent, but small, systematic differences of several percent between the two over the range of sweepback considered. The effect of Mach number and quarter-chord sweepback angle on the lift curve slope of the wing is illustrated in Figure C.20 for a wing with aspect ratio $A = 8$ and taper ratio $\lambda = 0.25$.

Table C.2 Coefficients $a_{\nu n}$ for the Lifting Surface Method of De Young and Harper (1948) for 0° Sweep

$\Lambda_{c/4}$	0	0	0	0
λ	0.25	0.5	0.25	0.5
A	8	8	10	10
a_{11}	12.61	11.65	13.44	12.16
a_{12}	−2.58	−2.96	−2.30	−2.75
a_{13}	−0.08	0.03	−0.18	−0.04
a_{14}	−0.10	−0.16	−0.07	−0.12
a_{21}	−1.58	−1.65	−1.43	−1.52
a_{22}	7.23	6.95	7.77	7.42
a_{23}	−1.61	−1.71	−1.42	−1.54
a_{24}	0.01	0.02	−0.04	−0.01
a_{31}	0.03	0.03	0.02	0.01
a_{32}	−1.41	−1.38	−1.29	−1.26
a_{33}	5.34	5.41	5.67	5.77
a_{34}	−1.28	−1.25	−1.17	−1.14
a_{41}	−0.16	−0.15	−0.15	−0.14
a_{42}	0.09	0.07	0.07	0.05
a_{43}	−2.61	−2.48	−2.45	−2.30
a_{44}	4.61	4.80	4.86	5.11

It is clearly seen here that the lift curve slope of the wing is diminished as the Mach number and quarter-chord sweepback angle increase. Though the unswept wing shows superior lift performance, it has been shown in Chapter 9 that the drag penalties incurred by unswept wings are too high to permit their practical use for high subsonic speeds.

The spanwise distributions of the loading coefficient for untwisted (flat plate) wings of geometries representative of commercial turbofan and turboprop aircraft, as computed by the method of De Young and Harper (1948), are shown in Figures C.21 and C.22. In Figure C.21 the result for a turbofan aircraft wing with a quarter-chord sweepback of 30° and taper ratio $\lambda = 0.25$ is contrasted with that for a turboprop aircraft wing with no sweepback and a larger taper ratio of 0.5; both wings have an aspect ratio $A = 8$. It is apparent that the differences between the two loading distributions are small but noticeable, with the swept back wing displaying higher loading outboard and lower loading inboard than the unswept wing. In Figure C.22 the result for a turbofan aircraft wing with a quarter-chord sweepback of 30° and taper ratio $\lambda = 0.25$ is contrasted with that for a turboprop aircraft wing with no sweepback and

Table C.3 Coefficients $a_{\nu n}$ for the Lifting Surface Method of De Young and Harper (1948) for 45° Sweep

$\Lambda_{c/4}$	45	45	45	45
λ	0.25	0.5	0.25	0.5
A	6	6	8	8
a_{11}	12.59	11.43	14.08	12.24
a_{12}	−4.21	−3.86	−4.72	−4.09
a_{13}	0.56	0.23	1.01	0.46
a_{14}	−0.50	−0.35	−0.70	−0.45
a_{21}	−0.93	−1.11	−0.47	−0.71
a_{22}	7.44	7.08	8.41	7.93
a_{23}	−2.84	−2.71	−3.20	−2.99
a_{24}	0.31	0.24	0.48	0.36
a_{31}	0.06	0.03	−0.02	−0.03
a_{32}	−0.95	−0.89	−0.62	−0.55
a_{33}	5.53	5.67	6.18	6.33
a_{34}	−1.84	−1.87	−1.97	−2.00
a_{41}	−0.12	−0.10	−0.09	−0.05
a_{42}	0.24	0.20	0.17	0.11
a_{43}	−2.12	−1.89	−1.75	−1.46
a_{44}	4.98	5.35	5.61	6.12

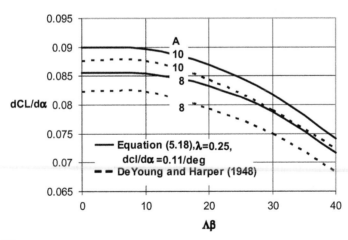

FIGURE C.19

The variation of the lift curve slope for wings of taper ratio $\lambda = 0.25$ and aspect ratio $A = 8$ and 10 is shown as a function of compressibility-corrected sweepback according to two different methods of calculation.

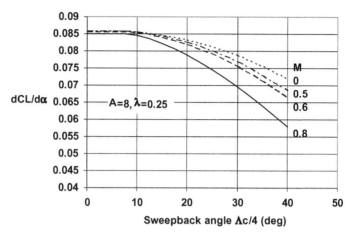

FIGURE C.20

The variation of the lift curve slope of a wing with $A=8$ and $l=0.25$ is shown as a function of quarter-chord sweepback angle with the flight Mach number appearing as a parameter.

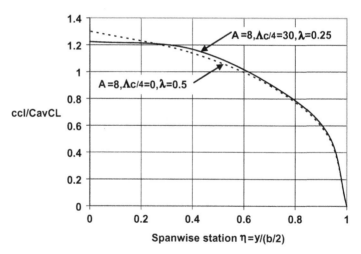

FIGURE C.21

The spanwise distribution of the loading coefficient for two untwisted wings of the same aspect ratio $A=8$ but different sweepback angle and taper ratio.

a larger taper ratio of 0.5; both wings have an aspect ratio $A=10$. Once again, the differences between the two distributions are small but distinctly noticeable. It may be noted that the spanwise distributions of the loading coefficients for the swept back wings of different aspect ratio are similar in appearance and the same is true for the unswept wings.

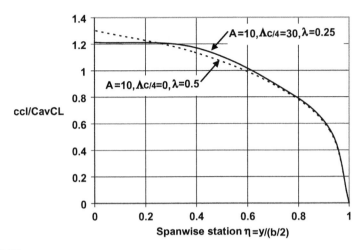

FIGURE C.22

The spanwise distribution of the loading coefficient for two untwisted wings of the same aspect ratio $A=8$ but different sweepback angle and taper ratio.

The differences in spanwise loading become apparent when we consider the distribution of the local lift coefficient c_l across the wing. We must now introduce the actual geometry of the wing and calculate

$$c_l = \left(\frac{cc_l}{c_{av}C_L}\right)\frac{c_{av}}{c}C_L = \left(\frac{cc_l}{c_{av}C_L}\right)\left(\frac{S}{b}\right)\frac{C_L}{c} \qquad \text{(C.86)}$$

The average chord of the wing $c_{av}=S/b$ and it is clear that the variation of the chord with the spanwise coordinate y can alter the spanwise distribution considerably. For example, in the case of wings with straight leading and trailing edges the ratio of the average chord to the local chord is given by

$$\frac{c_{av}}{c} = \frac{1+\lambda}{2\left[1-\eta\left(1-\lambda\right)\right]} \qquad \text{(C.87)}$$

The chord distribution is obviously known from the geometry of the wing being considered.

If we choose to consider the case of a wing with twist we must specify the twist distribution. Although a linear distribution of twist is not the most common, it is useful for descriptive purposes. Thus take $\varepsilon=\eta\varepsilon_t$ where ε is measured in radians and assume a tip twist of $1°$; positive twist increases the local angle of attack and is called "wash-in." For the wings with which we have been dealing the local lift coefficient distribution for 1 degree of wash-in and an angle of attack of $10°$ measured from the root chord zero angle of attack α_{r0} is shown in Figure C.23. The wing lift coefficient at the 10 degree of angle of attack is also given in Figure C.23 for each wing.

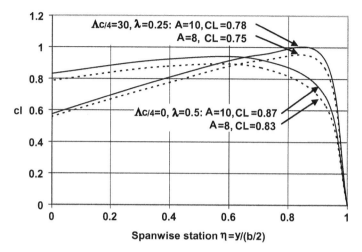

FIGURE C.23

Spanwise distribution of local lift coefficient for several wings with 1 degree of wash-in. All cases are shown for an angle of attack of 10 degrees.

The outboard portions of the swept wings are seen to be heavily loaded compared to the center section, while the unswept wings are loaded somewhat more heavily over the center section of the wing. The high local lift coefficient developed over the outboard section of the swept wing is likely to cause initial stall at the wing-tips, which is undesirable. This situation may be corrected by twisting down the tip sections, i.e., introducing "wash-out" at the tips. The effect of "washing-out" the wingtips is illustrated in Figure C.24 where the spanwise distribution of local lift coefficient for a swept wing with 1° of wash-in ($\varepsilon = \eta/57.3$) compared to that for the same wing with 5° of wash-out ($\varepsilon = -5\eta/57.3$). The tip loading has been diminished and the center section loading increased resulting in a more practical distribution of local lift coefficient. Black (1953) carried out a series of experiments on a swept wing with different twist distributions and reported that this simplification of the Weissinger method gave satisfactory results.

Aerodynamic wash-out may thus be useful for avoiding tip stall. The permissible amount of wash-out is often limited by the increase in induced drag, which is generally small for 1° or 2° of wash-out. The limiting wash-out may instead be that value which results in the tip section operating outside the low-drag range at the required high-speed lift coefficient, but such details are beyond the scope of this book. However, it should be kept in mind that a poor choice of planform may cause tip sections to be so heavily loaded that tip stalling is likely with any reasonable choice of wash-out. For example, high taper ratios and large amounts of sweepback are unfavorable in this respect and are particularly bad in combination and may promote longitudinal instability at the stall in addition to the usual lateral instability.

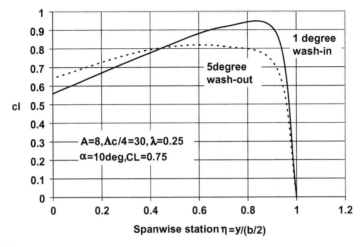

FIGURE C.24

Comparison of the spanwise distribution of local lift coefficient for a swept wing with 1° of wash-in compared to that for the same wing with 5° of wash-out.

C.6.6 Compressibility effects on the lifting surface method

The Prandtl-Glauert transformation of coordinates given by Equation (C.2) applied to an airliner wing of aspect ratio $A = 10$ in an $M = 0.8$ flow results in a transformed wing with $A = 6$ in a flow at the same speed, but with $M = 0$, as shown in Figure C.25. The span remains the same, as does the angle of attack, but the x-coordinate is stretched as the Mach number increases.

Thus the transformed wing aspect ratio decreases by a factor equal to β and the sweep of the quarter-chord line increases according to Equation (C.85) as the subsonic Mach number increases. The transformed wing may be analyzed with any span loading theory to find the load distribution at $M = 0$ and this will be the span load distribution on the actual wing at the flight Mach number. Obviously, the wing is continually distorted as the Mach number changes and a separate calculation would have to be carried out for each case. This is readily done, though tedious, once a suitable computational scheme is constructed. To facilitate more rapid estimates of the span load distribution several simplifications are presented in the next section.

C.6.7 Simplifications to the Weissinger method

The approach provided by DeYoung and Harper (1948) and described previously uses the Weissinger lifting surface analysis for subsonic wings. A generalized method for the loading distribution over wings in subsonic flow was presented by them, but it still is relatively tedious to carry out. To aid in rapid evaluation of different wing

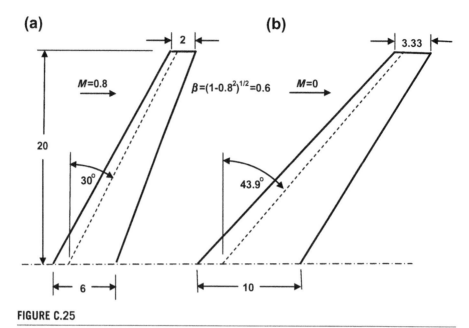

FIGURE C.25

Sample case of (a) subsonic wing ($M=0.8$, $A=10$) and (b) transformed wing ($M=0$, $A=6$).

designs they also provide individual graphs which permit the direct determination of the 16 influence coefficients based on the geometry of the wing with no further calculation. There is, of course, the necessity to sometimes interpolate or extrapolate the curves they present. However, one may then use these influence coefficients to determine all the loading information as discussed previously. Because we focus on the commercial airplane case we are typically concerned with airplanes whose wings have the following characteristics:

(a) Regional turboprops in the speed range $0<M<0.5$ with $A \sim 10$, $\Lambda_{c/4}=0$, and $\lambda \sim 0.5$.

(b) Turbofans in the speed range $0<M<0.85$ for $8<A<10$, $25° < \Lambda_{c/4}<35°$, and $0.25<\lambda<0.35$.

If we confine our attention to this approximate range of parameters we may extract the pertinent information for wing loading for straight tapered wings with no twist, that is, the so-called additional loading described in Section C.6.3, from the charts presented by DeYoung and Harper (1948). First, we note that the extensive calculations they performed illustrated that the taper ratio λ has a minor effect on the additional loading, particularly for the small range of values being considered. Curve fits to the loading coefficient for the range of variables listed are compiled as a function of the spanwise coordinate $\eta = 2y/b$ in Table C.4. To determine the span loading due to twist, that is, the so-called basic loading described in Section C.6.2, one must use the approach described therein or else use the simplified expression for the basic loading described in the next section.

C.6.8 The method of Diederich

Other simplified methods, which depend upon correlation graphs, have been presented by Schrenk (1940) and Diederich (1952). Basically, Diederich (1952) suggests the following relation for the additional loading function

$$\frac{cc_{l,a}}{c_{av}C_L} = C_1\frac{c}{c_{av}} + C_2\frac{4}{\pi}\sqrt{1 - \eta^2} + C_3 f \tag{C.88}$$

The coefficients C_1, C_2, and C_3 are presented on graphs as functions of the planform parameter

$$F = \frac{A}{\left(\frac{2\pi}{c_{l,\alpha}}\right)\cos\Lambda_{c/4}} \tag{C.89}$$

The variable $f = f(\Lambda_\beta, \eta)$ appearing in Equation (C.88) is also given by Diederich in the form of a graph of f as a function of spanwise coordinate η with Λ_β as a parameter. When there is no sweepback $\Lambda_\beta = 0$ and the function f is elliptic so that Equation (C.88) may be rewritten as follows:

$$\frac{cc_{l,a}}{c_{av}C_L} = C_1\frac{c}{c_{av}} + (C_2 + C_3)\frac{4}{\pi}\sqrt{1 - \eta^2} + C_3 f' \tag{C.90}$$

Table C.4 Approximations to the Results of De Young and Harper (1948) for the Additional Loading on Several Representative Wings

Spanwise station $\eta = y/(b/2)$	Turbofans[a] $A=10, \lambda=0.25$ Loading coefficient $cc_l/c_{av}C_L$	Turbofans[a] $A=8, \lambda=0.25$ Loading coefficient $cc_l/c_{av}C_L$	Turboprops $A=10, \lambda=0.5,$ $\Lambda_\beta=0$ Loading coefficient $cc_l/c_{av}C_L$
0	$1.42 - 0.0071\Lambda_\beta$	$1.40 - 0.0057\Lambda_\beta$	1.31
0.383	1.17	1.17	1.17
0.707	$0.82 + 0.0033\Lambda_\beta$	$0.83 + 0.0029\Lambda_\beta$	0.89
0.924	$0.46 + 0.30(\Lambda_\beta/60)^{1.5}$	$0.46 + 0.26(\Lambda_\beta/60)^{1.5}$	0.55
1	0	0	0

[a] The angle $\Lambda_\beta = \arctan(\tan\Lambda_{c/4}/\beta)$ is measured in degrees.

The coefficients are well behaved and may be represented by the following curve fits:

$$
\begin{aligned}
C_1 &= 0.079 F^{0.75} \\
C_2 &= 1 - \sin{(0.099F)} \\
C_3 &= 0.4 \sin{(0.112F)}
\end{aligned}
\tag{C.91}
$$

The shortcoming in applying this simple method is that the function f, and therefore the function f', is given only for $\Lambda_\beta = -60°, -45°, -30°, 0, 30°, 45°$, and $60°$, where the negative sign denotes a swept forward wing. Because the compressibility-corrected sweepback angle Λ_β varies with Mach number there is some question about the accuracy to be obtained by interpolation or extrapolation among the graphs presented. For example, a wing with $\Lambda_{c/4} = 30°$ will have $\Lambda_\beta = 33.7, 35.8, 39.0$, and $43.9°$ when the Mach number $M = 0.5, 0.6, 0.7$, and 0.8, respectively. However, a reasonable curve fit for f' may be made by using the plots of f as a function of spanwise coordinate η for $\Lambda_\beta = 0, 30°$, and $45°$. Such a curve fit may be written as follows:

$$
f' = 1.07 \left(\frac{\Lambda_\beta}{30}\right) \left[(\eta - 0.8)^2 - 0.24\right] + 0.53\eta^6 \sqrt{\left(\frac{\Lambda_\beta}{30}\right)}
\tag{C.92}
$$

This equation reproduces f' quite faithfully for the three values of Λ_β used except for the last 5% of the wing at the tip. However, this equation is merely a correlation based on the few curves for f given by Diederich (1952), and it has not been adequately tested, so one should be guarded in its use.

The basic loading function is given by Diederich in terms of the additional loading as follows:

$$
\frac{cc_{l,b}}{c_{av}} = k_1 C_{L,\alpha} (\alpha - \alpha_{av}) \left(\frac{cc_{l,a}}{c_{av} C_L}\right)
\tag{C.93}
$$

Here the terms k_1 and α_{av} are given by

$$
k_1 = \frac{\sqrt{1 + \left(\frac{2}{F}\right)^2} + \left(\frac{2}{F}\right)}{\sqrt{1 + \left(\frac{6}{F}\right)^2} + \left(\frac{6}{F}\right)}
\tag{C.94}
$$

$$
\alpha_{av} = \int_0^1 \alpha \left(\frac{cc_{l,a}}{c_{av} C_L}\right) d\eta = \int_0^1 (\alpha_r + \varepsilon) \left(\frac{cc_l}{c_{av} C_L}\right) d\eta
\tag{C.95}
$$

This formulation makes the lift of the basic loading equal to zero, as it should be, and that can be verified by integrating Equation (C.93) from $\eta=0$ to $\eta=1$. Then the total loading function

$$\frac{cc_l}{c_{av}} = C_{L,\alpha}\alpha_{av}\left(\frac{cc_{l,a}}{c_{av}C_L}\right) + \left(\frac{cc_{l,b}}{c_{av}}\right) \tag{C.96}$$

Diederich's method provides a rapid means of estimating the spanwise lift distribution using relatively simple equations involving solely the geometric characteristics of the wing. In Diederich's method the moment due to the basic loading, which is a pure couple, may be found by taking moments about an arbitrary axis. Selecting an axis in the y-direction and passing through the tip of the root chord leads to the following expression for the pitching moment M_r:

$$M_r = 2\int_0^{b/2}\left(x - x_r\right)c_l qc\,dy$$

Then the pitching moment due to the basic loading is the zero-lift pitching moment and is given by

$$C_{m0} = \frac{M_r}{qSc_{MAC}} = \frac{1}{2}A\tan\Lambda_{c/4}\frac{c_{av}}{c_{MAC}}\int_0^1\left(\frac{cc_{l,b}}{c_{av}}\right)\eta\,d\eta$$

The basic loading function is given in Equation (C.93) and for wings with straight leading and trailing edges

$$\frac{c_{av}}{c_{MAC}} = \frac{3}{4}\frac{(1+\lambda)^2}{\left(1+\lambda+\lambda^2\right)}$$

The spanwise location of the center of pressure for wings with straight leading and trailing edges is given in this method by

$$\eta_{c.p.} = \frac{1+2\lambda}{3(1+\lambda)}C_1 + \frac{4}{3\pi}C_2 + \left(0.425 + 1.067\Lambda_\beta\right)$$

C.6.9 Comparison of Diederich's method to that of De Young and Harper

Consider the spanwise distribution of the loading coefficient for an untwisted wing with aspect ratio $A=8$, sweepback angle $\Lambda=30°$, and taper ratio $\lambda=0.25$ as computed by the method of De Young and Harper (1948) and shown in Figure C.20. Using the simplified method of Diederich (1952) as described in the previous section we calculate the spanwise loading coefficient and compare it to that of Figure C.20 as shown in Figure C.26. It is seen that the simplified method tends to underestimate

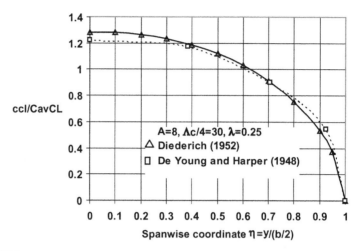

FIGURE C.26

Comparison of the spanwise loading coefficient as calculated by the Weissinger approach of De Young and Harper (1948) with the simplified method of Diederich (1952) for an untwisted swept wing.

the spanwise loading coefficient calculated by the method of De Young and Harper (1948) on the outboard stations and overestimate it on the inboard stations. Recall that the area under both curves is equal to unity because the lift coefficient of the wing is given by

$$C_L = \frac{L}{qS} = \frac{2}{qS} \int_0^{b/2} c_l q(c\,dy) = \frac{1}{c_{av}} \int_0^1 cc_l\,d\eta$$

We have called the span loading coefficient for an untwisted wing the "additional" loading because it is due solely to changes in angle of attack. We may now extend the comparison of the two methods to the case of the same wing with a wash-out of 5° at the tip. In this case we solve for the "basic" loading due to twist and add it to the additional loading. Doing this for the spanwise distribution of the section lift coefficient leads to the results shown in Figure C.27. Notice that the agreement between the two methods is quite good except at η around 0.9 where the lift coefficient predicted by the Diederich method is about 8% below that predicted by the De Young and Harper method.

C.7 The vortex lattice method

The lifting surface theory just described provides useful information about fairly general wings, but even greater flexibility was desired. In particular, in addition to the spanwise loading experienced by the wing, the chordwise loading is of substantial

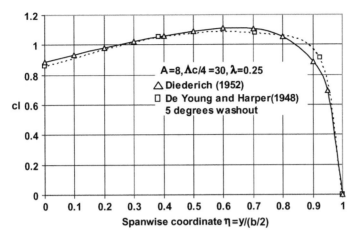

FIGURE C.27

Comparison of the spanwise distribution of the section lift coefficient as calculated by the Weissinger approach of Harper and Young (1948) with the simplified method of Diederich (1952) for a swept wing with 5° of washout at the tip.

interest. The logical extension of the simple lifting surface theory follows from the extension of the boundary conditions applied. In the simple lifting surface theory presented earlier, the actual wing was replaced by a flat plate with the same planform and the condition of no flow through the wing was satisfied at several spanwise points lying on the three-quarter-chord line. In the simplest extension, one may consider the flat plate wing to be made up of panels which each act as wings in that they produce a lifting line assumed to be acting on the quarter-chord line of the panel and a trailing vortex pair. On each of these panels the condition of zero net flow through the panel is enforced at the single control point. An illustration of the model for a single panel on the wing is shown in Figure C.28. The ability to treat a large number of panels by using a computer program to carry out the large number of calculations has made this approach now widespread. Margason and Lamar (1977) present a detailed development of such an approach and present a program which will handle up to 120 panels on a half-span.

In this simple flat plate wing case we may also consider that the wing and the vortices all lie in the plane $z=0$. Since typical angles of attack are small this is a reasonable approximation. One can make the approach more practical by considering twist and camber, etc. but at the expense of added complexity in the calculations. The degree of complexity carried out should be commensurate with the degree of accuracy required.

C.7.1 The induced velocity field

The induced velocity may be found in the same manner as was done previously. Consider the general nth panel shown in Figure C.28 in more detail, as illustrated in

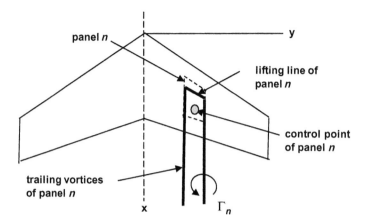

FIGURE C.28

The general nth panel on the wing showing the basic horseshoe vortex pattern. At the control point the normal component of the vector sum of the velocities induced by the lifting line and trailing vortex pair of every panel and the free stream velocity must balance such that the velocity normal to the wing is zero.

Figure C.29. The velocity induced by any segment of the vortex may be determined by using Equation (C.17):

$$dw = \frac{\Gamma}{4\pi r^3} R \, ds$$

Applying this equation to each leg of the nth vortex, AA', BB', and AB leads to

$$W_{AA'}(x_P, y_P) = \frac{\Gamma}{4\pi R_{AA'}} (\cos \alpha_1 - 1) \tag{C.97}$$

$$W_{BB'}(x_P, y_P) = \frac{-\Gamma}{4\pi R_{BB'}} (\cos \alpha_2 - 1) \tag{C.98}$$

$$W_{AB}(x_P, y_P) = \frac{\Gamma}{4\pi R_{AB}} (\cos \alpha_3 - \cos \alpha_4) \tag{C.99}$$

We may determine all the necessary quantities from the geometry in Figure C.29 as follows:

$$R_{AA'} = y_A - y_P \tag{C.100}$$

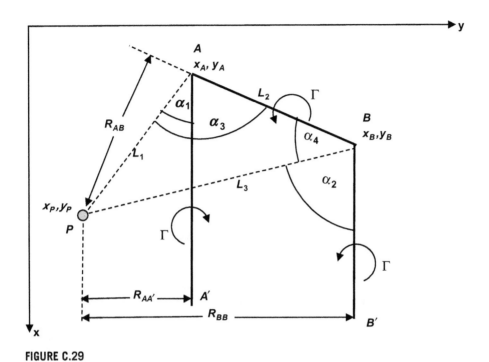

FIGURE C.29

Schematic of the details of the *n*th panel shown in Figure C.28.

$$R_{BB'} = y_B - y_P \tag{C.101}$$

$$R_{AB} = L_3\sqrt{\left[1 - \cos^2\alpha_4\right]} \tag{C.102}$$

$$\cos\alpha_1 = \frac{\sqrt{(x_A - x_P)^2}}{L_1} \tag{C.103}$$

$$\cos\alpha_2 = \frac{\sqrt{(x_B - x_P)^2}}{L_3} \tag{C.104}$$

$$\cos\alpha_3 = \frac{L_1^2 + L_2^2 - L_3^2}{2L_1L_2} \tag{C.105}$$

$$\cos \alpha_4 = \frac{L_2^2 + L_3^2 - L_1^2}{2L_1L_2} \qquad (C.106)$$

The lengths of the legs of the triangle ABP are shown in Figure C.29 and are defined as follows:

$$L_1 = \sqrt{(x_P - x_A)^2 + (y_P - y_A)^2}$$
$$L_2 = \sqrt{(x_B - x_A)^2 + (y_P - y_B)^2}$$
$$L_3 = \sqrt{(x_P - x_B)^2 + (y_P - y_B)^2}$$

The view shown in Figure C.29 is from above, looking down on the wing. The horseshoe vortex is producing downwash within its outline and we may consider downwash to be negative, that is, in the negative z-direction in the coordinate system shown. Thus at the point P the contribution to the induced velocity from vortex segment AA' is $W_{AA'} > 0$, from vortex segment AB it is $W_{AB} < 0$, and from vortex segment BB' it is $W_{BB'} < 0$. We note that $R_{AA'} > 0$ and $R_{BB'} > 0$ and, by definition, $R_{AB} > 0$. Note also that the terms $(\cos \alpha_1 - 1)$ and $(\cos \alpha_2 - 1)$ are bounded by -2 and 0 and $(\cos \alpha_3 - \cos \alpha_4)$ lies between 0 and 2. Though the value of Γ is the same along the entire horseshoe vortex, the sense of the circulation shown in Figure C.29 requires the negative sign for Γ in the vortex segment BB' so that the direction of the induced velocities in Equations (C.97)–(C.99) is correct.

The velocity induced at the general point P by the nth panel is then the sum of the velocities $W_{AA'}$, $W_{BB'}$, and W_{AB} as given by Equations (C.97)–(C.99), respectively. This may be written as

$$W_{Pn} = \frac{1}{4\pi} \left[\frac{\cos \alpha_1 - 1}{R_{AA'}} - \frac{\cos \alpha_2 - 1}{R_{BB'}} + \frac{\cos \alpha_3 - \cos \alpha_4}{R_{AB}} \right] \Gamma_n = w_{Pn}\Gamma_n \qquad (C.107)$$

Then the induced velocity at the point P due to N panels on the wing is

$$W_P = \sum_{n=1}^{N} w_{Pn}\Gamma_n \qquad (C.108)$$

Here the quantity w_{Pn} represents a matrix of influence coefficients describing the geometry of the control points with respect to the N vortex panels, each of which has a strength Γ_n. This set of N circulation values is unknown, but satisfying the boundary condition that the net flow normal to the plane of the wing at each control point is zero requires that

$$W_P = -V \sin \alpha \qquad (C.109)$$

C.7.2 The lift generated by the vortex lattice

With the free stream velocity, angle of attack of the planar wing, and the geometry of the panels and control points all known, Equation (C.108) provides N equations for the N unknowns, Γ_n. Of course, there is no need to restrict the wing to a flat plate with no camber or twist; the fixed angle of attack α in Equation (C.109) may be replaced by the N quantities $\alpha_P - \delta_P$ describing the difference between the local (small) angles of attack and camber, respectively, or $\alpha_P - \varepsilon_P$ describing the difference between the local (small) angles of attack and the washout angle, respectively. The total lift is found by summing the lift of each panel over all N panels to obtain

$$L = \rho V \sum_{n=1}^{N} \Gamma_n \left(y_B - y_A \right)_n = \frac{1}{2} \rho V^2 b^2 \sum_{n=1}^{N} \left(\frac{\Gamma_n}{bV} \right) \left(\eta_B - \eta_A \right) \qquad \text{(C.110)}$$

The lift coefficient then becomes

$$C_L = A \sum_{n=1}^{N} \left(\frac{\Gamma_n}{bV} \right) \left(\eta_B - \eta_A \right) \qquad \text{(C.111)}$$

The non-dimensional circulation may be rewritten as follows:

$$\frac{\Gamma_n}{bV} = \frac{C_L}{2A} \frac{c_l c}{c_{av} C_L} \qquad \text{(C.112)}$$

The quantity $(c_l c / c_{av} C_L)$ is the loading coefficient described in the previous sections of this Appendix so Equation (C.110) becomes

$$1 = \sum_{n=1}^{N} \frac{1}{2} \left(\frac{c_l c}{c_{av} C_L} \right)_n \left(\eta_B - \eta_A \right) \qquad \text{(C.113)}$$

Detailed descriptions of the vortex lattice method may be found in various textbooks, for example, Bertin and Smith (1998) and Katz and Plotkin (2001).

C.7.3 An application of the vortex lattice method

We may consider a wing with a sweepback angle of the quarter-chord $\Lambda_{c/4} = 30°$, an aspect ratio $A = 10$, and a taper ratio $\lambda = 0.25$. The wing is fitted with 10 horseshoe vortices and 10 control points as shown in Figure C.30. The characteristics of the wing and the location of all the vortex and control points are given in Tables C.5 and C.6. As an example we consider the wing to be an untwisted flat plate at an angle of attack $\alpha = 10° = 0.1745$ rad. When combined, Equations (C.108) and (C.109) yield the following array:

$$-4.22G_1 + 0.028G_2 + 0.023G_3 + 0.006G_4 - 0.006G_5 = -0.1745$$
$$0.213G_1 - 2.844G_2 - 0.024G_3 + 0.006G_4 - .0120G_5 = -0.1745$$
$$0.082G_1 - 0.039G_2 - 1.985G_3 - 0.067G_4 - 0.032G_5 = -0.1745 \quad \text{(C.114)}$$
$$0.039G_1 + 0.095G_2 - 0.224G_3 - 1.452G_4 - 0.163G_5 = -0.1745$$
$$0.027G_1 + 0.051G_2 + 0.127G_3 - 0.385G_4 - 1.421G_5 = -0.1745$$

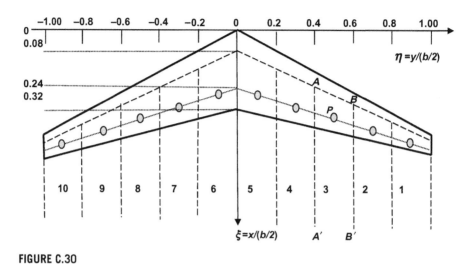

FIGURE C.30

A swept back wing with $N=10$ horseshoe vortices $A'ABB'$ and 10 control points P.

Because of the symmetrical nature of the wing and its loading the non-dimensional vortex strengths are symmetrical and therefore $G_1 = G_{10}$, $G_2 = G_9$, $G_3 = G_8$, $G_4 = G_7$, and $G_5 = G_6$. Here

$$G_n = \frac{\Gamma_n}{\left(\frac{b}{2}\right) V} \tag{C.115}$$

The values of G_n ($n=1, 2, 3, 4, 5$) which are solutions to Equations (C.114) are given in the second column of Table C.7. Using Equation (C.111) to find the contribution of each vortex to the total lift coefficient of the wing is given in the fourth column of Table C.7 and the sum yields $C_L = 0.783$. The relationship between the average chord c_{av} and the local chord is given by Equation (C.87) and the results for the loading coefficient $cc_l/c_{av}C_L$ are shown in the final column of Table C.7. The distribution of the loading coefficient as provided by the vortex lattice method with $N=10$ is compared to the Weissinger lifting surface method result for the same untwisted wing in Figure C.31. It is apparent that the loading coefficient predicted by the vortex lattice method with $N=10$ is not in very good agreement with that predicted by the lifting surface Weissinger method as applied by DeYoung and Harper (1948). The loading coefficient is overpredicted on the inboard sections of the wing and underpredicted on the outboard sections. The overall lift coefficient, on the other hand, is given as $C_L = 0.783$ by the vortex lattice method and 0.779 by the lifting surface method. In addition, the location of the center of pressure on the wing is reasonably close, as shown in Figure C.32 with $\eta_{c.p.} = 0.388$ according to the vortex lattice method and 0.424 for the lifting surface method, an under-prediction of about 8.5%.

Table C.5 Outboard Vortex Corner and Control Points for the Wing in Figure C.16

n	ξ_A	η_A	ξ_B	η_B	P	η_P	η_P
1	0.5419	0.8	0.6574	1.0	1	0.6517	0.9
2	0.4264	0.6	0.5419	0.8	1	0.6517	0.9
3	0.3110	0.4	0.4264	0.6	1	0.6517	0.9
4	0.1955	0.2	0.3110	0.4	1	0.6517	0.9
5	0.0800	0	0.1955	0.2	1	0.6517	0.9
6	0.1955	−0.2	0.0800	0	1	0.6517	0.9
7	0.3110	−0.4	0.1955	−0.2	1	0.6517	0.9
8	0.4264	−0.6	0.3110	−0.4	1	0.6517	0.9
9	0.5419	−0.8	0.4264	−0.6	1	0.6517	0.9
10	0.6574	−1.0	0.5419	−0.8	1	0.6517	0.9
1	0.5419	0.8	0.6574	1.0	2	0.5602	0.7
2	0.4264	0.6	0.5419	0.8	2	0.5602	0.7
3	0.3110	0.4	0.4264	0.6	2	0.5602	0.7
4	0.1955	0.2	0.3110	0.4	2	0.5602	0.7
5	0.0800	0	0.1955	0.2	2	0.5602	0.7
6	0.1955	−0.2	0.0800	0	2	0.5602	0.7
7	0.3110	−0.4	0.1955	−0.2	2	0.5602	0.7
8	0.4264	−0.6	0.3110	−0.4	2	0.5602	0.7
9	0.5419	−0.8	0.4264	−0.6	2	0.5602	0.7
10	0.6574	−1.0	0.5419	−0.8	2	0.5602	0.7
1	0.5419	0.8	0.6574	1.0	3	0.4687	0.5
2	0.4264	0.6	0.5419	0.8	3	0.4687	0.5
3	0.3110	0.4	0.4264	0.6	3	0.4687	0.5
4	0.1955	0.2	0.3110	0.4	3	0.4687	0.5
5	0.0800	0	0.1955	0.2	3	0.4687	0.5
6	0.1955	−0.2	0.0800	0	3	0.4687	0.5
7	0.3110	−0.4	0.1955	−0.2	3	0.4687	0.5
8	0.4264	−0.6	0.3110	−0.4	3	0.4687	0.5
9	0.5419	−0.8	0.4264	−0.6	3	0.4687	0.5
10	0.6574	−1.0	0.5419	−0.8	3	0.4687	0.5

The shortcoming of the vortex lattice method with $N=10$ is apparent in Figure C.32, in which the contribution of the 10 vortices is shown. Obviously, using a discrete number of vortices limits the spatial resolution of the lift distribution to the span of the individual vortices. Such is not the case for the lifting surface approach, which instead provides an (approximate) solution to the integral equation for the

Table C.6 Inboard Vortex Corner and Control Points for the Wing in Figure C.16

n	ξ_P	η_P	ξ_B	η_B	P	ξ_P	η_P
1	0.5419	0.8	0.6574	1.0	4	0.3772	0.3
2	0.4264	0.6	0.5419	0.8	4	0.3772	0.3
3	0.3110	0.4	0.4264	0.6	4	0.3772	0.3
4	0.1955	0.2	0.3110	0.4	4	0.3772	0.3
5	0.0800	0	0.1955	0.2	4	0.3772	0.3
6	0.1955	−0.2	0.0800	0	4	0.3772	0.3
7	0.3110	−0.4	0.1955	−0.2	4	0.3772	0.3
8	0.4264	−0.6	0.3110	−0.4	4	0.3772	0.3
9	0.5419	−0.8	0.4264	−0.6	4	0.3772	0.3
10	0.6574	−1.0	0.5419	−0.8	4	0.3772	0.3
1	0.5419	0.8	0.6574	1.0	5	0.2857	0.1
2	0.4264	0.6	0.5419	0.8	5	0.2857	0.1
3	0.3110	0.4	0.4264	0.6	5	0.2857	0.1
4	0.1955	0.2	0.3110	0.4	5	0.2857	0.1
5	0.0800	0	0.1955	0.2	5	0.2857	0.1
6	0.1955	−0.2	0.0800	0	5	0.2857	0.1
7	0.3110	−0.4	0.1955	−0.2	5	0.2857	0.1
8	0.4264	−0.6	0.3110	−0.4	5	0.2857	0.1
9	0.5419	−0.8	0.4264	−0.6	5	0.2857	0.1
10	0.6574	−1.0	0.5419	−0.8	5	0.2857	0.1

Table C.7 Loading Distribution for the Wing in Figure C.16

η	$2\Gamma/bV$	$\Delta\eta$	$\dfrac{A(\Gamma/bV)}{\Delta\eta}$	c	c_l	$cc_l/c_{av}C_L$
1	0	0.1	0	2.00	0.000	0.000
0.9	0.0402	0.2	0.040	2.60	0.774	0.506
0.7	0.0631	0.2	0.063	3.80	0.830	0.792
0.5	0.0828	0.2	0.083	5.00	0.828	1.040
0.3	0.1001	0.2	0.100	6.20	0.807	1.258
0.1	0.1053	0.2	0.105	7.40	0.711	1.323
−0.1	0.1053	0.2	0.105	7.40	0.711	1.323
−0.3	0.1001	0.2	0.100	6.20	0.807	1.258
−0.5	0.0828	0.2	0.083	5.00	0.828	1.040
−0.7	0.0631	0.2	0.063	3.80	0.830	0.792
−0.9	0.0402	0.2	0.040	2.60	0.774	0.506
−1	0	0	0	2.00	0.000	0.000

$$C_L = 0.783$$

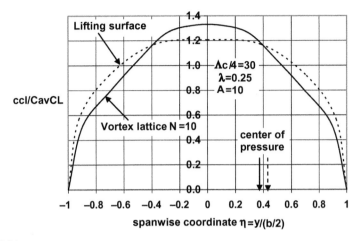

FIGURE C.31

Comparison between the results of the vortex lattice and lifting surface methods for the spanwise distribution of the loading coefficient for an untwisted wing with $\Lambda_{c/4}=30°$, $\lambda=0.25$, and $A=10$.

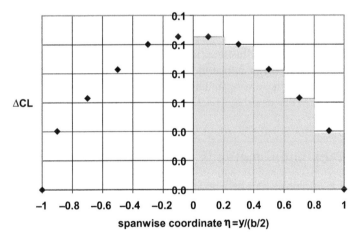

FIGURE C.32

Contribution of each of the 10 vortices to the total lift coefficient of the wing. The right-hand side of the figure is shaded to emphasize the discrete values of ΔC_L produced by the method.

induced velocity distribution. Early applications showed that $N=20$ vortices yield useful results, although more recent studies exploit available computer capability by using 50–100 vortices to achieve very detailed and accurate results. Margason and Lamar (1977) recommend 4 panels chordwise and 20 panels across the semi-span for

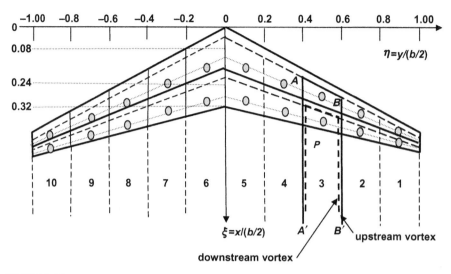

FIGURE C.33

A swept back wing with $N=20$ horseshoe vortices $A'ABB'$ and 20 control points P. For substantial accuracy about 160 horseshoe vortices with 160 control points are necessary.

reasonable accuracy. Just doubling the chordwise panels from those in Figure C.30 to those shown in Figure C.33 illustrates the increasing geometric complexity common to panel methods. The vortex lattice method may also be employed to handle lifting surface combinations, like a wing and horizontal tail, and a wing with pylons or other vertical surfaces, as discussed by Blackwell (1969).

C.8 Surface panel methods

The two-dimensional surface panel method for airfoils discussed previously may be extended to three-dimensional wings and bodies or combinations thereof. The simple case of a wing is shown schematically in Figure C.34. The first concern is to model the surface of the wing and this has generally been done by dividing the surface up into quadrilateral panels, thus the name panel method. Complex geometries can thereby be modeled, although doing so is a substantial and often tedious task.

Cebeci (1999) points out that the three-dimensional inviscid flow over an airplane may be acceptably modeled by the combination of a lifting wing with a sharp trailing edge from which an infinitesimally thin vortex wake is shed and a streamlined non-lifting slender fuselage, as shown in Figure C.35. The individual singularities that comprise the velocity potential are all solutions to Laplace's equation and satisfy the boundary condition that they vanish far from the body. Note that because Laplace's equation is linear, sums of any of these are also solutions. The boundary condition at the surface is that the component of velocity normal to it must be zero because

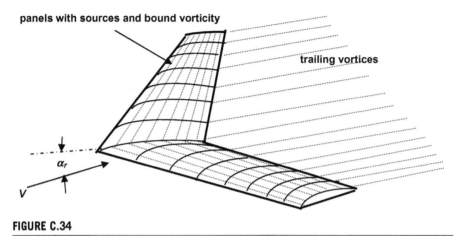

FIGURE C.34

Swept wing modeled by sources and bound vorticity on all panels.

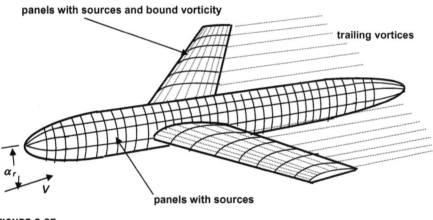

FIGURE C.35

Wing-body combination modeled by sources on all panels and bound vorticity on wing panels.

the body is a stream surface. Furthermore, if the body is a wing that produces lift, a Kutta condition must be applied along the trailing edge. The vortex wake shed from the trailing edge cannot support a pressure difference across it so its orientation must be determined as part of the complete solution for the flow field.

In this fashion whole airplanes may be modeled in great detail, and to do so may require 5000–10,000 panels for sufficient resolution. Obviously the computations are more extensive when the geometric details desired become more complex, but the equations developed can be handled readily by current computers. All airframe manufacturers have made substantial investments in panel methods for aircraft analysis and design and they are in routine use for those flight operations where the likelihood of flow separation is limited.

An online search shows there are a number of codes available, both publically and commercially. Drela (2013), for example, presents a program called AVL for the aerodynamic and flight-dynamic analysis of rigid aircraft of arbitrary configuration. It employs an extended vortex lattice model for the lifting surfaces, together with a slender-body model for fuselages and nacelles. General nonlinear flight states can be specified. AVL is released under the Gnu General Public License.

The surface panel method has been making increasingly important contributions to aircraft design each year since its first practical application over 40 years ago. Various computational methods applied to aerodynamics are presented in Henne (1990) including the early development of the panel method as described by Smith (1990), a pioneer in the field. The ability of the panel method to treat complete airplanes incorporating subsystems like geometrically complex multi-element airfoils has been partially responsible for reducing the demand for laboratory and flight test time during the development process, as described by Malik and Bushnell (2012).

The panel method has thus far been most successful in dealing with steady flight, like cruise, with little separated flow. In designing the Boeing 777, CFD was used extensively to refine the wing design for high-speed cruise and to optimize propulsion/airframe, as shown in Figure C.36. In addition, according to Johnson et al. (2005), CFD was used to design the fuselage cabin once the body diameter was fixed. They report that no further changes were necessary as a result of wind tunnel

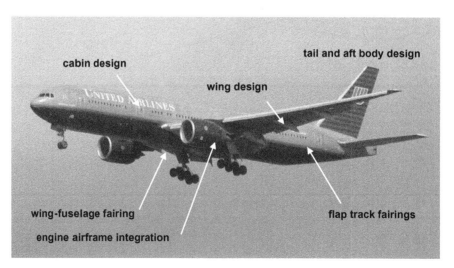

FIGURE C.36

Specific areas and systems of the B777 designed with the aid of CFD according to Johnson et al. (2005).

testing. CFD was combined with wind tunnel testing to finalize the design of the aft body and wing/body fairing shape and to aid in the design of the flap support fairings. CFD was used to determine important, yet small-scale, items like locations for static pressure ports and angle of attack vanes for the air data system, and environmental control system inlet and exhaust ports.

For more stressful flight conditions involving higher loading and more extensive and complex flow separation such as takeoff and landing more complex CFD methods must be employed. A major need in this regard is adequate physical modeling of the transitional and unsteady turbulent flow fields engendered by complex separated flows.

C.9 Computational fluid dynamics applied to airplane design

The development of computational fluid dynamics (CFD) for predicting flows over aircraft is concisely described by Jameson (1989), a pioneer and major contributor to the field. He points out that though the field of fluid dynamics had a comprehensive mathematical footing, the need for simplifying assumptions to achieve practical design solutions placed a heavy burden on experimental work. The capital and operating expense and the relatively long turn-around time in large-scale wind tunnel testing provided substantial incentives for airframe manufacturers to develop computationally based simulation techniques for dealing with large numbers of design iterations with increased speed and reduced costs.

The Navier-Stokes equations provide the fundamental basis for the analysis of fluid dynamic problems. Although the theoretical foundation for these equations is sound they are in general quite difficult to solve, being unsteady three-dimensional partial differential equations with variable coefficients. The entire history of fluid dynamics may be described as a collection of approximate approaches valid for particular geometries and conditions. The assumption that the viscosity of the fluid is vanishingly small ($\mu = 0$) leads to the Euler equations, a major simplification of the Navier-Stokes equations. For practical problems of aeronautical interest the effect of viscosity, particularly the lift, is typically found to be indeed small. Any such effects were observed to be confined to a thin region around a body, the boundary layer, and behind it, the wake. Of course, when the angle of attack of a body becomes large enough the boundary layer no longer remains thin and the flow separates from the body.

When the vorticity in the flow is confined to those same thin regions, then the remainder of the flow field may be considered irrotational and admits of a velocity potential ϕ which satisfies the potential equation

$$\left(1 - M^2\right) \frac{\partial^2 \phi}{\partial x^2} + \frac{\partial^2 \phi}{\partial y^2} + \frac{\partial^2 \phi}{\partial z^2} = 0 \qquad \text{(C.116)}$$

Here the free stream flow is aligned with the x-axis and the Mach number $M = \frac{u}{\sqrt{\gamma RT}}$. The velocity components in the x, y, and z directions are given, respectively, by

$$u = \frac{\partial \phi}{\partial x}, v = \frac{\partial \phi}{\partial y}, w = \frac{\partial \phi}{\partial z} \tag{C.117}$$

It is recognized that in the transonic range, $M = O(1)$, shock waves may form which introduce vorticity over a relatively large expanse of flow and the assumption of a velocity potential fails. Therefore, when the subsonic flow over a body is near neither separation nor sonic conditions the use of the potential equation can provide useful solutions to flow over a finite body such as a wing or a fuselage. Simplifications include the slender body approximation (small disturbance theory) where the velocity in the x-direction differs very little from the free stream velocity. As discussed in Section C.1, Equation (C.116) may be written as the compressible small disturbance equation:

$$(1 - M_\infty^2)\frac{\partial^2 \phi}{\partial x^2} + \frac{\partial^2 \phi}{\partial y^2} + \frac{\partial^2 \phi}{\partial z^2} = 0 \tag{C.118}$$

Here ϕ is the small disturbance potential and for incompressible flow $M_\infty = 0$ so that Equation (C.118) reduces to the Laplace equation. This is also the case if we transform the x-coordinate to the x_0-coordinate by introducing the Prandtl-Glauert factor $\sqrt{1 - M_\infty^2}$ to obtain $x_0 = \frac{x}{\sqrt{1-M_\infty^2}}$. The combination of a full or linearized potential flow solver coupled with a boundary layer code and a three-dimensional grid generation code provides one with a powerful tool for dealing with viscous flows about wings and entire airplanes under the flow assumptions described earlier. Note that these codes give information on the entire flow field, not just on the body surface like panel codes, so that global flow field and interference effects may be assessed.

The use of the coupled full potential and boundary layer codes is inappropriate when the irrotational flow condition is no longer applicable, as in the case where shock waves are present. If the boundary layers and wakes are still thin the inviscid Euler equations may be used in concert with boundary layer codes. Once there are significant regions of separated flow due to extreme angle of attack, shock waves, or vortex flows off sharp edges the boundary layer approximation, as well as the irrotational flow assumption, is no longer even approximately satisfied. In that case CFD techniques must be applied to the Navier-Stokes equations. As Malik and Bushnell (2012) point out, "The challenges that are faced by CFD (e.g., unsteady separation, boundary-layer transition) are such that they cannot be resolved by the mere availability of faster machines. Research is needed for the development of more accurate numerical schemes, advanced solver technology, grid adaptation, error estimation, physics modeling, and schemes for efficiently exploiting the capabilities of future massively parallel machines. The full potential of ever-increasing computer power cannot be realized without strategic investments in the computational infrastructure."

C.10 Nomenclature

A	aspect ratio or area
A_n	nth term in a series
a	lift curve slope
a_{vn}	influence coefficients for lifting surface method
b	wingspan
c	chord length
C_D	wing drag coefficient
c_d	airfoil drag coefficient
c_l	airfoil lift coefficient
$c_{l\alpha}$	airfoil lift curve slope $= a$
C	constant or path contour
C_L	wing lift coefficient
$C_{L\alpha}$	wing lift curve slope
C_m	wing moment coefficient
C_p	pressure coefficient
D	drag
F	planform parameter, Equation (C.89), or force
f'	span loading curve-fitted function, Equation (C.92)
G	normalized circulation Γ/bV
k	constant, Equation (C.8)
k_1	planform factor, Equation (C.94)
L	lift
L'	lift per unit span
M	Mach number
N	number of terms
n	index for a series of terms or for vortex locations
p	pressure
q	dynamic pressure
R	perpendicular distance from vortex
r	distance from a vortex to a point in space
Re	Reynolds number
S	wing planform area
s	path length
U	velocity parallel to the x-axis
V	velocity parallel to the y-axis
W_P	velocity induced at point P, Equation (C.108)
w	downwash velocity
x	chordwise distance from wing leading edge

y	spanwise distance from centerline
z	distance normal to the x-y plane
α	angle of attack
α^0	angle of attack where a departs from linearity
α_{av}	average angle of attack, Equation (C.94)
β	Prandtl-Meyer factor $(1-M^2)^{1/2}$
ε	downwash angle or wing twist angle
Φ	velocity potential
ϕ	perturbation velocity potential or dummy variable, Equation (C.40)
Γ	circulation
γ	circulation per unit length
η	normalized coordinate $y/(b/2)$
κ	$c_{l\alpha}/2\pi$
Λ	wing sweepback angle
Λ_β	compressibility-corrected sweepback angle, Equation (C.85)
λ	wing taper ratio
ν	index for wing control points
ρ	density
θ	dummy variable, Equation (C.40)
ξ	normalized coordinate $x/(b/2)$
ω	vorticity

C.10.1 Subscripts

a	additional loading
av	average
b	basic loading
c	chord
$c.g.$	center of gravity
$c.p.$	center of pressure
$crit$	critical
$c/4$	quarter-chord position
i	induced
LE	leading edge
l	lower surface
MAC	mean aerodynamic chord
max	maximum
n	normal or index for a series of terms
p	pressure
r	root
ref	reference
TE	trailing edge
t	tip
u	upper surface
0	zero-lift conditions or incompressible flow conditions

∞	free stream conditions
$(...)'$	perturbation quantity

References

Abbott, H., et al. 1945. Summary of Airfoil Data. NACA Technical Report No. 824.

Abbott, H., Von Doenhoff, A.E., 1959. Theory of Wing Sections. Dover, NY.

Ashley, H., 1992. Engineering Analysis of Flight Vehicles. Dover, NY.

Bertin, J.J., Smith, M.L., 1998. Aerodynamics for Engineers, third ed. Prentice-Hall, New Jersey.

Bisplinghoff, R., Ashley, H., Halfman, R., 1983. Aeroelasticity. Dover, New York.

Black, J., 1953. An Experimental Investigation of the Effects of Deformation on the Aerodynamic Characteristics of A Swept-Back Wing. Aeronautical Research Council Reports and Memoranda No. 2938.

Blackwell Jr., J.A., 1969. A Finite-Step Method for Calculation of Theoretical Load Distributions for Arbitrary Lifting-Surface Arrangements at Subsonic Speeds. NASA TN D-5335.

Cebeci, T., 1999. An Engineering Approach to the Calculation of Aerodynamic Flows. Springer, New York.

De Young, J., Harper, C.W., 1948. Theoretical Symmetric Span Loading at Subsonic Speeds for Wings Having Arbitrary Planforms. NACA Report 921.

Diederich, F.W., 1952. A Simple Approximate Method for Calculating Spanwise Lift Distributions and Aerodynamic Influence Coefficients at Subsonic Speeds. NACA TN 2751.

Drela, M., 1989. XFOIL: an analysis and design system for low reynolds number airfoils. In: Mueller, T.J. (Ed.), Lecture Notes in Engineering: Low Reynolds Number Aerodynamics, vol. 54. Springer-Verlag, New York. <http://web.mit.edu/drela/Public/web/xfoil/>.

Drela, M., 2013. <http://web.mit.edu/drela/Public/web/avl/>.

Eppler, R., Somers, D.M., 1980. A computer program for the design and analysis of low-speed airfoils. NASA TM 80210, August 1980. <http://www.pdas.com/eppler.html>.

Henne, P. (Ed.), 1990. Applied computational aerodynamics, Progress in Aeronautics and Astronautics, vol. 125. American Institute of Aeronautics and Astronautics, Reston, VA.

Jameson, A., 1989. Computational aerodynamics for aircraft design. Science 245, 361–371.

Johnson, F.T., Tinoco, E.N., Jong Yu, N., 2005. Thirty years of development and application of CFD at Boeing Commercial Airplanes, Seattle. Computers & Fluids 34, 1115–1151.

Katz, J., Plotkin, A., 2001. Low-Speed Aerodynamics, second ed. NY, Cambridge.

Liepmann, H.W., Roshko, A., 2001. Elements of Gasdynamics. Dover, NY.

Malik, M.R., Bushnell, D.M., 2012. Role of Computational Fluid Dynamics and Wind Tunnels in Aeronautics R&D. NASA/TP-2012-217602.

Margason, R., Lamar. J., 1977. Vortex-Lattice FORTRAN Computer Program for Estimating Subsonic Aerodynamic Characteristics of Complex Planforms. NASA TN D-6142.

von Mises, R., 1959. Theory of Flight. Dover, New York.

Schrenk, O., 1940. A Simple Approximate Method for Obtaining Spanwise Lift Distributions. NACA TM 948.

Smith, A.M.O., 1990. The panel method: its original development. In Henne (1990). pp. 3–20.

Stevens, V.I., 1948. Theoretical Basic Span Loading Characteristics of Wings with Arbitrary Sweep. NACA Technical Note 1772.

Weissinger, J., 1947. The Lift Distribution of Swept-back Wing. NACA TM 1120.

Graphs of Critical Mach Number

Critical Mach numbers predicted for various NACA 6-series of airfoils are shown as a function of lift coefficient in curves presented by Abbott et al. (1945). The major effects are that both thickness and camber tend to reduce the critical Mach number. The use of these curves in predicting drag effects due to compressibility is discussed in Section 9.14.

Commercial Airplane Design Principles. http://dx.doi.org/10.1016/B978-0-12-419953-8.00021-8

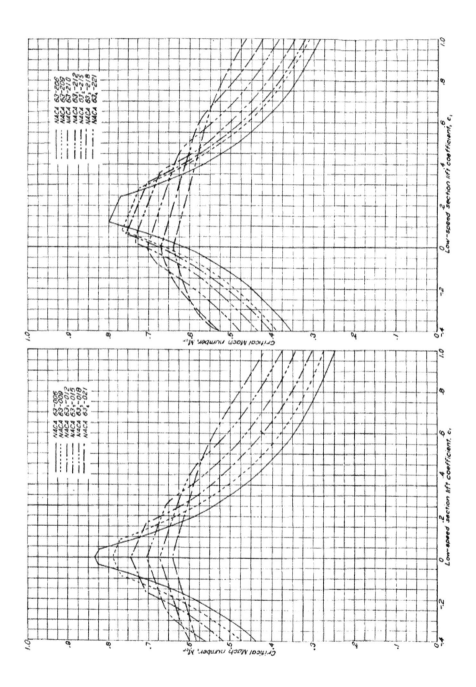

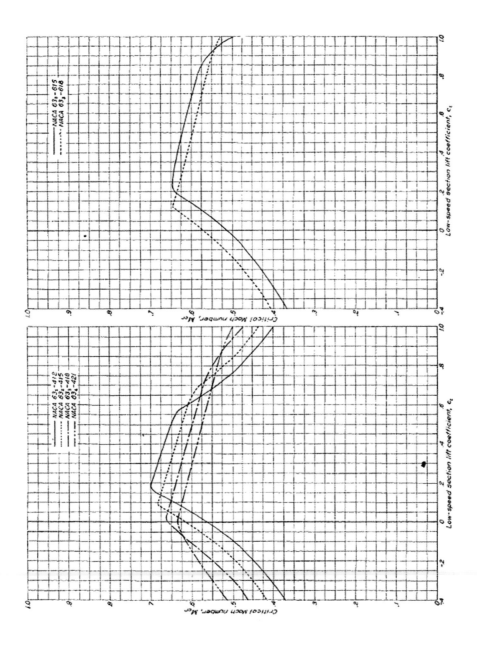

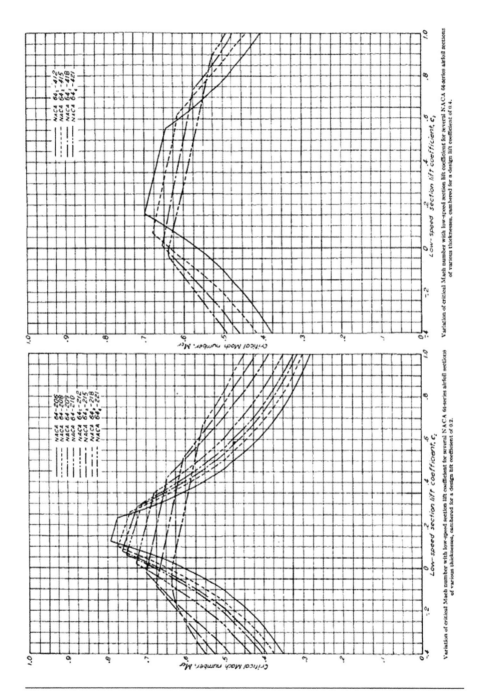

Variation of critical Mach number with low-speed section lift coefficient for several NACA 64-series airfoil sections of various thicknesses, cambered for a design lift coefficient of 0.4.

Variation of critical Mach number with low-speed section lift coefficient for several NACA 64-series airfoil sections of various thicknesses, cambered for a design lift coefficient of 0.2.

Reference

Abbott I.H. et al., 1945. Summary of Airfoil Data, NACA Technical Report No. 824.

Units and Conversion Factors

The English system of units continues to be the system of choice in the aircraft industry although SI units are entering the field, but slowly. Choosing a system of units to use is primarily a matter of taste and past practice with often very little to logically encourage adherence to one system or the other. In its daily operations the commercial aircraft industry tends to employ nautical miles and nautical miles per hour, or knots, to describe range and speed, while pounds and gallons remain favored to denote weight or thrust and volume of liquids. Engineering usage in the industry tends to more fundamental units within the English system; this is called the foot-pound-second or "fps" system. The major difference is that the "fps" engineering system favors force as a basic unit while the science system (SI) favors mass as the basic unit, and it is this difference between the two systems that often leads to confusion. This is obviously the case because the time units are equivalent in the English and SI systems and the relationship between the length units is simply a multiplicative constant. In the space-related sector of aerospace engineering the mass-based system has substantial merit in application while in the aeronautics-related arena, where gravitational and relativistic variations are of little importance, the force-based system has no shortcomings in application. Therefore, the fact that the English system of units is still in wide use and that the historical background of the aviation industry is steeped in its use makes it a requisite for understanding that system when researching the field.

The commercial aviation industry is firmly based in subsonic flight in the lower atmosphere so that our attention is focused on constant composition atmospheric air as the medium of interest. For reference, atmospheric properties of interest are shown in Table E.1 for English, SI, and other units while Table E.2 presents conversions between the systems.

E.1 The error function

The error function $erf(x)$ and its complement are given by

$$erf(x) = \frac{2}{\sqrt{\pi}} \int_0^x e^{-\xi^2} \, d\xi = 1 - erfc(x)$$

Commercial Airplane Design Principles. http://dx.doi.org/10.1016/B978-0-12-419953-8.00022-X

Table E.1 Properties of the Atmosphere and the Earth at Sea Level

Property	English Units	SI Units	Other Units	Other Units
Pressure	2116.22 lb/ft^2 (psf)	101.325 kN/m^2 (kPa)	14.696 psi	1.0133 bar
Density	0.002378 slug/ft^3	1.225 kg/m^3		
Temperature	518.67 R	288.15 K	59°F	15°C
Dynamic viscosity	3.74E-07 lb-s/ft^2	1.7894E-05 kg/m-s		
Kinematic viscosity	1.5728E-04 ft^2/s	1.4607E-05 m^2/s		
Sound speed	1116.3 ft/s	340.26 m/s	760.94 mph	660.83 kts
Specific heat	0.240 Btu/lb-R	1.004 kJ/kg-K		
Molecular weight	28.96 lb/lb-mol	28.96 kg/kg-mol		
Specific heat ratio	1.4	1.4		
Gas constant	1716 ft^2/s^2-R	287 m^2/s^2-K	53.37 ft/R	0.287 kJ/kg-K
Gravity	32.174 ft/s^2	9.8067 m/s^2		
Earth radius	3757 mi	6357 km		

Table E.2 Conversion Factors for English and SI Units[a]

	Multiply	**By**	**To Obtain**
Pressure	lb/ft^2 (psf)	0.047880	kN/m^2 (kPa)
	lb/in^2 (psi)	6.8948	kN/m^2 (kPa)
	atmospheres (atm)	101.33	kN/m^2 (kPa)
	bar	1.00E+05	kN/m^2 (kPa)
	in. of Hg at 32 °F	3.3864	kN/m^2 (kPa)
	kN/m^2 (kPa)	20.842	lb/ft^2 (psf)
	kN/m^2 (kPa)	0.14504	lb/in^2 (psi)
	atmospheres (atm)	2116.20	lb/ft^2 (psf)
	atmospheres (atm)	14.696	lb/in^2 (psi)
	bar	2,088.60	lb/ft^2 (psf)
	mm of Hg at 0 °C	2.7845	lb/ft^2 (psf)
Density	lbm/ft^3 (slug/ft^3)	16.018	kg/m^3
	kg/m^3	0.068948	lbm/ft^3 (slug/ft^3)
Temperature	R	0.55556	K
	F+459.67	0.55556	C+273.15
	K	1.8	R
	C+273.15	1.8	F+459.67
Length	feet (ft)	0.30480	m
	inches (in.)	0.025373	m
	miles (mi)	1.6093	km
	nautical miles (nm)	1.8520	km
	meters (m)	3.2808	ft
	meters (m)	39.370	in
	kilometers (km)	0.59160	mi
Area	square feet (ft^2)	0.092903	m^2
	square inches (in^2)	6.4516E-04	m^2
	square meters (m^2)	10.764	ft^2
	square cm (cm^2)	0.15500	in^2

(Continued)

Table E.2 (Continued)

	Multiply	By	To Obtain
Volume	cubic feet (ft³)	0.028317	m³
	gallon (gal)	3.7854E-03	m³
	gallon (gal)	3.7854E+00	l
	cubic meter (m³)	35.315	ft³
	cubic meter (m³)	264.17	gal
	liter (l)	0.264	gal
Mass	slug (lbm)	0.45359	kg
	kilogram	2.2046	lbm
Energy	foot-pounds (ft-lb)	1.3558	J (N-m)
	Btu	1.0551	kJ (kN-m)
	Joule (J)	0.73756	ft-lb
	kilojoule (kJ)	0.9478	Btu
Force	pound (lb)	4.4482	N
	Newton (N)	0.22481	lb
Power	horsepower (hp)	0.74570	kW
	ft-lb/s	1.35580	W
	kilowatt (kW)	1.3410	hp
	Watt (W)	0.73756	ft-lb/s
Velocity	ft/s (fps)	0.3048	m/s
	mi/hr (mph)	1.6093	km/hr
	nautical mi/hr (kts)	0.5148	m/s
	nautical mi/hr (kts)	1.852	km/hr
	m/s	3.2808	ft/s
	km/hr (kph)	0.62139	mi/hr (mph)
	m/s	1.9425	nm/hr (kts)
	km/hr (kph)	0.53996	nm/hr (kts)

[a] *Quantities in parentheses are other commonly used forms of the units shown.*

Table E.3 Values for the Error Function and its Complement

x	erfc(x)	erf(x)
0	1	0
0.05	0.944	0.056
0.10	0.888	0.112
0.20	0.777	0.223
0.40	0.572	0.428
0.60	0.396	0.604
0.80	0.258	0.742
1.00	0.157	0.843
1.50	0.0339	0.966
2.00	0.00468	0.995

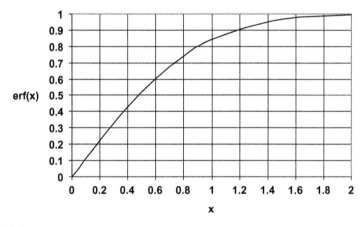

FIGURE E.1

The variation of *erf*(*x*) with its argument, *x*.

A table of values for both functions is given in Table E.3 and a plot of *erf*(*x*) as a function of *x* is shown in Figure E.1.

General Database for Commercial Aircraft

Throughout this book reference is often made to a database of airplane characteristics when correlations are developed from theoretical considerations and practical industrial experience. This database is a collection of a number of general characteristics of commercial airplanes assembled from the various sources cited in this book. The data that have been gathered and presented here should be considered representative of the aircraft cited because there are many variants of each aircraft and no one set should be considered authoritative. Instead, the information in this database is offered as a comparative tool to aid in guiding the development of airplane configurations and acting as a check on their practicality. The tables have been divided into those for aircraft with gross weights less than 100,000 lb and those with gross weights greater than 100,000 lb. This arbitrary characteristic was selected to divide the collection into those aircraft that serve primarily regional markets and those that serve continental and intercontinental markets. The regional market division includes both turboprop-powered and turbofan-powered aircraft while the continental and intercontinental division includes only turbofan-powered aircraft.

F.1 Weight data

The aircraft manufacturer and aircraft designation are noted to identify each entry in the database. The number of passengers N_p is representative of the typical passenger capacity in mixed classes of service and the range R, in statute miles, is the nominal range for the indicated number of passengers. Tables F.1 and F.2 display data for the maximum takeoff weight W_{to}, the nominal empty weight W_e, the maximum cargo weight W_c, and the maximum landing weight W_l. The aircraft are listed in order of increasing takeoff weight and all weights are given in pounds.

Commercial Airplane Design Principles. http://dx.doi.org/10.1016/B978-0-12-419953-8.00023-1

Table F.1 Selected Weight Data for Aircraft of Aircraft of Gross Weight Under 100,000 lbs

Manufacturer	Aircraft	Np	R (mi)	Engine	W_{to}	W_e	W_c	W_l
Embraer	EMB-120	30	1088	2XPWC PW118A/B (TP)	26,433	16,711	1,543	25,794
Saab	S340	35	1075	2XGE CT7-9B2 (TP)	29,000	18,600	2,100	28,500
Bombardier	Q100	38	1174	2XPWC PW123C/D (TP)	36,300	22,941	9,000	33,900
Avions de T.R.	ATR42-500	42	966	2XPWC PW127E (TP)	41,005	24,800	12,015	40,344
Embraer	ERJ135LR	37	2015	2xRR AE3007	44,092	25,342	2,205	40,785
Fokker	F50	50	1070	2XPWC PW125B (TP)	45,900	27,600	4,225	43,650
Embraer	ERJ145LR	50	1900	2xRR AE3007	46,275	26,694	2,646	42,549
P.R. of China	Y-7	52	1239	2XWJ-5A-1 (TP)	48,050	31,317	15,000	46,740
Avions de T.R.	ATR72-500	68	947	2XPWC PW127F (TP)	48,500	28,549	15,542	48,170
Bombardier	CRJ200LR	50	2307	2xCF34-3B1	53,000	30,500	13,500	47,000
Boeing	DC9-10	72	1380	2xPW JT8D	90,700	53,000	6,600	81,700
Fokker	F70	79	2072	2xRR Tay 620	92,000	50,230	4,510	81,000
BAE Systems	ARJ70	70	1911	4xASE LF507-1F	95,000	52,700	5,000	83,500

Table F.2 Selected Weight Data for Aircraft of Gross Weight Over 100,000 lbs

Manufacturer	Aircraft	Np	R (mi)	Engine	W_{to} (lbs)	W_e (lbs)	W_c (lbs)	W_l (lbs)
Boeing	B737-500	110	2752	2xCFM56-3C-1	115,500	70,440	10,060	110,000
Boeing	B717-200	106	2389	2xRR BR700-715	121,000	68,500	10,800	110,000
Boeing	B737-300	126	2600	2xCFM56-3C-1	138,500	72,360	7,440	116,600
Boeing	B737-600	110	2384	2xCFM56-7BE	145,500	82,480	9,670	120,500
Airbus	A318	124	3713	2xCFM56-5B/PW6000	149,910	86,650	12,700	123,500
Boeing	B737-400	147	3696	2xCFM56-3C-1	150,000	76,760	7,420	124,000
Boeing	B737-700ER	126	3961	2xCFM56-7BE	154,500	84,990	11,810	129,200
Airbus	A319-100	124	4214	2xCFM56-5B6/IAE V2524-A5	166,500	89,500	15,100	134,500
Airbus	A320-200	150	3455	2xCFM56-5B4/IAE V2527-A5	169,800	92,800	14,900	142,200
Boeing	B737-800	162	3587	2xCFM56-7BE	174,200	92,190	13,910	146,300
Boeing	B737-900	177	3160	2xCFM56-7B	174,200	94,740	10,160	146,300
Boeing	B737-900ER	180	3720	2xCFM56-7BE	187,700	98,190	14,790	157,300
Airbus	A321-200	185	3455	2xCFM56-5B3/IAE V2533-A5	206,100	106,300	19,600	171,500
Boeing	B757-200	200	4490	2xRR RB.211-535E4	255,000	130,440	15,560	210,000
Boeing	B757-300	243	3909	2xRR RB.211-535E4	272,500	141,690	19,710	224,000
Airbus	A310-300	220	5978	2xGE CF6-80C2	361,600	178,700	28,650	273,400
Boeing	B787-3	296	3512	2xGEnx-1B	364,000	182,000	–	280,000
Boeing	B767-200ER	181	7560	2xGE CF6-80C2	395,000	189,900	34,690	300,000
Boeing	B767-300ER	218	6880	2xGE CF6-80C2	412,000	203,500	49,020	320,000
Boeing	B767-400ER	245	6477	2xGE CF6-80C2	450,000	228,900	52,150	350,000
Boeing	B787-8	242	9442	2xGEnx-1B/Trent 1000	503,500	242,000	–	380,000

(Continued)

Table F.2 (Continued)

Manufacturer	Aircraft	Np	R (mi)	Engine	W_{to} (lbs)	W_e (lbs)	W_c (lbs)	W_f (lbs)
Airbus	A330-300	335	7427	2xCF6-80E1/PW4000/Trent700	513,670	274,650	58,150	412,300
Airbus	A330-200	253	8809	2xCF6-80E1/PW4000/Trent700	513,670	263,700	56,200	396,200
Boeing	B777-200	305	9788	2XGE90-77B	545,000	302,400	53,450	445,000
Boeing	B787-9	259	9788	2XGEnx-1B	553,000	254,000	–	425,000
Airbus	A350-800	270	6034	2xTrent XWB-74	571,000	246,000	–	425,000
Airbus	A350-900	315	9557	2xTrent XWB-83	591,000	255,000	–	452,000
Airbus	A340-200	240	9212	4XCFM56-5C	606,270	287,160	44,180	407,855
Airbus	A340-300	295	8285	4XCFM56-5C	609,580	288,500	50,500	423,280
Boeing	B777-200ER	301	10,589	2XGE90-94B	656,000	320,800	57,090	470,000
Boeing	B777-300	368	8867	2xPW4098/Trent892	660,000	342,500	75,220	524,000
Airbus	A350-1000	369	6915	2xTrent XWB-92	679,000	339,000	–	514,000
Boeing	B777-200LR	301	10,818	2xGE90-115B	766,000	346,300	53,190	492,000
Boeing	B777-300ER	365	9131	2xGE90-115B	775,000	372,100	80,250	554,000
Boeing	B747-400	416	8320	4XGE CF6-80C2	875,000	398,800	68,840	652,000
Boeing	B747-8I	467	9212	4xGEnx-2B67	975,000	466,700	69,730	682,000
Airbus	A380-800	525	9212	4xGE/PW GP7270/Trent 970	1,234,600	611,000	112,000	851,000

F.2 Scale data

The aircraft manufacturer and aircraft designation are noted to identify each entry in the database. The aircraft are listed in order of increasing takeoff weight, in pounds, and the range R, in statute miles, is also listed. Tables F.3 and F.4 collect data on the wing planform area S, the wingspan b, and the fuselage length L is used to form the aspect ratio $A = b^2/S$ and the slenderness ratio b/L.

F.3 Wing loading data

The aircraft manufacturer and aircraft designation are noted to identify each entry in the database. The aircraft are listed in order of increasing takeoff weight, in pounds, and the range R, in statute miles, is also listed. Tables F.5 and F.6 use data on the wing planform area S, to form the wing loading in takeoff W_{to}/S, the wing loading in landing W_l/S, and the ratio of the takeoff weight to the maximum landing weight, W_{to}/W_l.

F.4 Thrust and power data

The aircraft manufacturer and aircraft designation are noted to identify each entry in the database. The aircraft are listed in order of increasing takeoff weight, in pounds. Tables F.7 and F.8 use data on the maximum power or takeoff thrust to form the takeoff thrust loading F_{to}/W_{to} or the takeoff power loading P/W_{to}, which is measured in horsepower per pound. The nominal takeoff and landing distances reported in the literature for the aircraft are also listed.

Table F.3 Selected Scale Data for Aircraft of Gross Weight Under 100,000 lbs

Manufacturer	Aircraft	R (mi)	W_{to} (lbs)	S(sq.ft)	b (ft)	A	L (ft)	b/L
Embraer	EMB-120	1088	26,433	424	64.9	9.93	65.6	0.99
Saab	S340	1075	29,000	450	70.3	10.98	64.8	1.08
Bombardier	Q100	1174	36,300	585	85.0	12.35	73.0	1.16
Avions de T.R.	ATR42-500	966	41,005	586	80.6	11.09	74.3	1.08
Embraer	ERJ135LR	2015	44,092	551	65.8	7.86	86.5	0.76
Fokker	F50	1070	45,900	754	95.2	12.02	82.8	1.15
Embraer	ERJ145LR	1900	46,275	551	65.8	7.86	98.0	0.67
P.R. of China	Y-7	1239	48,050	810	96.1	11.40	77.7	1.24
Avions de T.R.	ATR72-500	947	48,500	657	88.7	11.98	89.1	1.00
Bombardier	CRJ200LR	2307	53,000	520	69.7	9.34	87.1	0.80
Boeing	DC9-10	1380	90,700	934	89.4	8.56	104.4	0.86
Fokker	F70	2072	92,000	1,006	92.1	8.43	101.4	0.91
BAE Systems	ARJ70	1911	95,000	832	86.4	8.97	85.8	1.01

Table F.4 Selected Scale Data for Aircraft of Gross Weight Over 100,000 lbs

Manufacturer	Aircraft	W_{to}	S (sq.ft.)	b (ft)	A	L (ft)	b/L
Boeing	B737-500	115,500	980	94.8	9.17	101.8	0.93
Boeing	B717-200	121,000	1001	93.3	8.70	124.0	0.75
Boeing	B737-300	138,500	980	94.8	9.17	109.6	0.86
Boeing	B737-600	145,500	1341	112.6	9.45	102.5	1.10
Airbus	A318	149,910	1320	111.0	9.3	103.0	1.08
Boeing	B737-400	150,000	980	94.8	9.17	119.6	0.79
Boeing	B737-700ER	154,500	1341	117.5	10.30	110.3	1.07
Airbus	A319-100	166,500	1320	111.1	9.35	111.0	1.00
Airbus	A320-200	169,800	1320	111.1	9.35	123.3	0.90
Boeing	B737-800	174,200	1341	117.4	10.28	129.5	0.91
Boeing	B737-900	174,200	1341	117.5	10.30	138.2	0.85
Boeing	B737-900ER	187,700	1341	117.4	10.28	138.2	0.85
Airbus	A321-200	206,100	1320	111.1	9.35	146.0	0.76
Boeing	B757-200	255,000	1951	124.0	7.88	155.3	0.80
Boeing	B757-300	272,500	1951	124.8	7.98	178.6	0.70
Airbus	A310-300	361,600	2360	144.0	8.79	153.1	0.94
Boeing	B787-3	364,000	3735	170.0	7.7	186.0	0.91
Boeing	B767-200ER	395,000	3050	156.1	7.99	159.2	0.98
Boeing	B767-300ER	412,000	3050	156.1	7.99	180.3	0.87
Boeing	B767-400ER	450,000	3130	170.3	9.27	201.3	0.85
Boeing	B787-8	503,500	3735	197.0	10.4	186.0	1.06
Airbus	A330-300	513,670	3890	197.8	10.06	208.1	0.95

(Continued)

Table F.4 (Continued)

Manufacturer	Aircraft	W_{to}	S (sq.ft.)	b (ft)	A	L (ft)	b/L
Airbus	A330-200	513,670	3892	197.8	10.1	189.0	1.05
Boeing	B777-200	545,000	4605	199.9	8.68	209.1	0.96
Boeing	B787-9	553,000	3735	197.0	10.4	206.0	0.96
Airbus	A350-800	571,000	4740	213.0	9.57	199.0	1.07
Airbus	A350-900	591,000	4740	212.0	9.48	220.0	0.96
Airbus	A340-200	606,270	3890	197.8	10.06	194.9	1.01
Airbus	A340-300	609,580	3890	197.8	10.06	208.9	0.95
Boeing	B777-200ER	656,000	4605	199.9	8.68	209.1	0.96
Boeing	B777-300	660,000	4605	199.9	8.68	242.3	0.83
Airbus	A350-1000	679,000	4750	213.0	9.55	242.0	0.88
Boeing	B777-200LR	766,000	4702	212.6	9.61	209.1	1.02
Boeing	B777-300ER	775,000	4702	212.6	9.61	242.3	0.88
Boeing	B747-400	875,000	5650	211.4	7.91	231.8	0.91
Boeing	B747-8I	975,000	5958	224.6	8.47	250.0	0.90
Airbus	A380-800	1,234,600	9104	261.8	7.53	239.0	1.10

Table F.5 Selected Wing Loading Data for Aircraft of Gross Weight Under 100,000 lbs

Manufacturer	Aircraft	R (mi)	W_{to}	W_l	S (sq.ft)	W_{to}/S	W_l/S	W_{to}/W_l
Embraer	EMB-120	1088	26,433	25,794	424	62.3	60.8	1.02
Saab	S340	1075	29,000	28,500	450	64.4	63.3	1.02
Bombardier	Q100	1174	36,300	33,900	585	62.1	57.9	1.07
Avions de T.R.	ATR42-500	966	41,005	40,344	586	70.0	68.8	1.02
Embraer	ERJ135LR	2015	44,092	40,785	551	80.0	74.0	1.08
Fokker	F50	1070	45,900	43,650	754	60.9	57.9	1.05
Embraer	ERJ145LR	1900	46,275	42,549	551	84.0	77.2	1.09
P.R. of China	Y-7	1239	48,050	46,740	810	59.3	57.7	1.03
Avions de T.R.	ATR72-500	947	48,500	48,170	657	73.8	73.3	1.01
Bombardier	CRJ200LR	2307	53,000	47,000	520	101.8	90.3	1.13
Boeing	DC9-10	1380	90,700	81,700	934	97.1	87.5	1.11
Fokker	F70	2072	92,000	81,000	1,006	91.5	80.5	1.14
BAE Systems	ARJ70	1911	95,000	83,500	832	114.2	100.4	1.14

Table F.6 Selected Wing Loading Data for Aircraft of Gross Weight Over 100,000 lbs

Manufacturer	Aircraft	R (mi)	W_{to} (lbs)	W_l (lbs)	S (sq.ft)	W_{to}/S	W_l/S	W_l/W_{to}
Boeing	B737-500	2752	115,500	110,000	980	117.9	112.2	0.952
Boeing	B717-200	2389	121,000	110,000	1001	120.9	109.9	0.909
Boeing	B737-300	2600	138,500	116,600	980	141.3	119.0	0.842
Boeing	B737-600	3713	145,500	120,500	1341	108.5	89.9	0.828
Airbus	A318	3696	149,910	123,500	1320	113.6	93.6	0.824
Boeing	B737-400	2384	150,000	124,000	980	153.1	126.5	0.827
Boeing	B737-700ER	3961	154,500	129,200	1341	115.2	96.3	0.836
Airbus	A319-100	4214	166,500	134,500	1320	126.1	101.9	0.808
Airbus	A320-200	3455	169,800	142,200	1320	128.6	107.7	0.837
Boeing	B737-800	3587	174,200	146,300	1341	129.9	109.1	0.840
Boeing	B737-900	3160	174,200	146,300	1341	129.9	109.1	0.840
Boeing	B737-900ER	3720	187,700	157,300	1341	140.0	117.3	0.838
Airbus	A321-200	3455	206,100	171,500	1320	156.1	129.9	0.832
Boeing	B757-200	4490	255,000	210,000	1951	130.7	107.6	0.824
Boeing	B757-300	3909	272,500	224,000	1951	139.7	114.8	0.822
Airbus	A310-300	5978	361,600	273,400	2360	153.2	115.8	0.756
Boeing	B787-3	3512	364,000	280,000	3735	97.5	75.0	0.769

(Continued)

Table F.6 (Continued)

Manufacturer	Aircraft	R (mi)	W_{to} (lbs)	W_l (lbs)	S (sq.ft)	W_{to}/S	W/S	W_l/W_{to}
Boeing	B767-200ER	7560	395,000	300,000	3050	129.5	98.4	0.759
Boeing	B767-300ER	6880	412,000	320,000	3050	135.1	104.9	0.777
Boeing	B767-400ER	6477	450,000	350,000	3130	143.8	111.8	0.778
Boeing	B787-8	9442	503,500	380,000	3735	134.8	101.7	0.755
Airbus	A330-300	7427	513,670	412,300	3890	132.0	106.0	0.803
Airbus	A330-200	8809	513,670	396,200	3892	132.0	101.8	0.771
Boeing	B777-200	6034	545,000	445,000	4605	118.3	96.6	0.817
Boeing	B787-9	9788	553,000	425,000	3735	148.1	113.8	0.769
Airbus	A350-800	9788	571,000	425,000	4740	120.5	89.7	0.744
Airbus	A350-900	9557	591,000	452,000	4740	124.7	95.4	0.765
Airbus	A340-200	9212	606,270	407,855	3890	155.9	104.8	0.673
Airbus	A340-300	8285	609,580	423,280	3890	156.7	108.8	0.694
Boeing	B777-200ER	8867	656,000	470,000	4605	142.5	102.1	0.716
Boeing	B777-300	6915	660,000	524,000	4605	143.3	113.8	0.794
Airbus	A350-1000	10,589	679,000	514,000	4750	142.9	108.2	0.757
Boeing	B777-200LR	10,818	766,000	492,000	4702	162.9	104.6	0.642
Boeing	B777-300ER	9131	775,000	554,000	4702	164.8	117.8	0.715
Boeing	B747-400	8320	875,000	652,000	5650	154.9	115.4	0.745
Boeing	B747-8I	9212	975,000	682,000	5958	163.6	114.5	0.699
Airbus	A380-800	9212	1,234,600	851,000	9104	135.6	93.5	0.689

Table F.7 Selected Thrust and Power Data for Aircraft of Gross Weight Under 100,000 lbs

Manufacturer	Aircraft	Engine	F_{to} (lbs)	P (hp)	W_{to}(lbs)	F_{to}/W_{to} or P/W_{to}	S_{to} (ft)	S_l (ft)
Embraer	EMB-120	2XPWC PW118A/B	–	1800	26,433	0.136	5105	4527
Saab	S340	2XGE CT7-9B2	–	1870	29,000	0.128966	3830	3258
Avions de T.R.	ATR42-500	2XPWC PW121	–	2150	36,300	0.118457	3196	2605
Embraer	ERJ135LR	2XPWC PW127E	–	2400	41,005	0.117059	3822	3691
Bombardier	Q100	2XRR AE3007-A1/3	8917	–	44,092	0.404472	5774	4462
Embraer	ERJ145LR	2XPWC PW127B	–	2750	45,900	0.119826	2920	3337
Fokker	F50	2XRR AE3007-A1	8917	–	46,275	0.385392	6070	4528
Avions de T.R.	ATR72-500	2XDONGAN WJ5-A	–	2900	48,050	0.120708	1791	2034
Bombardier	CRJ200LR	2XPWC PW127F	–	2750	48,500	0.113402	4012	3438
P.R. of China	Y-7	2XGE CF34-3B1	9220	–	53,000	0.347925	6017	4847
Fokker	F70	2XPW JT8D-7	14000	–	90,700	0.30871	6150	5080
BAE Systems	ARJ70	2xTay620	27700	–	92,000	0.301087	3545	3855
Boeing	DC9-10	4XHoneywell ALF502R-5	6970	–	95,000	0.293474	3610	3550

Table F.8 Selected Thrust Data for Aircraft of Gross Weight Over 100,000 lbs

Manufacturer	Aircraft	Engine	F_{to} (lbs)	W_g (lbs)	F_{to}/W_g	S_{to} (ft)	S_l (ft)
Boeing	B737-500	2xCFM56-3C-1	40,000	115,500	0.346	8700	4500
Boeing	B717-200	2xBR700-715	42,000	121,000	0.347	5750	5000
Boeing	B737-300	2xCFM56-3C-1	44,000	138,500	0.318	7600	4700
Boeing	B737-400	2xCFM56-3C-1	47,000	138,500	0.339	8880	5050
Airbus	B737-600	2xCFM56-7BE	39,000	145,500	0.268	6180	4380
Boeing	A318	2xCFM56-5B/PW6000	47,600	149,910	0.318	4200	4200
Boeing	B737-700ER	2xCFM56-7BE	54,600	154,500	0.353	5500	4690
Airbus	A319-100	2xCFM56-5B6/IAE V2524-A5	47,000	166,500	0.282	4800	4700
Airbus	A320-200	2xCFM56-5B4/IAE V2527-A5	53,200	169,800	0.313	5900	4800
Boeing	B737-800	2xCFM56-7BE	52,600	174,200	0.302	7330	5440
Boeing	B737-900	2xCFM56-7BE	54,600	174,200	0.313	7900	5450
Boeing	B737-900ER	2xCFM56-7BE	52,600	187,700	0.280	8970	5200
Airbus	A321-200	2xCFM56-5B3/IAE V2533-A5	64,000	206,100	0.311	7100	5200
Boeing	B757-200	2xRB.211-535E4/PW2037/2040	86,200	255,000	0.338	7750	5100
Boeing	B757-300	2xRB.211-535E4/PW2037/2040	86,200	272,500	0.316	8650	5700
Airbus	A310-300	2xCF6-80C2/PW4000	118,000	361,600	0.326	7400	4950
Boeing	B787-3	2xGEnx-1B54	106,400	364,000	0.292	–	–
Boeing	B767-200ER	2xCF6-80C2/PW4000	115,800	395,000	0.293	8150	5300
Boeing	B767-300ER	2xCF6-80C2/PW4000	121,600	412,000	0.295	8900	5500
Boeing	B767-400ER	2xCF6-80C2/PW4000	121,600	450,000	0.270	10800	6200

(Continued)

Table F.8 (Continued)

Manufacturer	Aircraft	Engine	F_{to} (lbs)	W_g (lbs)	F_{to}/W_g	S_{to} (ft)	S_l (ft)
Boeing	B787-8	2xGEnx-1B64	144,600	503,500	0.287	–	5676
Airbus	A330-300	2xCF6-80E1/PW4000/Trent700	140,000	513,670	0.273	8700	5873
Airbus	A330-200	2xCF6-80E1/PW4000/Trent700	142,200	513,670	0.277	8700	5723
Boeing	B787-9	2xGEnx-1B70	139,600	540,000	0.259	–	–
Boeing	A350-800	2xTrent XWB-74	190,000	540,120	0.352	–	–
Airbus	B777-200	2xGE90-77B/PW4077/Trent877	154,400	545,000	0.283	8450	5100
Airbus	A350-900	2xTrent XWB-83	190,000	584,225	0.325	–	–
Airbus	A340-200	4xCFM56-5C	136,000	606,270	0.224	10043	6115
Airbus	A340-300	4xCFM56-5C	136,000	609,580	0.223	10450	6432
Boeing	A350-1000	2xTrent XWB-92	190,000	650,360	0.292	–	–
Boeing	B777-200ER	2xGE90-94B/PW4090/Trent895	187,400	656,000	0.286	10000	5350
Airbus	B777-300	2xPW4098/Trent892	183,200	660,000	0.278	12250	6050
Boeing	B777-200LR	2xGE90-110B-115B	220,000	766,000	0.287	9700	5250
Boeing	B777-300ER	2xGE90-115B	231,080	775,000	0.298	10550	5850
Boeing	B747-400	4xCF6-80C2/PW4056/RB211-524	232,360	875,000	0.266	9900	7150
Boeing	B747-8I	4xGEnx-2B67	266,000	975,000	0.273	10700	6800
Airbus	A380-800	4xGE/PW GP7270/Trent 970	280,000	1,234,600	0.227	9350	6200

Index

Printed and bound by CPI Group (UK) Ltd, Croydon, CR0 4YY

03/10/2024

01040323-0010